GENETIC IMPROVEMENT OF CATTLE AND SHEEP

Geoff Simm

First published 1998

ISBN 0 85236 351 6

A catalogue record for this book is
available from the British Library

Published by Farming Press
Miller Freeman UK Ltd
Wharfedale Road, Ipswich, IP1 4LG
United Kingdom

Distibuted in North America
by Diamond Farm Enterprises
Box 537, Bailey Settlement Road
Alexandria Bay, NY 13607, USA

Cover and book design by Ian Garstka
Page layout by Hannah Berridge
Printed and bound in Great Britain by
Biddles Ltd, Guildford and King's Lynn

CONTENTS

Foreword

The development and use of breeding programmes based on sound scientific principles have had a major impact on animal production and on the quantity, price and quality of the food we eat. We have only to consider the revolutions in dairy cattle improvement brought in by the use of artificial insemination, progeny testing and BLUP, and the opportunities to increase rates of progress by effective use of multiple ovulation and embryo transfer. We realise that the rate of growth of lean meat of broiler chickens has changed by several per cent for over four decades, such that poultry meat has long ceased to be a luxury product. It is futile to argue whether genetic improvement or health care or nutrition are most important in animal production; each needs the other.

The theory of genetic improvement is soundly based on Mendelian genetic principles, but because the important traits such as growth rate and milk yield do not fall into two or three classes but are continuous, the genetic composition of an animal cannot be deduced simply by looking at it. In order to select those animals likely to have the best progeny animals have to be reared and recorded in an organised way, and the data analysed using quite complicated mathematical and statistical procedures. Students and breeders rarely find the concepts easy. In this book Dr Geoff Simm outlines the principles and shows how they are utilised. The arguments are made clearly and there is a minimum of mathematics, so the ideas and practice of genetic improvement are made available to student and practitioner alike.

There are no fundamental differences between the methodologies applied in different livestock species; but because of the actual structure of the industries, the longer generation turnover and lower reproductive rates, less effective breeding programmes operated for ruminants, particularly in the more extensive beef cattle and sheep systems, than for non-ruminants. Yet much can be done, and Dr Simm and his colleagues have successfully introduced modern programmes and evaluation procedures for cattle and sheep in the UK. This experience adds much to the book.

Many of the methods have a long-standing scientific base. But new molecular genetic technologies are being developed and used which will help us better to understand the genetic basis of the important production traits and may revolutionise the way animals are selected. The ideas are described in a straightforward way and evaluated here. Together with the developments being made in the more traditional quantitative approaches, Geoff Simm shows that genetic improvement of livestock is an exciting topic.

William G Hill, FRS
Professor of Animal Genetics
University of Edinburgh

Preface

For over 10,000 years the fortunes of the human race have been closely intertwined with the husbandry of livestock. For much of this time, whether knowingly or unknowingly, livestock have been changed genetically by subjective means. Some of the scientific foundations for more objective genetic improvement methods were laid over 100 years ago, and others followed from the 1900s to the 1930s. However, it is only over the past 50 years or so that these have been applied to any great extent in livestock improvement. These objective methods are based on measurement of performance in economically important traits. To date they have been used most widely in pig, poultry and dairy cattle breeding and to a lesser extent in beef cattle and sheep breeding. The application of scientific methods in animal breeding has led to major improvements in the output, cost and quality of food over the last few decades. The wider use of existing scientific methods in animal breeding, and the development of new ones, could help to meet the challenge of feeding a dramatically increased human population over the next few generations.

This book concentrates on genetic improvement of dairy cattle, beef cattle and sheep. It does so partly because scientific methods have been applied less widely in ruminants than in non-ruminants, and so there is a challenge in promoting the wider use of them in these species. Also, my own interests are in cattle and sheep breeding, so I feel better equipped to write about these subjects. Despite this, the principles discussed in the first few chapters apply equally well to other species. I hope that the book will be a useful introduction to the application of scientific methods in animal breeding for students and teachers at a range of levels, and for advisers, consultants and farmers specialising in livestock breeding.

The first chapter in the book sets the scene for modern livestock breeding, by looking at the origins of today's livestock breeds. The next four chapters deal with the scientific principles of livestock improvement. Chapter 2 outlines some of the basic principles in genetics, and attempts to illustrate the link between genes and the performance of individual farm animals, or populations of them. In Chapter 3 the main strategies for genetic improvement are discussed. The factors which affect responses to within-breed selection, and some tools for more effective within-breed selection, are discussed in Chapters 4 and 5. The next three chapters deal with the application of these principles in practical livestock breeding. Chapters 6, 7 and 8 deal with dairy cattle, beef cattle and sheep breeding respectively. Finally, Chapter 9 discusses the role of new technologies in animal breeding, especially reproductive and molecular technologies. It also explores the ethical implications of the use of some of these technologies. There is a glossary of technical terms at the end of the book.

The fairly wide use of statistics in modern methods of animal breeding is often offputting to both students and practitioners. I have tried to keep the use of statistics to a minimum, but a little understanding of some of the basics goes a long way. So, some statistical methods are described, but little or no prior knowledge is assumed. New techniques from molecular biology could have a potentially important role in animal breeding in the future. But, here too, there is a risk that the new technology and new language could widen the gulf between developer and user. I have tried to outline the potential applications and value of these new techniques without going into too much detail on the molecular methods themselves.

I have tried to illustrate as many of the principles of animal breeding as possible by the use of practical examples. For reasons of convenience and familiarity, many of these examples are from work done by my colleagues and myself, but there are many other examples in the scientific literature which are just as relevant. To keep the book as readable as possible, I have kept direct references to published work to a minimum. Where these references are helpful, or where Figures or Tables are drawn directly or modified from them, a number in square brackets is used to indicate the source publication. The full list of sources appears at the end of each chapter. Where possible I have tried to use recent reviews of a subject area, rather than individual papers. If you want to know more about a subject, these reviews are a good place to start. Also, I have listed examples of further reading at the end of the chapters.

Geoff Simm
SAC
Edinburgh

Acknowledgements

Many people have helped in many ways in the production of this book. In particular, I am very grateful indeed to the following friends and colleagues who read all or parts of the book and provided many constructive suggestions: Professor Bill Hill (University of Edinburgh; who also very kindly wrote the Foreword), Professor Robin Thompson (IACR/Roslin Institute), Dr Beatriz Villanueva, Dr Ron Lewis (both SAC), Dr Peter Amer (formerly SAC, now AgResearch, Invermay, New Zealand), Dr Jennie Pryce (SAC/University of Edinburgh), Dr Jon Mercer (Independent Breeding Consultants), David and Judy Hiam, Phil Beatson (formerly Lincoln University, now Holstein Friesian Association (Inc), New Zealand), Dr Chris Haley (Roslin Institute), Dr Peter Visscher (University of Edinburgh), Derrick Guy (formerly MLC), Dr Brian McGuirk (Genus), Gordon Swanson and Dr Raphael Mrode (both ADC), John Southgate (Signet), Dr Ron Crump (formerly Roslin Institute, now University of New England, Australia), Dr Roel Veerkamp (formerly SAC, now ID-DLO, Netherlands), Professor John Robinson (SAC), Dr Kevin Atkins (NSW Agriculture, Australia), Dr Tom Broad (AgResearch, Invermay, New Zealand), Stewart Hall (MLC) and Dr Chris Brown (formerly MLC, now Marks and Spencer plc), Libby Henson and Dr Basil Wolf (Welsh Institute of Rural Studies).

Many colleagues also kindly gave me access to their unpublished data, or helped by producing data, figures etc. for examples. In particular, I thank Dr Ron Crump, Dr Jennie Pryce, Dr Peter Amer, Dr Beatriz Villanueva, Dr Sue Brotherstone (University of Edinburgh), Gordon Swanson, Dr Raphael Mrode, Mike Coffey, Ann Hardy (both Holstein Friesian Society of Great Britain and Ireland), Dr Georgios Banos (INTERBULL), Dr George Wiggans (USDA), Dr Brian McGuirk, Dr Michael Cowan (Genetic Visions Inc., Madison, Wisconsin), Claire Heley (Signet), Dr Chris Brown, Stewart Hall, Dr Dick Esslemont (University of Reading/DAISY), Dr Chris Haley, Dr Naomi Wray, Dr Francis Barillet, Dr François Ménissier (both INRA), Dr Ray Peterson (University of British Columbia), Mark Jeffries (Australian Dairy Herd Improvement Scheme), Dr Kevin Atkins, Dr Laurie Piper and Dr Hans-Ulrich Graser (both University of New England, Australia), Dr Brian Wickham (formerly Livestock Improvement Corporation Ltd, New Zealand), Dr Arnold Bryant (Dairy Research Corporation, Hamilton, New Zealand), Dr Jean-Noël Bonnet (Eléveurs de Bovins Limousins, France) and Dr Larry Cundiff (USDA) in this respect. I also thank Associate Professor Frank Nicholas (University of Sydney) for his friendly advice in general, and on several diagrams in Chapter 2 in particular. I am also grateful to Dr Alan Archibald (Roslin Institute) for his help in using the Anubis Mapping software.

I thank Dr Sandra Eady (CSIRO, Armidale, Australia), Professor Dave Thomas (University of Wisconsin, Madison, US), Antonello Cannas (Cornell University, US), Mrs Pat Stanley, Dr Kreg Leymaster and Dr Gary Snowder (both USDA), Professor Ian Wilmut, Dr John Williams, Dr Alan Archibald (all Roslin Institute), Dr Luca Ferretti (IDVGA-CNR, Milan, Italy), Dr Tom Broad, Dr H A Ansari, Mrs Ùna Cochrane, Carey Coombs, Peter Johnson (Elite Texel Sires Ltd), Dr Peter Amer, Professor John Robinson, Dr Mark Young and Dr Tony Waterhouse (SAC), the Royal Agricultural Society of England, Iona Antiques, Genus, and David Gaunt (formerly British Simmental Cattle Society), for allowing me to uwse their excellent photos.

I have been fortunate to have many good teachers and colleagues in Edinburgh and elsewhere who first stimulated, and continue to stimulate, my interest in animal breeding. Many of these are listed above, but I am deeply indebted to the late Professor Charles Smith, who supervised my postgraduate studies at ABRO in Edinburgh. I am very grateful indeed to the many colleagues who provided support in other ways, particularly Gillian Ramsay for additional typing and help with permissions, and other colleagues who tolerated my book 'diversions'. I also thank Ross McGinn, Allan McBride and Jackie Calder (SAC) for computing support, line drawings and photographic assistance respectively. I also thank my employers, SAC, for their support in this venture. I am grateful to Roger Smith, Claire Newbery and Elizabeth Ferretti of Farming Press for encouraging me to write this book, for their patience in waiting for it and for their editorial input. I thank the funders of our research in Edinburgh, especially SOAEFD, MAFF, MLC, MDC, BBSRC and HFS. I also thank the many breeders and breeding organisations who have worked with us, for their interest and support in getting research into practice.

I am also very grateful to friends and members of my family who provided support in many ways, and particularly to my wife, Marie, who read many drafts of the book and provided lots of good advice, produced most of the graphs, tolerated many invasions of our leisure time and regularly exaggerated the rate of progress being made.

Many authors and publishers kindly gave their permission to reproduce direct or amended versions of their material, or to reproduce material for which they are copyright holders. These original sources are noted in the table or figure captions concerned, and full details are given in the reference list at the end of each chapter. In particular I thank the following:

The Natural History Museum, London (extracts from Clutton-Brock, 1987, in Chapter 1), Dr Tom Broad and his colleagues associated with SheepBase (Figures 2.3 and 9.8), US Congress, Office of Technology Assessment (Figures 2.4, 2.5, 9.5 and 9.7), the Biotechnology and Biological Sciences Research Council, Swindon, UK (Figure 2.6).

Figures 2.8, 2.9, 2.14, 2.15, 3.1, 3.10, 4.7, 4.8 and Tables 2.3, 2.4, 3.9, 3.10 and 4.6 are reproduced or adapted from *Veterinary Genetics* by F W Nicholas (1987) by kind permission of Oxford University Press. Figures 2.12, 2.14, 2.15 and several formulae in Chapter 4 are reproduced or adapted from *Introduction to Quantitative Genetics* by D S Falconer and T F C Mackay (1996) by kind permission of Addison Wesley Longman.

Similarly, I thank Blackwell Science Ltd (Table 2.6); Dr A H Visscher on behalf of the publishers of the *Proceedings of the Working Symposium on Breed Evaluation in Crossing Experiments with Farm Animals*, IVO (1974) (Table 3.9); the British Society of Animal Science (Tables 3.1, 3.3, 3.4, 3.7, 3.8, 5.9, 6.3, 6.6, 6.8, 6.11, 6.12, 6.16, 6.17, 6.22, 6.24, 6.26, 6.27 and 7.16); United States Department of Agriculture (Figures 3.4 and 3.10); Professor W G Hill on behalf of the publishers of the *Proceedings of the Fifth World Congress on Genetics Applied to Livestock Production*, 1990 (Figures 3.8 and 7.3 and Tables 7.4 and 8.2); the Meat and Livestock Commission and Signet (Figures 3.9, 7.6, 8.7, 8.13 to 8.15 and Tables 7.1, 7.4, 7.6, 7.7, 8.1, 8.3 to 8.5 and 8.11); Professor W A Becker and

Academic Enterprises (Table 4.3, and the expanded version of this table given at the end of the book); Genus (Figure 6.5, Tables 6.4 and 6.11); the Food and Agriculture Organisation of the United Nations (Figures 3.8, 6.1 and 6.2, 7.1 and 7.2, 8.1 to 8.6); INTERBULL and the UK Animal Data Centre (many tables and graphs as indicated in Chapters 5 and 6); the Holstein Friesian Society of Great Britain and Ireland (Figures 6.7, 6.12 and 6.13 and Table 6.11); the Residuary Milk Marketing Board of England and Wales and the National Dairy Council (Tables 6.2 and 7.5).

I gratefully acknowledge Springer-Verlag GmbH and Co. KG, for their permission to reproduce Plate 5 from 'Resolving ambiguities in the karyotype of domestic sheep (Ovis aries)' by H A Ansari, P D Pearce, D W Maher, A A Malcolm and T E Broad in *Chromosoma*, Vol. 102, pp. 340-347 (1993), and Plate 64 and part of the cover design from 'Chromosomal localization and molecular characterization of 53 cosmid-derived bovine microsatellites' by A Mezzelani, Y Zhang, L Redaelli, B Castiglioni, P Leone, J L Williams, S Solinas Toldo, G Wigger, R Fries and L Ferretti in *Mammalian Genome*, Vol. 6, pp. 629-635 (1995).

The formula for the prediction of a bull's BV from single records on his progeny, which is cited in Chapter 5, is from *Genetics for the Animal Sciences* by Van Vleck, Pollak, Oltenacu © 1987 by W H Freeman and Company, and is used with permission. Tables 2.11, 6.9 and 6.10 are reproduced from the publication cited in the text by kind permission of Publications Elsevier, France. Table 6.1 is reprinted from *Livestock Production Science,* Vol. 11, pp. 401-15 (1984) 'Holstein Friesians, Dutch Friesians and Dutch Red and Whites on two complete diets with a different amount of roughage: performance in first lactation' by J K Oldenbroek and Figure 8.11 is reprinted from *Livestock Production Science* Vol 21, pp. 223-33 (1989), 'Selection indices for lean meat production in sheep' by G Simm and W S Dingwall, both with kind permission from Elsevier Science – NL, Sara Burgerhartstraat 25, 1055 KV Amsterdam, the Netherlands.

I thank CAB International for their permission to reproduce Figures 8.12, 9.1, 9.2, 9.3, 9.6, 9.9 and Tables 7.11 to 7.14, 8.7 to 8.10, 8.14 and 8.15. I am grateful to the Federation of the American Society for Experimental Biology for their permission to reproduce Figure 9.7 and to Elsevier Trends Journals for their permission to reproduce Figure 9.10. I am grateful to Professor L B Schook and the University of Minnesota for their permission to reproduce Figure 9.4. The extracts from the Banner Committee Report in Chapter 9 are Crown copyright; Crown copyright is reproduced with the permission of the Controller of Her Majesty's Stationery Office. Finally, I thank SAC for their permission to reproduce many diagrams and excerpts from tables, as indicated in the text.

CHAPTER 1

The origins of today's livestock breeds

Natural selection and evolution

Animals come in an incredible variety of sizes, shapes and colours. Just why this is so has occupied many great minds over the centuries. Most biologists today believe that the species that we farm, keep as pets, observe in the wild or watch on TV, and those now extinct, evolved by the process of **natural selection**. Although several nineteenth century naturalists and philosophers contributed ideas, Charles Darwin is usually credited with putting together a cohesive theory of natural selection in 1838. He later explained this in his book *The Origin of Species*, first published in 1859.

The main principles of Darwin's theory of natural selection are that species can change over time, and that their survival or success depends on how well they fit their environment – what one of Darwin's supporters, Thomas Huxley, described as "the survival of the fittest". The key to this process is variation between individuals. Darwin recognised the "many slight differences which appear in the offspring from the same parents"; it is on these differences which natural selection, or breeders of domestic livestock, can act. Those individuals which have favourable attributes stand a higher chance of surviving and reproducing than those which do not.

This theory provides an explanation of how the huge variety of animal, and other, species has arisen from the earliest primitive forms of life. Chance variations in the size, shape or functioning of animals, over the course of millions of years, have allowed them to adapt to particular environments, or niches. Some species have been able to adapt to and profit from major changes in the environment while others, such as the dinosaurs, have not and have become extinct. Many new species have emerged, usually as a result of physical isolation of part of a population – for example when continents drifted apart – or due to isolation of other kinds, such as increasing dependence on a particular type of food. One of the most famous examples of natural selection comes from Darwin's visit to the Galapagos Islands. Here he noticed the wide variety in the size and shape of the beaks of different species of finches, which was related to the way in which each species obtained its food. Although Darwin's work is usually linked to evolution and natural selection of wild animals, he was well aware of, and greatly influenced by, the changes in domestic animals brought about by the 'artificial' selection of early livestock breeders.

Darwin's theory was highly controversial at the time, but there is now overwhelming evidence, from many branches of science, that it is substantially correct. For more details of the background to Darwin's work, and developments from it, see references [1, 3] listed at the end of this chapter.

Domestication

Over the last 250,000 years the human population has increased from an estimated three million to about 4,000 million [1]. (During that time the life expectancy of humans has also more than doubled.) This increase in population size has occurred in three main surges. The first was stimulated when our early ancestors first learned to use tools and make fire – attributes which allowed them to spread from the tropical regions in which they first evolved to inhabit colder, northern areas. The second surge occurred about 12,000 to 10,000 years ago, when humans began to cultivate plants and domesticate animals after the end of the Ice Age. The third surge in population size began with increased industrialisation. For example, in Britain the population increased from about 5.5 million to about 10.75 million in the eighteenth century, with a particularly rapid rise in the second half of the century [2]. The world's population is expected to double again in the next generation and so, despite food surpluses in affluent nations, the challenges facing agriculture globally have by no means diminished.

When humans first spread northwards they were primarily hunters who adapted their own lifestyle to that of their prey [1]. Some groups, such as the indigenous people of North America following herds of bison, or those of Lapland following reindeer, continued this lifestyle. Others learned to modify the behaviour of some of their prey species, and so began the process of domestication.

Of the very large number of animal species, very few have been successfully domesticated. Francis Galton (Darwin's cousin) wrote an essay on domestication in 1865, suggesting that the process of domestication happened by trial and error. He reasoned that "...a vast number of half-unconscious attempts have been made throughout the course of ages, and that ultimately, by slow degrees, after many relapses, and continued selection, our several domestic breeds became firmly established." Galton also identified six conditions for successful domestication of a species of animals. These are given below, together with more modern interpretations [1].

1. **"They should be hardy."** The animals had to be able to cope with earlier weaning than normal, to adapt to artificial feeding and husbandry, and probably to cope with new disease challenges.

2. **"They should have an inborn liking for man."** They had to be social animals, with a hierarchy, able to imprint on humans and accept them as 'leader' in later life. Galton also stressed the importance of being able to understand the behaviour of a species, and communicate with it, for domestication to be successful.

3. **"They should be comfort-loving."** That is, they should not be highly adapted for instant flight, but more amenable to being herded and closely confined.

4. **"They should be found useful to the savages."** The primary function of the first animals to be domesticated was to provide a reliable source of food for humans (other uses, such as provision of clothing, a means of transport and draught power, and uses as religious,

ritual or status symbols came later).

5. **"They should breed freely."** The ability to breed in captivity is perhaps the most important attribute for domestication; failure to do so has prevented the domestication of many species.

6. **"They should be easy to tend."** The animals should be reasonably placid, have versatile feeding habits, and tend to herd together.

Sheep and goats were probably the first of our current farm livestock species to be domesticated, about 10,000 years ago, though domestication of the dog began about 2,000 years earlier. They were followed by cattle and pigs, and later still by horses. Domestication has led to many differences from the wild ancestors, either as a direct result of domestication itself, by chance, by natural selection or, directly or indirectly, as a result of artificial selection. Usually there is a much greater diversity in appearance of the domestic strains of livestock than in their wild counterparts. In many cases the domestic strains differ from the wild type in physical appearance, including a shortening of the facial region of the skull and the jaws, longer ears, longer or curling tails, differences in the size or shape of horns, differences in coat colour, and a wide range in the thickness of the coat, depending on the local climate. Most domestic species have a higher concentration of fat in their carcass than their wild ancestors (Plate 1). Also, most breeds of domestic sheep have lost the characteristic of completely shedding wool in the summer.

Modern livestock breeders are sometimes accused of producing maladapted breeds or individuals (of which more in Chapters 6 to 9), but it is interesting to note that as long ago as 450 BC there were strains of fat-tailed sheep in Arabia with tails so long and fat that shepherds crafted small wooden trailers, harnessed to the sheep, to prevent their tails from dragging on the ground [1].

Plate 4 Leicester wethers, belonging to GL Foljambe of Osberton Hall, near Retford, Nottinghamshire, UK, which won the first prize of £20 and silver and gold medals at the Smithfield Club Cattle Show in December 1859. (Courtesy of Mrs Pat Stanley.)

Origins of livestock breeds

A breed of livestock is basically a recognised interbreeding group of animals of a given species. In most cases, animals which belong to the same breed are of fairly uniform appearance. This appearance is inherited, and usually distinguishes the breed concerned from other breeds. However, in other cases animals are considered to belong to the same breed by virtue of their geographical location, and there is quite wide variation in their appearance. Breeds have been created by 'reproductive isolation' – that is, the formation of separate groups of animals, where matings occur within the groups but not usually between them. This is analogous to the way in which new species or subspecies of wild animals have evolved, except that the reproductive isolation of wild animals has often occurred due to geographical dispersion, whereas reproductive isolation of domestic breeds has often been imposed by humans. Within many breeds there is further subdivision into strains or lines. These share a common ancestry, but have become reproductively isolated, to varying extents, as a result of physical separation or pursuit of different breeding objectives. For example, many of the strains of black and white Friesian or Holstein dairy cattle around the world originate from importations of Dutch Black Pied animals, mostly in the nineteeth century. In many countries these local strains have been crossed, over the last few decades, to North American Holstein strains to increase yield.

Distinct breeds of dogs, cattle and sheep had been developed by ancient Egyptian civilisations by the second millennium BC [1]. In northern Europe there is little evidence of the presence of distinct breeds until the Roman period. The Romans are thought to have made positive efforts to improve their stock, possibly by selective breeding of favoured animals.

There was probably a slow evolution of livestock breeds, with strong regional ties, until a couple of centuries ago. In Britain the impetus for more rapid livestock improvement came in the second half of the eighteenth century with the dramatic increase in the size of the human population and the need for greater food production. By the earlier part of that century there were already some notable changes in agriculture which increased production, including the enclosure of land into fields, drainage of land, felling of woodland, improvement of roads, the use of root crops as winter feed for livestock, and the start of mechanisation [2].

Robert Bakewell (1725–1795) is usually regarded as the pioneer of livestock improvement as we know it (Plate 2). He and his followers employed **linebreeding** – the mating of closely related animals – which had been practised quite effectively in racehorses already. He also appears to have made widespread use of comparisons of growth and feed intake to help his selection decisions, and to have used both measurements and preserved specimens to allow him to chart his progress. Bakewell was particularly associated with the improvement of the Leicester breed of sheep and the Longhorn breed of cattle (Plates 3 and 4). His Dishley or New Leicester breed of sheep was created from other sheep breeds in Leicestershire and Lincolnshire, and was of uniform appearance by 1770. He selected for improved mutton production and, as a result, sheep of his own breed were claimed to reach market a year sooner than the usual three or four years of age [2]. (Incidentally, much of this early selection in meat breeds, together with husbandry methods of the time, produced extremely fat animals. Fat was in great demand at that time, at least in the high proportion of the human population which was engaged in physically demanding work. Today in most Western societies, with a more sedentary lifestyle, reversing the propensity of animals to fatten is a major preoccupation of livestock breeders.)

Bakewell began letting out rams for hire for the mating season from about 1760. He is reputed to have ridden around his customers' fields comparing the progeny of different rams, and using the best of these rams at home – an early example of **progeny testing**. Bakewell established the Dishley Society which laid down the rules for ram letting for Leicester breeders, and was, in many respects, a forerunner for today's pedigree breed societies. Anyone who is shocked by the spiralling price records in some of today's pedigree sheep sales will be interested to know that in 1789 Bakewell made 1,200 guineas from letting a single ram for one season [2]. That is equivalent to about £500,000 at today's prices, and it must have caused a few raised eyebrows at the time! Many British sheep breeds are thought to have some New Leicester ancestry, but the Border Leicester and Leicester Longwool are comparatively pure descendants [2].

Bakewell had less success in improving the meat qualities of the Longhorn breed of cattle, but his followers did achieve success with the Shorthorn breed. The *Coates Herd Book* for this breed was founded in 1822 (some 31 years after the first stud book for thoroughbred horses), and herd or flock books for many other breeds of cattle, pigs and sheep around the globe followed [2].

For many decades afterwards, in many breeds, the new profession of pedigree breeding in Britain enjoyed an enviable reputation for producing quality purebred livestock. The evidence for that is still visible today in the wide geographical distribution of British breeds and, less obviously, in the contribution they made to the formation of many other breeds around the globe. However, there are few of these British breeds which still enjoy this reputation today. To some extent this is a result of the pre-eminence of a small number of breeds in today's more specialised agricultural industry. Those breeds which have been selected for current market needs, such as efficient production of lean meat, milk solids, or wool of specific fibre diameter, have expanded at the expense of those which have not. This process has been stimulated by vast improvements in international communication and transport, and accelerated by techniques like artificial insemination and embryo transfer. Many people believe that we now depend too heavily on a small number of very specialised breeds and that we need to do more to protect those which are declining in numbers. The subject of conservation of genetic resources is very topical, and some of the issues involved are discussed in more detail in Chapter 3.

Summary

- The huge variety of animal and other species that we see today, together with those now extinct, evolved by the process of natural selection.

- The key to natural selection, and to the artificial selection practised by breeders, is the variation in most characteristics that exists between individual animals.

● Domestication of animals began 12,000 to 10,000 years ago. Whether or not it has been done knowingly, artificial selection has been practised in domestic animals ever since then. (Natural selection has continued to occur too.)

● Although distinct breeds or strains of cattle and sheep existed long before then, the practices of pedigree recording and selection of related animals with the aim of breed improvement date from the mid-1700s. The formation of herd books began early in the following century.

References

1. Clutton-Brock, J. 1987. *A Natural History of Domesticated Mammals.* British Museum (Natural History), London.

2. Hall, S.J. and Clutton-Brock, J. 1989. *Two Hundred Years of British Farm Livestock.* British Museum (Natural History), London.

3. Leakey, R.E. 1986. *The Illustrated Origin of Species* by Charles Darwin. Abridged and introduced by Richard E. Leakey. Faber and Faber, London.

Further reading

Clutton-Brock, J. 1987. *A Natural History of Domesticated Mammals.* British Museum (Natural History), London.

Hall, S.J. and Clutton-Brock, J. 1989. *Two Hundred Years of British Farm Livestock.* British Museum (Natural History), London.

Leakey, R.E. 1986. *The Illustrated Origin of Species* by Charles Darwin. Abridged and introduced by Richard E. Leakey. Faber and Faber, London.

Mason, I.L. 1996. *A World Dictionary of Livestock Breeds, Types and Varieties.* 4th edn. CAB International, Wallingford.

Ryder, M.L. 1983. *Sheep and Man.* Duckworth, London.

Stanley, P. 1995. *Robert Bakewell and the Longhorn Breed of Cattle.* Farming Press, Ipswich.

CHAPTER 2

Genes, genetic codes and genetic variation

Introduction

The last chapter outlined the long history of livestock improvement by subjective means. Some of the scientific foundations for more objective genetic improvement of livestock were laid over 100 years ago, and others followed from the 1900s to the 1930s. However, it is only over the past 50 years or so that these methods of selection (i.e. based on measurements of performance) have become fairly widely used in pig, poultry and dairy cattle breeding and, to a much lesser extent, in beef cattle and sheep breeding. Where they have been applied, these scientific methods have helped to increase the output, decrease the price and improve the quality of food and other animal products. The wider use of existing scientific methods in animal breeding, and the development of new ones, could help to meet the challenge of feeding a dramatically increased human population over the next few generations.

There have been contributions to objective methods of livestock improvement from several branches of science, and many of these are explained in later chapters. However, it is from the science of genetics that some of the most important contributions have come. School lessons on Mendel and his experiments with peas, or laboratory practicals studying the genetics of fruit flies or mice, can seem a far cry from the business of breeding better sheep or cattle. However, most of the principles of genetics established in these laboratory species, and which most of us have encountered at some time, apply equally to farm livestock. A little understanding of these principles goes a long way towards understanding modern approaches to livestock improvement, which are described later in the book. Hence, the purpose of this chapter is to introduce some of these principles, and to begin to explain the link between basic genetics and the genetic improvement of economically important traits in livestock. The first part of the chapter deals with chromosomes and genes – the means by which animal characteristics are passed on, or inherited, from one generation to the next. The next section deals with the structure and function of genes in a little more detail. Then the influence of genes on animal performance and some of the different types of gene action are discussed. Finally, the genetic basis of traits of economic importance in farm livestock is examined.

With any specialist subject, one of the most off-putting things for a newcomer is the language used by insiders. Although it is not always apparent, the main function of technical terms is to provide a sort of shorthand means of communication. Many of the technical terms in widespread use in genetics and livestock improvement are introduced in this chapter, but remember there is nothing mystical about these terms, they are meant to aid communication, not make it harder! There is a glossary at the end of the book if you need a reminder of the meaning of some of these technical terms.

Chromosomes and genes

What was not clear in Bakewell's or Darwin's time was just how characteristics were passed on, or **inherited**, from one generation to the next. Although we are still learning about this today, some major advances have been made. Ironically, it was during Darwin's lifetime that the first seeds were sown (no pun intended!) by the monk and naturalist, Gregor Mendel, though the significance of Mendel's work was not appreciated widely until much later. Mendel worked in a monastery in what is now Brno, in the Czech Republic. Much of his work was with peas and, in 1866, he reported that some characteristics such as flower colour, height of plants, seed shape and texture followed particular patterns of inheritance. This he attributed to 'elements', which were inherited from each parent. We now call these elements **genes**.

By the early 1900s a great deal of effort had gone into studying the process of cell division using the microscope. These studies revealed structures, which were named **chromosomes**, present in the nuclei of all cells, and which these early biologists suspected were involved in the process of inheritance (See Figure 2.1). We now know that these suspicions were true and that genes occur in sequences, often likened to a string of beads, on the chromosomes. In higher animals there are thousands of genes on each chromosome.

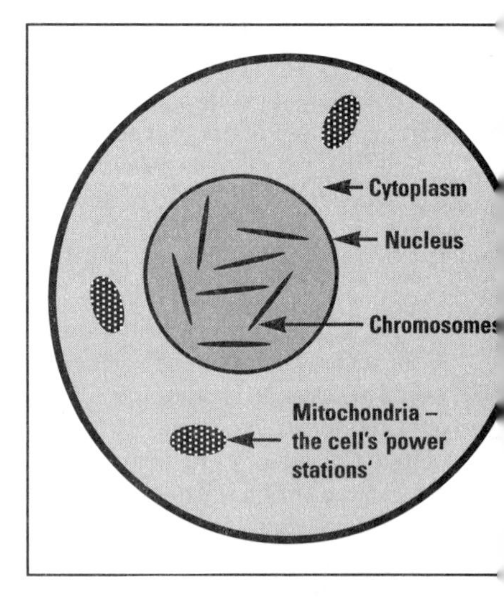

Figure 2.1 Diagram of a cell showing chromosom[es in] the nucleus (not all chromosomes are shown).

The chromosomes can be seen clearly with the aid of a microscope only at certain stages of cell division, at other times they are difficult to distinguish. (The complete set of chromosomes is known as a **karyotype** (see Plate 5).) In all cells except the **gametes** – that is, the sperm and eggs – the chromosomes occur in pairs, with the number of pairs being a characteristic of the species. For example, humans have 23 pairs of chromosomes in each of the body cells, cattle have 30 pairs and sheep have 27 pairs (this is known as the **diploid** number of chromosomes). When sperm and eggs are created in the reproductive organs (testes and ovaries) each sperm and each egg receives only a single set of chromosomes. In other words, the sperm or eggs of humans, cattle and sheep carry 23, 30 and 27 single chromosomes, respectively (this is known as the **haploid** number of chromosomes). When a sperm fertilises an egg, the resulting embryo carries the original number of paired chromosomes.

In some species, particular pairs of chromosomes can be distinguished from one another, under a microscope, by differences in their size or shape, and by visible differences in the patterns of banding after staining (see Figures 2.2 and 2.3). These patterns of banding are nature's equivalent of the bar codes now found on just about every item on supermarket shelves. Differences in the size, shape and banding

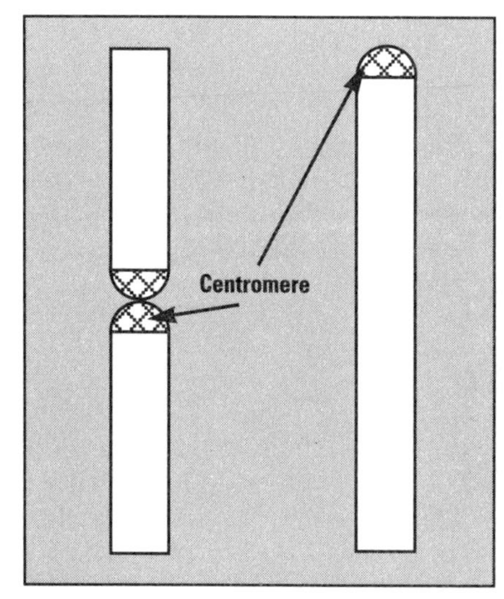

Figure 2.2 Chromosomes have different shapes. Some chromosomes have a constricted area called the centromere at or near the middle; others have the centromere at one end. Most sheep chromosomes and all cattle chromosomes, apart from the sex chromosomes, are of the second type.

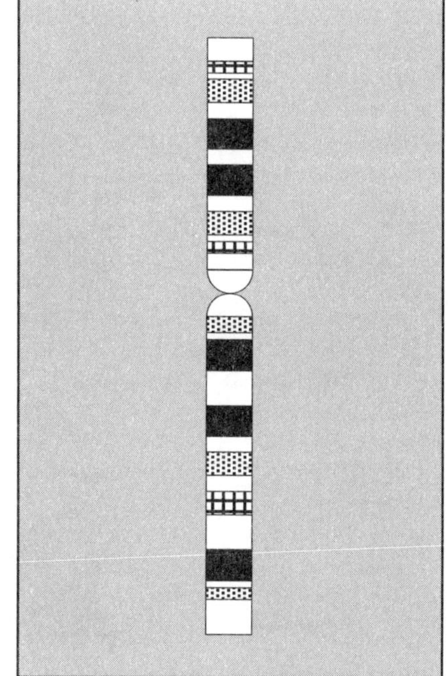

Figure 2.3 The patterns of banding on sheep chromosome number 3, after a particular method of staining. Darker bands in the diagram represent greater intensity of staining. The patterns are used to distinguish pairs of chromosomes from one another. The bands on each chromosome of a species have been given internationally agreed numbers, for ease of reference. ([2]; redrawn from a map produced from the SheepBase World Wide Web site using Anubis mapping software – see Chapter 9 for more details.)

patterns have allowed identification and numbering of each pair of chromosomes in some species, but in others, such as cattle and sheep, this is difficult because some pairs are very similar in appearance. Although different pairs of chromosomes may differ in size, the two members of each pair of chromosomes are the same size, with the exception of one pair, the **sex chromosomes**.

In the body cells of female mammals there are two sex chromosomes of similar size, termed X chromosomes. In the body cells of male mammals, there is one X chromosome, and a smaller Y chromosome (see Plate 5). Hence, in mammals it is the sperm which determines the sex of the resulting embryo. This is because all eggs carry an X chromosome, but sperm carry either an X or a Y sex chromosome. Eggs fertilised by sperm carrying an X chromosome develop into females, while eggs fertilised by sperm carrying a Y chromosome develop into males. (A single gene on the Y chromosome causes the development of males, from what would otherwise become females.) The reverse is true in birds: sperm carry the same sex chromosomes, but eggs carry either male or female sex chromosomes. All chromosomes, apart from the sex chromosomes, are called **autosomes**, and the genes on these chromosomes are said to be **autosomal**. Genes on the sex chromosomes are said to be **sex-linked**. (For more details on chromosomes see [9, 10].)

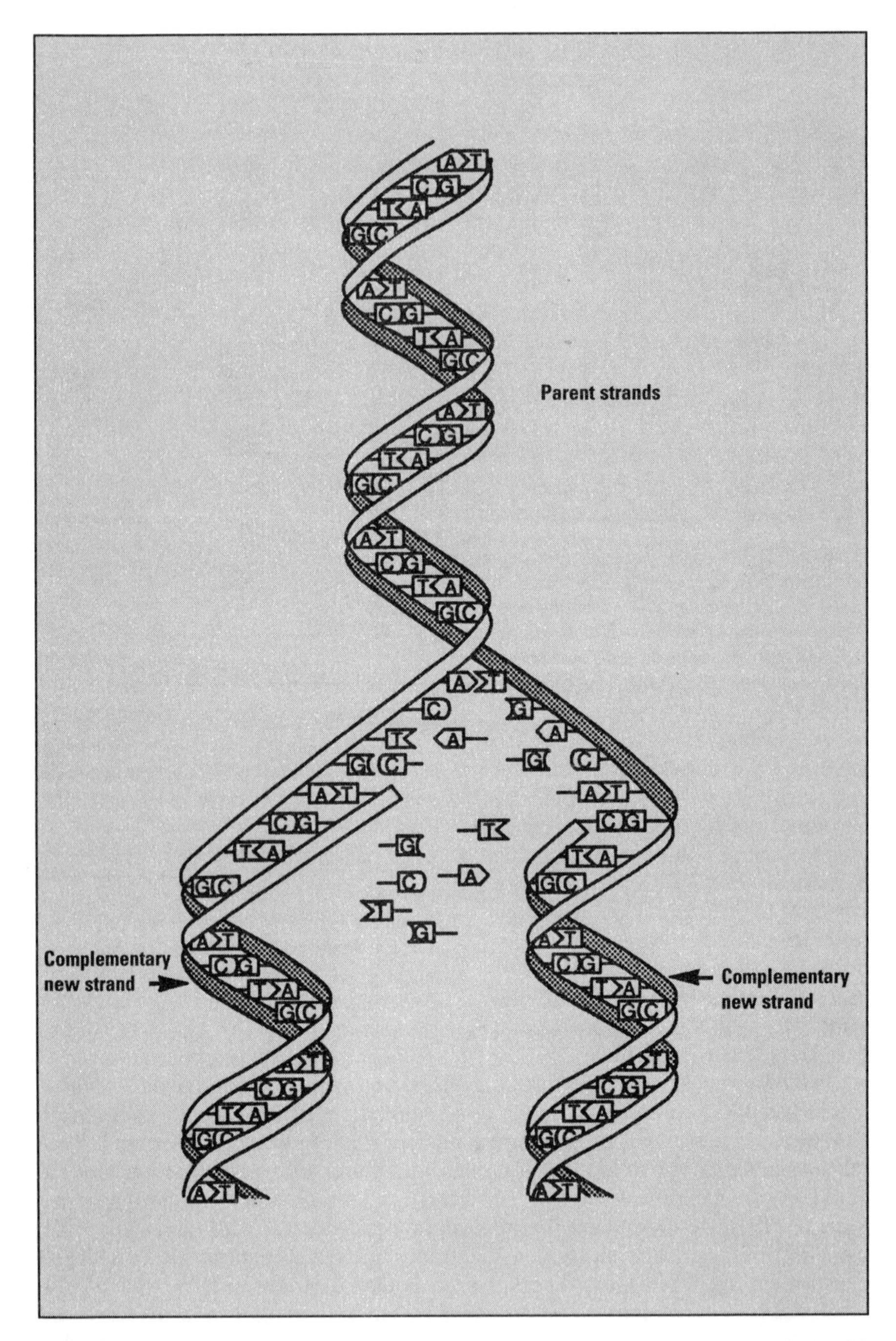

Figure 2.4 Diagram illustrating the structure of DNA, especially the pairing of the bases adenine (A) with thymine (T) and guanine (G) with cytosine (C). The diagram also shows how the DNA molecule unwinds during replication, with each single strand acting as a template for the synthesis of a new complementary strand [14, 15].

The structure and function of genes

DNA and genetic codes

One of the most significant scientific events of the twentieth century was the discovery of the chemical structure of the material which chromosomes and genes are made of: **deoxyribonucleic acid** or **DNA** for short. This discovery was made by James Watson and Francis Crick in 1953. Although the detail of the structure of DNA is outside the scope of this book, there are a few points about it that help to explain how genes function and how genetic variation arises among farm animals, or any other form of life. DNA molecules are made up of two strands, or backbones, a bit like the sides of a very long ladder, with cross linkages between these strands, equivalent to the rungs on the ladder (see Figure 2.4). The cross linkages are made up of pairs of different chemical substances called **bases**. There are four different bases which make up the cross linkages: adenine (abbreviated to A), thymine (T), guanine (G) and cytosine (C), and they form pairs in a particular way. Thymine always pairs with adenine, and guanine always pairs with cytosine. This means that if the sequence of bases on one strand of the DNA molecule is known, then the sequence on the complementary strand can be worked out. The whole molecule of DNA (the long ladder) is twisted into a corkscrew shape, which gives rise to the commonly used description of a double helix. Genes are effectively segments of the ladder, made up of long sequences of these base pairs.

Genes are both the design plans and the instruction manuals for all living creatures. They provide the instructions for a newly fertilised egg to grow into an adult, and for the body to go about its business at all times. Also, genes are the means by which these instructions are passed from one generation to the next. Each of these functions rests heavily on the structure of DNA and the fact that the bases which make up DNA always pair in the same way.

Genes are made up of sequences of bases at particular locations in DNA molecules, but they account for only about ten per cent of the total DNA in chromosomes (the entire **genome**). The rest is involved in controlling the expression of genes, or it has other functions which are currently unknown, or it has no function at all. Genes function by coding for the production of proteins. Many of these proteins are themselves 'signals' which control biochemical processes elsewhere in the body (e.g. the enzymes involved in controlling digestion, or the hormones controlling growth, reproduction and lactation). Others are 'structural' proteins which are used directly in the growth or repair of body tissues such as muscle, internal organs, skin or hair, or they are secreted in milk.

Wherever a gene occurs in a DNA molecule, the sequences of bases on one side of the strand form a template for the production of a series of amino acids. Amino acids are the building blocks from which proteins are made. Sets of three bases together in DNA are called a **triplet**, and each triplet either forms the code for producing a particular amino acid, or it signals where a gene stops or starts. Since there are 64 possible combinations of the four bases into triplets, but only 20 amino acids, it follows that several triplets code for the same amino acid. For example, the triplets TTT and TTC both code for the amino acid phenylalanine, and the triplets TCT, TCC, TCA and TCG all code for the amino acid serine. (See [10] for the full set.)

Hence, genes are long sequences of these triplets. On average, the informative

Unit of genetic information	Relative size (number of bases or base pairs)
a triplet	**3 bases**
an average gene (informative part)	**about 3,000 bases**
a typical chromosome (for species like cattle with 30 pairs)	**about 100,000,000 base pairs**
the entire genome (all the genetic information contained on a single member of each pair of chromosomes)	**about 3,000,000,000 base pairs**

Table 2.1 The relative size of different units of genetic information in mammals [10, 15].

part of each gene is about 3,000 bases or 1,000 triplets long (see Table 2.1). This sequence of triplets makes up a code or template for the production of particular proteins. In order for the template to be 'read', the double strand of DNA unwinds and the base pairs become separated temporarily (equivalent to unzipping a zip fastener) leaving a sequence of unpaired bases. A single strand sequence of complementary bases is then formed from this template by a substance closely related to DNA, called **messenger RNA (mRNA)**. Groups of three bases in mRNA, which code for the production of an amino acid, are known as **codons**. Messenger RNA strands leave the nucleus and move to the ribosomes,

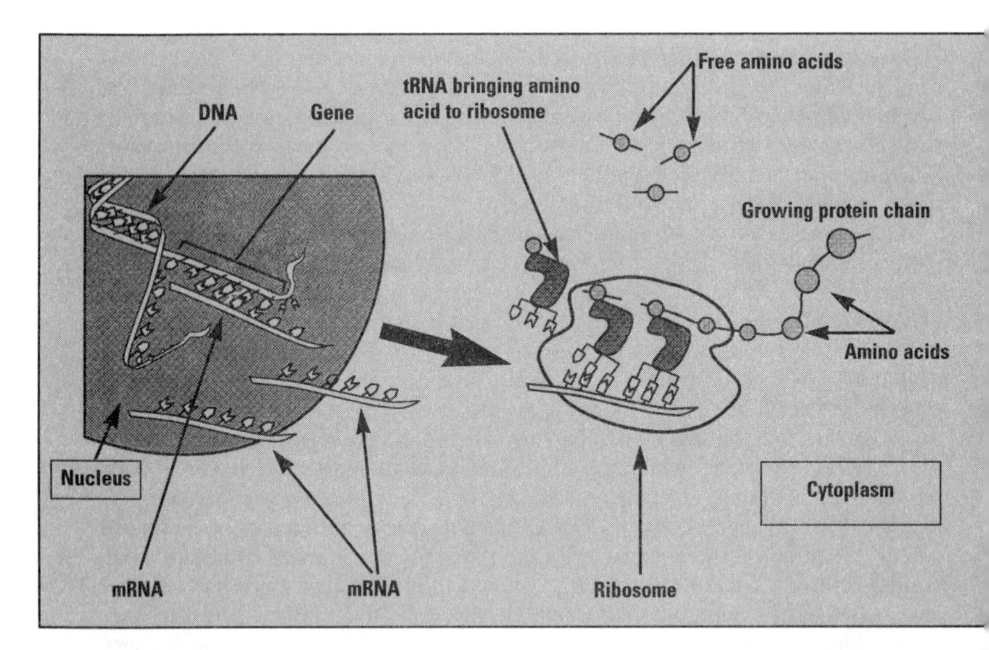

Figure 2.5 How genes function. Genes form templates for the production of a series of amino acids which make up a protein. In order for the template to be 'read', the double strand of DNA unwinds and the base pairs become separated temporarily to leave a sequence of unpaired bases. A single strand sequence of bases is then formed from this template by messenger RNA (mRNA) – which is very similar to DNA. This messenger RNA leaves the nucleus and moves to a ribosome in the cytoplasm of the cell. These ribosomes are the cell's protein factories. Another type of RNA (transfer RNA or tRNA) is involved in transporting free amino acids in the cell to the ribosomes for assembling proteins [14, 15].

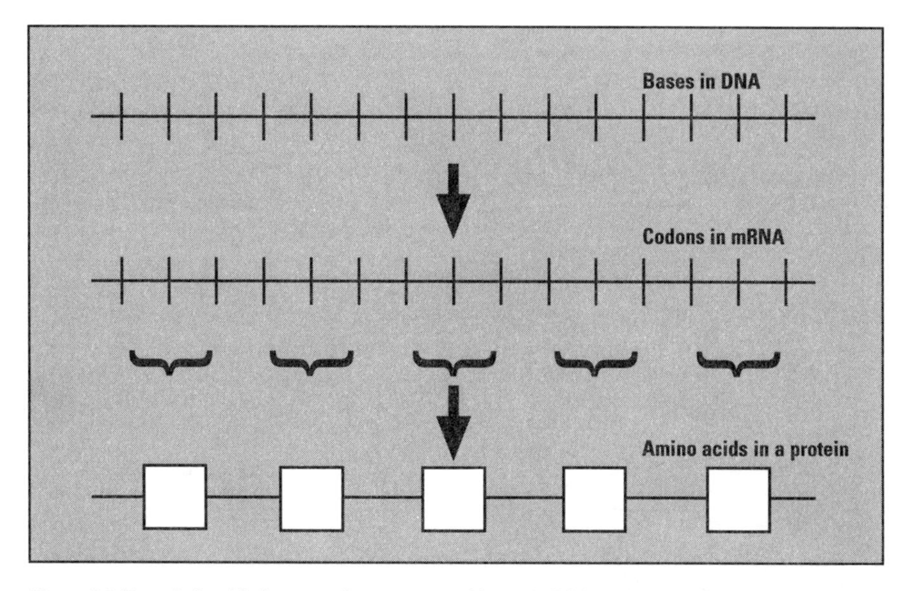

Figure 2.6 The relationship between the sequence of bases in DNA, codons in mRNA and amino acids in a protein [4].

the cell's 'protein factories', where the appropriate protein is synthesised. (See Figures 2.5 and 2.6 for more details).

Although all body cells in the same animal carry the same genetic code, different cells have very different functions. Just why some cells develop into a particular organ or tissue is still largely a mystery. What is amazing is that the same triplets appear to code for the same amino acids in all animals, and virtually the same genes are present in all animals. It is simply the presence or absence of a few triplets, or different combinations of these, arranged over a different number of chromosomes, which distinguish a frog from an elephant or a Suffolk ewe from a Belted Galloway cow.

Loci and alleles

As described above, the chromosomes occur in pairs in the body cells of all animals, with one member of each pair coming from the father via the sperm and one from the mother via the egg. Genes occur in a particular sequence along the chromosome. So, for instance, as a result of the gene mapping activities which are outlined in Chapter 9, the gene affecting the presence or absence of horns in many cattle breeds is now known to be on chromosome number 1. (Animals which have horns, and those which are naturally hornless, are called horned and polled respectively.) This gene will be located at the same position on each member of the pair for chromosome number 1 – the one from the father and the one from the mother – and the gene will be in this location for all animals of the breed or species. This location, or the site of a gene on a chromosome is called a **locus** (the plural is **loci**) (see Figure 2.7). Each locus is in the same position on the

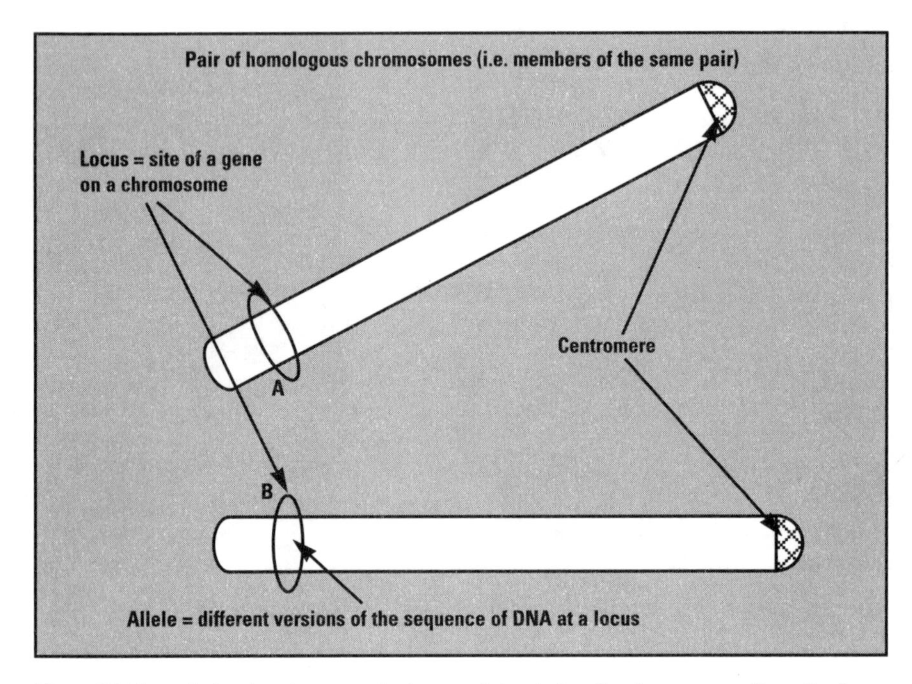

Figure 2.7 A locus is the site where a particular gene is located on the chromosome; alternative forms of DNA sequence at any locus are termed alleles. In this example the two members of a pair of homologous chromosomes (i.e. members of the same pair) are carrying different alleles, one abbreviated to A and the other to B. In a given breed, or in the population as a whole, there may be many other alleles possible at this locus (e.g. C, D, E), but individual animals can only carry either two copies of the same allele, or one copy of two different alleles. Often alleles differ only because of the substitution of a single base pair by another alternative pair of bases in their DNA sequence.

chromosome for all animals of a species. However, at that locus the two members of a pair of chromosomes may have identical or slightly different segments of DNA present. Often segments of DNA at a single locus differ only by the substitution of one base pair for another. In a population of animals there may be several alternative segments of DNA present at a given locus, but an individual animal can carry a maximum of two different versions – one on each chromosome. It is the particular form of the segment of DNA on a chromosome which is called a gene or, more properly, an **allele**. (The word gene gets used interchangeably to mean the site, or locus, and the actual form of DNA present, or allele.) As an analogy, think of the two members of any pair of chromosomes as being streets of the same length, each with the same number of houses, but with houses of various designs. The locus on a chromosome is equivalent to the house number or address, while the allele describes the type of house that is built there.

Transfer of genetic codes during cell division

Cell division is a vital part of both the growth and development of animals from a single cell to an adult containing millions of cells, and the production of the sex cells or **gametes** (sperm and eggs). There are different types of cell division involved in growth and reproduction. These are called **mitosis** and **meiosis** respectively, and they are outlined in the following sections.

Mitosis

During growth and development, the purpose of normal cell division is to create more cells which carry identical genetic information. This type of cell division is called **mitosis** or **multiplication division**. It creates two 'daughter cells' from one original cell, with the nucleus of each daughter cell containing a copy of the complete original set of chromosomes. There are several steps involved in producing copies of each of the chromosomes; the most important ones for our purpose are illustrated in Figure 2.8. The first step involves duplication of each chromosome. This is achieved by the DNA molecule unwinding from its tightly coiled shape. Then the two strands of DNA separate, i.e. the ladder is pulled apart down the middle of each rung. Next, new bases, which are present in the nucleus, move in to join up with each strand of DNA in the usual formation – thymine pairing with adenine, and guanine with cytosine (as shown already in Figure 2.4). When this process is complete for each of the original strands, there are two identical molecules of DNA (two ladders). At this stage the two copies of each original chromosome, called **chromatids**, are still joined at a point called the **centromere**. They then line up in the centre of the cell and separate, so that one copy ends up in the nucleus of each daughter cell.

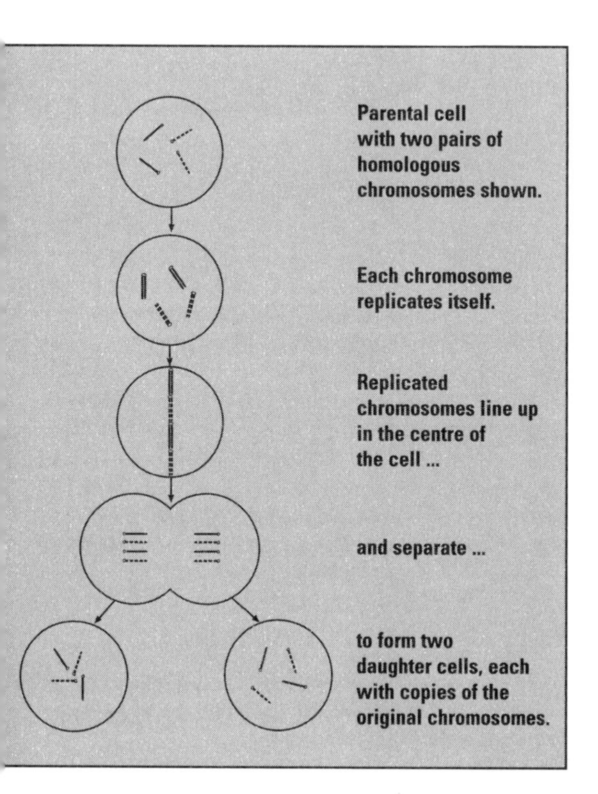
Parental cell with two pairs of homologous chromosomes shown.
Each chromosome replicates itself.
Replicated chromosomes line up in the centre of the cell ...
and separate ...
to form two daughter cells, each with copies of the original chromosomes.

Figure 2.8 Mitosis. This is the type of cell division which occurs during normal growth, development and tissue repair. Each of the daughter cells produced receives a copy of the complete original set of chromosomes [9].

Meiosis

Sperm and eggs are produced by specialised cells in the reproductive organs. The type of cell division involved here is called **meiosis** or **reduction division**. It is illustrated in Figure 2.9. In the early stages of sperm and egg production the process of cell division in the reproductive organs is similar to that which occurs during the growth of an animal. However, there are some very important differences in the later stages.

The first of these is a phenomenon known as **recombination** or **crossing over**, which occurs just before the division of cells during the production of sperm and eggs. Once the chromosomes have been copied, but while the two chromatids are still attached to each other, segments of one of the paternally-derived chromatids

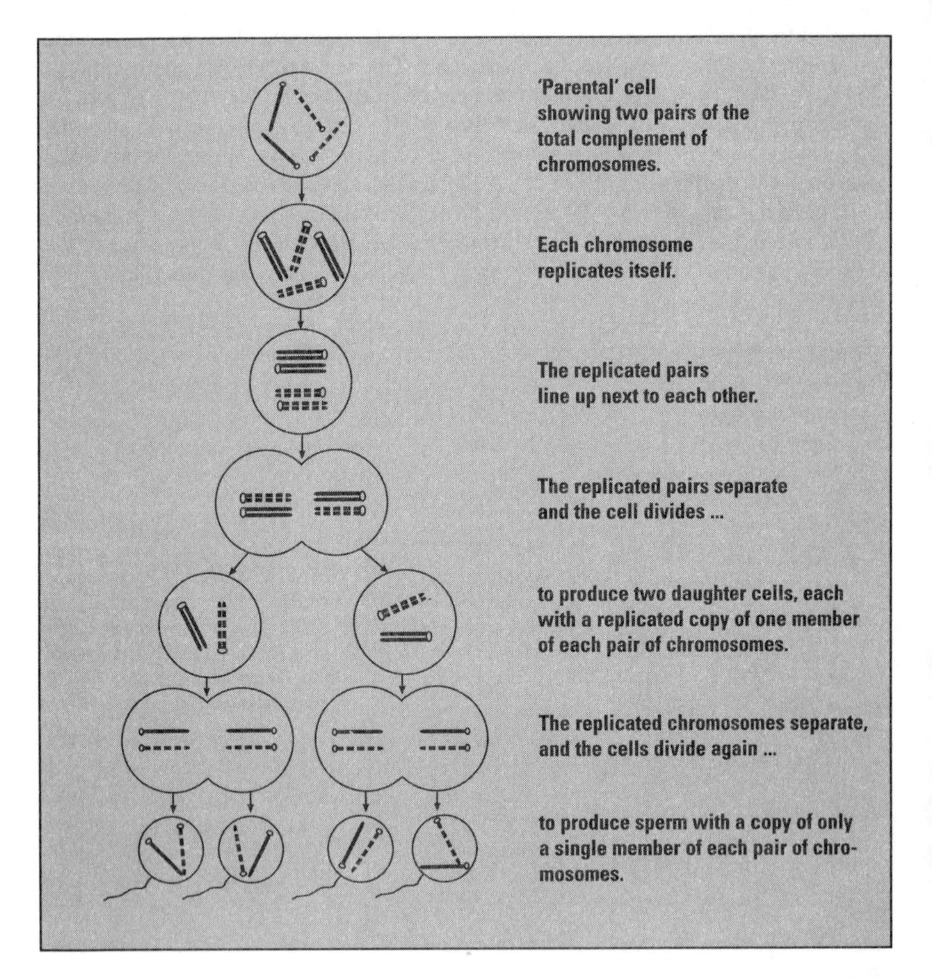

'Parental' cell showing two pairs of the total complement of chromosomes.
Each chromosome replicates itself.
The replicated pairs line up next to each other.
The replicated pairs separate and the cell divides ...
to produce two daughter cells, each with a replicated copy of one member of each pair of chromosomes.
The replicated chromosomes separate, and the cells divide again ...
to produce sperm with a copy of only a single member of each pair of chromosomes.

Figure 2.9 Meiosis. This is the type of cell division involved in the production of the sex cells or gametes. This example shows the production of sperm in the male, but a similar process is involved in producing eggs in the female. The resulting cells get only a single set, rather than pairs of chromosomes. The normal complement of pairs of chromosomes is restored at fertilisation [9].

may swap with a corresponding segment on a maternally-derived chromatid. This is illustrated further in Figure 2.10. The segments of chromosome which cross over correspond exactly in location, so there is no change in the size of chromosomes after crossing over. Recombination simply leads, as the name suggests, to new combinations of genes on each of the two chromosomes involved – some of these came from the animal's father, and others from its mother. (Recombination also occurs between the two identical chromatids, but since the two segments which swap over are identical, there is no effect on the genetic make-up of the resulting chromosomes.) Typically there are two or three recombination events per chromosome during the formation of sperm and eggs. This means that chromosomes in sperm or eggs are made up of a patchwork of segments originating from the father and mother of the animal producing the sperm or egg. Because segments of the chromosome are involved in crossing over, genes which are closer together on the same chromosome will tend to cross over together more often than genes which are far apart on the same chromosome. This is known as **linkage**, and it is one reason why some characteristics appear to be inherited together.

Meiosis also differs from mitosis in that the cells which become sperm or eggs receive only a single set of chromosomes. For example, the sperm or eggs of cattle carry 30 individual chromosomes, rather than the 30 pairs found in other cells. Whether a sperm or egg receives a copy of the maternally- or the paternally-derived member of a particular pair of chromosomes is an entirely chance event. On average, each sperm and each egg will carry half paternally-derived and half maternally-derived chromosomes, but there will be quite wide variation in this, purely by chance. As explained above, recombination causes a further mixing of maternally- and paternally-derived genes. As a result of the chance sampling of chromosomes, and of recombination, sperm and eggs contain a largely independent assortment of the genes of the animals producing them. Exceptions occur in the case of genes which happen to be located close together on the same chromosome and so tend to be inherited together.

Mutation

Another event associated with cell division which has important consequences for genetic variation is **mutation**. There are two main types of mutation. The first, called **chromosomal mutation**, involves changes in the number or structure of chromosomes. This often has a profound effect, and is believed to be a major source of embryonic mortality (see [9] for other examples of the effects of this type of mutation). The second type of mutation is termed a **gene** or **point mutation**. Whether cell division is occurring during growth, or in the production of sperm and eggs, the copying of DNA is usually correct, so that each new strand is identical to the original strand from which it was copied. However, occasionally mistakes are made, and the sequence of bases in the new copy of the DNA strand is slightly different from that in the original strand. Some of these gene mutations are corrected by special repair enzymes within the cell. If uncorrected mutations occur in any part of the body other than the reproductive organs they may interfere with the body functions of that animal, but they will not affect the genetic make-up of offspring. If, on the other hand, gene mutations occur in cells responsible for producing sperm or eggs in the reproductive organs, they lead to a change in the genetic code of offspring. For any particular gene, between one in a

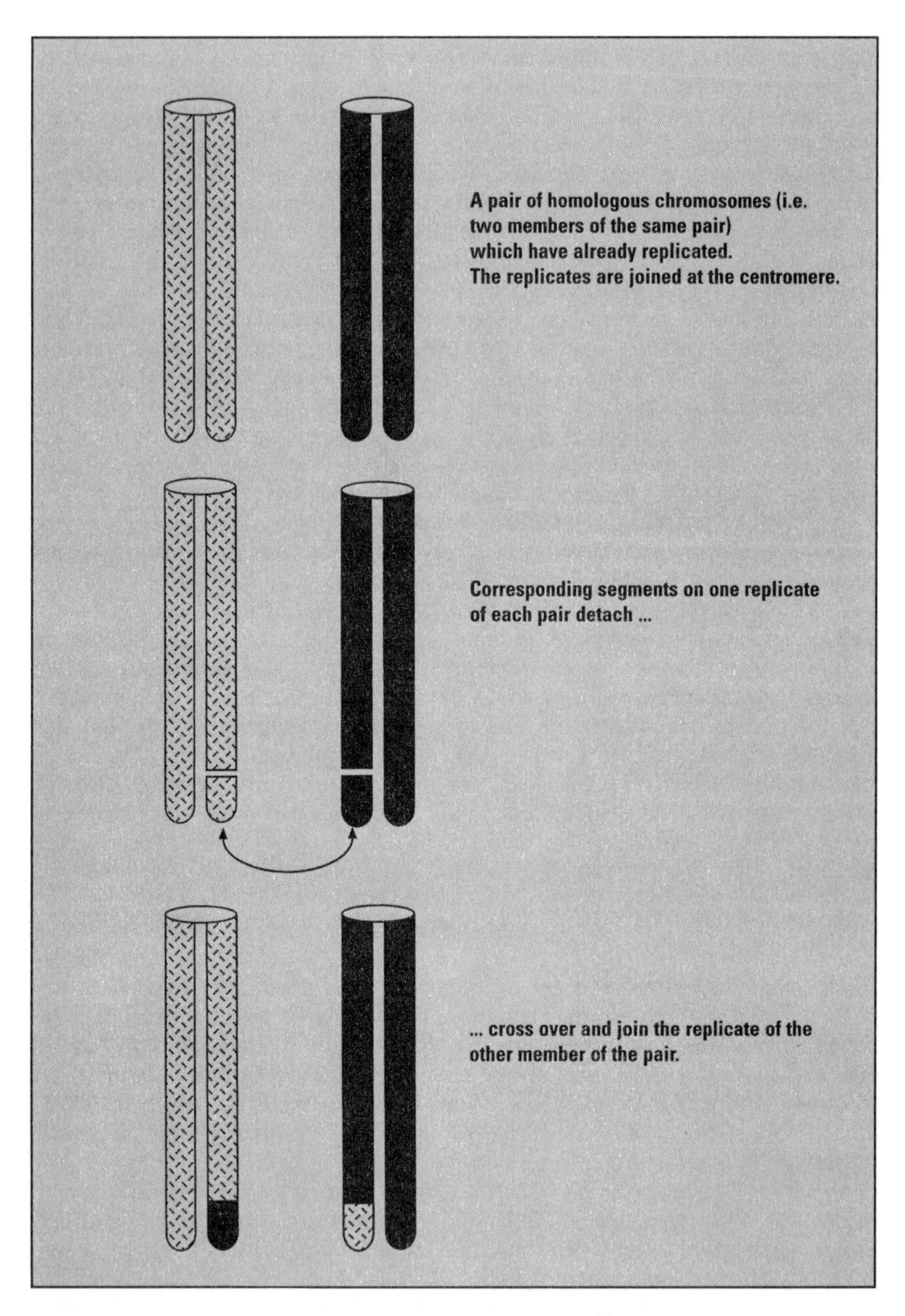

A pair of homologous chromosomes (i.e.
two members of the same pair)
which have already replicated.
The replicates are joined at the centromere.
Corresponding segments on one replicate
of each pair detach ...
... cross over and join the replicate of the
other member of the pair.

Figure 2.10 The process of recombination or crossing over. This involves segments on the maternally-derived and paternally-derived members of a pair of chromosomes swapping over during meiosis, prior to formation of sperm and eggs. Characteristics which are controlled by genes close together on the same chromosome tend to be inherited together. These genes are said to be linked. Characteristics controlled by genes further apart on the same chromosome will be inherited together less often, as a result of recombination.

hundred thousand and one in a million sperm or eggs carry a newly mutated version of the gene [7].

Variation among alleles at any locus is a direct result of gene mutations at some time in the past. The name mutation conjures up images of freaks of nature. In fact the majority of gene mutations which occur have little or no effect on an animal. But some do have a negative effect. For example, there are quite a few genetic disorders of farm animals which have arisen because of a mutation in a sequence of bases coding for important enzymes (see [10] for more details). Other gene mutations may have a positive effect. For example, the variation in colour, shape or function of wild animals which allows them to exploit new habitats or sources of food is partly a result of mutations. Similarly, much of the variation among farm animals which is exploited by selection is a result of mutations occurring over many generations both before and after domestication.

Tracking genes across generations

Tracking the contribution of chromosomes and genes across generations is quite confusing, but it is important in order to understand the degree of similarity between relatives (we will return to this in Chapters 4 and 5). It is easiest to follow if we consider three generations of animals. In this example the oldest generation are the grandparents. The middle generation of animals, the ones producing the sperm or the eggs in this example, are the parents. The animals resulting from the sperm and eggs produced by the parents are the offspring. Each of the parents got half their genes from their father and half from their mother (the grandparent generation). *On average* the sperm and eggs produced by the parent generation will carry equal numbers of genes from each grandparent. However, any particular gene carried by the sperm or egg must come from only one of the two possible grandparents. So, by chance, an individual sperm or egg may have more or less than half its genes from a particular grandparent. This happens because of the separation and chance sampling or **segregation** of paternally- and maternally-derived genes during meiosis in the production of a sperm or egg. So, all offspring get exactly half their genes from each parent, but individuals may get more or less than a quarter of their genes from each grandparent (though across the offspring generation as a whole, animals will get an average of one quarter of their genes from each grandparent). It is worth emphasising that this *potential* inequality of genetic contributions is from the grandparents, and not the parents of the offspring which results from the sperm or egg concerned. Figure 2.11 illustrates this process.

Genes, genotypes and phenotypes

It will be easier to understand more about the way that genes or alleles affect the appearance or performance of animals by considering the inheritance of coat colour as an example. In some livestock breeds there are several different versions of a gene affecting colour which can occur at a single site or locus. However, individual animals will have a maximum of two of these versions present at a locus – one on each chromosome. (In some breeds there is more than one locus

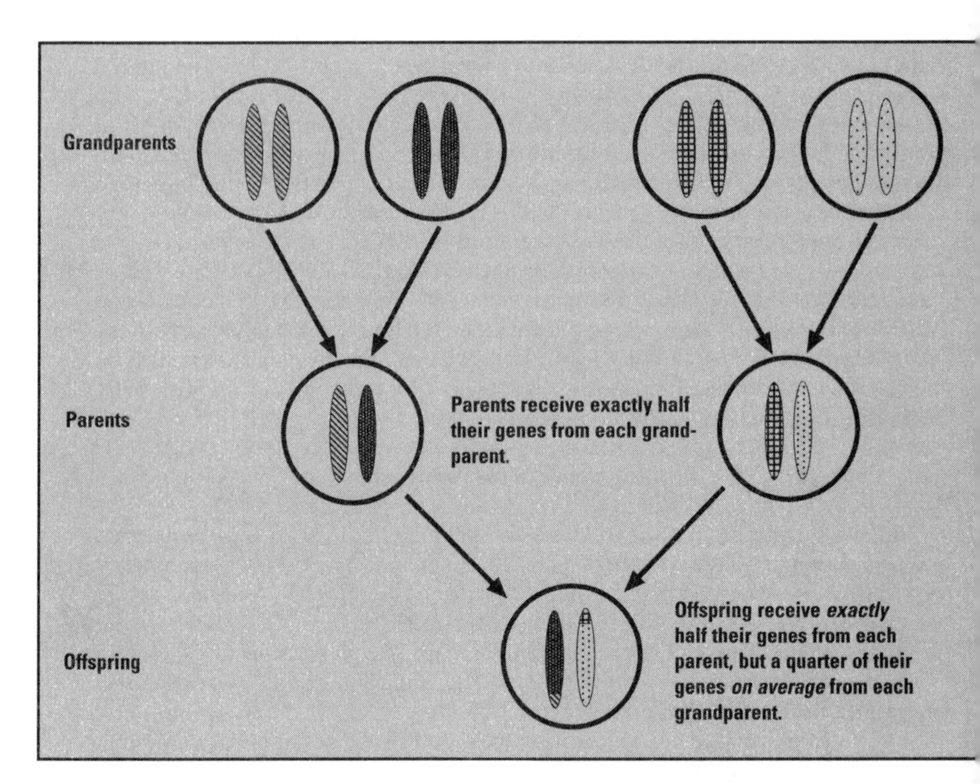

Figure 2.11 The transmission of genes from grandparents to grandoffspring. This illustrates that, although offspring get exactly half their genes from each parent, the contributions from grandparents can vary. On average one quarter of an animal's genes will come from each grandparent, but by chance individual animals may receive greater or smaller contributions from particular grandparents. In this example only a single pair of chromosomes is shown in each generation. These are equivalent pairs in each animal, but they are shown in different patterns to make it easier to follow their transmission. For any particular pair of chromosomes, offspring could potentially inherit one chromosome each from a single grandparent – so only two of the four grandparents' genes would be represented. However, in practice, recombination mixes the contributions of grandparents to particular chromosomes in offspring. Also, when *all* chromosomes are considered, the contribution of each grandparent is likely to be more equal (i.e. closer to 25%).

involved in determining colour, but for simplicity only a single locus is considered here.) The effect of different genes can be seen clearly in the coat colour of the Shorthorn breed of cattle. In this breed there are three main colours observed – red, white and roan. Combinations of these colours can also appear – for example, some animals have patches of both solid red and solid white – but for simplicity only the three single colours are considered here. What we observe when we look at animals, or measure them in some way, is referred to as the **phenotype**, so in Shorthorns there are three main phenotypes for coat colour – red, white and roan (see Plate 6). These coat colours are determined at a single locus with two versions of the gene, or two alleles, involved. One of these genes codes for red coat colour (abbreviated R here) and the other one codes for white (abbreviated W here). All Shorthorn calves get a copy of one of these coat colour genes from their father and one from their mother. So there are three possible combinations of these genes in the calves. These are: two copies of the red gene (RR) – these calves are red; two copies of the white gene (WW) – these calves are white; or one copy of each of the genes (RW) – these calves are roan. The particular combination of genes or alleles which an animal inherits is called its **genotype**. In this case there are three possible genotypes for coat colour – RR, WW and RW – although there are only two genes or alleles involved, R and W. Except in the case of a few rare genes, it does not matter which parent contributes which allele to the offspring, the effect is the same. That is, a calf which receives an R gene from its sire and a W gene from its dam will be the same colour as one which gets the R gene from its dam and the W gene from its sire. In both of these cases the genotype is abbreviated in the same way (RW). In breeds where there are three genes or alleles at a single locus for coat colour, six different combinations of these are possible. That is, six different genotypes are possible. If there are four genes involved at a single locus, then there are ten different genotypes, and so on.

Animals which carry two copies of the same gene or allele, for example the RR or WW Shorthorn cattle, are said to be **homozygous** for that particular gene or allele. Those animals which carry different copies of a gene, such as the RW Shorthorns, are said to be **heterozygous**. (The corresponding nouns are **homozygote** and **heterozygote**.)

Predicting the outcome of matings among different genotypes

Obviously, breeders of farm livestock would like to be able to predict the outcome of matings between different pairs of parents so that they can make the best mating decisions. Unfortunately, the coat colour example used above is one of the few cases where we can actually tell the genotype of the animal by looking at it. Most of the traits of economic importance in farm animals are controlled by genes at many different loci, and have many 'shades' of performance, rather than discrete categories like colours. Despite this, understanding what happens at a single locus with visible effects is an important step towards understanding the more complex situation when many genes are affecting a trait.

Although genes or alleles are present in pairs in the nuclei of body cells – one on each member of a pair of chromosomes – this pairing of alleles is broken during the production of sperm or eggs. That is, the alleles which were paired in parents pass to the next generation singly. This 'unpairing' and mixing of alleles or genes is termed segregation, as mentioned earlier. Mendel was the first to demonstrate this process, and to link it to visible effects such as seed type and flower colour in plants. The study of single genes and their effects is often termed Mendelian genetics in recognition of this. The ratios of different offspring genotypes expected from matings between particular genotypes of parent are often called the **Mendelian segregation ratios**. (For more details on Mendel's work, see the Further Reading section at the end of the chapter.)

Table 2.2 shows how the outcome of matings between particular genotypes can be predicted for a single locus of interest. The main principles are that at any locus an individual animal has two alleles present, one derived from each parent. If the animal is a homozygote these two alleles will be of the same type. If the

(a) Mating between RW bull and RW cow

Type of egg from RW cow	Type of sperm from RW bull ½ R	 ½ W
½ R	**¼ RR**	**¼ RW**
½ W	**¼ RW**	**¼ WW**
	Expected offspring genotypes and their frequencies	

Expected genotype frequency of offspring: ¼ RR, ½ RW, ¼ WW.

(b) Mating between RW bull and RR cow

Type of egg from RR cow	Type of sperm from RW bull ½ R	 ½ W
½ R	**¼ RR**	**¼ RW**
½ R	**¼ RR**	**¼ RW**
	Expected offspring genotypes and their frequencies	

Expected genotype frequency of offspring: ½ RR, ½ RW. The same outcome is expected from matings between RR bulls and RW cows.

(c) Mating between RR bull and WW cow

Type of egg from WW cow	Type of sperm from RR bull ½ R	 ½ R
½ W	**¼ RW**	**¼ RW**
½ W	**¼ RW**	**¼ RW**
	Expected offspring genotypes and their frequencies	

Expected genotype frequency of offspring: all RW.

(d) Mating between RR bull and RR cow

Type of egg from RR cow	Type of sperm from RR bull ½ R	 ½ R
½ R	**¼ RR**	**¼ RR**
½ R	**¼ RR**	**¼ RR**
	Expected offspring genotypes and their frequencies	

Expected genotype frequency of offspring: all RR. Similarly, all the offspring from matings between WW bulls and WW cows will be WW.

Table 2.2 Calculating the expected offspring genotype frequencies at a single locus, such as that determining coat colour in Shorthorn cattle, following matings between parents of different genotype.

animal is a heterozygote the two alleles will be different. Normally, when animals produce sperm or eggs, the two alleles at any locus will be represented equally, i.e. half the sperm or eggs will carry copies of one of the alleles, and half will carry copies of the other. For example, half the sperm from an RW Shorthorn bull will carry the R allele, and half will carry the W allele. So, the first step in predicting the outcome of a mating is to 'split' the genotype of each parent down into the two individual alleles or genes which their sperm or eggs will carry. Which particular egg a female sheds, and which particular sperm fertilises that egg are entirely random or chance events. So, we have to calculate the **probability** or **expected frequency** of all possible offspring genotypes which could result from a given mating.

To do this it is easiest to write down the two types of sperm and eggs on adjacent sides of a grid consisting of four squares. This grid is the shaded part of Table 2.2. Next, all four combinations of alleles from the sperm and eggs are entered into the appropriate part of the grid. On average, a quarter of the offspring from a particular type of mating are expected to have the genotype in each of the four grid squares. If there are only two different alleles involved then some of the grid squares will contain the same offspring genotype, so the frequencies of these are simply added together. For example, as shown in Table 2.2, matings between RW Shorthorn bulls and RW Shorthorn cows are expected to result in ¼ RR, ½ RW and ¼ WW calves. On average half the RW calves get their R gene from their father, and half get their R gene from their mother.

It is worth stressing that these are *expected* genotype frequencies, and that the *actual* frequencies observed can depart quite widely from these expectations, especially when the outcomes of only a few matings are considered. For example, a single Shorthorn calf born from a mating between an RW bull and an RW cow must fall into only one of the three genotypes expected, and so the expected ratio of colours cannot be reached. In fact, if we look at any less than four calves born from matings between RW Shorthorn bulls and cows, we cannot possibly reach the expected ratio of genotypes. Even with four calves from this type of mating there is less than a 20% chance of getting the expected frequency of genotypes. In practice we would need many more observations to get close to the expected ratio of genotypes. It is easy to demonstrate this by putting a ball marked R and a ball marked W in a box. These represent the alleles produced by one parent. Then two balls, representing the alleles produced by the other parent, are put in a second box. The boxes are shaken to mix up the balls, and then one ball is drawn from each box, representing the chance combination of alleles from the two parents at fertilisation. The combination of the two letters corresponds to the new animal's genotype. The balls are then replaced and the whole exercise is repeated a number of times to simulate repeated matings.

So, it is worth remembering that expected genotype frequencies tell us what to expect on average. Actual observations can depart quite widely from these expectations, especially when the number of observations is small. (See [16] for more details of how to calculate the probability of getting particular genotypes from different numbers of matings.)

Genotype and gene frequencies

The method described above for predicting the outcome of matings is useful if there is only a small number of animals involved. However, it can be extended to larger numbers of animals, as long as we know how many animals there are of each genotype in the herd or flock, or in the whole breed. The proportion of animals of each genotype is called the **genotype frequency**. It is simple to calculate if the different genotypes can be distinguished – it is just the number of animals of each genotype present, expressed as a fraction or percentage of the total (i.e. the total genotype frequency must add up to 1 or 100%). For example, if we had a herd of 100 Shorthorn cows comprising 30 red, 50 roan and 20 white cows, the genotype frequency in the herd would be 30% RR, 50% RW and 20% WW (or 0.3, 0.5 and 0.2 respectively). These genotype frequencies can be used to calculate the expected outcome of matings to bulls of a particular genotype. If only a single bull is used, the genotype frequencies can be calculated in three steps.

1. Calculating the expected outcome of matings between this bull genotype and each of the cow genotypes (as shown in Table 2.2). For example, if a red bull was used the expected outcome of the matings to these three cow genotypes would be:

Bull	Cows	Offspring		
		RR	RW	WW
RR	RR	1 (100%) RR	—	—
RR	RW	0.5 (50%) RR	0.5 (50%) RW	—
RR	WW	—	1 (100%) RW	—

2. 'Weighting' the results according to the genotype frequencies of cows in the herd. In this case there are 30% RR cows, 50% RW cows and 20% WW cows, so the weighted proportions of expected offspring genotypes as a result of mating each cow genotype to an RR bull is:

Cow genotype	Proportion of herd of this genotype	Weighted proportion of offspring of each genotype following mating to an RR bull (column 2 in this table multiplied by the proportion of each offspring genotype expected, taken from the table above)		
		RR	RW	WW
RR	0.3 (or 30%)	0.3 x 1 = 0.3 (30%) RR	—	—
RW	0.5 (or 50%)	0.5 x 0.5 = 0.25 (25%) RR	0.5 x 0.5 = 0.25 (25%) RW	—
WW	0.2 (or 20%)	—	0.2 x 1 = 0.2 (20%) RW	—

3. Totalling the expected proportion of calves of each genotype. In this case the overall proportion of different calf genotypes is 0.55 (or 55%) RR plus 0.45 (or 45%) RW.

If more than one genotype of bull is involved, then this process has to be repeated for each genotype, and the results weighted according to the genotype frequency of bulls. This approach soon gets laborious when there are several genotypes of each sex involved. Although it is useful to know the genotype frequency in an initial group of animals, once the numbers are large, and we consider the results of several generations of matings, it is often more convenient to think of **gene (or allele) frequency**. As the name suggests, gene frequency is based on a count of the genes or alleles present in a herd, flock or breed, rather than the genotypes. This is also more logical as it is individual genes or alleles, not genotypes, that get passed from parent to offspring.

To calculate the gene frequency we have to split each different genotype down into its two constituent alleles, and then weight these according to the proportion of the herd or flock carrying these alleles. For example, RR Shorthorn cattle carry only R genes; WW animals carry only W genes; RW animals carry half R genes and half W genes. Hence, the frequency of the R gene can be calculated as:

(1 x RR genotype frequency) + (0.5 x RW genotype frequency)

So, in our herd of 100 cows made up of 30% RR, 50% RW and 20% WW animals, the frequency of the R gene is:

(1 x 0.3) + (0.5 x 0.5) = 0.55

Similarly, the frequency of the W gene can be calculated as:

(1 x WW genotype frequency) + (0.5 x RW genotype frequency)

= (1 x 0.2) + (0.5 x 0.5)
= 0.45

Or more conveniently, since only two genes are involved, the frequency of the W gene is simply:

1 – frequency of R gene

= 1 – 0.55
= 0.45

If there is no selection for or against particular alleles at a locus, no mutation and no movement of animals into or out of a large random-mating population of animals, then the gene and genotype frequencies in that population tend to stay the same from one generation to the next. This is called the **Hardy-Weinberg equilibrium**, after the mathematician and the physician who demonstrated it independently in the early 1900s. So, without forces such as selection operating, the variation in a population is maintained rather than disappearing over time. In farm animal breeding the conditions for the Hardy-Weinberg equilibrium are rarely met, and so we can expect gene and genotype frequencies to change. For a start, the populations are usually relatively small (e.g. individual herds or flocks, or even all herds or flocks which interbreed in a particular country), and so

ur by chance alone. These chance changes in gene frequency are
ift. Also, there are often imports or exports of breeding stock to
lar breeding population. Finally, most breeding programmes are
videspread use of parents with favourable characteristics, and
m mating. So, selection in farm animals usually changes gene and genotype frequencies, but this is not usually obvious, unless selection is for something as visible as coat colour. For example, if the herd of cows mentioned above was bred entirely to WW (white) bulls, then the frequency of the W allele would be higher in the calves than it was in the original herd of cows. If female calves from this new generation were in turn bred to a white bull, then the frequency of the W allele would increase further in their offspring. If matings were always made to WW bulls, the R gene would eventually disappear altogether from the herd. In this case the frequency of the W allele would be 1, and that of the R allele would be 0. The W allele is then said to be **fixed** in this population, and the R allele **lost** from it.

If we know the frequency of a particular allele of interest, we can estimate how many generations of selection it would take to increase or decrease the frequency by a certain amount. This is useful in planning eradication of genetic diseases controlled by single genes, or increasing the frequency of a favourable single gene in a breed, for example a single gene affecting the cheese-making properties of cows' milk, or one affecting fertility in sheep. It is generally easier to alter the frequency of a gene which is already at an intermediate frequency, than to alter the frequency of a gene which is already at either a very high or very low frequency. (For more details on these calculations, and their application, see [7, 9]).

Sources of genetic variation between animals

Genetic improvement depends on genetic variation. If there is no genetic variation between animals in traits of interest, there can be no improvement. To put it another way, if there are no animals genetically better than the rest, there is no way of picking better parents to breed from. The most important sources of genetic variation between individuals or groups of animals have been mentioned in the preceding sections, but genetic variation is so central to genetic improvement that it is worth discussing them in more detail here. These sources of variation are:

- differences in gene (allele) frequencies
- segregation of genes (alleles)
- recombination
- mutation

Differences in gene (allele) frequencies are a major source of variation between groups of animals of a particular species. If we look far enough back into history, then all animals of the same species have common ancestors. Offspring get half their genes from each parent, and this continues down the generations. So, at first sight, we might expect distant ancestors and current animals to have copies of the same genes. However, as mentioned earlier, there is an element of chance as to which half of each parent's alleles the offspring receive. This means that some alleles carried by parents may never appear in offspring, while others may appear often. If you toss a coin often enough you expect it to land on heads 50% of the time, and on tails 50% of the time, but with a short sequence of tosses you often get several heads in a row, or several tails in a row. So it is with the transfer of alleles from one generation to the next. A parent which is heterozygous at any locus is expected to produce half of its sperm or eggs carrying each of the two alleles. However, because only a few offspring are produced, by chance these may all carry copies of the same allele. This can lead to changes in gene frequency. In the case of domestic livestock species, the populations have expanded enormously, and have become subdivided. So there may be differences between subpopulations which have occurred by chance alone. Also, these species have been subject to many generations of natural and artificial selection. As a result, particular alleles present in the original population may have been lost altogether or, at the other extreme, all the animals may carry two copies of the same allele at a particular locus. Between these two extremes, different breeds or groups of animals may have different gene frequencies.

Perhaps it is easiest to visualise this by considering genes for coat colour again. In some breeds of cattle or sheep all the animals are the same colour, whereas in others there are several different colours. Coat colour in all mammals appears to be controlled by a series of at least six loci, with several alleles possible at each locus [9]. The reason why some breeds have little or no variation in colour, while others show wide variation, is that some of the alleles at some loci have been lost in the less variable breeds, either by chance or by deliberate selection. In the Icelandic breeds of cattle, sheep and horses, and a few other breeds, there has been little or no systematic selection for or against particular colours, and so there is a huge range in coat colour (see Plates 7 to 9). In contrast, in many other cattle breeds coat colour is effectively determined at a single locus. This may be because there is only variation at a single locus (i.e. there are two identical alleles at each of the other loci), or because any variation at other loci is masked by the alleles present at a single locus. So, in breeds where all the animals are the same colour, it is likely that the allele for this colour has become fixed, while those which produced other colours in ancestors have been lost, or that the effect of any remaining alleles for other colours is masked. (This process of masking is discussed in the next section.)

Within a population of animals, genetic variation also occurs because of segregation. As explained earlier, this is a consequence of parents carrying two copies of each gene, one on each member of the pair of chromosomes concerned, but sperm and eggs carrying only a single copy of each gene at any locus. When animals reproduce, the sperm or eggs they produce carry half their genes, but the particular sample of genes which any sperm or any egg carry is down to chance. Chance is also involved in which particular egg is shed by a female, and which particular sperm successfully fertilises this egg. Similarly, recombination helps to mix the genes present in a breed, herd or flock, and to create new combinations of genes, which contributes to genetic variation.

The different genes present at any locus have arisen because of mutations at some time in the past. However, until recently, new gene mutations were considered to be a minor source of genetic variation, and usually to have negative effects. It is now recognised that they make an important contribution to genetic variation, and that their effects are sometimes beneficial. New gene mutations are a particularly important source of variation in populations which have been under selection for a long time. They are largely responsible for most selected populations continuing to change, even after many generations of selection. Without mutation as a source of new variation, very highly selected populations would reach a plateau or limit in the character under selection.

To use a card-playing analogy, the first three sources of genetic variation mentioned above – differences in gene frequency, segregation and recombination – all cause variation by shuffling and splitting the existing pack. However, mutation can create completely new cards.

Types of gene action

Additive and non-additive gene action

So far we have considered only the simplest type of gene action – where the heterozygous animals show characteristics of both homozygous types. In the Shorthorn coat colour example discussed earlier, for instance, the heterozygous animals had both red and white hairs, whereas the homozygous types had either red or white hairs. In this case there is said to be **co-dominance**, or **no dominance**. The equivalent type of gene action when performance is measured on some scale occurs when heterozygotes are exactly intermediate to the two homozygous types. In these cases the type of gene action is described as **additive**. However, there are several other types of gene action, collectively called **non-additive** types of gene action. These are **complete dominance**, **partial** (or **incomplete**) **dominance** and **overdominance**. The distinctions between these types of gene action, and some practical examples, are discussed in the following sections. These types of gene action are also illustrated in Figure 2.12.

Co-dominance, no dominance or additive gene action

The inheritance of red, white and roan coat colour in Shorthorn cattle, described above, is a good example of co-dominance. (As mentioned before, as well as occurring singly, these colours can occur in combination, but for simplicity this is ignored here.) In this breed, the segment of DNA which we call the gene R leads to the production of red pigment in the coat hairs. In animals which have two copies of this gene all the hairs will be red. The gene W leads to no pigment production, and the hairs appear white. Animals with two copies of this gene appear all white. Heterozygous animals get one copy of each gene, and so have some pigmented hairs and some unpigmented hairs, and they appear roan. Similarly, this type of gene action is often seen when the products of a gene – such as enzymes or blood types – can be detected directly. Advances in molecular biology are now allowing direct analysis of the sequence of bases seen in alternative alleles at many loci in farm animals. Sequences of DNA used as markers or potential indicators of improved performance are often co-dominant. That is, the two alternative sequences seen in homozygotes can both be detected in heterozygotes (see Chapter 9).

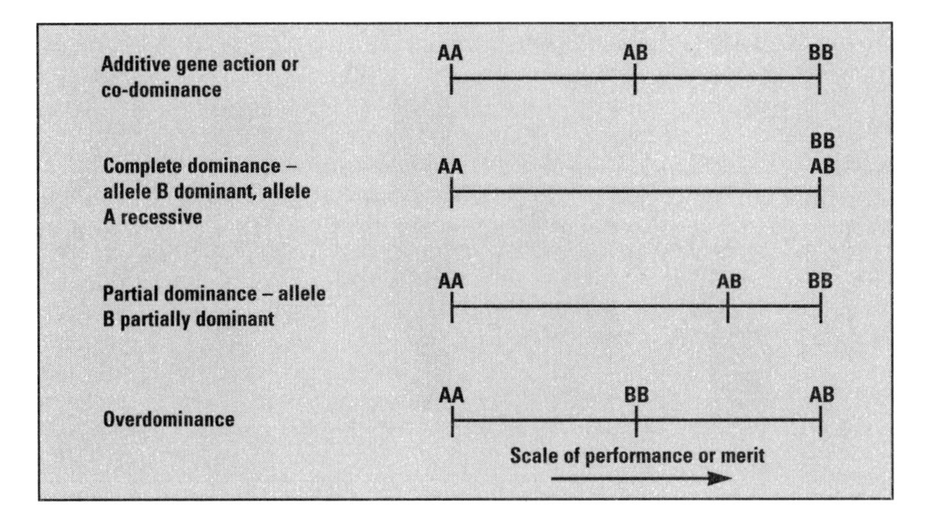

Figure 2.12 Four different types of gene action. The horizontal scale shows some visible or measurable 'performance' characteristic in animals. This could be a trait like colour or blood group with few alternatives, or a trait like milk yield or live weight measured on a continuous scale. Co-dominant gene action gives heterozygotes which show characteristics of both homozygous types, e.g. they have both red and white hairs in the case of Shorthorn coat colour, or they have two blood haemoglobin types, whereas homozygotes have only one. The equivalent type of gene action when performance is measured is called additive gene action. Additive gene action gives heterozygotes with performance midway between the two homozygotes. With complete dominance, heterozygotes have the same level of performance as one of the homozygotes. The performance of heterozygotes is less extreme with partial dominance, but still closer to that of one of the homozygotes. Overdominance leads to heterozygotes which are more extreme in performance than either homozygote [7].

The Booroola gene in sheep provides a good example of the equivalent form of gene action – additive gene action – in characteristics which are counted or measured, rather than showing obvious visible differences like colour. The Booroola gene was discovered in the late 1970s in Merino sheep in Australia, but it has been introduced since then to a number of other breeds. The Merino breed normally has a relatively low number of lambs, but animals carrying this gene were noticed because of their high fecundity. Closer examination of the results of matings between fecund and normal animals showed that a single gene was involved, which had an approximately additive effect on ovulation rate. In other words the performance of heterozygous animals is close to the average of that of the homozygous normal animals and those homozygous for the Booroola gene. The genotype of normal animals is denoted ++, the genotype of those carrying one copy of the Booroola gene is denoted F+, and the genotype of those carrying two copies of the Booroola gene is denoted FF. In one well-documented experiment the ++, F+ and FF Merino ewes had ovulation rates of about 1.40, 2.82 and 4.38 ova respectively, so each copy of the F gene added an extra 1.4 to 1.6 ova ([11]; see Figure 2.13). The term additive refers to the fact that there is a direct link between the number of copies of a gene and some characteristic of the animal – in this case ovulation rate.

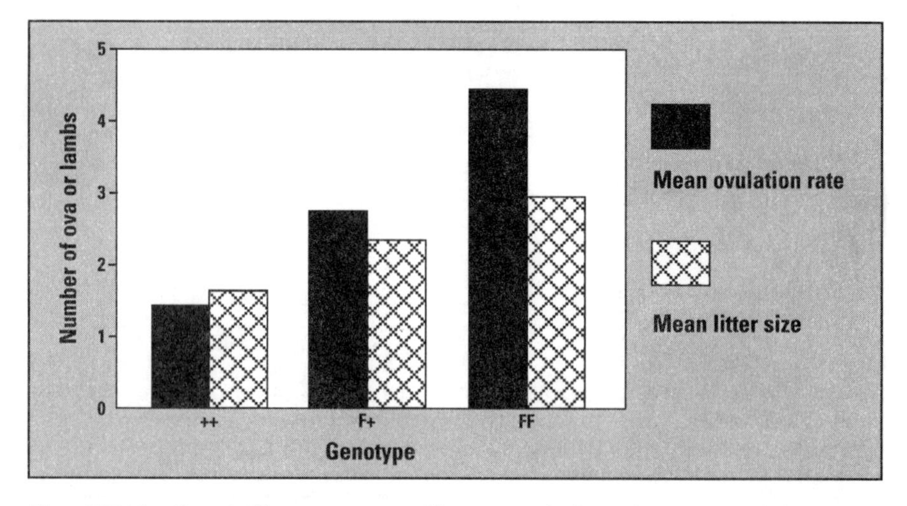

Figure 2.13 The effect of different genotypes, with respect to the Booroola gene, on ovulation rate and litter size of Merino sheep. In this study, the ++, F+ and FF ewes had mean ovulation rates of 1.40, 2.82 and 4.38 ova, so each copy of the F gene adds an extra 1.4 to 1.6 ova. The three genotypes gave birth to 1.48, 2.17 and 2.66 lambs, so each copy of the F gene adds 0.5 to 0.7 lambs (litter size for ++ animals is higher than predicted from ovulation rate, probably because not all of the ovulation and litter size records refer strictly to the same oestrous cycle). Hence, additive gene action appears to apply to ovulation rate. The effects on mean litter size appear to be partially dominant, probably as a result of higher prenatal mortality in larger litters. ([11]; Dr LR Piper.)

Complete dominance

Unfortunately, with non-additive gene action it is not usually this easy to tell what the genotype of an animal is from its phenotype. To illustrate this, let us look at another coat colour example. Most Aberdeen Angus cattle are black, but there is also a red gene in the breed. Animals which carry two copies of the black gene appear black (this genotype is abbreviated to BB here), and those that carry two copies of the red gene appear red (this genotype is abbreviated bb here; the rationale for these abbreviations follows shortly). However, when these two types of animals are mated to each other, all the resulting calves are black (they have the genotype Bb). So in this case, there are still two genes and three genotypes involved, but only two phenotypes – black or red. It is not possible to tell from looking at the black animals whether or not they carry the red gene (i.e. whether they are BB or Bb). This type of gene action is called complete dominance. It occurs when the effect of one allele is completely masked, or overridden, by the presence of another. In this case a single copy of the red gene is completely masked by the presence of a single copy of the black gene. The black gene is called a **dominant gene**, while the red gene is called a **recessive gene**. The expected genotype frequencies of offspring from different matings, the segregation ratios, can be calculated as shown in Table 2.2. The results for all possible matings between BB, Bb and bb animals are summarised in Table 2.3. (The black or red coat colour seen in Holstein Friesians and several other breeds is inherited in the same way as that described here for the Angus breed.)

Type of mating	Expected genotype frequency (segregation ratio) among offspring			Corresponding phenotypes	
	BB	Bb	bb	Black	Red
BB x BB	1	0	0	1	0
BB x Bb	$^1/_2$	$^1/_2$	0	1	0
BB x bb	0	1	0	1	0
Bb x Bb	$^1/_4$	$^1/_2$	$^1/_4$	$^3/_4$	$^1/_4$
Bb x bb	0	$^1/_2$	$^1/_2$	$^1/_2$	$^1/_2$
bb x bb	0	0	1	0	1

Table 2.3 Expected offspring genotype frequencies, or segregation ratios, and corresponding phenotypes following matings between parents of different genotype for coat colour in the Aberdeen Angus cattle breed [9].

If a gene acts co-dominantly, and it also has a readily visible effect, then the heterozygotes can be distinguished by their appearance or phenotype, as in the case of roan Shorthorn cattle. If there is complete dominance, as in the case of coat colour in the Aberdeen Angus or Holstein Friesian breeds, then the heterozygotes cannot be distinguished by their appearance. These heterozygotes are also called **carriers** of a recessive gene. In some cases it is possible to detect carriers from a blood test, for example when carriers of a genetic disease produce different levels of a particular enzyme from homozygous normal animals, but often there are still animals whose genotype remains uncertain. In other cases, until the recent development of DNA-based tests, the only way of detecting heterozygous or carrier animals when a recessive gene is involved was to do test matings. This involves mating animals of unknown genotype to those of known genotype, and making an inference about the unknown genotype based on the frequency of offspring of different types. For example, if a black Angus bull is mated to red Angus cows we would expect to get no red calves at all if the bull is a homozygote (BB). But, if he is a heterozygote or carrier (Bb), we would expect to get 50% red calves, on average.

There are a number of genetic diseases in farm animals which are under the control of a single recessive gene and where test matings of this type have been used to prevent carriers being used inadvertently. For example, there is a condition called tibial hemimelia syndrome which occurs in the Galloway cattle breed. Homozygous recessive calves have severely shortened and twisted hind limbs and a large abdominal hernia. Affected calves are usually born alive but die soon after birth. In the mid-1970s a screening programme was operated by the Galloway Cattle Society and the East of Scotland College of Agriculture, to check that Galloway bulls were not carriers of the gene concerned, before they were used for artificial insemination (AI). Since the condition is lethal, and so no homozygous recessive cows were available, a herd of known carrier cows was established, and test matings were made to these cows.

Table 2.4 shows the number of offspring from test matings of this type required to achieve different probabilities of detecting carriers of a recessive gene. The table also shows the numbers of offspring required from matings to recessive homozygous animals, when these are available, and for matings to offspring of the

Animal under test mated to	Number of offspring required to reach this probability of detecting carriers		
	95%	99%	99.9%
Homozygous recessive (i.e. affected) animals	5	7	10
Known heterozygotes (carriers)	11	16	24
Its own offspring	23	35	52

Table 2.4 Number of offspring required from different types of test mating to achieve probabilities of 95%, 99% or 99.9% of detecting a carrier of a recessive gene [9].

parent under test. Test matings for recessive genes should be made with these minimum numbers of offspring in mind. For example, to have a 95% chance of detecting a bull carrying a copy of the gene causing a defect like the tibial hemimelia syndrome, the bull under test would need to be mated to enough known carrier cows to produce eleven progeny. To have a 99% chance of detecting a carrier bull, sixteen progeny would be needed. Of course, whenever a single affected offspring is produced the test is complete, and the animal under test is known to be a carrier.

One of the spin-offs from the development of gene maps in farm livestock (maps describing the location of functional genes, or other sequences of DNA on the chromosomes of the species concerned; see Chapter 9 for more details) is that DNA-based tests for carriers of recessive genes affecting colour, polledness, genetic diseases, and many other characteristics of interest, are becoming available. When the exact location of the gene of interest is known, then these tests can detect carriers with 100% certainty. However, if the exact location of the gene of interest is unknown, but the location of a closely linked gene is known, this can be used as a marker. Because the gene of interest and the marker occur close together on the same chromosome, they will usually be inherited together. Recombination may separate them occasionally, but this will occur less frequently the closer together the loci for the marker gene and the gene of interest. So a test for a particular marker genotype may indicate the presence of the favoured genotype for the trait of interest. A useful feature of many DNA markers is that, even if the gene acts non-additively at the level of the trait of interest (e.g. colour or polledness) markers at the DNA level are co-dominant, and so heterozygotes are readily detectable.

Whether the location of the gene of interest is known, or whether it is the location of a marker which is known, blood or other tissue tests could be used to detect carriers as an alternative to expensive test mating programmes such as that described above. Such DNA tests are already in use for some genetic disorders in dairy cattle. For example, BLAD (bovine leucocyte adhesion deficiency) and DUMPS (deficiency of uridine monophosphate synthase) are both genetic diseases of cattle, which are controlled by single genes. To prevent the inadvertent widespread use of carriers through AI, and hence the risk of producing homozygous affected animals in later generations, most young dairy bulls of affected breeds are screened using simple blood tests prior to entering progeny tests, and homozygous affected or carrier animals are excluded. Tests are also available now for cattle which are carriers of the horned gene and the red coat colour gene.

Partial dominance

Partial dominance occurs when the effect of one allele is partially, but not completely, masked by the presence of another allele. In other words, the 'performance' of heterozygotes is closer to one type of homozygote than the other. Although the effect of the Booroola gene is additive with respect to ovulation rate, the gene appears to show partial dominance with respect to litter size. In this case the F+ genotype is slightly closer to the FF genotype in litter size than to the ++ genotype (i.e. the F gene appears to be partially dominant – see Figure 2.13).

In some cattle breeds there is a single gene present which causes extreme muscularity or **double muscling** (muscular hypertrophy). This is particularly common in the Belgian Blue and Piedmontese breeds. Although the results are somewhat inconsistent, from some studies it has been suggested that the double muscling gene is partially dominant to the gene for normal muscularity. In other words heterozygotes are closer to homozygous double-muscled animals in appearance and performance than to the homozygous normal animals [3].

Overdominance

Overdominance occurs when heterozygotes show more extreme performance than either homozygote. This type of gene action is probably quite rare, at least for traits of economic importance. The Inverdale gene, which affects fertility in sheep, provides a good example of overdominance. The gene was discovered in Romney sheep which had been screened, on the basis of high fertility, into a research flock at the Invermay Agricultural Centre in New Zealand. Romney ewes carrying a single copy of the Inverdale gene have litter sizes about 0.58 lambs higher than normal Romney ewes. (Because of higher neonatal lamb mortality, this is reduced to about 0.17 extra live lambs at one day of age.) However, ewes carrying two copies of the Inverdale gene are completely infertile, as they have undeveloped ovaries [6].

Other types of gene action and inheritance

Sex-linked genes

As mentioned earlier, genes are either located on the autosomes (i.e. on any of the chromosomes except the sex chromosomes), or they are located on the sex chromosomes, in which case they are said to be **sex linked**. Since there is only one pair of sex chromosomes in each cell, and many more pairs of autosomes, it follows that the majority of genes are autosomal. However, there are a few important or interesting examples of sex-linked genes in farm livestock. These include the Inverdale fertility gene in sheep which was mentioned above. The Inverdale gene is located on the X chromosome, so females can carry 0, 1 or 2 copies of the gene (abbreviated ++, I+ and II respectively), while males, which have only one copy of the X chromosome, carry either 0 or 1 copy of the gene (abbreviated + and I). Hence, calculating segregation ratios for sex-linked genes is a bit more complicated than it is for autosomal genes. The key is to calculate genotype frequencies separately for male and female offspring. This is illustrated

Type of mating		Expected genotype frequencies in offspring				
		Female offspring			Male offspring	
Sire	Dam	SS	S+	++	S	+
S	SS	1	0	0	1	0
S	S+	½	½	0	½	½
S	++	0	1	0	0	1
+	SS	0	1	0	1	0
+	S+	0	½	½	½	½
+	++	0	0	1	0	1

Table 2.5 Expected offspring genotype frequencies for a hypothetical sex-linked gene S, on the X chromosome, following matings between parents of different genotype. (The Inverdale fertility gene in sheep follows this pattern of inheritance, except that females homozygous for the Inverdale gene (II) – equivalent to the SS genotype in this example – are infertile, so not all types of matings listed here produce offspring.)

for a hypothetical X-linked gene in Table 2.5 (few genes have been identified on Y chromosomes). An added complication in the case of the Inverdale gene is that females homozygous for the Inverdale gene (II) are infertile, so not all types of matings listed produce offspring.

Sex-limited genes

Sex linkage is not the only route for the sex of an animal to affect gene expression. Sex-linked genes occur on the sex chromosomes. But autosomal genes may be **sex limited**, even though two copies of the genes are carried by both sexes, if the trait they affect is only shown by one sex, e.g. milk production or ovulation rate.

Sex-influenced genes

The expression of genes may be **sex influenced**, without being sex linked or sex limited. For example, in some sheep breeds, including the Merino and the Welsh Mountain, the males are usually horned and the females usually polled even though they have the same genotype. The polled locus is not on one of the sex chromosomes, so it is not sex linked. Similarly, there is no basic biological reason for horn growth to be sex limited – the fact that both the males and females of many other sheep breeds are horned proves the point. So there must be another mechanism, such as the level of circulating sex hormones, which alters the expression of the horned gene in some breeds, depending on the sex of the animal. The coat colour of some cattle breeds is darker in males than females, probably for similar reasons.

Genomic imprinting

In most cases, it does not appear to matter which sex of parent an animal inherits an allele from – the effects are the same. For instance, in the Shorthorn coat colour example mentioned earlier, it does not matter whether roan calves get their R allele from their sire or their dam, they still turn out to be roan. However, there are a few known cases (and probably many more unknown ones) where this rule does not hold, and the effects of a particular genotype on the appearance or function of an animal depend on which parent contributed each allele [13]. This is known as **genomic imprinting**. Most of the known examples of this type of gene action are in laboratory animals or humans, but there is a suspected case of this in sheep. This is the so-called **callipyge gene**, which causes extreme muscularity in sheep [5] (see Plates 10 and 11). Initially this was thought to act like an autosomal dominant gene, but more recent studies on the pattern of inheritance suggest that lambs which inherit the callipyge gene from their dam do not show extreme muscularity.

Pleiotropy

Often an allele at a single locus has an influence on more than one animal characteristic. This is known as **pleiotropy**, and a gene which affects several characteristics is said to have a **pleiotropic** effect. For instance, a single gene is believed to be responsible for double muscling in several cattle breeds, and this condition is often associated with reduced fertility, lower calf survival and sometimes with increased stress susceptibility [3]. The homozygous polled condition in goats leads to the development of intersex animals, rather than females [16]. Also, there are associations between the Merle coat-colour gene and eye and other defects in dogs, and between the Overo coat-colour gene and lethal intestinal obstructions in horses [10]. There are many other less dramatic and less obvious cases of pleiotropy in traits of economic importance in cattle and sheep.

Epistasis

We have already seen that the effect of one allele at a single locus can be influenced by the presence of another allele at that locus on the other member of a pair of chromosomes – this is termed dominance. When the presence of an allele at one locus masks the effect of an allele *at a different locus*, this is termed **epistasis**. While it is likely that epistasis affects many traits of economic importance in farm animals, once again it is easiest to visualise the effects of epistasis from a coat colour example. The example which is probably familiar to most people is that of coat colour in Labrador retriever dogs. Colour is controlled by two alleles at each of two loci, called the B and E series. In the presence of the dominant E allele (either one or two copies), one or two copies of the dominant allele B leads to black animals (i.e. BB and Bb are both black), while homozygous recessive animals (bb) are chocolate coloured. All animals which are homozygous for the recessive e allele are yellow, regardless of their genotype at the B locus (though their noses are different colours, just to make life even more interesting!). These combinations of genotypes, and their effect on colour, are summarised in Table 2.6. A similar example occurs in some cattle breeds, such as the Simmental. In Europe most Simmentals are either deep red, or a lighter yellow colour, in each case with a white face and markings. In fact both types carry two copies of the red allele, but the expression of these is influenced by the presence or absence of an allele at a different locus which causes dilution of coat colour. Without the dilution allele, the cattle appear red, but with either one or two copies of the dilution allele, they appear yellow [12].

Genotypes				Phenotype
BBEE	BbEE	BBEe	BbEe	Black
bbEE	bbEe			Chocolate
BBee	Bbee			Yellow (black nose)
bbee				Yellow (liver nose)

Table 2.6 The epistatic effect of coat colour genes in Labrador retrievers [18].

Incomplete penetrance and variable expressivity

There are some genetic disorders in animals which are known to be the result of a single recessive gene, but not all homozygous recessive animals show the disease. This is termed **incomplete penetrance**. A good example of incomplete penetrance occurs in the malignant hyperthermia syndrome in pigs. Animals affected by this disorder show increased body temperature, muscle rigidity and respiratory complications, leading to death, when they are exposed to mild stress. Fortunately a test has existed for many years which allows pig breeders to check whether prospective breeding animals are susceptible. The test involves allowing pigs to breathe halothane vapour. Those which react by showing a stiffening of the muscles are classified as susceptible (the majority show a complete recovery from the test). However, a proportion of animals are misclassified as reactors or non-reactors. Although the frequency of misclassifications is reduced with increasing experience of the person conducting the test, there is still less than 100% detection of affected animals.

A similar situation arises with a few other genetic disorders caused by a single gene. In these cases all animals of the affected genotype show the disease, as expected, but they show it to varying extents. A good example is the condition known as mulefoot which occurs in cattle and pigs. It is believed to be caused by a single gene which is recessive in cattle but dominant in pigs. Affected animals have solid hooves, rather than the usual cloven hooves. However, animals with the mulefoot genotype can have from one to four feet affected. This is known as **variable expressivity** [16].

There are many situations in animal breeding where the inheritance of particular phenotypes cannot be explained in terms of the animal's genotype at a single locus. In a few cases, such as those outlined above, this may be due to incomplete penetrance or variable expressivity. However, in the vast majority of cases, it is much more likely that the phenotypes observed are the result of the action of genes at more than one locus, or that the phenotype is affected by non-genetic factors (e.g. feeding and management) as well as by genes at one or more loci. These situations are discussed in later sections in this chapter.

Cytoplasmic or mitochondrial inheritance

For generations livestock breeders have claimed that particular maternal lineages have special attributes. Dairy cattle breeders in particular stress the importance of 'cow families'. Until recently these claims were often dismissed by scientists as being chance effects, or wishful thinking. However, there is now some evidence of a real effect of cow families, albeit small. So far the discussion of the transmission of genetic material has focused solely on DNA in the nucleus of cells. While the majority of the DNA is in the nucleus, there are small amounts outside the nucleus, in the cytoplasm. In particular, there is DNA in the **mitochondria** (see Figure 2.1). These are small structures in the cytoplasm of cells which are responsible for energy generation – the cell's 'power stations'. Because it is only eggs, and not sperm heads, which contain mitochondria, it is possible that mitochondrial DNA, which gets passed from mother to daughter, is responsible for the apparent superiority of some cow families. Proving or disproving this from records of performance is incredibly difficult. For example, mothers and daughters usually occur in the same herd, and it is difficult to distinguish between preferential treatment of favoured families and true mitochondrial inheritance. Nonetheless, several studies indicate that up to five per cent of the total variation in performance could be due to cytoplasmic effects [8].

Names of genes and their abbreviations

At this point it is worth mentioning the way genes are named and abbreviated, because this is often based on their type of gene action. Dominant, co-dominant or additive genes are sometimes abbreviated by using an upper case letter (e.g. R and W in the Shorthorn examples above, and B for the black gene in the Angus breed) while recessive genes are abbreviated by using the lower case version of the letter used to abbreviate the dominant gene. For example, b was the name used for the red gene in the Angus example above; the gene was called b rather than r to denote that red is recessive to the black gene.

In other cases, one or more letters are used to denote the name of the locus. These are often taken from the character affected by the locus, the name of the person who discovered it, or the place of discovery. Superscripts may be used to denote the alleles that are found there. This is particularly useful if there are more than two alleles at a locus. For example N^d, N^t, and n are the names of three alleles affecting medullation, or hairiness, of the fleece in Romney sheep; the n comes from Nielson, the name of the owner of the property on which a ram with an exceptionally hairy fleece was first found. (The N^d and N^t alleles – and a third allele, N^J, discovered in a part Border Leicester part Romney flock – increase medullation compared to the normal type. This increases the value of the fleece for carpet manufacture, and has led to the formation of three new breeds, the Drysdale, the Tukidale and the Carpetmaster, based on these three alleles respectively; see [9] for more details.) In cases where an allele is a rare or unusual version of a normal type, the symbol + is sometimes used to denote the normal allele, and a letter is used to denote the unusual version (as with the Booroola gene). For sex-linked genes, the genotype is often presented as a superscript on the letter X or Y, depending on which chromosome the gene is located. For example, the full description of the genotype of a ewe heterozygous for the Inverdale gene is Fec X^{I+}. In this case Fec is an abbreviation of fecundity.

International committees oversee the naming of newly discovered genes in an attempt to maintain consistent methods and unique names and to improve communication (see [1]). This is particularly important now that many more genes are being identified, as a result of gene mapping. However, for genes whose effect has been known for a long time, such as those affecting coat colour, there are often many alternative names in use for the locus and its alleles. Also, similar loci and alleles occur across species, and they may have different names in each.

The nature of traits of interest in farm animals

Qualitative and quantitative traits

So far, most of the examples of types of gene action in farm animals have concerned coat colour. Characteristics like coat colour and polledness, where the animals can be divided into discrete types with no intermediates, are called **qualitative** characteristics. It was this type of characteristic in plants, such as flower colour and seed coat type, which Mendel studied. Many of these characteristics are controlled by single genes. Also, these traits are often entirely controlled by genes and are unaffected by non-genetic influences. For example, except in rare cases of nutrient deficiency, the coat colour of animals is controlled entirely by genes and is unaffected by the type of feed offered.

However, most traits of economic importance in livestock are either expressed in units which can be counted (e.g. numbers of lambs born or weaned, number of services per conception) or, more often, measured on a continuous scale (e.g. kg of live weight, litres of milk, mm of fat), rather than falling into discrete categories. These are called **quantitative** characteristics. Most of these quantitative traits are affected by many genes, rather than single genes. Additionally, most of these traits are affected, to a greater or lesser extent, by non-genetic influences. These non-genetic influences are collectively called the **environment**. This includes factors such as the standard of feeding and management of the animal, geographic or climatic influences on performance, together with a host of other chance influences, such as exposure to disease agents. For example, there is an important genetic component to the milk yield of dairy cows, but the actual yield achieved by cows depends not only on their genetic merit, but also on how well they are fed and managed.

The bridge between Mendel's work on qualitative characters and the genetic basis of measured or quantitative traits was made in the first couple of decades of the 1900s by R A Fisher in Britain and Sewall Wright in the USA. The application of these ideas in animal breeding was pioneered by J L Lush in the USA in the 1930s.

Genetic and environmental influences

An important concept in animal breeding is that the genotype of an animal confers on it a particular value for a given trait. Across the whole population of animals, the average performance of animals will be a reflection of their average genetic merit. However, the performance of individual animals may appear better or worse than their true genetic merit, depending on whether they receive a favourable or unfavourable environment. For instance, the milk yield of a cow of very high genetic merit may be quite low if she happens to be in a herd in which she is fed very small quantities of concentrates and poor quality silage. Conversely, a cow of lower genetic merit may produce much higher yields than the cow mentioned previously as a result of being kept in a herd where larger quantities of concentrates are fed and where the silage is of very high quality.

To summarise, for most quantitative traits, the **phenotype** (**P**), or observed performance of an animal, depends on the **genotype** (**G**) it inherits and the **environment** (**E**) it 'receives'. This is often abbreviated to [7]:

$$\mathbf{P = G + E}$$

The genotype of an animal can be split further into additive effects (i.e. the combined effects of all genes which act additively on the trait of interest, abbreviated to **A**), plus non-additive effects, due to dominance and epistasis, abbreviated to **NA**, so:

$$\mathbf{G = A + NA}$$
and
$$\mathbf{P = A + NA + E}$$

For most quantitative characteristics it is the additive genetic part of the animal's genotype which we try to measure and alter by selection – this part is also termed its **breeding value**. There are reasonably good methods for predicting the breeding value, or additive genetic merit of individual animals, and for predicting changes in additive genetic merit of a herd or flock from one generation to the next. These are discussed in the next three chapters. However, it is very difficult to assess the effects of dominance and epistasis on quantitative traits. They are treated largely as 'noise' in within-breed genetic improvement programmes, though they are important in crossbreeding. It is simplest to think of the non-additive effects acting alongside the additive part of the animal's genotype, either to enhance or diminish the performance of the animal in a way which is difficult to predict. For example, live weight in cattle is probably affected by genes at tens or even hundreds of loci, some with additive gene action and some with non-additive gene action (i.e. there is dominance or epistasis or both). Selection on predicted breeding values (i.e. additive genetic merit) is expected to give a fairly smooth increase in the average weight of animals in the herd, year by year. But, in practice, non-additive gene action can create irregularities in response. For instance, one of the many loci affecting live weight might have two alternative alleles (H and h), one (H) showing partial dominance over the other (h). Let us assume that, on average, the HH animals are 10 kg heavier than the hh animals and the Hh animals are 8 kg heavier than the hh animals. If we select for higher live weight, we will tend to get more HH animals and fewer hh animals – we are increasing the frequency of the H allele and decreasing the frequency of the h allele. However, the increments in performance are not equal as we substitute h alleles by H alleles – there is a bigger increment in response from replacing an hh animal by an Hh animal (8 kg on average) than from replacing an Hh by an HH (2 kg on average). So, non-additive gene action such as this, especially if it is occurring at many loci, contributes to unevenness in response to selection (see Figure 2.12).

The fact that there are so many grades or levels of performance in quantitative traits, as a result of many genes and the environment acting together, means that it is usually impossible to observe the Mendelian segregation ratios seen in qualitative characteristics. However, this segregation is still happening and the principles of inheritance of quantitative traits are exactly the same as those already described for traits like colour – it is simply harder to see what is happening. Most of the following chapters are about the methods that are used to surmount the problem of not being able to detect different genotypes easily to make genetic improvement in quantitative traits.

Traits affected by single genes

There are a number of examples of traits of importance where a single gene is known or thought to be involved. In cattle and sheep these include:

- coat colour (although several loci are involved, in many breeds there is no variation at some of these loci, or this is masked by variation at other loci, so control of colour is often effectively at a single locus)
- polledness (absence of horns; the polled gene P is dominant to the horned gene p in *Bos taurus* breeds; a different single gene controls polledness in *Bos indicus* breeds)
- muscular hypertrophy, or double muscling, in cattle
- muscularity in some sectors of the Poll Dorset sheep breed in the US, and now other breeds which have been mated to Poll Dorset animals carrying the gene concerned (the callipyge gene)
- milk protein types (κ-casein and ß-lactoglobulin) affecting cheese-making properties and protein yield of cows' milk
- a number of genetic diseases in cattle, including tibial hemimelia, 'amputated' lethal or 'bulldog calf' syndrome, BLAD and DUMPS
- 'spider syndrome', or chondrodysplasia, in sheep
- susceptibility to scrapie in sheep
- wool fibre diameter in Romney sheep, and breeds derived from it
- fecundity in Merino sheep, and other breeds which have been crossed to the Merino (the Booroola gene)
- fecundity in Romney sheep (the Inverdale gene)

Many of the traits listed above are qualitative (i.e. they have discrete phenotypes) but, as discussed earlier, the Booroola gene provides a good example of a single gene which influences a quantitative characteristic.

Traits affected by many genes

Usually it is impossible to equate the appearance or performance of an animal with its genotype at a single locus in the way we have done already for coat colour and the Booroola gene. This is because several or many genes, as well as non-genetic factors, are influencing the phenotype, or measured performance of the animals. (These traits are sometimes said to be under **polygenic** control.) Even so, what we see when we measure the performance of animals is the net result of Mendelian segregation at many different loci.

Figure 2.14 illustrates this. If we consider a hypothetical (but quite feasible) example of a single locus affecting milk yield additively, with only two alleles possible at that locus, A and B, then there are three possible genotypes: AA, AB and BB. If the A allele has no effect on average lactation yield, but each copy of the B allele adds 50 kg milk to the average yield, then there will be three discrete levels of performance due to variation at this locus. AA cows will produce no extra milk, AB cows will have 50 kg higher average yield (i.e. + 50 kg from the single copy of the B allele), and BB cows will produce an extra 100 kg of milk (two

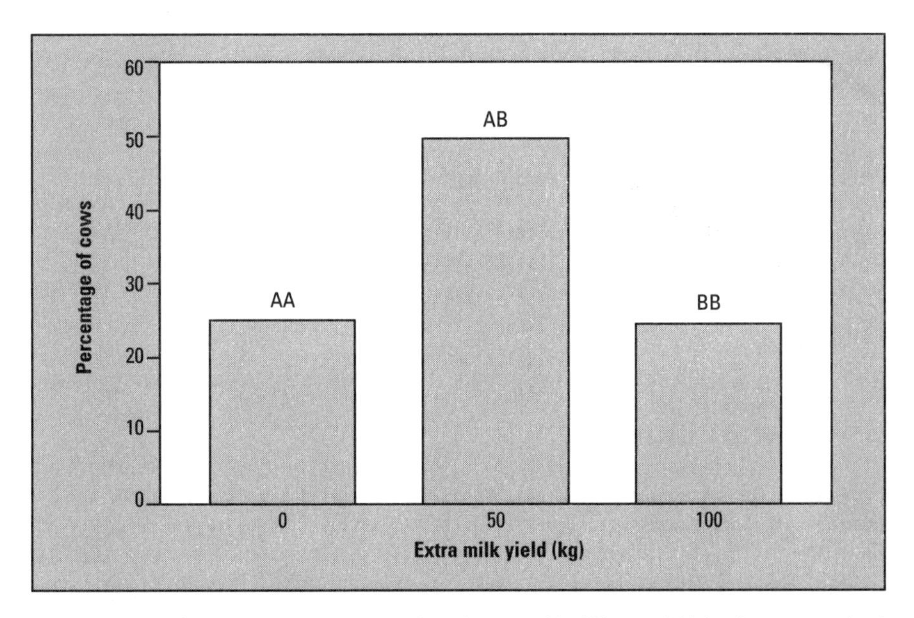

Figure 2.14 Histogram showing the proportion of cows with different yield levels as a result of variation in genotype at a single locus. The horizontal scale shows the yield level, increasing from left to right. In this example there are three yield levels: 0 kg extra for the AA cows, 50 kg extra for the AB cows and 100 kg extra for the BB cows. The vertical scale shows the proportion of cows which fall into each yield level. In this example the gene frequency is 0.5, so there are 25% AA cows, 50% AB cows and 25% BB cows [7, 9].

Genotype at second locus, genotype frequency and effect on yield	Genotype at first locus, genotype frequency and effect on yield		
	AA 25% 0 kg	AB 50% +50 kg	BB 25% +100 kg
JJ 25% 0 kg	AAJJ 6.25% 0 kg	ABJJ 12.5% +50 kg	BBJJ 6.25% +100 kg
JK 50% +100 kg	AAJK 12.5% +100 kg	ABJK 25% +150 kg	BBJK 12.5% +200 kg
KK 25% +200 kg	AAKK 6.25% +200 kg	ABKK 12.5% +250 kg	BBKK 6.25% +300 kg

Table 2.7 The proportion of cows with different yield levels as a result of variation in genotype at two independent loci. The grid shows all possible combinations of genotypes at the two loci, the expected frequency of these combined genotypes given gene frequencies of 0.5 at each locus (this is obtained by multiplying together the genotype frequencies at the first and second loci – for example, if the frequency of the AA genotype is 0.25 and the frequency of the JJ genotype is 0.25, the expected frequency of the AAJJ genotype is 0.25 x 0.25 = 0.0625 or 6.25%), and the expected extra yield as a result of the B allele at the first locus increasing yield by 50 kg per copy, and the K allele at the second locus increasing yield by 100 kg per copy.

copies of the B allele, each worth 50 kg). If the frequency of the A and B allele is the same (0.5), then 25% of the cows will be AA, 50% will be AB, and 25% will be BB. This leads to the distribution shown in the figure.

If we consider a second locus affecting yield, also with two alleles acting additively (J and K) and independently from the first locus (i.e. they are not linked), then there are nine genotypes possible for the two loci considered together. These are shown in Table 2.7. If the J allele has no effect on yield, but each copy of the K allele adds 100 kg, then the three genotypes at this locus will produce 0 kg (JJ), 100 kg (JK) and 200 kg (KK) extra milk. Although there are nine different genotypes, several of these have the same 'value', so only seven different levels of performance are observed. Figure 2.15 shows the expected distribution of yield deviations assuming gene frequencies of 0.5 at each locus, and thus genotype frequencies of 25% AA, 50% AB and 25% BB, and 25% JJ, 50% JK and 25% KK.

Comparing Figures 2.14 and 2.15 shows that as the number of loci affecting performance increases, so too does the number of 'levels' of performance. As the number of loci increases, eventually these levels merge together and the distribution of performance in a population of animals follows a smooth bell-shaped distribution. The performance of most animals falls near the middle of this distribution and relatively few animals have performance at either extreme. The fact that feeding, management and other non-genetic factors influence performance also helps to produce a smooth curve. This curve is termed the **normal distribution**, and it is characteristic of many traits of importance in farm animals which are influenced by many genes. Figures 2.16 and 2.17(a) show distributions of the actual live weights of Suffolk ram lambs from a single flock

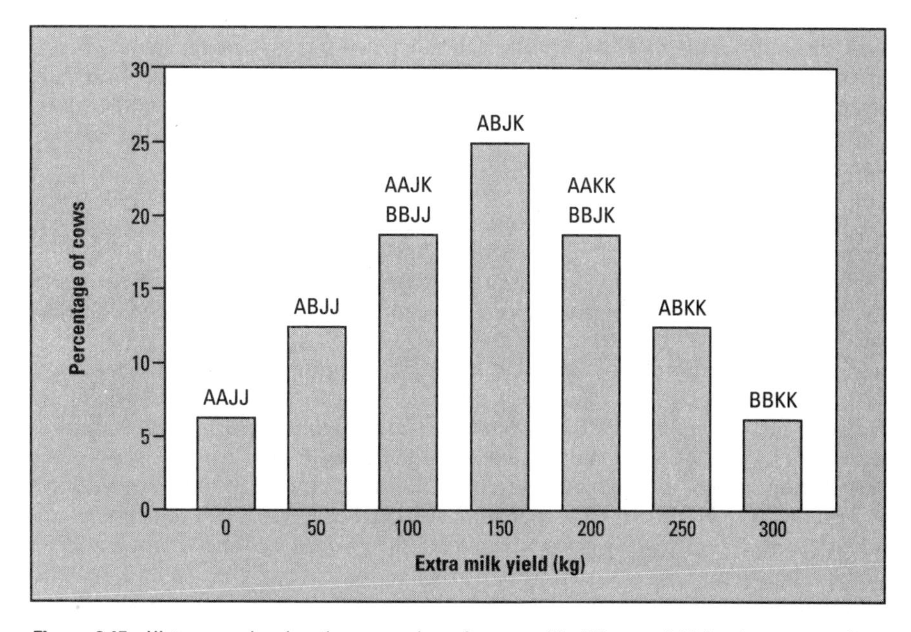

Figure 2.15 Histogram showing the proportion of cows with different yield levels, expressed as deviations from the average for the herd, as a result of variation in genotype at the two independent loci described in the text and Table 2.7. The horizontal scale shows the yield level, increasing from left to right. In this example there are seven yield levels. The vertical scale shows the proportion of cows which fall into each yield level. The genotype frequencies are 25% AA, 50% AB and 25% BB, and 25% JJ, 50% JK and 25% KK [7, 9].

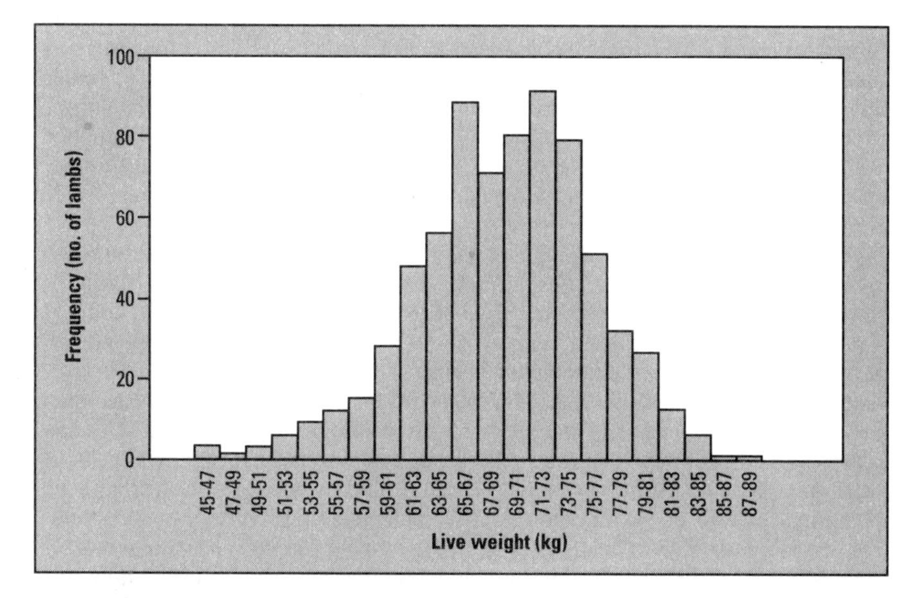

Figure 2.16 Distribution of the 21-week live weight of over 700 ram lambs, recorded over nine years, in the SAC Suffolk flock.

and the milk yields of Holstein Friesian dairy cows from several herds. These distributions are already fairly smooth, but if we repeated this exercise with data from many more animals the distributions would become even smoother. (As explained later, the distributions also get smoother when some of the known environmental influences on performance are accounted for.)

The distributions such as those shown in Figures 2.16 and 2.17(a) are quite useful in animal breeding, but there are several additional steps which can increase their value further. The shape of a distribution can be influenced by the number of 'bands' of performance which are used to group animals. For instance, if we chose to count the number of cows with total lactation milk yields in only three bands, say between 0–4,000, 4,001–8,000 and 8,001–12,000 kg, we would get a rather narrow, spiky distribution. If we went to the other extreme, and chose 120 bands, in increments of 100 kg, then we would end up with a fairly flat distribution. However, if the number of bands is chosen carefully, comparing the shapes of distributions can be very informative. For instance, Figure 2.17(b) shows the distribution of milk protein percentages for the animals whose yields were shown in Figure 2.17(a). In both cases performance has been grouped in increments of about 5% to 6% of the mean performance (300 kg milk and 0.2% protein). The graphs show that there is a narrower distribution, or less variation, in milk protein percentage than in milk yield. Although the amount of variation in a trait can be influenced by management, such as quantity and quality of feed offered, it is also a biological characteristic of the trait concerned. Measuring the total amount of variation, and apportioning this to genetic and non-genetic effects is central to livestock improvement. The next section describes how this variation is measured; how the measurements are used in improvement programmes is described in Chapters 4 and 5.

At this point it is worth mentioning another group of traits of interest in animal breeding. The animals' phenotypes for these traits fall into only two, or a small number, of categories. However, they are not inherited in a simple Mendelian way like coat colour etc., but are influenced by many genes and the environment, just like the normally distributed traits described above. Ovulation rate and litter size in most sheep breeds (excluding those with the Booroola gene, which has a big impact on litter size), survival and **multifactorial** diseases such as mastitis are good examples.

Although there are many genes and environmental factors influencing ovulation rate and litter size, most ewes have either 0, 1, 2 or 3 lambs. Similarly, cows or ewes are either unaffected or affected by mastitis (often coded 0 and 1 respectively). In these cases it is still useful to think of an underlying normal distribution, but the distribution now has one or more **thresholds** which determine the phenotype. For example, within breeds of sheep, ovulation rate, and hence litter size, is partly influenced by an underlying distribution of body condition. Below a certain threshold body condition, a ewe may shed only a single egg, while above it she may shed two. Above an even higher threshold body condition she may shed three eggs, and so on. It may be easier to understand the concept of thresholds from an analogy – that of hen eggs sold on the basis of weight. The weight of eggs follows a normal distribution, but in many countries eggs are sold according to a small number of bands (e.g. small, medium, large, extra large). The weight of an egg in relation to a series of thresholds determines which band it falls into.

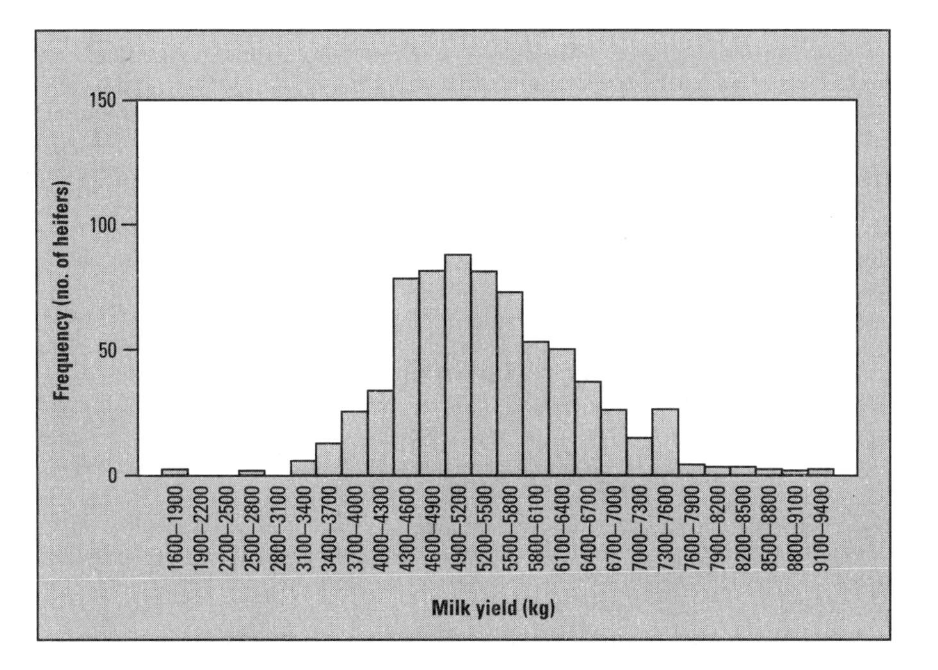

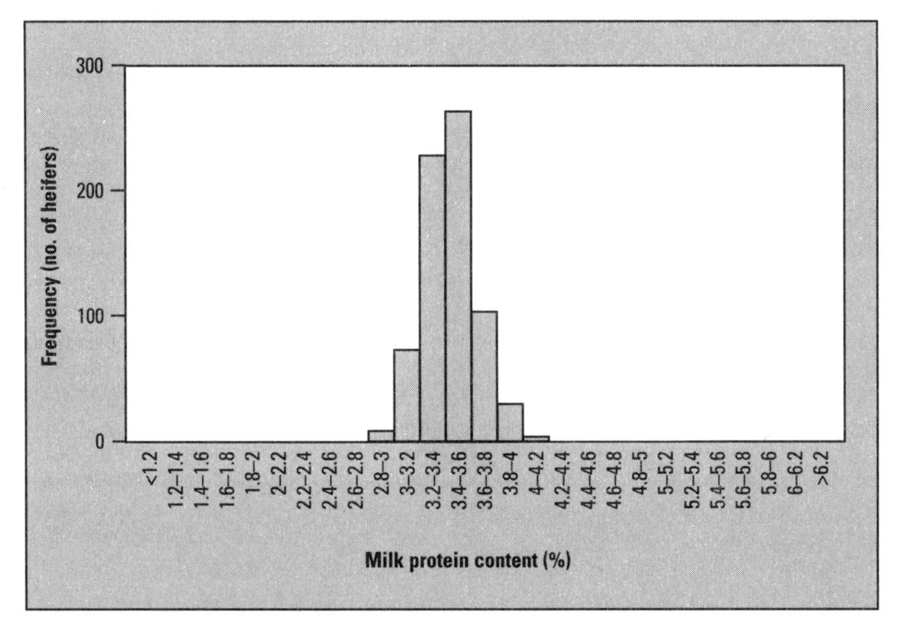

Figure 2.17 (a) (above) Distribution of the first lactation yields of about 700 Holstein Friesian heifers from a sample of English herds; and (b) (below) distribution of milk protein contents for the same heifers. In both cases performance has been grouped in increments of about 5% to 6% of the mean performance (300 kg milk and 0.2% protein). (Source: Dairy Information System (DAISY), University of Reading/National Milk Records.)

In the case of multifactorial diseases, it is common to think of an underlying normal distribution of **liability** to be affected by the disease. At one end of this distribution there are animals which have a favourable genotype and a favourable environment, and they have a low liability of being affected by the disease. At the other end of the distribution, there are animals which have the worst possible combination of genes and environment, and they are much more liable to be affected by the disease. Dealing with this sort of trait is more complex than dealing with traits like milk yield, and special statistical methods have been developed for the purpose (see [7] for more details).

Measuring variation in performance

In the last section we saw that, even when distributions are drawn on a comparable scale, there can be differences in the spread of the distributions, which are characteristics of the traits concerned. For example, there was a wider spread in the distribution of cows' milk yields than in the distribution of their milk protein concentrations. Distributions for the same trait may also vary, depending on the range of genotypes and the range of environments present. For example, we would expect a wider distribution of yields in a herd using bulls of both very good and very bad genetic merit for yield, than in a herd which only used good bulls. Similarly, we would expect a wider distribution of yields in a herd which fed cows *ad libitum* than in a herd where all the cows received a fixed quantity of feed. This is because under *ad libitum* feeding there is likely to be a wider range of feed intakes than under restricted feeding, and feed intake affects yield.

So far, we have only described the shape of these distributions or amounts of variation in words, but it is important to actually measure the spread of distributions of performance. This is achieved using a statistic called the **variance** (defined as the mean squared deviation from the mean) or its square root which is called the **standard deviation**. The use of statistics often evokes emotions normally reserved for a trip to the dentist. You do not have to be a mathematical genius to understand what follows, and a small amount of statistics goes a long way in helping to understand the techniques involved in genetic improvement. So please persevere!

Variances can only be calculated for a group or population of animals. The first step in doing so is to prepare a list of performance records for that group of animals, such as live weights, milk yields or fleece weights. The variance is then calculated in four further steps:

1. Calculate the mean, or average, of the performance records (i.e. the sum of the performance records for all animals divided by the number of records);

2. Calculate the difference, or deviation, between each individual record and the mean. If the records follow a normal distribution, half will be higher than the mean, and half will be lower. The deviations for records which are higher than the mean are recorded as positive (i.e. they have a + sign in front); deviations for records which are lower than the mean are recorded as negative (they have a – sign in front);

3. Square each of these differences (i.e. multiply each deviation by itself). Note that a negative number becomes positive when it is squared e.g. –3 squared is +9;

4. Add up these squared deviations, and then divide by n–1, where **n** is the total number of records (this is the same as calculating the mean squared deviation, except that the total is divided by n–1 rather than n, for statistical reasons that do not need to concern us here). This is the variance of our sample of observations. To calculate the standard deviation, take the square root of the variance. Variances are measured in squared units e.g. kg squared or litres squared (abbreviated kg^2 or l^2), and these are difficult to interpret. Hence the popularity of standard deviations which are expressed in the same units that the records were measured in, for example kg or litres.

Table 2.8 gives an example of how these steps are applied to calculate the variance and standard deviation of 20-week weight for a group of lambs. At first sight it may seem odd to go through all this squaring and 'unsquaring'. However, the reason for it is quite simple. If we calculated the deviations from the mean, and averaged these without squaring, they would sum to zero – try it yourself with the values in Table 2.8. Squaring is a way of getting rid of the negative signs, while still measuring the size of deviations from the mean.

As mentioned earlier, many of the traits of interest in farm livestock follow a normal distribution. Calculating variances and standard deviations for these traits, as described above, allows some important predictions to be made about the proportions of animals expected in different bands of performance. These predictions are based on some very useful characteristics of normal distributions

Animal identity	20-week weight (kg)	Deviation from mean (kg)	Squared deviation (kg^2)
1	70	+5	+25
2	73	+8	+64
3	61	–4	+16
4	58	–7	+49
5	63	–2	+4
6	67	+2	+4
7	60	–5	+25
8	68	+3	+9
9	69	+4	+16
10	61	–4	+16
Total	650	0	228 kg^2
Mean	65 kg	**Variance** (total from column 4, divided by 9 (n=10, so n–1 = 9)	25.33 kg^2
		Standard deviation	$\sqrt{25.33\ kg^2}$ = 5.03 kg

Table 2.8 Calculation of the variance and standard deviation of lamb weights. The unshaded cells show the records, the calculations are shown in the shaded areas.

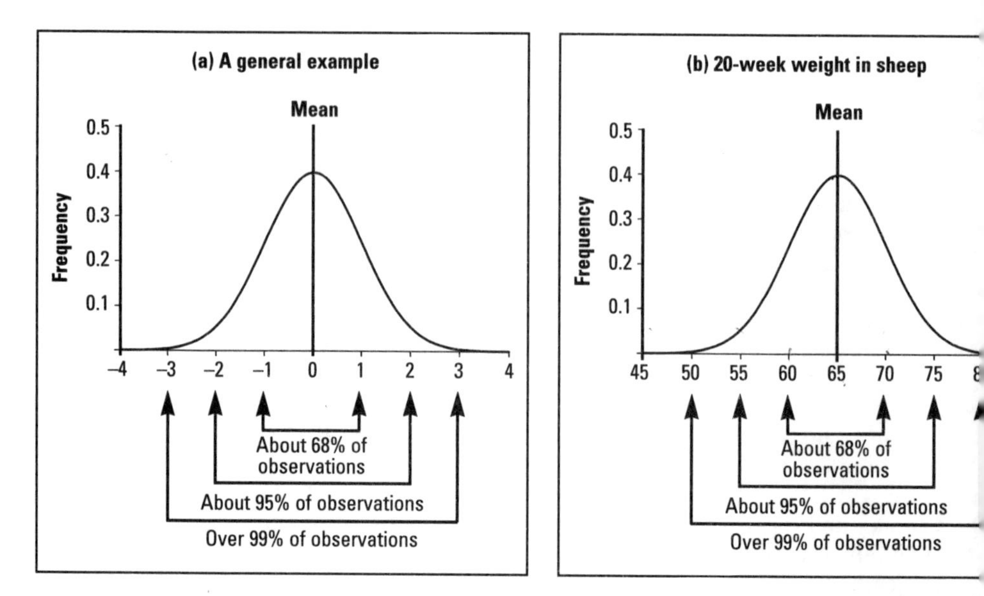

Figure 2.18 Some properties of the normal distribution, which are useful in predicting the proportions of animals which fall in different 'bands' of performance. (a) A general example; the vertical axis shows the expected frequency of observations, and the horizontal axis is in standard deviation units. (b) An example of 20-week weight in sheep, which is assumed to have a standard deviation of about 5 kg; the horizontal axis is in kg.

which are worth noting here (see Figure 2.18(a)).

- The mean is in the middle of the distribution – half the records will be above the mean, and half below.

- About 68% of the records fall in the range –1 to +1 standard deviations from the mean – a characteristic which is complex to explain, but very useful in practice. For example, if the standard deviation (**sd**) of lamb 20-week weight is 5 kg, and the mean is 65 kg, then about 68% of the lamb weights are expected to fall in the range 60–70 kg (the mean minus 1 sd is 60 kg, the mean plus 1 sd is 70 kg). This is illustrated in Figure 2.18(b).

- About 95% of the records fall in the range –2 to +2 standard deviations from the mean. So, in the lamb weight example, we expect about 95% of lamb weights to fall between 55 kg and 75 kg.

- Over 99% of the records fall in the range –3 to +3 standard deviations from the mean. So, in the lamb weight example, we expect over 99% of lamb weights to fall between 50 kg and 80 kg. In other words, when traits are normally distributed the performance of most animals will be close to the average. Progressively fewer animals will have performance at either extreme.

Trait	Mean	Standard deviation	Coefficient of variation = sd/mean
Live weight (kg)	69.4	7.1	0.10 (10%)
Fat depth (mm)	7.3	1.3	0.18 (18%)
Muscle depth (mm)	28.5	2.3	0.08 (8%)

Table 2.9 Means, standard deviations and coefficients of variation of live weight, ultrasonic fat depth and ultrasonic muscle depths in a group of SAC Suffolk ram lambs.

Variances or standard deviations provide a valuable tool for comparing variation in the same trait in different groups of animals, e.g. animals in different herds or flocks, or belonging to different breeds. A higher variance or standard deviation indicates a greater amount of variation, which may be a result of a wider variation in genetic merit or a wider variation in environment or both. However, it is difficult to make a direct comparison between variances or standard deviations for different traits, such as growth rate and fatness. This can be done by calculating the **coefficient of variation** – this is the standard deviation divided by the mean for the trait concerned. The coefficient of variation is often expressed as a percentage, i.e. multiplied by 100. Table 2.9 shows the means, standard deviations and coefficients of variation for 20-week weights and ultrasonic fat and muscle depths in a group of SAC Suffolk ram lambs. This table shows that, proportionately, there is most variation in fat depth, and slightly more variation in weight than in muscle depth.

The principle of splitting the individual animal's phenotype into genotype and environment was explained earlier. The same principle can be extended to groups of animals, so that the total variation observed (**phenotypic variance**, or V_P for short) may be split into a component due to variation in the genotype of the animals (**genetic variance, V_G**), and a component due to variation in their environment (**environmental variance, V_E**). (The fact that variation from different sources can be split and added together in this way is another reason why variances are such a useful tool in animal breeding.) To summarise, the total phenotypic variance in a population of animals, for a particular trait, is the sum of the genetic and environmental variance[7]:

$$V_P = V_G + V_E$$

The split between genetic and environmental variation is illustrated in Figure 2.19. The narrower, taller distribution in the centre of the curve represents the genetic variation in this particular population. The environmental variation adds to this genetic variation, to create the wider, flatter curve, which represents the total phenotypic variation. In practice, what we see and measure directly when we record animal performance is the wider distribution, or phenotypic variation. However, it is important to remember that this results from underlying distributions of genetic and environmental variation.

As before, the genetic part of the variation can be split into parts due to additive genetic variance, or variance in breeding values, and variance due to the non-additive effects of dominance and epistasis:

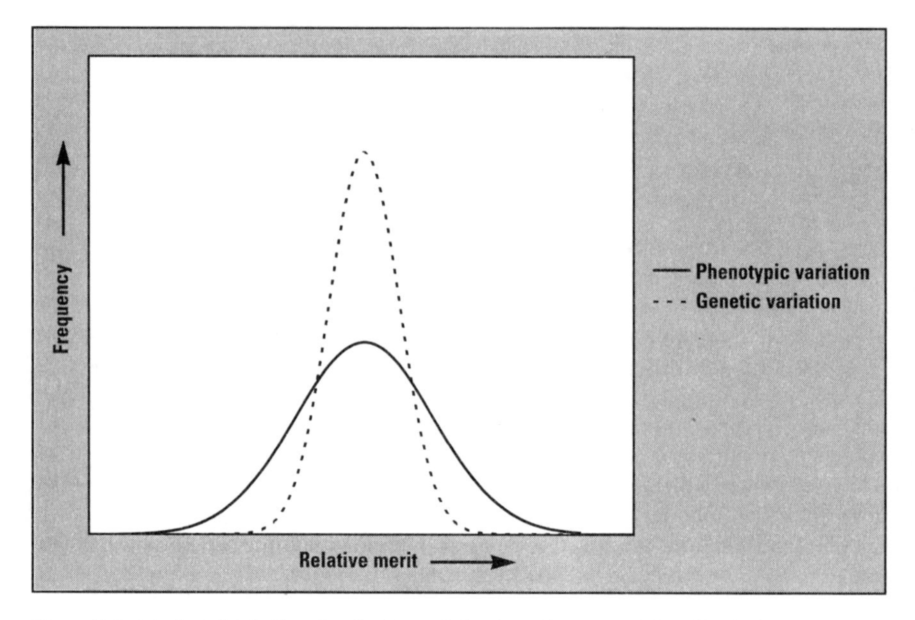

Figure 2.19 A typical distribution showing the variation in performance of farm livestock.

$$V_G = V_A + V_{NA}$$
and
$$V_P = V_A + V_{NA} + V_E$$

How we attempt to split the distribution into its components, and the role these have in practical genetic improvement, is discussed in more detail in later chapters. However, Figure 2.20 gives a simple example. This diagram shows distributions of estimated genetic merit of bulls for milk yield, based on the performance of their daughters, along with distributions of the yield of their daughters in a sample of herds.

Measuring associations between traits and between animals

Within breeds of farm livestock there are often associations between traits. For example, higher yielding dairy cows tend to have lower milk protein percentage, faster growing meat animals tend to be fatter at a given age, and sheep with higher body weight tend to have slightly larger litter sizes. These associations are caused mainly by pleiotropy – that is, the same gene (or genes) influences the two associated traits – or by environmental conditions which affect one trait also affecting the other. Linkage of genes can also cause an association between traits, but this association is soon broken down by recombination, unless the genes concerned occur very closely together on the same chromosome.

It is often easiest to see any association between traits by plotting a graph of the measurements of two characteristics for each animal. Figure 2.21 illustrates

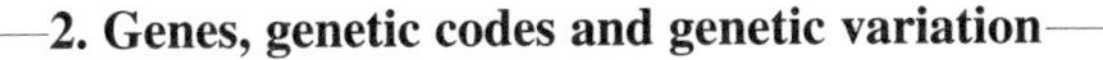

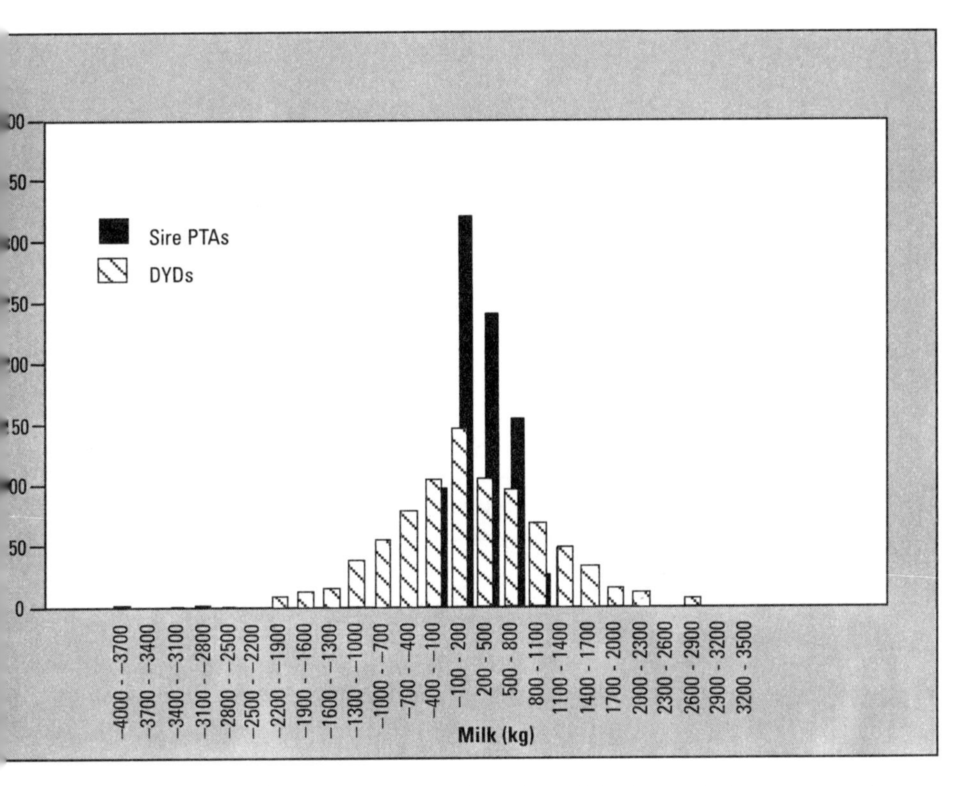

Figure 2.20 Distribution of the first lactation yield of about 700 Holstein Friesian heifers from a sample of English herds, together with a distribution of the estimated genetic merit of their sires for milk yield. The heifers' yields have been expressed as deviations from the mean yield of heifers calving in the same herd, year and season, to remove some of the known sources of environmental variation. Note that this distribution of '**daughter yield deviations**' (**DYD**s) is much smoother than that of the 'raw' yield data from the same animals shown in Figure 2.17 (a). The distribution of yields estimates V_P, though some environmental effects have been removed by expressing yields as DYDs. The sires' genetic merit has been estimated from the performance of their daughters in many herds. Sires' merit is expressed as '**predicted transmitting abilities**' (**PTA**s) – the expected deviation in daughter's yield. The distribution of sire PTAs is less smooth than that of daughters' yields because there are relatively few sires. The distribution of sires' PTAs estimates $\frac{1}{4}V_A$. (In practice sire PTAs are likely to underestimate this, as (i) only the very best bulls get widely used, and so they are not a truly representative sample of the population as a whole, and (ii) PTAs are predictions of additive genetic merit.) (Source: ADC; Dairy Information System (DAISY), University of Reading/National Milk Records.)

this. When there is a positive association between two characters, an upward-sloping line (from left to right) through the points produces the best fit. If there is no association between the traits, the points appear to be scattered randomly all over the graph and it is impossible to see any pattern at all. With a negative association, the points appear to be gathered around a downward-sloping line.

It is important to be able to measure the direction and strength of associations between traits for a number of purposes in animal breeding. For instance, if we select on one trait we might want to know how other traits will change. Or, we may wish to choose between alternative measurements as predictors of the trait of

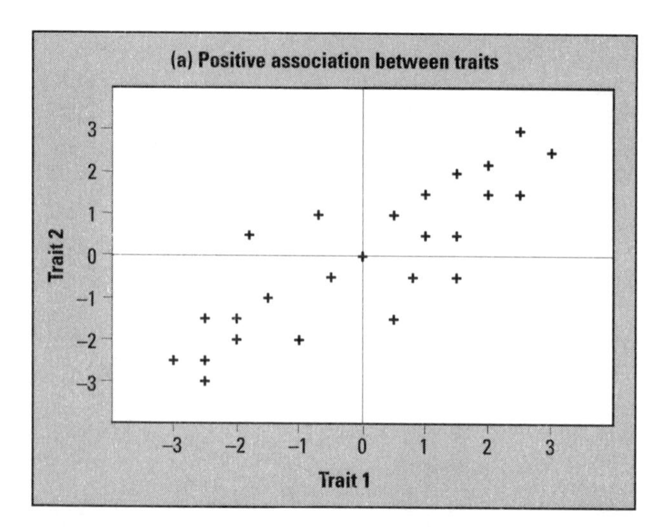

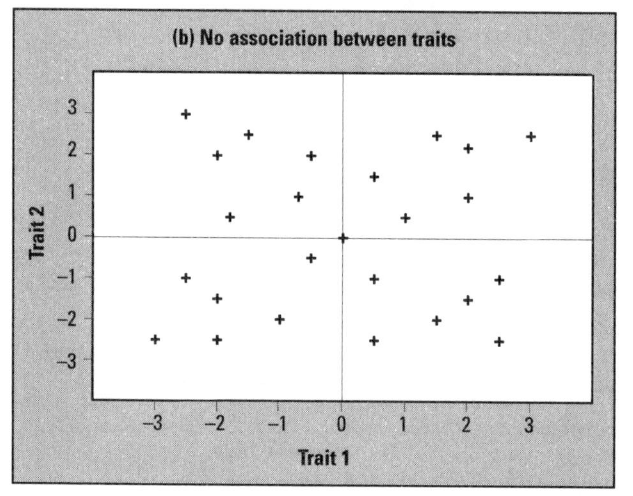

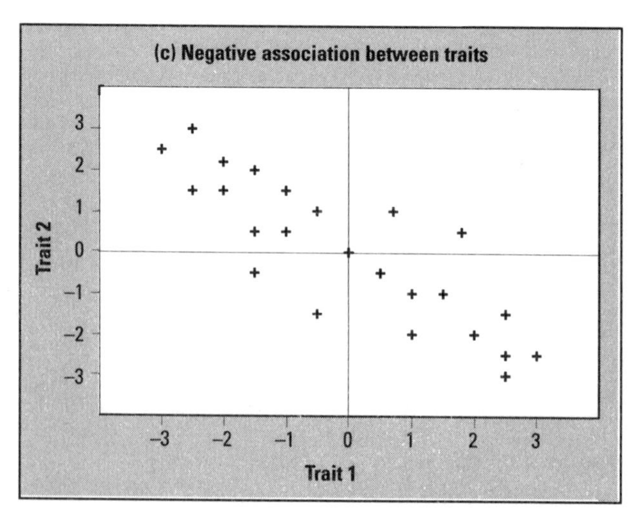

Figure 2.21
Scatter diagrams illustrating
(a) positive,
(b) zero and
(c) negative associations between traits.

interest, e.g. different live animal measurements as predictors of carcass composition. Also, measuring the degree of similarity in performance among related animals provides valuable information on the extent to which the measured character is under genetic control. The degree of association between two characters, or between the same character at different times, or between the same character on related animals, is usually measured by one of two closely related statistics – the **regression coefficient** and the **correlation coefficient** (**regression** and **correlation** for short).

Regressions measure the extent to which changes in one character are associated with changes in another, in the units of measurement. For example, a regression coefficient would express the association between milk protein % and milk yield in terms of the percentage unit change in protein per kg increase in milk yield. This is useful if we want to predict an animal's milk protein % from its yield. It also tells us about the direction of the association between traits (slightly negative in this case – protein % is going down as yield goes up), but it is not easy to tell at a glance whether the association between the two traits is strong or weak. Similarly, the regression of offspring performance on parent performance is often used to estimate the heritability of a trait, or the extent to which it is under genetic control.

Correlation coefficients also measure the association between traits, but they do so on a scale from –1, through 0, to +1, rather than in units of measurement. Correlations are more useful than regressions if we want to tell at a glance the strength (and direction) of the association between two traits, or if we want to compare the strength of associations between different pairs of traits. A correlation which is greater than zero indicates that as one character increases, the other generally increases too (as in the example of live weight and fatness in growing animals). A correlation of zero indicates that there is no apparent association between the two characters. A negative correlation indicates that as one character increases, the other generally decreases (as in the example of milk yield and protein percentage). The size of the correlation, whether it is positive or negative, indicates how closely the individual points are clustered around the line drawn through them. If the correlation is +1 or –1 then all the points will be on the line. With weaker positive or negative correlations then the points are more spread out around the line.

Regression and correlation coefficients are calculated from the variance of one or both of the two characters concerned, and the **covariance** between the two characters. The covariance is calculated in a similar way to the variance, as shown in the last section. However, instead of calculating squared deviations from the mean for a single trait, the deviations from the means of the two traits measured on the same animal are multiplied together to give **cross products**. Unlike squared deviations, which are always positive, cross products (and covariances) can be positive or negative. To calculate the covariance these cross products are summed, and then divided by n–1, where n is the number of animals. Table 2.10 gives an example of the calculation of a covariance. In this case it is the covariance between fat depth and live weight, for the ten sheep included in Table 2.8. (Note that these examples are for illustration – substantially larger numbers of animals are needed to get reliable estimates of variances and covariances.)

The regression of one trait on another is calculated from the covariance between the two traits and the variance of one of them:

Animal identity	20-week weight (kg)	Deviation from mean (kg)	Ultrasonic fat depth at 20 weeks (mm)	Deviation from mean (mm)	Cross products (column 3 x column 5; kg x mm)
1	70	+5	8.0	+1.1	+5.5
2	73	+8	8.7	+1.8	+14.4
3	61	–4	5.7	–1.2	+4.8
4	58	–7	5.6	–1.3	+9.1
5	63	–2	6.1	–0.8	+1.6
6	67	+2	6.4	–0.5	–1.0
7	60	–5	7.3	+0.4	–2.0
8	68	+3	7.8	+0.9	+2.7
9	69	+4	6.2	–0.7	–2.8
10	61	–4	7.2	+0.3	–1.2
Total	650	0	69	0	31.1
Mean	65 kg		6.9 mm	**Covariance** (total from column 6 divided by 9 (n–1)	3.456 (approx.)

Table 2.10 An example showing the calculation of the covariance between two traits (lamb weight and ultrasonic fat depth at 20 weeks of age). The unshaded cells show the records, the calculations are shown in the shaded areas.

$$\textbf{regression of trait 2 on trait 1} = \frac{\textbf{covariance between traits 1 and 2}}{\textbf{variance of trait 1}}$$

This expresses the change in trait 2, in units of measurement, for each unit change in trait 1. (Conversely, the regression of trait 1 on trait 2 is simply the covariance between the two traits divided by the variance of trait 2.) For example, using the information in Tables 2.8 and 2.10, we can calculate the regression of ultrasonic fat depth on 20-week weight in this group of sheep as:

$$\textbf{regression of fat depth on live weight} = \frac{\textbf{covariance between fat depth and live weight}}{\textbf{variance of live weight}}$$

$$= \frac{3.456}{25.33}$$

$$= \textbf{0.136 mm fat per kg live weight (approx.)}$$

In other words, in these sheep, fat depth increases, on average, by about 0.136 mm for each 1 kg increase in 20-week weight.

Correlation coefficients are calculated from the covariance between two traits and the standard deviations of each of them:

$$\textbf{correlation between traits 1 and 2} = \frac{\textbf{covariance between traits 1 and 2}}{\textbf{sd trait 1 x sd trait 2}}$$

If we repeat the calculations shown in Table 2.8 for the fat depth measurements shown in Table 2.10 we get a value of 1.11 mm^2 (approx.) for the variance of fat depth, which gives a standard deviation of 1.05 mm. Substituting this into the formula above gives:

$$\textbf{correlation between weight and fat depth} = \frac{3.456}{5.03 \times 1.05}$$

$$= +0.65$$

Correlations express the relationship between two characters in standard deviation units. For example, a correlation of +1 indicates that an animal which is 1 sd above average in one trait is expected to be 1 sd above average in the second. Similarly, an animal 2 sd better than average in one trait is expected to be 2 sd better in the second. A correlation of +0.65 such as that calculated above, indicates that an animal which is 1 sd above average in one trait is expected to be 0.65 sd above average in the second. Similarly, a correlation of –1 or –0.5 indicates that animals which are 1 sd above average in one trait are expected to be 1 sd or 0.5 sd below average in the other, and so on.

Phenotypic, genetic and environmental associations

There are three types of correlation which are widely used in animal breeding: **phenotypic**, **genetic** and **environmental correlations** (abbreviated $\mathbf{r_P}$, $\mathbf{r_G}$ and $\mathbf{r_E}$ respectively). There are also equivalent regressions, of which the phenotypic and genetic regressions are the most commonly used ($\mathbf{b_P}$ and $\mathbf{b_G}$).

Phenotypic correlations or regressions measure the direction and strength of the association between observed performance, or phenotype, in two characters, e.g. the correlation between live weight and fat depth measured on the same animals. The example calculations in the last section were for phenotypic correlations or regressions.

Genetic correlations or regressions measure the direction and strength of the association between genetic merit or breeding values for the two characters – strictly speaking these are **additive genetic correlations** or **regressions**, abbreviated $\mathbf{r_A}$ or $\mathbf{b_A}$. Genetic correlations and regressions are calculated from covariances and variances of breeding values rather than covariances and variances of phenotypic measurements. A genetic correlation (or regression) between two traits implies that they are affected directly or indirectly by the same genes (either positively or negatively). Often the phenotypic and genetic correlations between two traits are similar, e.g. for many growth traits. They are less similar for traits separated across time, e.g. milk production traits in first and later lactations. Figure 2.22 illustrates the phenotypic association between milk yield and milk protein % in dairy cattle and the association between breeding values for milk yield and milk protein %.

A positive environmental correlation indicates that environmental conditions which are favourable for one character are also favourable for the second. For instance, feeding a high energy diet *ad libitum* may favour both weight gain and fat deposition and so create an environmental correlation between them. A

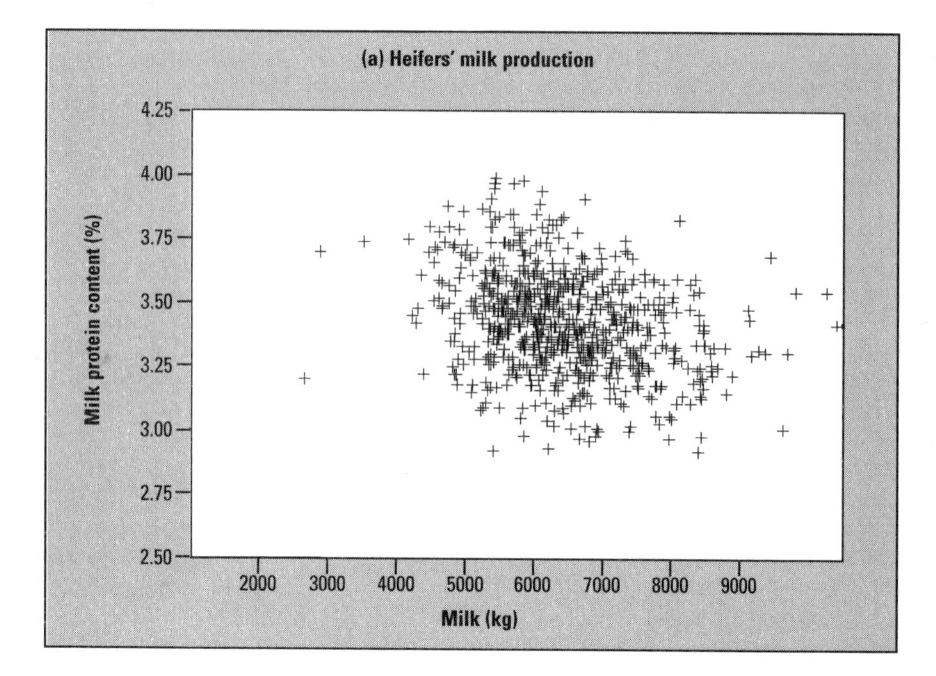

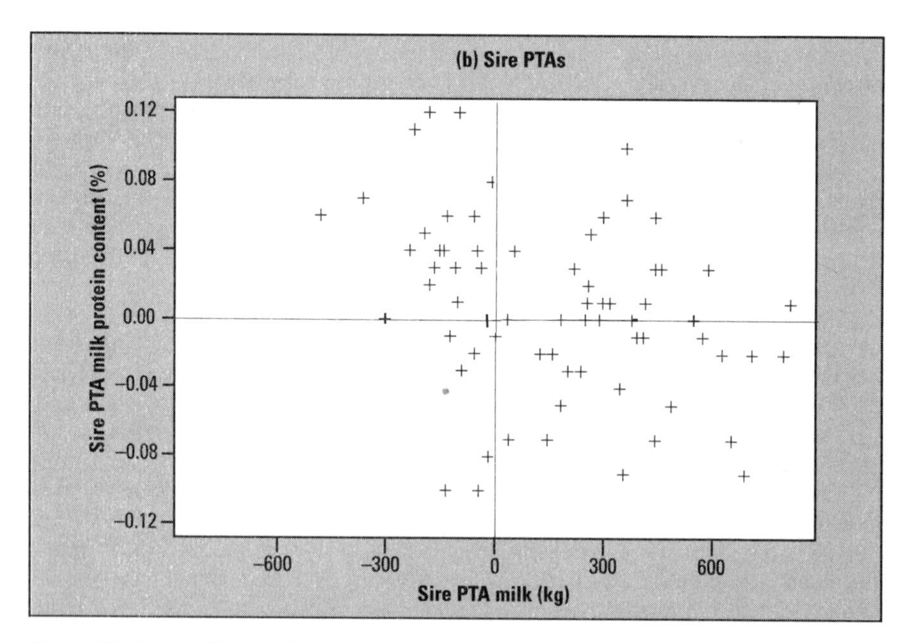

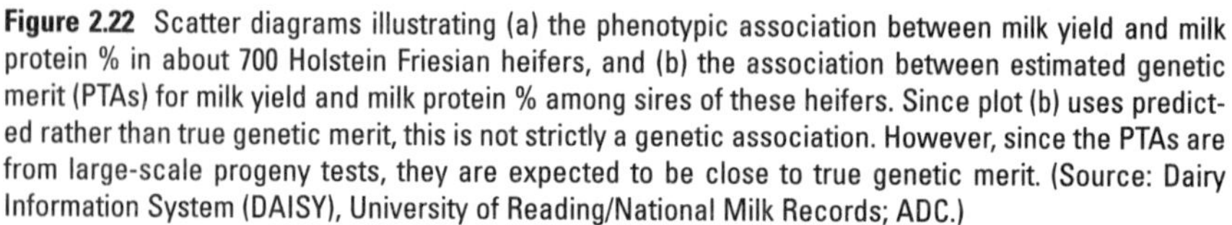
Figure 2.22 Scatter diagrams illustrating (a) the phenotypic association between milk yield and milk protein % in about 700 Holstein Friesian heifers, and (b) the association between estimated genetic merit (PTAs) for milk yield and milk protein % among sires of these heifers. Since plot (b) uses predicted rather than true genetic merit, this is not strictly a genetic association. However, since the PTAs are from large-scale progeny tests, they are expected to be close to true genetic merit. (Source: Dairy Information System (DAISY), University of Reading/National Milk Records; ADC.)

Correlations between	r_P	r_A	r_E
Milk yield and protein % (1st lactation)	–0.42	–0.50	–0.36
Milk yield in 1st and 2nd lactations	0.58	0.87	0.42

Table 2.11 Examples of phenotypic, genetic and environmental correlations in dairy cattle [17].

negative environmental correlation indicates that environmental conditions which favour one character are unfavourable for the second. For example, there is a positive genetic correlation but a negative environmental correlation between mature size and litter size within several species. The negative environmental correlation arises because the growth of females which are born in large litters is often permanently stunted because of the competition they face for space in the uterus as they develop from an embryo and the competition they face after birth for a limited supply of milk from their mother. So, although they carry the genes for both high body size and high litter size, they are prevented from fully expressing their genes for body size because of the 'poor' environment they experienced in early life. It is not always easy to define causes of environmental correlations, and they do not often act in the opposite direction to genetic correlations, as in the weight and litter size example. However, it is important to recognise that they exist, and that they lead to differences in the scale, and occasionally the direction, of genetic and phenotypic correlations.

With estimates of any two of these types of correlations, and of the heritabilities of the traits concerned, it is possible to calculate the third (e.g. see [7]– the phenotypic correlation is *not* simply the sum of the genetic and environmental correlations). Some examples of the three types of correlation in dairy cattle are shown in Table 2.11. More examples of phenotypic and genetic correlations in dairy cattle, beef cattle and sheep are given in Chapters 6, 7 and 8.

Correlations or regressions are used in animal breeding to:

- predict the changes in one trait which are expected following selection on another

- construct selection indexes which allow simultaneous selection for several characters

- predict genetic merit or breeding value for single or multiple traits (when there is a positive or negative genetic correlation between two traits, records on one trait can help in predicting genetic merit in the other, e.g. if we had a dairy bull with a high predicted breeding value for milk production, but no records on his daughters' protein percentage, we would predict that he would have a relatively poor breeding value for protein percentage, because of the negative genetic correlation between milk yield and protein percentage)

- measure the accuracy of prediction of breeding values for individuals

- measure the repeatability of a trait (the correlation between repeated records on the same animal)

- measure the accuracy of selection in a population (the accuracy of selection is the correlation between the criterion on which selection is based e.g. the animal's own performance, or average performance of a group of siblings or progeny, and the true breeding value)

The first of these uses is discussed in Chapter 4. The others are discussed in more detail in Chapter 5.

Summary

- Modern methods of livestock improvement are founded on the principles of genetics. Hence, an understanding of some of these principles helps in understanding approaches to the genetic improvement of livestock.

- The unit of inheritance is the gene. Genes occur in sequences on the chromosomes which are present in the nuclei of cells. In body cells the chromosomes occur in pairs; the number of pairs is a characteristic of the species concerned. However, the sex cells or gametes – sperm and eggs – contain only a single member of each pair of chromosomes. Hence, every animal receives one member of each pair of chromosomes from its mother, via the egg, and the other from its father, via the sperm. As a result, progeny receive half their genes from each parent.

- Chromosomes and genes are made up of deoxyribonucleic acid, or DNA. DNA has a ladder-like structure, with rungs of the ladder formed by pairs of bases. This structure, together with the fact that these bases always pair in the same way, provides a very reliable mechanism for copying DNA during cell division – this occurs as part of the growth and development of an animal (mitosis) or during the formation of gametes (meiosis).

- Genes are made up of sequences of bases down one side of the 'ladder', which form a template for the production of proteins. These proteins are either used directly in growth and repair of the body tissues, or they act as signals controlling body functions. Genes account for only about ten per cent of the total DNA in chromosomes – the rest is involved in controlling the expression of genes, or has an unknown function or no function.

- The site or 'address' of a gene on a chromosome is known as a locus (though the words gene and locus get used interchangeably). Alternative sequences of bases at a locus are termed alleles or genes. There may be many alternative alleles at a locus, but any individual animal will either have two copies of the same allele (one on each member of the pair of chromosomes concerned) or will have a copy of two different alleles.

- Although sperm and eggs receive only one member of each pair of chromosomes, each of these carries genes from both the father and mother of the animal producing the sperm or egg (i.e. the grandparents of the animal which is created when a sperm fertilises an egg), as a result of recombination. Unless loci occur very close to each other on the same chromosome (i.e. they are linked), genes at different loci are effectively inherited independently, because of recombination and the independent assortment of chromosomes at meiosis.

- Usually DNA gets copied faultlessly during cell division, or errors are corrected by special repair enzymes. However, occasionally mistakes are made which do not get corrected. If these gene mutations occur in cells producing sperm and eggs they lead to changes in the genetic code of offspring. Often these mutations have little or no effect, but sometimes they have favourable or deleterious effects. Mutations which have occurred over many generations are responsible for much of the genetic variation we see today. New mutations are largely responsible for most selected populations continuing to change, even after many generations of selection.

- The particular combination of alleles or genes which an animal inherits is called its genotype (e.g. RR, RW or WW). Animals which have two copies of the same allele (e.g. RR or WW) are homozygous, while those with a copy of two different alleles (e.g. RW) are heterozygous. What we observe when we look at animals or measure them in some way is called the phenotype (e.g. red, roan or white coat colour). In some cases we can infer the genotype from the phenotype (e.g. as with Shorthorn coat colour). In many others we cannot do so, either because of the type of gene action or because many genes, or many genes plus non-genetic effects influence the phenotype for the character concerned.

- The outcome of matings between two animals of known genotype can be predicted by drawing a 2x2 grid. The two types of alleles carried by sperm are noted at one side of the grid, and the two types of alleles carried by eggs are noted on the adjacent side of the grid. The four combinations of alleles possible in offspring are then entered inside the grid. These are the genotype frequencies expected in offspring. The actual genotype frequencies observed can depart quite widely from these expectations, especially when the number of matings is small. At a herd, flock or breed level it is easiest to work with gene frequencies rather than genotype frequencies.

- In large, random-mating populations with no selection, no movement of animals in or out of the population, and ignoring mutation, gene frequencies are expected to stay the same from one generation to the next. However, these conditions are rarely met in farm animals. For instance, chance changes in gene frequency can occur because populations are subdivided into smaller interbreeding groups. Also, the frequency of genes which have a favourable effect on the trait of interest is expected to increase as a result of selection, while the frequency of those with an unfavourable effect is expected to decrease.

- Genetic improvement depends on genetic variation. The most important sources of genetic variation between individuals or groups of animals are differences in gene frequencies, segregation of genes, recombination and mutation.

- Genes or alleles can act in several different ways. With co-dominance, heterozygotes show characteristics of each of the homozygous types. The equivalent type of gene action on measured traits, when heterozygotes are exactly intermediate to homozygotes, is called additive gene action. Complete dominance occurs when the presence of one allele completely masks the effect of another allele at the same locus. Partial dominance occurs when the performance of heterozygotes is closest to that of animals homozygous for one of the alleles concerned. Overdominance occurs when the performance of heterozygous animals exceeds that of both homozygous types.

- The action of some genes depends on the sex of the animal. If the locus is on one of the sex chromosomes it is said to be sex linked; if the trait of interest is only ever seen in one sex, but the locus is not on one of the sex chromosomes, it is said to be sex limited; other traits, which are neither sex linked nor sex limited, may be sex influenced. Genomic imprinting occurs when the expression of a gene in offspring depends on which sex of parent contributed the gene. Pleiotropy occurs when an allele at a single locus influences more than one characteristic of the animal. Epistasis occurs when the presence of an allele at one locus masks the effect of an allele at another locus. When a character is entirely controlled by a single gene, but not all animals of a given genotype show the expected phenotype, this is termed incomplete penetrance. The equivalent type of gene action when animals show the expected phenotype, but to varying degrees, is called variable expressivity. Small amounts of genetic information are passed from mothers to daughters via mitochondrial DNA. This may influence animal performance to a small extent. This is termed cytoplasmic or mitochondrial inheritance.

- Many traits of interest in farm animals are influenced not only by genes, but also by the environment in its widest sense (including feeding and management). For these traits it is useful to think of an animal's phenotype being comprised of its genotype (which can be further subdivided into an additive genetic component, termed its breeding value, and a non-additive genetic component) and an environmental component. Modern methods of livestock improvement attempt to disentangle these components as far as possible. Selection between and within breeds acts largely on additive genetic merit, while crossbreeding may be used to benefit from additive or non-additive genetic differences between animals, or both of these.

- There are quite a few traits of interest in farm animals under the control of single genes (e.g. coat colour, polledness, many genetic disorders). However, most traits of interest are affected by genes at many different loci, as well as by non-genetic factors. Although Mendelian segregation is at work at each of these loci, it is difficult to distinguish different phenotypes. Instead the performance of animals tends to show continuous variation. Often the performance of animals follows a bell-shaped curve, or normal distribution, with most animals near the middle, and only a few at either extreme.

- Many of thc 'tools' used in livestock improvement rest on properties of this bell-shaped distribution of performance. The variation in performance in a group of animals can be measured using a statistic called the variance, or its square root the standard deviation. From these statistics we can predict the proportions of animals with different levels of performance. The variance in performance in a group of animals can also be split into additive genetic, non-additive genetic and environmental components. This allows comparisons of the relative importance of these different sources of variation, and is useful when deciding on a strategy for genetic improvement and for predicting responses to selection.

- There are often associations between traits in livestock. These associations are caused mainly by pleiotropy or by environmental conditions which affect one trait also affecting the other. Linkage of genes can also cause an association between traits, but this association is soon broken down by recombination, unless the genes concerned occur very closely together on the same chromosome.

- It is important to be able to measure the direction and strength of associations between traits. This is achieved using regression or correlation coefficients, which are calculated from the covariance between the two traits and the variances of one or both traits. Regressions express the association between two traits in units of measurement. Correlations fall in the range –1 to +1. The sign indicates the direction of the association between traits. The size of the coefficient indicates the strength of the association – the closer to 1 the correlation, the stronger the association between the traits.

- There are three types of correlation which are widely used in animal breeding: phenotypic correlations, genetic correlations and environmental correlations (abbreviated r_P, r_G and r_E). Similarly, phenotypic and genetic regressions are widely used (b_P, b_G). Phenotypic correlations or regressions measure the degree of association between observed performance, or phenotype, in two characters. Additive genetic correlations or regressions measure the degree of association between breeding values for two characters (r_A or b_A). A positive environmental correlation indicates that environmental conditions which are favourable for one character are also favourable for the second. A negative environmental correlation indicates that environmental conditions which favour one character are unfavourable for the second.

- Correlations or regressions are used in animal breeding to predict the changes in one trait which are expected following selection on another, to construct selection indexes which allow simultaneous selection for several characters, to predict breeding values, and to measure the accuracy of selection or the accuracy of predicting breeding values.

References

1. Andresen, E., Broad, T., Di Stasio, L. et al. 1991. 'Guidelines for gene nomenclature in ruminants 1991.' *Genetics, Selection, Evolution,* 23:461–6.
2. Ansari, H.A., Pearce, P.D., Maher, D.W., et al. 1993. 'Resolving ambiguities in the karyotype of domestic sheep (*Ovis aries*).' *Chromosoma*, 102:340–7.
3. Arthur, P.F. 1995. 'Double muscling in cattle: a review.' *Australian Journal of Agricultural Research*, 46:1493–1515.
4. Biotechnology and Biological Sciences Research Council. 1996. *The New Biotechnologies – Opportunities and Challenges*. BBSRC, Swindon.
5. Cockett, N.E., Jackson, S.P., Shay, T.L., et al. 1996. 'Polar overdominance at the ovine callipyge locus.' *Science*, 273:236–8.
6. Davis, G. H., Dodds, K.G., McEwan, J.C. and Fennessy, P.F. 1993. 'Liveweight, fleece weight and prolificacy of Romney ewes carrying the Inverdale prolificacy gene ($FecX^I$) located on the X-chromosome.' *Livestock Production Science*, 34:83–91.
7. Falconer, D.S. and Mackay, T.F.C. 1996. *Introduction to Quantitative Genetics*. 4th edn. Longman, Harlow.
8. Gibson, J.P., Freeman, A.E. and Boettcher, P.J. 1997. 'Cytoplasmic and mitochondrial inheritance of economic traits in cattle.' *Livestock Production Science,* 47:115–24.
9. Nicholas, F.W. 1987. *Veterinary Genetics*. Oxford University Press, Oxford.
10. Nicholas, F.W. 1996. *Introduction to Veterinary Genetics*. Oxford University Press, Oxford.

11. Piper, L.R., Bindon, B.M. and Davis, G.H. 1985. 'The single gene inheritance of the high litter size of the Booroola Merino.' In R.B. Land and D.W. Robinson (eds), *Genetics of Reproduction in Sheep*, pp.115–25. Butterworths, London.

12. Schalles, R.R. 1986. *The Inheritance of Color and Polledness in Cattle.* American Simmental Association, Bozeman, Montana.

13. Surani, M.A. and Allen, N.D. 1990. 'Genomic imprinting: epigenetic control of gene expression, phenotypic variations and development.' *Proceedings of the 4th World Congress on Genetics Applied to Livestock Production*, Vol XIII, pp. 27–34.

14. US Congress, Office of Technology Assessment. 1988. *Mapping Our Genes – The Genome Projects: How Big, How Fast?* OTA-BA-373, US GPO, Washington DC.

15. US Department of Energy. 1992. *DOE Human Genome Program. Primer on Molecular Genetics.* US Department of Energy, Office of Energy Research, Office of Health and Environmental research, Washington DC.

16. Van Vleck, L.D., Pollak, E.J. and Oltenacu, E.A.B. 1987. *Genetics for the Animal Sciences.* W.H. Freeman and Company, New York.

17. Visscher, P.M. and Thompson, R. 1992. 'Univariate and multivariate parameter estimates for milk production traits using an animal model. I. Description and results of REML analyses.' *Genetics, Selection, Evolution,* 24:415–30.

18. Willis, M.B. 1991. *Dalton's Introduction to Practical Animal Breeding.* 3rd edn. Blackwell Scientific Publications, Oxford.

Further reading

Lush, J.L. 1945. *Animal Breeding Plans.* 3rd edn. Iowa State College Press, Ames, Iowa.

Nicholas, F.W. 1987. *Veterinary Genetics.* Oxford University Press, Oxford.

Nicholas, F.W. 1996. *Introduction to Veterinary Genetics.* Oxford University Press, Oxford.

Orel, V. 1996. *Gregor Mendel: the first geneticist.* Oxford University Press, Oxford.

Van Vleck, L.D., Pollak, E.J. and Oltenacu, E.A.B. 1987. *Genetics for the Animal Sciences.* W.H. Freeman and Company, New York.

CHAPTER 3

Strategies for genetic improvement

Introduction

For thousands of years humans have attempted to alter populations of animals to make them more suitable for the production of food or fibre, or as providers of transport, draught power etc. These attempts have been increasingly effective over the last couple of centuries. Generally, improved breeds of livestock produce food, fibre or other products which are of higher quality or are better matched to modern requirements than their predecessors. Also, improved breeds usually have higher efficiency of production than unimproved breeds and so the relative cost of their produce is lower. For example, in many countries genetic improvement of pigs and poultry has contributed to the change in status of pig and poultry meat from being luxury foods to being the cheapest meats available. In many countries the price of milk has fallen in real terms over the last few decades, partly as a result of genetic improvement of yield and overall efficiency of production. Selection for reduced fibre diameter and increased fleece weight in specialised sheep breeds has been important in allowing wool to continue to win a share of the market for clothing fabrics. Genetic improvement is particularly valuable because it is permanent, it is cumulative when selection is continuous, and it is usually highly cost-effective.
The aim of this chapter is to describe the strategies which are employed to achieve genetic improvement. Some more detailed examples of both the benefits and, importantly, the possible negative consequences of selection and how these can be avoided or reduced, are discussed in later chapters. Traditionally, three main strategies have been used for the genetic improvement of livestock. These are:

- **Selection between breeds or strains** – substituting one breed or strain for another

- **Selection within breeds or strains** – choosing better parents within a particular breed or strain

- **Crossbreeding** – mating parents of two or more different breeds, strains or species together

These strategies are discussed in more detail below. In future, new strategies may become available as a result of developments in molecular biology. These may include the ability to transfer genes within or between species, and the ability to regulate or modify the expression of existing or introduced genes. Some of these new possibilities are discussed in more detail in Chapter 9.

For any genetic improvement strategy to be effective it is important to have a clear view of what the economically important animal characteristics (traits) are.

Then it is logical to choose the most appropriate breed or cross, based on their performance in these traits. It is then sensible to consider whether this pure breed, or component breeds of the cross, can be improved further by within-breed selection. In practice, the availability of information on the performance of different breeds and crosses, the financial and physical resources available on the farm, the current breeds or crosses in use, local traditions, local market demands and personal preferences also influence the choice of breed or cross, and the extent to which within-breed selection is practised, at least in the short term.

Before discussing the different strategies for genetic improvement in detail, it will be helpful to consider the structure of the livestock breeding industries in which they are applied.

The structure of livestock breeding industries

The structure of livestock breeding industries in most industrialised nations is often described schematically as a pyramid with **elite** (or **nucleus**, **seedstock** or **stud**) breeders at the top, one or more middle tiers of **purebred** (**straightbred**) or **crossbred multipliers**, and a final tier of **commercial herds** or **flocks**, or end users (see Figure 3.1). In this generalised model, the elite breeders' role is to produce breeding stock, particularly males, for use within the top tier and in multipliers' herds or flocks. The elite breeders are the main focus for genetic improvement. The main role of multipliers, as the name suggests, is to take improved stock from the tier above and to create larger numbers of animals for sale to the tier below. In most industries there are purebred multipliers involved in producing greater numbers of purebred animals, particularly males, for the tiers below. However, in some industries there are also crossbred multipliers, who produce crossbred animals, particularly females, for use in the commercial tier. Usually, flocks or herds in the commercial tier are primarily involved in the production of meat, milk or fibre, and have little or no involvement in selling stock for further breeding. Hence, genetic improvement flows down the pyramid, or is disseminated from elite, through multiplier, to commercial tiers. The

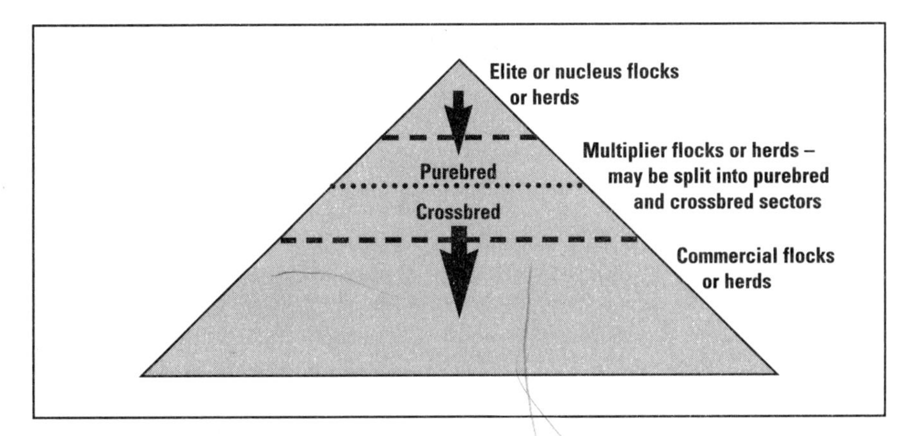

Figure 3.1 The structure of livestock breeding industries in many industrialised nations [27].

difference in genetic merit of different tiers of the pyramid is termed the **improvement lag**. (The use of the words 'pedigree' and 'commercial' to describe different tiers of the industry offends some pedigree breeders, as they feel it implies that their activities are not commercial. However, I am going to stick to those terms here, for the want of better terms to use.)

In the pig and poultry industries in many countries the elite and multiplier tiers are mainly in the hands of a small number of multinational breeding companies. They maintain elite purebred or synthetic lines in which most of the selection is practised. Breeding stock from these lines then moves to the company's own multiplier herds or flocks, or to herds or flocks contracted by them, to produce both males and females for sale to commercial producers. These pigs and poultry produced for the commercial tier are usually crossbred.

The very widespread use of artificial insemination (AI) in the dairy industries of most countries has effectively removed the middle tier from the breeding pyramid. That is, the commercial tier has direct access to elite animals via AI. The elite tier in dairy cattle breeding is usually owned by a mixture of private breeders and breeding companies – the former own most of the elite cows, and the latter own most of the elite bulls. Unlike the situation in the pig and poultry industries, in most countries the commercial tier in the dairy industry is made up largely of purebred animals.

In most countries all tiers of the pyramid in the sheep and beef cattle breeding industries are in the hands of individual breeders, rather than breeding companies. However, the structures of the breeding industries vary quite markedly from country to country. For instance, the Australian finewool industry is based on purebred Merino sheep in all tiers. The middle and lower tiers of the Merino industry also act as crossbred multipliers for some sectors of the lamb meat industry by producing halfbred Merinos (crosses between Merino ewes and, usually, Border Leicester rams). These in turn are mated to rams from **terminal sire** or specialised meat breeds (especially the Poll Dorset). The New Zealand sheep industry is based largely on purebred dual-purpose ewes in all tiers. These include the Romney, Coopworth, Perendale and Corriedale breeds, which have been selected to produce both meat and wool. In commercial flocks a proportion of these pure dual-purpose ewes are often crossed to terminal sires. In the UK, hill sheep flocks in all tiers are usually purebred. As in the Merino example, many of these UK hill flocks act as crossbred multipliers, producing crossbred ewes for the commercial tier of the sheep industry in the lowlands and uplands. These are usually produced from **draft** hill ewes – those which have had about four lamb crops on the hill and have been moved to better ground for crossing, usually to rams from one of the longwool or crossing breeds. The resulting crossbred ewes are generally mated to purebred rams from a terminal sire breed to produce lambs for meat production.

The pastoral beef industries of many temperate countries or areas are based on purebred cows of the traditional British beef breeds – Hereford, Aberdeen Angus and Shorthorn, or crosses between them. Increasingly, these are mated to purebred bulls from the larger continental European breeds, such as the Charolais, Simmental or Limousin. Purebred animals are the norm in all tiers of the French beef industry. In contrast, much of the commercial tier in the UK and Eire has been based traditionally on beef x dairy crossbred cows. At least until recently the popularity of these crossbred cows was partly because of the plentiful supply of

Tier	Dairy cattle	Beef cattle	Sheep
1. Elite or nucleus breeders	Pedigree Holstein Friesian breeders, selling young bulls to AI companies for progeny testing, or privately testing young bulls. The same breeders selling heifers to other breeders in this tier, or to the tier below. A few companies operating nucleus breeding schemes (see Chapter 6 for more details).	Pedigree terminal sire (or other) beef breeders, selling bulls to other elite pedigree herds (tier 1), and purebred multiplier herds (tier 2).	Pedigree terminal sire or longwool sheep breeders, selling rams to other elite pedigree flocks (tier 1) and multiplier pedigree flocks (tier 2). Purebred 'stud' breeders of Merinos selling rams within this tier and to purebred multipliers.
2. Purebred multipliers	Pedigree or other breeders producing heifers for sale to commercial herds.	Pedigree terminal sire herds buying bulls from tier 1, and selling bulls for crossing in commercial herds (tier 4).	Pedigree terminal sire flocks buying rams from tier 1, and selling rams for crossing in commercial flocks (tier 4). Pedigree longwool breeders selling rams to crossbred multipliers.
3. Crossbred multipliers		Dairy herds buying beef bulls from tier 2 or beef semen from tier 1 or 2, and selling beef x dairy heifers to suckler herds in tier 4. Pure beef herds (e.g. Galloway) crossing to another beef breed (e.g. White Shorthorn) and selling crossbred heifers (e.g. Blue Grey) to suckler herds in tier 4.	Draft ('retired') hill ewes (e.g. Scottish Blackface) crossed to longwool sires (e.g. Bluefaced Leicester) to produce crossbred breeding females (e.g. Scottish Mule) for use in tier 4. Draft Merino ewes crossed to Border Leicester rams to produce crossbred females for tier 4.
4. Commercial or end-users	Purebred Holstein Friesian (or other) dairy herds, using AI with semen from bulls in tier 1.	Crossbred suckler cow herds (e.g. Hereford or Simmental x Holstein Friesian; Blue Grey) buying replacement females from tier 3, and bulls from tier 2).	Crossbred ewe flock (e.g. Scottish Mule, Border Leicester x Merino) buying replacement females from tier 3, and rams from tier 2.

Table 3.1 Examples of sheep and cattle enterprises in various tiers of the 'breeding pyramid'. Several tiers may be present on a single farm [30].

replacement beef x dairy heifers from dairy herds – a by-product of the practice of crossing dairy heifers, and those cows not required for breeding replacement dairy heifers, to a beef bull. In these cases, dairy herds act as multipliers of cross-bred cows for the beef industry. (These cows kept for rearing calves for beef production are called **suckler** cows.) Table 3.1 gives some examples of where some typical sheep and cattle enterprises fit on the breeding and dissemination pyramid.

The terms purebred and pedigree require further explanation. Purebred or straightbred animals are those produced from several or many generations of matings between animals of the same breed. Strictly speaking any purebred animal whose parentage, or more distant ancestry, is known can be described as a pedigree animal. However, in this context pedigree means pedigree registered – in other words an animal whose ancestry, and other details, are officially recorded by a breed society or other organisation set up for this purpose. In Britain, most sheep in the elite tier, except in some hill breeds, and virtually all beef and dairy cattle in the elite tier are pedigree registered. This is also true for a significant proportion of 'commercial' dairy cattle. Most purebred beef cattle and sheep in the multiplier tier are pedigree registered; others have unofficial records of ancestry but are not registered with a breed society. In the commercial tier most beef and sheep are crossbred, though a few are purebred but not pedigree registered. Traditionally great status has been attached to animals which are officially pedigree registered. In many cases pedigree registered animals have a higher monetary value than non-registered animals, though this does not always reflect their genetic merit. Breed societies have generally provided a valuable service in maintaining accurate records of animals' pedigrees – a service which is valued by both breeders and potential buyers of pedigree livestock. However, until recently, relatively few societies have actively promoted performance recording or the use of both pedigree and performance records to assist selection decisions. The combination of performance and pedigree records is one of the factors which makes modern methods of genetic improvement so effective. The more enlightened breed societies have recognised this and they have broadened their services to provide detailed information on performance as well as ancestry and have improved their members access to, and understanding of, selection tools.

In the idealised model of a livestock breeding industry, strong market signals ought to go up this pyramid from one tier to the next, resulting in a clearly defined breeding goal in each tier. This is probably a reasonably accurate reflection of what happens in most pig and poultry breeding industries, and to a lesser extent in most dairy cattle breeding industries. However, in many beef cattle and sheep breeding industries breeding goals are often set 'top down' instead of 'bottom up', except between the lower tiers. In other words, breeding goals for elite animals are influenced much more by market signals from within the top tier, rather than from tiers below. Frustration with this situation led to the formation of **group breeding schemes** in several sheep and beef cattle breeds in New Zealand in the 1960s [29]. The farmers concerned were not satisfied that the tiers above were producing breeding stock relevant to their needs, and so they established cooperative breeding schemes to produce their own replacement breeding stock, especially males. In most cases this involved the formation of **nucleus flocks** or **herds**. These are elite flocks or herds formed by screening the best cows or ewes from

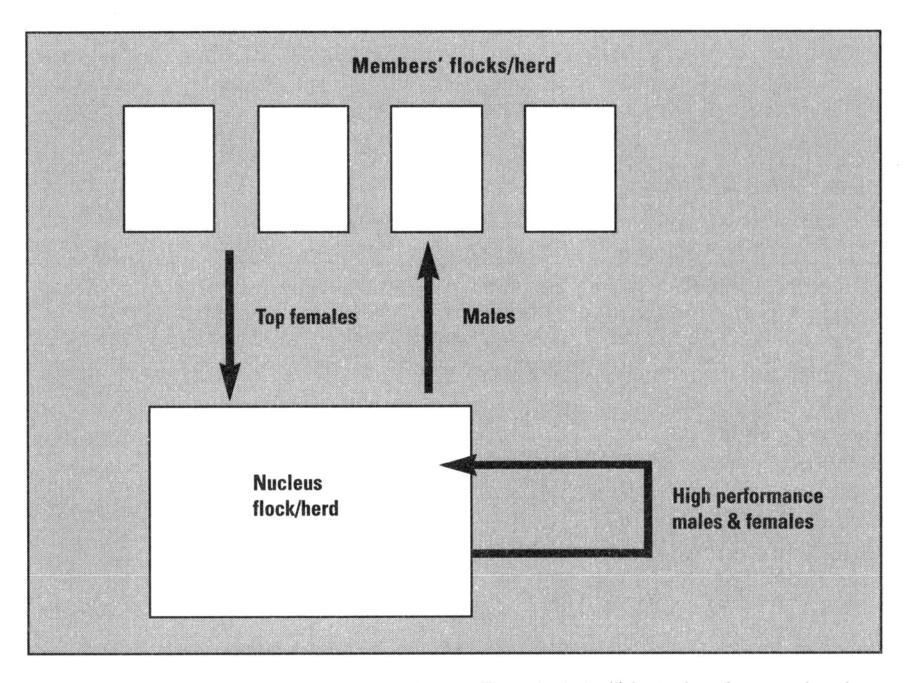

Figure 3.2 The principles of group breeding schemes. These include: (i) formation of a central nucleus herd or flock, from the best females available in members' own herds or flocks; (ii) comprehensive recording and rigorous selection for commercially important traits in the nucleus; (iii) the best males produced in the nucleus are retained there for breeding, whilst the next best group (or ex-nucleus males) are the main source of breeding males for members' herds or flocks – this is the main route for dissemination of genetic improvement from the nucleus to the base herds or flocks; (iv) elite females may continue to be 'promoted' to the nucleus from base herds or flocks, on the basis of genetic merit (this only takes place in 'open' nucleus schemes – in some cases 'closed' nucleus schemes are preferred, for example to minimise risks of transfer of disease to the nucleus, or because of a lack of recording in base herds or flocks).

cooperators' own herds or flocks. Comprehensive recording and rigorous selection in the nucleus results in genetic improvement of commercially important traits in the nucleus animals, which is then disseminated to members' own flocks or herds when they obtain replacement rams and bulls (or females) from the nucleus (see Figure 3.2).

A number of factors are involved in diluting the signals between tiers in livestock industries. The many links in the chain connecting primary producers to consumers often leads to imperfect communication, and so the signals reaching commercial producers may be weak in the first place. In some countries direct or indirect subsidies may mask 'true' market signals. There may be a lack of understanding of, or low availability of, objective information on the genetic merit of breeds, crosses or individual animals within breeds. Also, social factors are often important – for instance, some breeders may prefer to breed show winners than to follow strict market signals. For a small number of elite breeders there may be niche markets which are based mainly on show performance. At least in the short term it is probably more profitable for this minority of breeders to breed for show performance, than to breed for the traits of most economic importance in the tiers below. Unfortunately, many more breeders get lured by the potentially big rewards from these small niche markets and neglect the larger markets. A further complication occurs if the animals produced in some tiers are by-products of other primary activities. For example, the main concerns of most dairy farmers crossing to a beef bull is the bull's effect on calving ease in their dairy herd, and then the potential sale value of the calf. They are less concerned about the possible use of some daughters of the beef bull as suckler cows, and so most of them pay little attention to the performance traits which might be useful in that role (e.g. intermediate mature size and good maternal performance) when choosing a beef bull.

It will be apparent now that the number of tiers present in the breeding pyramid, and their relative importance, will vary between species and countries. The most relevant strategy for genetic improvement for any individual breeder will depend on the breeder's position in the breeding pyramid. Most elite breeders and purebred multipliers will be interested in within-breed selection, and occasionally in between-breed selection if the market changes substantially and their current breed becomes less relevant. Crossbred multipliers will be interested mainly in crossbreeding. Commercial producers will probably be interested in several or all of the strategies for genetic improvement. Although they are users, rather than creators, of genetically improved stock, a knowledge of the techniques involved should allow them to make better decisions on which breed, cross or individual animals to use.

Selection between breeds

Selection between breeds or strains can achieve dramatic and rapid genetic change when there are large genetic differences between populations in characteristics of economic importance. However, this benefit is achieved only once, unlike the improvement brought about by continuous within-breed selection. It is obviously costly to replace whole flocks or herds of breeding females at once. In practice, changes are often made more gradually. If replacement females (i.e. females brought into the breeding herd or flock to replace those which have died of natural causes, or have been culled) are usually purchased, then the switch to the new breed can be made gradually by buying replacements from the new breed. If replacement females are normally homebred, then the composition of the herd or flock can be changed gradually by **grading up** or repeated **backcrossing** to the new breed. Grading up or backcrossing involves repeated mating of the current females, and subsequently their female offspring, to sires of the new breed. Table 3.2 shows the proportion of genes coming from each of two breeds following successive generations of backcrossing – on average each generation of crossing halves the proportion of genes from the original breed, compared to that in the previous generation. The table also shows the minimum time required to achieve particular proportions of genes from each breed. Many pedigree breed societies allow registration of crossbred animals in a separate register, with eventual acceptance as purebred animals once they are $^7/_8$ or $^{15}/_{16}$ purebred.

The first important criterion for choosing between breeds, strains or crosses is that the choice ought to be made on the basis of objective comparisons of performance *in the relevant environment* (and at a relevant endpoint). There are some

Breed composition of			Number of years after initial cross to produce offspring of the composition shown, assuming females first leave offspring at		
Sire	Dam	Offspring	1 year old (some sheep breeds)	2 years old (sheep and some cattle breeds)	3 years old (cattle)
A	B	$^1/_2$ A, $^1/_2$ B	–	–	–
A	$^1/_2$ A, $^1/_2$ B	$^3/_4$ A, $^1/_4$ B	1	2	3
A	$^3/_4$ A, $^1/_4$ B	$^7/_8$ A, $^1/_8$ B	2	4	6
A	$^7/_8$ A, $^1/_8$ B	$^{15}/_{16}$ A, $^1/_{16}$ B	3	6	9

Table 3.2 The proportion of genes from each of two breeds following successive generations of backcrossing, and the time taken to achieve these proportions in cattle and sheep. The letters A and B are used as abbreviations for the two breeds here (not as abbreviations for alleles, as in the previous chapter). In the first cross exactly half of each animal's genes come from each of the two breeds. In subsequent generations of backcrossing the proportions of genes shown are the averages expected – individual animals may have greater or smaller proportions than shown because of segregation and recombination.

dramatic examples of the cost of ignoring this rule: high performing temperate breeds of livestock have been introduced often to the tropics without this sort of trial, and have then succumbed to diseases or to nutritional deprivation, to which local breeds were tolerant. For example, in the tropics, European dairy breeds often have lower survival and herdlife than pure zebu animals, and crosses with them. The reproductive rate of the pure European breeds is often too low to maintain herd sizes [11].

Less dramatically, there are important economic benefits to be gained, in any country, from matching breeds or crosses to particular production systems. The concept that genotypes do not always rank the same in different environments, or that the advantage to a particular genotype in one environment may be smaller or greater in another, is an important one in livestock improvement. In general this is called a **genotype x environment interaction**, or in this particular case, a **breed x production system interaction**. Figure 3.3 illustrates two types of genotype x environment interaction. The genetic correlation between the performance of related animals in two environments can also be used to detect genotype x environment interactions. In theory, if the genetic correlation between performance in the two environments is less than 1, then there is an interaction. However, correlations have to be substantially less than 1 before an interaction is of practical importance.

A great deal of research has been done on comparing the performance of different genotypes (breeds, strains, crosses, or animals of different genetic merit within a breed) in different environments (countries, production systems, feeding levels) to help livestock producers choose the most appropriate breed for their circumstances. At first sight the results of these studies are often in conflict: many of them appear to deny the existence of interactions, while others firmly support their existence. It is probably fair to say that if there is a big enough difference in performance between the genotypes, or a big enough difference between the

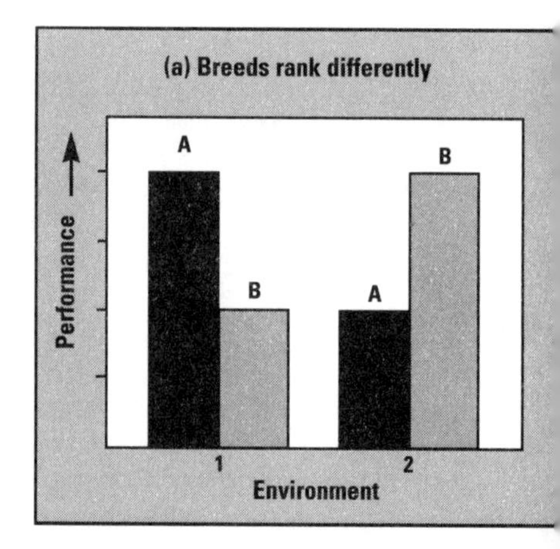

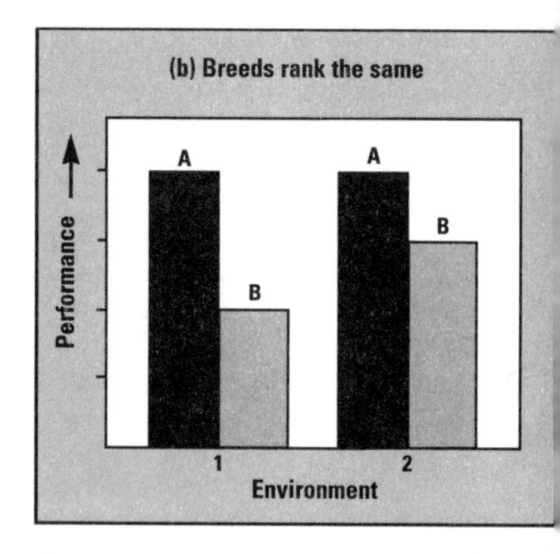

Figure 3.3 Two types of genotype x environment interaction. (a) Breeds A and B rank differently in the two environments. (b) Breeds A and B rank the same in the two environments, i.e. breed A is best in both environments, but the advantage to breed A is much less in environment 2 than in environment 1.

environments in which they are compared, or both, then there will be an interaction. For example, several studies indicate that small or medium-sized beef and sheep breeds have higher overall productivity or profitability in extensive grazing systems than larger breeds. But larger breeds tend to do best in more intensive systems with high levels of concentrate feeding [4, 13]. There are several possible explanations for this. The more extensive the production system the greater the effort animals will have to make to harvest the resources they require for maintenance and growth. Hence, smaller breeds may have an advantage in extensive systems because they have lower requirements for maintenance and growth than larger breeds, and so their requirements are more in balance with the effort they expend in meeting them. An alternative explanation is that the larger sheep and beef breeds are probably the ones which have been subjected to the greatest artificial selection. At the same time, there may have been modifications to their environment, which have reduced natural selection for adaptation to extensive conditions. As a consequence of these two factors, favourable attributes for extensive systems (e.g. high mobility, particular patterns of grazing behaviour, cold tolerance, disease resistance and strong maternal behaviour) may have been lost, or may have diminished, in larger, more intensely selected sheep and beef cattle breeds. (The results of some experiments investigating interactions are given in later chapters.)

Another important criterion for breed or strain comparisons is that the samples of animals compared should be sufficiently large and representative of the breeds concerned to allow inferences to be drawn about these breeds in general, and not just the sample itself. In other words, the breed comparisons need plenty of animals from different sires (and dams, though this usually happens automatically in the less prolific species), and these sires themselves need to be sampled randomly, or in a balanced way, from several different flocks or herds. This criterion is often violated: newly introduced breeds are often promoted on the basis of very flimsy comparisons, or no objective comparisons at all. The size of the sample required depends on the minimum difference between breeds that is considered to be of economic importance. It also depends on the amount of variation that exists in the trait concerned.

For example, let us assume that a 25 kg advantage in the 18-month weight of Charolais-cross beef calves was the minimum that would justify a producer switching from Simmental-crosses. If the standard deviation of 18-month weight is about 50 kg in both crosses, then we would need to compare about 70 animals of each cross to be 90% sure that Charolais crosses were really better, and about 85 animals of each cross to be 95% sure. (See reference [34] or other standard statistics textbooks for details of how to calculate the numbers of animals required for this sort of test).

Generally, the higher the degree of genetic control of a trait (i.e. the higher the **heritability** – this is the ratio of additive genetic to total phenotypic variation in the trait of interest), the greater the number of sires that should be used in a breed comparison. As a rough guide, five or more sires should be used if the traits of interest are of low heritability, and ten or more should be used if the trait is highly heritable. (This seems odd at first sight, but there is proportionally more variation between sire families for highly heritable traits than for low heritability traits. This means that there is a greater risk of mistakenly choosing one breed over another as a result of the chance sampling of one or two exceptional sires, when the heritability of the trait is high.) Ideally, sires should be from different flocks or herds. References to examples of well-designed breed comparisons are given at the end of this chapter. Tables 3.3 and 3.4 summarise the results of two of these. These results are presented here for illustration only. It is important to remember that breeds change over time as a result of within-breed selection. So, the results of breed comparisons can become outdated fairly quickly, particularly if some breeds are pursuing effective improvement programmes and others are not, or if breeds are selecting for different characteristics.

There are many recent examples of selection between breeds in the livestock industries. In Britain many dairy herds of Shorthorn, Ayrshire and Channel Island breeds (Jersey and Guernsey) changed to Friesians three or four decades ago. Over the last decade or two, the local strains of black and white dairy cattle in many

	Sire breed mean			
Trait	Oxford Down	Southdown	Suffolk	Texel
Age at slaughter (days)	258^{c}	214^{ab}	228^{bc}	262^{c}
Carcass weight (kg)	19.7^{c}	17.0^{a}	18.6^{b}	19.2^{c}
Daily gain of carcass lean (g)	46^{bc}	50^{c}	49^{c}	46^{bc}
Daily gain of carcass fat (g)	23^{bc}	25^{cd}	23^{bc}	20^{ab}

Table 3.3 Results of a comparison between different sire breeds of sheep, mated to Mule (Bluefaced Leicester x Swaledale) ewes. These results are from just a sample of the breeds involved in a large trial run by the Meat and Livestock Commission in Britain. The results here are from 'late flocks' producing lambs off grass, forage crops and roots. The sire breed means are for lambs at the same estimated level of subcutaneous fatness. The superscripts show whether or not sire breeds differ significantly. Within a row, sire breeds with different superscripts differ significantly; those with the same superscript do not differ significantly, e.g. Oxford Down cross lambs were significantly older at slaughter than Southdown crosses, but were not significantly different in age from Suffolk or Texel crosses [21].

Trait	Sire breed mean				
	Aberdeen Angus	Charolais	Hereford	Limousin	Simmental
Age at slaughter (days)	477^{c}	520^{a}	492^{b}	517^{a}	517^{a}
Carcass weight (kg)	205^{e}	268^{a}	214^{de}	247^{c}	258^{b}
Conformation (15 point scale)	9.9^{b}	11.2^{a}	8.7^{cd}	11.0^{a}	9.9^{b}
Saleable meat in carcass (g/kg)	725^{bc}	727^{b}	719^{cd}	733^{a}	720^{cd}
Fat trim in carcass (g/kg)	96^{abc}	90^{c}	97^{abc}	92^{bc}	93^{abc}

Table 3.4 Results of a comparison between different sire breeds of beef cattle, mated to Hereford x Friesian and Blue Grey suckler cows. These results are from just a sample of the breeds involved in a large trial run by the Meat and Livestock Commission in Britain. The results here are from winter fattening systems. The sire breed means are for steers at the same estimated level of subcutaneous fatness. The superscripts show whether or not sire breeds differ significantly. Within a row, sire breeds with different superscripts differ significantly; those with the same superscript do not differ significantly e.g. Charolais crosses were older at slaughter than Aberdeen Angus or Hereford crosses, but were not significantly older than Limousin or Simmental crosses [20, 35].

European and other countries have been partly or wholly substituted by North American Holstein strains (see Plate 12). (Most of these different black and white strains originate from importations of Dutch Black Pied animals at some time in the last few hundred years, but they have been selected for different objectives, with different rates of progress in different countries since then.) In each case the breed substitutions have been prompted by the higher total milk yield of the incoming breed or strain. In many temperate beef-producing countries the traditional British beef breeds have been replaced, at least as terminal sires, by the larger, leaner continental European breeds such as the Charolais, Limousin and Simmental. In the British sheep industry, many rams from the Border Leicester breed have been replaced by rams from the more prolific Bluefaced Leicester breed for mating to draft hill ewes to produce crossbred breeding females for commercial flocks.

Selection within breeds

Selection within breeds involves comparing animals of the same breed and mating the preferred animals to produce the next generation. This process is usually repeated each generation, and as long as there is genetic variation in the characters under selection this produces changes in the following generation (compared to between-breed selection, where the changes occur only once).

In any population of farm (or wild) animals, the size of the population can only be maintained if adults produce enough offspring in their lifetime to replace themselves when they die or are culled. The capacity to produce potential replacements varies between species and between sexes. For instance, pigs produce far more potential replacements per adult than cattle, and in all species there are generally far fewer replacement males needed than females, since each male is usually mated to several females. It is this production of more replacements than are needed to maintain population size which provides the opportunity for selection within breeds.

Whenever a group of animals is reproductively isolated and some are allowed to breed and others are not, then within-breed selection is being practised. As discussed in Chapter 1, this has occurred for many centuries during and following the creation of distinct breeds. However, the use of within-breed selection as a tool for genetic improvement became widely recognised following its successful application in improving racehorses in the seventeenth century and farm livestock breeds in the eighteenth century.

At this time it was common to mate closely related animals in an attempt to concentrate desirable characteristics in the offspring. For example, strains of animals would be created from a series of father-daughter or mother-son matings. This practice of mating close relatives, called linebreeding, is much less common in livestock breeding today. Linebreeding is a rather dramatic form of **inbreeding** – the mating of animals which have one or more ancestors in common. Eventually inbreeding occurs whenever selection is practised in a closed population, but steps are usually taken to limit it by deliberately avoiding matings between close relatives.

Related animals have more genes in common than unrelated animals, and the closer the relationship, the more genes they have in common. This means that animals of outstanding genetic merit, for any characteristic, are likely to have relatives which have higher merit than average. For these reasons, it appears entirely logical to breed from closely related animals, but there is a downside. As well as having more favourable genes in common, related animals also have more unfavourable genes in common on average. For instance, there are believed to be many recessive genes which cause genetic diseases, or adversely affect reproduction, survival or the overall functional fitness of animals. Because these genes generally act recessively, they only cause problems in animals that carry two copies of the gene (homozygous recessive animals). Across a breed as a whole, there may be very few animals which are homozygous for these particular genes, but the genes are still there, undetected, in heterozygous or carrier animals. Matings between related animals are more likely to produce offspring which are homozygous for some of these genes than matings between unrelated animals. So in some cases breeders may be lucky, and they may only see the benefits of mating related animals of high genetic merit. However, in other cases there may be very unfavourable consequences because of infertility or genetic disease in the offspring.

Today we tend to hear only about the past success stories which resulted from linebreeding or close inbreeding, but for each of these there were probably several failures. In plant breeding, and to a lesser extent in poultry breeding, breeders can afford to try linebreeding or close inbreeding in many different lines, and only continue to breed from those which show favourable responses. However, in other livestock species, because of the longer generation intervals, the lower reproductive rate, and the higher value of the individual breeding animals, it is not practical to keep many different lines, and the risks from linebreeding usually outweigh the potential benefits. A more gradual, but inevitable and cumulative increase in inbreeding occurs whenever selection is practised in a closed population (e.g. a breed, or a closed herd or flock) over a long period of time. This too can result in the same unfavourable effects as linebreeding. The effects of inbreeding, and methods of controlling inbreeding in selection programmes are discussed further in Chapter 4.

Requirements for within-breed selection

The choice of animals to breed from may be based solely on their appearance, or on a subjective assessment of their own, or their relatives' performance. For example, selection may be based on visual appraisal of the breed characteristics of an animal, its shape, colour etc., or on subjective estimates of the amount of wool or milk produced, or the size of animals. Alternatively, objective records of the performance of the animal itself, or its relatives, may be used. That is, actually weighing the fleeces or the animals, or measuring the volume of milk produced rather than guessing the production. Often a combination of objective and subjective information is used. Objective methods of within-breed selection (i.e. those which rely on recording performance, rather than making subjective judgements about the merit of animals) have become widely used in pig, poultry and dairy cattle breeding over the last few decades. They have been used to a much lesser extent in sheep and beef cattle breeding, but this situation is changing.

Objective selection within breeds or strains is intended to increase the average level of additive genetic merit (termed **breeding value**) of the population. Ideally, the steps involved are:

1. Deciding what to improve – termed the **breeding goal**. As mentioned before, it is sensible to identify the breed or cross which has the highest merit in this trait or combination of traits, and to decide on the most appropriate breeding strategy (e.g. pure- or crossbreeding) before embarking on within-breed selection in the best breed, or in component breeds of the best cross. Also, there must be genetic variation in this trait or combination of traits within the breed if any improvement is to be made – in other words, there need to be differences between animals in each of these traits, and at least part of these differences need to be inherited.

2. Deciding what to measure and select on within breed (the **selection criterion**) in order to make improvements in the breeding goal. In some cases the selection criterion or criteria may be the same as the goal; in others some indirect measurement is needed, for example when the goal trait can only be measured in one sex or after slaughter.

3. Designing the breeding programme (e.g. numbers of males and females to be selected annually, ages at mating).

4. Implementing the programme, i.e. doing the routine recording, evaluation and mating of animals.

5. Monitoring progress and redesigning the programme if necessary.

These steps are summarised in Figure 3.4. In practice, most breeding schemes evolve in a less planned way than this, but it is useful to have this ideal structure in mind when attempting to improve them.

In most circumstances unique and permanent identification of individual animals is a requirement for successful within-breed selection. This may sound obvious, but it is harder to achieve than it first seems – particularly when the identities of animals have to be unique across herds, flocks, years or countries. There are several reasons why unique identification is important. Firstly, it is important to be able to assign records of performance to the right animal each time its performance is recorded. Secondly, modern methods of genetic evaluation make full use of information from relatives, so it is important to be able to establish how recorded animals are related to each other. Thirdly, having recorded and evaluated animals and decided which ones to use for breeding, it is important to be able to find them again! In some special cases, there is justification for simplified recording of identification. For example, in some extensive production systems animals may be identified only as members of a particular sire family, rather than individually identified.

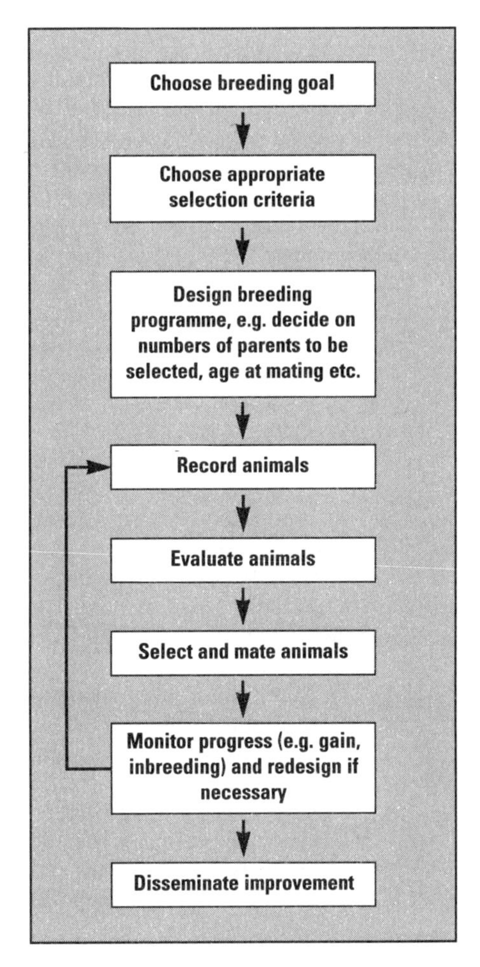

Figure 3.4 The steps ideally involved in a within-breed improvement programme based on objective measurement of performance. Once the breeding goal is established, different breeds and crosses should be compared in this character, and then within-breed improvement should be used in the best breed, or in component breeds of the best cross [18].

Objective selection depends on having records of performance on the candidates for selection, or their relatives, or both. In principle, there is no reason why each breeder should not devise his or her own method for recording performance in traits relevant to their own breeding goal. Most major pig and poultry breeding companies have done this. However, in practice in most Western countries the majority of cattle and sheep breeders use recording schemes which are operated by regional or national agencies which specialise in recording and evaluation. (The difference between the species has probably arisen for a number of reasons. These include the relatively small number of pig and poultry breeders in the elite tier compared to the situation in cattle and sheep breeding, so there are fewer 'products' for commercial producers to be aware of. Also, many pig and poultry companies aim to fulfil both elite and multiplier roles, so the breeding companies can communicate directly with their final customers and there is less need for a nationally recognised recording system.) For example, in many countries there are milk recording agencies which visit dairy farms at regular intervals to record the yield of individual cows and to sample the milk for analysis of fat and protein content. This information is used to assist management decisions, and when accumulated and analysed it is used to provide estimates of the genetic merit of cows in the herd and their sires. In Britain, for example, for the past couple of decades, the Meat and Livestock Commission (MLC), and subsequently Signet (a joint MLC/SAC company), have performed equivalent functions for pedigree beef cattle and sheep breeders. Similar government or private agencies operate in many other countries. Increasingly records such as dates of birth, sex and live weights of young animals are collected by the breeders themselves, while others, such as ultrasonic measurements of fatness and muscularity, are made by field staff. In many countries the recording agencies adopt the recognised international standards of the International Committee on Animal Recording (ICAR) for recording and evaluation methods.

There are several advantages to using these regional or national systems:

- they encourage discipline and consistency in which traits are recorded, and how and when they are recorded

- they often provide some type of authentication of the records of performance – either by supervising some of the recordings or, more commonly now, simply by checking that performance records fall into expected ranges, and that details from different sources (e.g. breeder and breed society) match

- they usually allow access to more sophisticated methods of data storage, processing and evaluation than is feasible on most farms

- depending on the method of evaluation and the population structure, they can allow direct comparison of the estimated genetic merit of animals across many herds or flocks; this increases the pool of animals available for selection by breeders, and also allows customers to identify individual animals, or flocks and herds of high genetic merit

- there are often benefits when it comes to selling breeding stock, because large recording schemes can achieve a stronger 'identity' and better education about their purposes and products than individual breeders could achieve.

There are also disadvantages to membership of large regional or national recording schemes. There is usually a greater (apparent) cost than with home-grown schemes and often there is less flexibility for tailoring breeding goals to individual breeder's requirements. However, on balance, the advantages of coordinated regional or national schemes usually far outweigh the disadvantages.

Factors affecting rates of improvement

As discussed in Chapter 2, many traits of economic importance in farm livestock are under the control of a large number of genes and show continuous variation. Annual rates of genetic improvement in these polygenic traits depend on four main factors:

1. The **selection intensity** achieved. This is related to the proportion of animals selected to become parents, based on their own performance or that of their relatives. The lower the proportion of animals selected, the higher the selection intensity and hence the better the selected animals will be, on average.

2. The **accuracy** with which genetic merit in the trait of interest is predicted. This itself depends on the extent to which the trait is under genetic control: traits such as growth and wool production tend to be more strongly influenced by the animal's genetic make-up (they have a higher heritability) than traits associated with reproduction and survival. The accuracy also depends on the extent to which performance information on relatives is used in helping to pick animals for breeding. The more information available on relatives, the more accurately an individual's genetic merit or breeding value can be predicted.

3. The amount of **additive genetic variation** in the trait of interest. Again, this is largely a characteristic of the trait concerned. For example, there is relatively more additive genetic variation in growth rate than in carcass lean content in sheep and beef cattle. Also, there is relatively more additive genetic variation in milk yield than in milk protein concentration in dairy cattle.

4. The **generation interval**, which depends on the average age of parents when their offspring are born. Essentially this regulates the speed with which selected animals contribute their better genes to the flock or herd, via their offspring.

These factors are discussed in more detail in Chapter 4. Generally speaking, the higher the selection intensity, accuracy and genetic variation, and the lower the generation interval, the higher the annual rate of genetic improvement. The main opportunities for breeders to accelerate rates of improvement are through choice of the most accurate methods of predicting breeding values and by maintaining high selection intensities and low generation intervals. However, there are biological limits to the extent to which selection intensity and generation interval can be altered. It is possible to achieve much higher selection intensities in species or breeds with a high reproductive rate. Similarly, shorter generation intervals can be achieved in those species or breeds which reach sexual maturity at a younger age. These reproductive characteristics of different species are contrasted in Table 3.5. Largely because of biological advantages in these reproductive characteristics, higher rates of genetic change are possible in pigs and poultry than in ruminants (see Figure 3.5). However, the reproductive rate of female cattle and sheep can be altered dramatically by the use of reproductive technologies such as multiple ovulation and embryo transfer. This effectively makes cattle and sheep reproductively more like poultry and pigs, and enables higher rates of genetic gain. The use of these technologies is discussed in more detail in Chapter 9, and specific examples are given in Chapters 6 to 8.

Species	Typical age of females at birth of first offspring (limits generation interval)	Typical annual number of offspring per female (limits selection intensity)
Poultry	7 months	120–280
Pigs	9–12 months	20–24
Sheep	12–24 months	1–3
Cattle	24–36 months	1

Table 3.5 Comparison of the reproductive characteristics of farm livestock species. The values shown are for natural reproduction. Those for ruminants can be increased dramatically by the use of reproductive technologies such as embryo transfer, as discussed in Chapter 9.

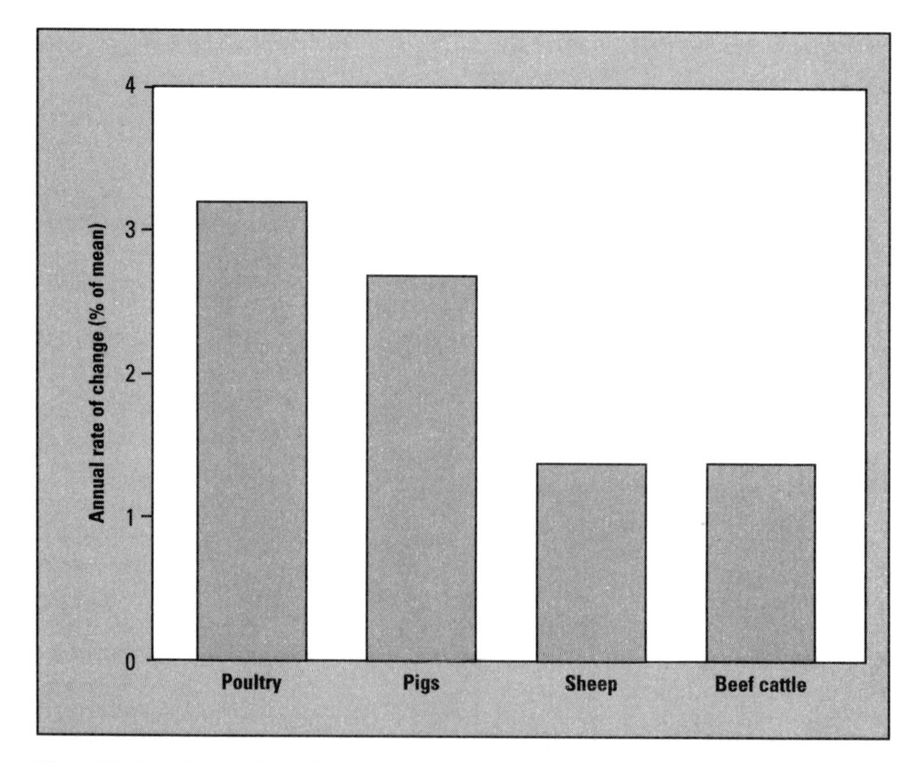

Figure 3.5 Annual rates of genetic change possible in growth rate in different livestock species with natural reproduction. The rates of genetic change are expressed in percentage units of the average growth rate of the species concerned. The differences in rates of change between species are largely due to differences in the natural reproductive characteristics of the species [33].

Selection for more than one trait

In most livestock production systems profitability depends on several different animal characteristics rather than on any single trait. For instance, in dairy cattle, income often depends not only on yield but also on fat and protein content of the milk and often on somatic cell count too. Feed costs, health and rebreeding costs are important variable costs. In meat animals, returns per animal depend not only on carcass weight, but also on measures of carcass quality, such as fatness and conformation, and on feed costs. In genetic improvement programmes it is important to reflect the fact that several traits influence profit, and so animals are usually selected (whether objectively or subjectively) on a combination of traits. This can be achieved in a number of ways.

One approach, called **tandem selection** involves selection for one trait for one or more generations, followed by selection for a second trait for one or more generations, possibly followed by selection on more traits, eventually returning to selection on the first, and so on. It can be useful in some circumstances; for instance, if there is an urgent need to improve one trait and other traits need only very minor improvement, it might be appropriate to select only on the first trait until some economically important threshold was reached, and then to turn to the others. However, if the associations between traits of importance are unfavourable, the selection goes round in circles, improving one trait in one generation, and then wholly or partly cancelling that improvement in the next generation when selection is for a second trait.

A more consistent method, called **independent culling levels**, is to set minimum qualifying standards, or thresholds, in several traits of interest. To qualify for selection, animals must then surpass the qualifying standard in each trait. There are methods available to allow the most appropriate qualifying standards, or cut-off points, to be calculated. Figure 3.6 (a) illustrates this method of selection. In this example beef bulls are only selected if they achieve a 400-day weight above 560 kg, and a muscling score of over 12 points (on a scale of 1 to 15). An informal version of this approach is used by most breeders when they check candidates for selection for breed type or any 'functional' defects (such as overshot or undershot jaws, locomotion faults or small testicles) before breeding from them. This approach is probably a very efficient way of dealing with functional defects which occur irregularly but are suspected of having a genetic component, or traits of minor importance. A version of independent culling levels, or sequential selection, may also be useful when one trait of interest is very difficult or expensive to measure. For example, it may not be cost-effective to record and select on feed intake or feed efficiency on all animals in a selection programme, but it may well be effective to do so, say, on those with weaning weights in the top 25% of those tested.

In theory, the optimal method of selection when there are several traits of economic or functional importance is to calculate a **selection index**. This is a score of overall genetic merit for each of the animals available for selection, based on their own or their relatives' performance in the traits of interest. In some respects index selection is similar to independent culling levels in that animals are selected for more than one trait simultaneously and altering the emphasis on a trait in a selection index is broadly similar to moving the cut-off point for culling with

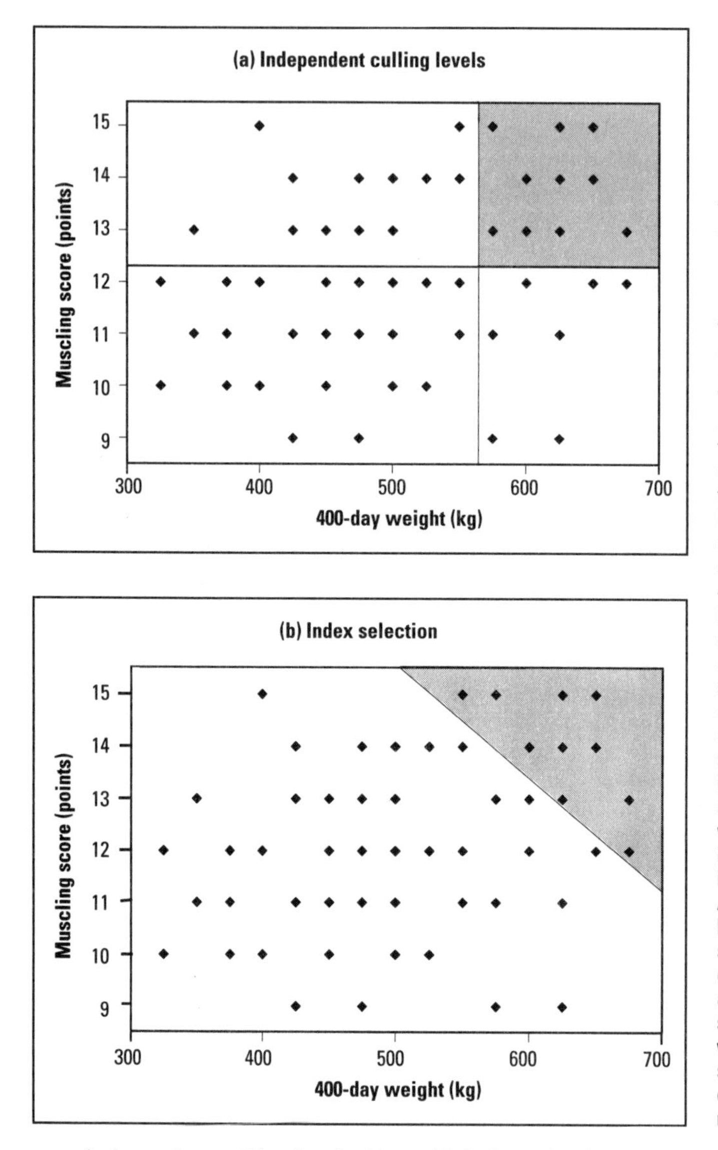

Figure 3.6 The use of independent culling levels or index selection to select simultaneously for increased 400-day weight, and increased muscling score. Each point on the two graphs shows the 400-day weight and muscling score of a single beef bull. (a) Independent culling levels involves setting minimum qualifying standards, or thresholds, in each trait of interest. To qualify for selection, animals must then surpass the qualifying standard in each trait, represented by the vertical and horizontal lines on the graph. In this example, bulls are only selected if they achieve a 400 day weight above 560 kg, and a muscling score greater than 12 points (on a scale of 1-15). (b) A selection index scores animals based on all traits of interest, but there are no fixed cut-off points for individual traits – an animal can be selected with lower performance in one trait, providing that it excels in another. Animals above the diagonal line in the graph would be selected. In this example eight of the ten selected animals would be the same whichever method was used; two of the ten selected animals would differ, depending on the method used.

independent culling levels. But with index selection an animal can compensate for poor performance in one trait by excelling in another. Although the individual animals which are selected may not have the ideal combination of characters, this is the most efficient way of moving a whole population in the desired direction. A good analogy for index selection is the entrance system for many colleges or universities, where candidates have to accumulate a minimum points score to be accepted, where this score is calculated from the examination grades they have achieved in several subjects. Index selection is illustrated in Figure 3.6(b).

Table 3.6 gives a simple example of how an index can help in selection. The table shows milk fat and protein yields for three dairy cows. If selection was based solely on protein yield, then cow A would be selected. If selection was based

Cow	Protein yield (kg)	Fat yield (kg)	Protein + fat yield (kg)
A	**230**	280	510
B	200	**300**	500
C	225	295	**520**

Table 3.6 An example of the use of a simple index to select cows on protein plus fat yield. The yields underlined in bold type show which cow would be selected if selection was for protein yield only, fat yield only, or protein plus fat yield respectively.

solely on fat yield, then cow B would be selected. However, if selection was based on fat plus protein yield – a simple index – then cow C would be selected. In this case, the index gives equal emphasis to protein and fat yields, so the two are simply added together. If the desired emphasis on protein:fat was 2:1, then animals could be selected on twice their protein yield, plus their fat yield, and so on. In practice, the traits of interest are often measured in different units, and they have different heritabilities, and so arriving at the most appropriate weighting factors is a bit more complicated than it seems at first sight. To derive these weighting factors it is necessary to know: (i) how much additive genetic variation there is in the traits of interest; (ii) the direction and strength of associations among these traits and (iii) their relative economic importance. Obtaining reliable estimates of the genetic variation in traits and associations among them requires comprehensive recording of hundreds or thousands of animals. When the traits of interest are already recorded in a large regional or national scheme then there should be plenty of records to obtain these estimates. However, it will be more difficult and costly to obtain them for traits which are not already recorded, which can be a major limitation to the wider use of index selection. Selection indexes are discussed further in Chapter 5.

Crossbreeding

Reasons for crossing

Crossbreeding involves mating animals of different breeds, lines or species. It is usually used for one or more of the following reasons:

- **To improve the overall efficiency of a production system by crossing breeds which have high genetic merit in different traits (*complementarity*).** The use of specialised sire and dam breeds or lines is a good example of this [32]. It is common for commercial beef herds and sheep flocks to be made up of breeding females from small or medium-sized breeds or crosses which have relatively low maintenance costs and good reproductive and maternal characteristics. These are mated to males from much larger terminal sire breeds, with faster growth rate and better carcass characteristics. The use of different breed types with complementary characteristics usually results in production systems with much higher overall efficiency and profitability than those based entirely on small breeds with good reproductive and maternal traits, or those based entirely on large breeds with good growth and carcass traits [13].

- **To produce individual animals of intermediate performance between that of two more extreme parent breeds**. At first sight this is similar to the use of crossbreeding to exploit complementarity. However, the emphasis here is on creating individual animals of intermediate performance, rather than matching different breeds with different roles in a crossing system. Beef x dairy suckler cows, such as Hereford or Simmental x Holstein Friesians, provide a good example of this use of crossing – the crossbred cows have higher beef merit than pure dairy cows, but higher milk yield than pure beef cows. Also, beef x beef crosses, such as the Blue Grey, combine the hardiness of the Galloway with the faster growth and likely higher milk yield of the Shorthorn.

- **For *grading up* to a new breed or strain**. As explained earlier, this involves mating animals (usually females) of an existing breed or strain to those (usually males) from a new breed or strain (see Table 3.2). The progeny from these matings are themselves mated to the new breed or strain, and this process continues for several generations. Many pedigree herds of continental European beef cattle breeds were established or expanded in importing countries in the 1960s and 1970s by grading up, or repeated crossing, using imported bulls or semen. Similarly, the spread of North American Holsteins to many temperate countries has been brought about largely by grading up indigenous populations of black and white (or other) dairy cattle. This approach is probably more common in cattle than sheep because the higher average value per head of pedigree cattle than pedigree sheep makes it more expensive to start afresh with purchased pedigree cattle of a new breed.

- **As an intermediate step in the creation of a new *synthetic* or *composite* breed**. Creating a synthetic breed usually involves crossing two or more different breeds. If more than two breeds are involved, then this step can take several generations. For instance, if four breeds are involved (A,B,C and D), two different crossbred types could be produced in the first generation (AB and CD). In the second generation, these two crossbred types could be mated to each other, to produce offspring which, on average, have a quarter of their genes from each of the four original breeds. Males and females of these fourway crosses could then be mated to each other in subsequent generations.

When two breeds are crossed, the offspring (called the **F1 generation**) are relatively uniform, since exactly half of each animal's genes come from each parent breed. However, if F1 males are mated to F1 females (i.e. if the F1 generation is **interbred**), the resulting offspring, called the F2 generation, show huge variation in appearance and performance.

This is because segregation and recombination have now occurred, leading to a wide variation in the proportion of genes which individual animals inherit from the original breeds (as explained in Chapter 2). For any particular characteristic, some offspring will resemble one of the original parent breeds, while others will resemble the other parent breed, and most will be somewhere in the middle. Following this explosion of variation in the F2 generation, it takes several generations of selection for the desired characteristics before the variation in performance and appearance is reduced, and a recognisable breed type emerges.

Most synthetic lines of pigs and poultry have been created in this way, although they are not regarded as new breeds since, in most cases, these lines are still open to the introduction of yet more breeds or other synthetic lines. However, there are several examples of synthetic breeds in cattle and sheep. The Luing cattle breed was created from crosses between Shorthorn and Highland cattle, subsequently interbred. There are several synthetic dairy cattle breeds in the tropics which have been created by crossing more productive *Bos taurus* dairy breeds with indigenous heat- and disease-tolerant *Bos indicus* breeds (e.g. the Australian milking zebu, formed from the Sahiwal, Red Sindi and Jersey breeds and the Jamaica Hope formed from the Jersey, Friesian and Sahiwal breeds [38]). The Coopworth dual-purpose sheep breed from New Zealand was created by crossing Border Leicester and Romney sheep, followed by interbreeding the crosses with intense selection for numbers of lambs born or weaned, weaning weight, fleece weight and quality and 'easy care' characteristics, such as lambing ease [2] (see Plate 13). The Meatlinc is a synthetic terminal sire sheep breed created in Britain by crossing animals from several breeds, including the Suffolk, Dorset Down, Ile de France, Berrichon du Cher and Charollais, and selecting for growth and carcass traits [14]. The ABRO Damline was developed in the 1960s and 70s as a synthetic crossing breed from the Finnish Landrace, East Friesland, Border Leicester and Dorset Horn breeds. The Cambridge is another synthetic crossing breed created in Britain by selecting foundation ewes which had given birth to three consecutive sets of triplets. The ewes were from several breeds, including the Clun Forest, Llanwenog, Lleyn and Radnor breeds. These were mated to Finnish Landrace rams and the offspring generation was interbred [28].

- **To introduce new variation to numerically small breeds**. In many numerically small breeds it is difficult for breeders to find enough unrelated animals of sufficiently high genetic merit to sustain a genetic improvement programme. Often, these problems are exacerbated because the population is inbred and genetic defects have emerged which need to be controlled. In these cases it is quite common for animals from other breeds to be introduced, either officially or unofficially! This is similar to the creation of a synthetic breed, except that the long-term aim is to have a much lower proportion of genes from the new breed.

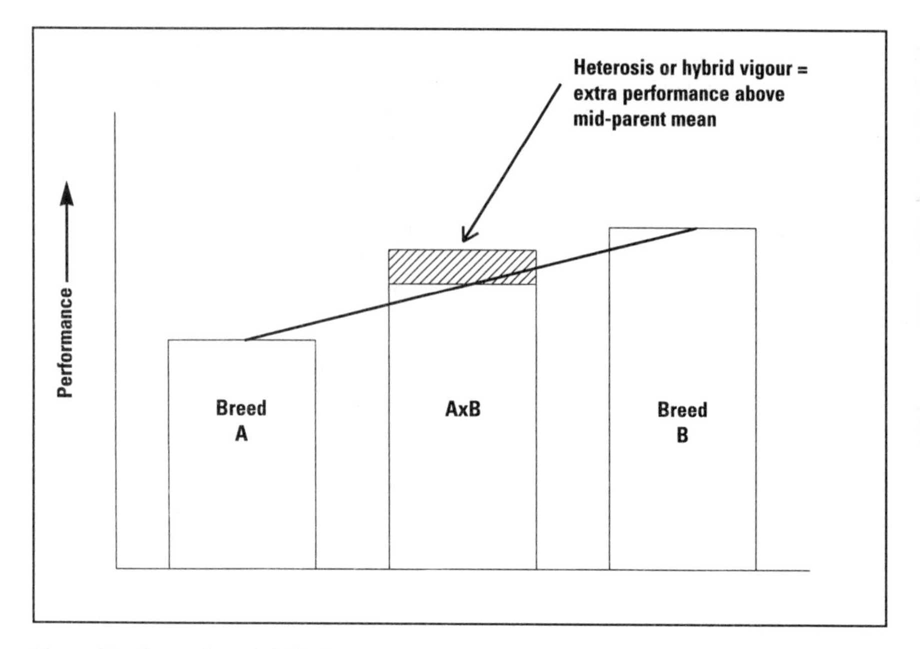

Figure 3.7 Heterosis or hybrid vigour. Heterosis is defined as the advantage in performance of crossbred animals above the mid-parent mean of the two parent breeds. In this case the amount of heterosis is the shaded part of the middle bar in the graph. Heterosis is most useful when it leads to the average performance of the crossbred animals exceeding that of the *best* parent breed in a single important trait or a combination of important traits.

- **To introduce a single gene for a favourable characteristic to an existing breed (*introgression*).** Crossing is sometimes used to introduce a single gene for a favourable characteristic to an existing breed. Good examples include the introgression of the gene controlling polledness into naturally horned cattle breeds, and introgression of the Booroola gene affecting fecundity, originally found in Merino sheep, into other sheep breeds. Unlike the creation of a synthetic breed, the aim of introgression is usually to introduce and retain only the desired gene from the new breed, and to remove the other genes contributed by this breed by successive generations of backcrossing to the original breed. It is important in this case to ensure that as many as possible of the females retained for backcrossing carry the desirable gene. After several generations of backcrossing, carrier animals can be mated to each other to create offspring which are homozygous for the newly introduced gene but in other respects are similar to the original breed. (See Chapter 9 for more details on how molecular genetic techniques can assist in this process.)

- **To exploit *heterosis* or *hybrid vigour*.** When two breeds are crossed, intuitively we expect the performance of the crossbred offspring to fall midway between that of the parent breeds. However, in practice the performance of crossbreds is often better than we expect. This advantage in performance above the mid-parent mean is called heterosis or hybrid vigour (see Figure 3.7). It is measured either in the units in which the trait was originally measured, or as a percentage increase over the mid-parent mean. For example, if two sheep breeds have average litter sizes of 1.0 and 2.0 lambs respectively, we expect crossbred ewes to have an average litter size of 1.5 (the mid-parent mean). If, in fact, the crosses have an average litter size of 1.6 lambs, this indicates that there is heterosis in litter size of 0.1 lamb (1.6 minus 1.5) or 6.7% (0.1/1.5 x 100). Heterosis is usually greatest in traits associated with reproduction, survival and overall fitness. There are many important examples of heterosis in farm animals, but heterosis is not confined to these. The fact that streetwise mongrel dogs and city pigeons appear to thrive so well, and procreate so freely, compared to their pampered purebred cousins is at least partly due to heterosis. Heterosis is also important in plant breeding as well as in animal breeding. The beneficial effects of heterosis occur for exactly the opposite reason that the detrimental effects of inbreeding occur. In other words, when two breeds or lines are crossed there is a much lower proportion of offspring which are homozygous for recessive genes affecting reproduction, survival and overall fitness, or causing genetic disease, than if animals of the same breed are mated. Crossbreeding creates animals which are heterozygous at more loci, whereas purebreeding creates animals which are homozygous at more loci.

Selection between breeds, selection within breeds and the first four applications of crossbreeding listed above all exploit differences in additive genetic merit between populations or individual animals. However, heterosis is the result of non-additive gene action. That is, it is a result of dominance at individual loci, or epistasis between loci, or both (see Chapter 2 for details). The fact that heterosis is a result of non-additive gene action means that it is difficult to predict the amount of heterosis to expect when particular breeds are crossed. Some crosses result in substantial heterosis and others do not. When a particular cross produces a large amount of heterosis, the parent breeds are sometimes said to **nick** well, or show good **combining ability**. Although it is difficult to predict the level of heterosis that will arise from crossing any two breeds, it is usually greater for crosses between genetically diverse breeds. For example, heterosis is usually greater in beef x dairy cattle breed crosses than in crosses between two dairy breeds, and it is usually greater in crosses between *Bos taurus* and *Bos indicus* breeds than in crosses between two *Bos taurus* breeds. This is probably because the more distantly related the two breeds, the greater the proportion of loci at which different alleles are fixed in the two breeds, and hence the greater the number of loci at which the crossbred offspring are heterozygous. As mentioned above, heterosis is usually greatest for traits affecting reproduction, survival and overall fitness, and it is usually least for production traits like growth and milk yield. Figure 3.8 shows the relative milk yield of indigenous *Bos indicus* cattle breeds, *Bos taurus* dairy breeds, F1 crosses between indigenous and exotic dairy breeds, and subsequent generations of interbred animals from a review of many dairy cattle crossbreeding experiments in the tropics.

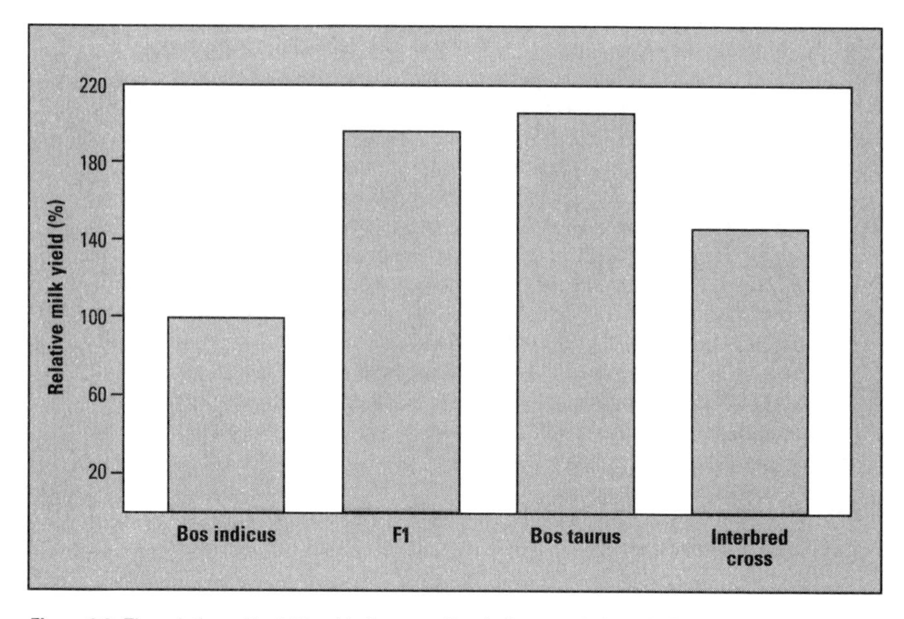

Figure 3.8 The relative milk yields of indigenous *Bos indicus* cattle breeds, F1 crosses between *Bos indicus* and *Bos taurus* dairy breeds, *Bos taurus* dairy breeds, and subsequent generations of interbred animals from a review of many dairy cattle crossbreeding experiments in the tropics. Yields are expressed in percentage units relative to that of the *Bos indicus* breeds [8, 9].

Heterosis is a useful bonus if there are other primary reasons for crossbreeding, e.g. to exploit complementarity of breeds. However, crossbreeding solely to exploit heterosis is only really justified if the heterosis is sufficient to make the crossbred animals better, on average, than the best parent breed. In other words, having substantial heterosis is of no net benefit if the crossbred animals are still inferior to purebred animals of one of the parent breeds. So, any evaluation of crossbreeding as a strategy for genetic improvement needs to take into account the additive genetic merit of the pure breeds, as well as the non-additive bonus which occurs when they are crossed.

There are numerous examples of experiments to compare both pure breeds and crosses of cattle and sheep. The results from one of these, an experiment to compare the reproductive and maternal performance of three different types of crossbred sheep in Britain, are shown in Table 3.7. Others include a large 'multi-breed' experiment, established at the Animal Breeding Research Organisation (ABRO) in Edinburgh in 1970, to estimate the extent of between-breed differences in traits affecting overall efficiency. The experiment involved 25 cattle breeds of different mature size and different levels of milk production. The results showed that at about one year of age, between-breed variation accounted for about 70% of the total variation in body weight, and 60% of the total variation in cumulated food intake, which highlights the importance of selecting the most appropriate breed [36]. Over the last couple of decades a series of large-scale experiments have been done at the US Meat Animal Research Center in Nebraska to compare the growth, carcass characteristics and reproductive performance of many beef breeds and crosses, to compare crossbreeding systems and to compare different synthetic

Trait	Crossbred ewe type		
	Border Leicester X	Bluefaced Leicester X	ABRO Damline X
Ewes lambing per ewe mated in first year of life	0.62[a]	0.73[b]	0.78[b]
Total number of lambs born per ewe mated (avg. of first 3 matings)	1.29[a]	1.40[b]	1.54[c]
Number of lambs born alive per ewe mated (avg. of first 3 matings)	1.16[a]	1.28[b]	1.37[c]
Estimated litter weight at 10 weeks, per ewe mated (kg)	25.94	31.10	28.68
Productivity relative to Border Leicester cross (= 100) taking into account estimated ewe feed requirements	100	114	117

Table 3.7 Results of a comparison between three different crossing breeds of sheep: the Border Leicester, the Bluefaced Leicester and the ABRO Damline. The table shows the average performance of crossbred ewes produced by mating rams from these three breeds to ewes of several hill breeds. The crossbred ewes were mated in their first year, which partly explains the relatively low levels of performance. Matings were to a terminal sire breed. The superscripts show whether or not the crossbred ewe types differ significantly. Within a row, crossbred means with different superscripts differ significantly; those with the same superscript do not differ significantly e.g. significantly fewer Border Leicester cross ewes lambed in their first year, compared to the other two crosses, but there was no significant difference in the proportion of Bluefaced Leicester cross or Damline cross ewes lambing in their first year [3].

strains [6, 7, 16]. A similar experiment was established in the early 1990s at the Australian Beef Cooperative Research Centre in Armidale, with the emphasis on investigating differences in meat quality between breeds and crosses.

In most temperate dairy industries there is very little systematic use of crossbreeding. A major reason for this is that the milk production of crossbreds rarely exceeds that of the pure Holstein Friesian, although there is heterosis for several important traits. In contrast, there is widespread use of systematic crossing, especially between Jerseys and Friesians, in the extensive dairy industry of New Zealand. (In 1995/96 Holstein Friesians, Jerseys and crosses between them accounted for 57%, 17% and 17% of dairy cows in New Zealand [22].) Both of these breeds have a long history of successful within-breed selection in New Zealand, and so have high additive genetic merit for production from grass. Additionally, crossbred animals appear to be particularly favoured when comparisons are made on the basis of milk production per hectare. Since heterosis for production is higher than that for live weight, crossbred animals achieve extra production above the parental mean with a relatively small 'penalty' as a result of higher live weight, and hence higher maintenance costs [17].

Types of heterosis

When parents of two different breeds are mated, the crossbred progeny may show heterosis in a range of characteristics. These might include the time taken to get up and suck after birth, neonatal survival, and early growth. Once the crossbred animals themselves mature and reproduce, heterosis may be evident in another set of traits associated with fertility and maternal ability. At this stage, some of the benefits of heterosis accrue to the offspring of the crossbred female, rather than to the female herself. So it is useful to distinguish between:

- **Individual heterosis**: influencing performance as a result of animals themselves being crossbred.

- **Maternal heterosis**: influencing the reproductive and other maternal performance characteristics of crossbred females. The benefits in this case are often measured in the offspring (e.g. extra weight of offspring weaned by crossbred cows, compared to purebreds). Maternal heterosis occurs as a result of dams being crossbred.

Species/type of animal	Breeds crossed	Trait	Type of heterosis	Amount of heterosis in units of measurement or as a % of the mid-parent mean	Source
Sheep	Galway, Border Leicester, Cheviot and Blackface	Fertility	Maternal	7.3%	[37]
		Litter size	"	2.3%	"
		Lambs/ewe mated	"	10.0%	"
		Lamb wt/ewe mated	"	11.8%	"
		Lamb mortality	"	2.0%	"
	Several breeds	Lamb daily gain	Individual	2–4%	[25]
Beef cattle	Hereford, Angus, Shorthorn	Wt calf weaned per cow exposed	Maternal	14.8%	[6]
			Individual	8.5%	"
	Several breeds	Age at puberty	"	–9.4%	[7]
		Post-weaning gain	"	11.0%	"
		Carcass weight	"	15.0%	"
		Fat thickness	"	0.1%	"
Dairy cattle	Holstein Friesian, Jersey	Milk	Individual	129 litres (3.9%)[1]	[17]
		Fat	"	6.8 kg (4.1%)[1]	"
		Protein	"	5.0 kg (4.1%)[1]	"
		Survival from 1st to 2nd lactation	"	4.7%	"
		Live weight	"	7.2 kg (1.9%)[2]	"

[1] % heterosis derived from mean performance of the two pure breeds in 1995/96 [22].

[2] % heterosis derived from assumed mid-parent mean live weight of 375 kg.

Table 3.8 Some examples of heterosis in traits of economic importance in sheep and cattle. Values shown are specific for the combination of breeds concerned [30].

- **Paternal heterosis**: influencing the reproductive performance of crossbred males. Paternal heterosis occurs as a result of sires being crossbred.

Although there may be paternal heterosis for traits such as libido and fertility, individual and maternal heterosis are of most practical value. Table 3.8 gives some examples of individual and maternal heterosis in economically important traits in crosses between temperate breeds of sheep, beef and dairy cattle. There are several good examples of industry structures, or breeding schemes, which make good use of both individual and maternal heterosis, as well as complementarity. There is widespread, organised use of crossing in the pig and poultry industries. The aim is to produce breeding females for commercial herds and flocks which, as well as having high additive genetic merit for reproduction and associated characteristics, show maternal heterosis. Mating these females to males of a different breed or strain then maximises individual heterosis in the offspring. For example, in the pig industry in the UK and elsewhere, commercial F1 females are produced typically by crossing strains based on the Large White and Landrace breeds, which have been selected for high maternal performance. In commercial herds these sows in turn are mated to boars from breeds, or synthetic lines based on several breeds, which have been selected for growth, carcass traits and feed conversion efficiency.

Crossbreeding schemes in sheep and beef cattle have probably evolved over time in a less planned way than in the pig and poultry industries. For instance, some production systems based on crossbred animals have probably been

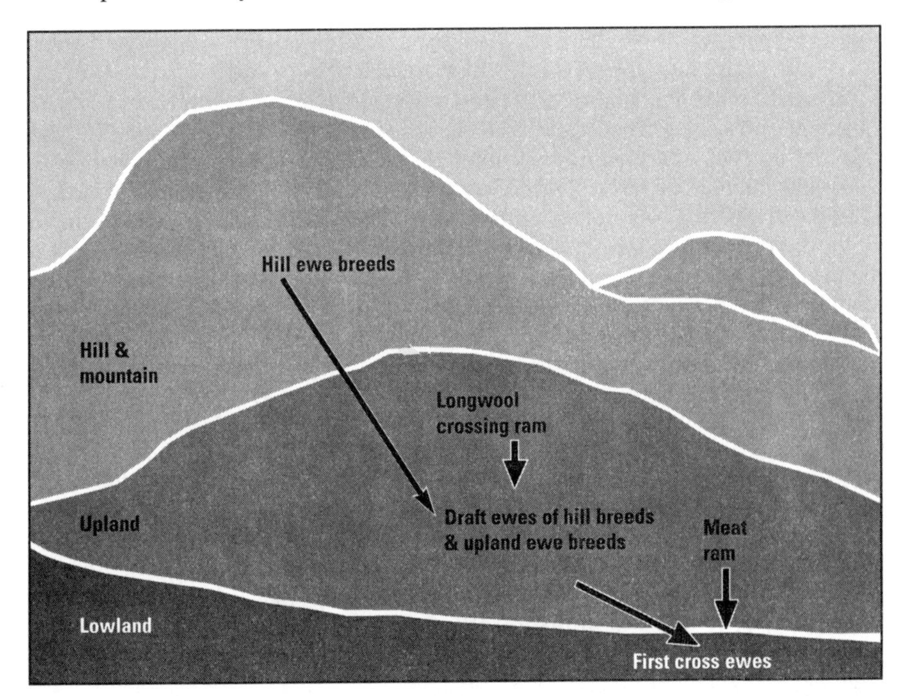

Figure 3.9 The stratified system of sheep breeding in the UK (source: MLC).

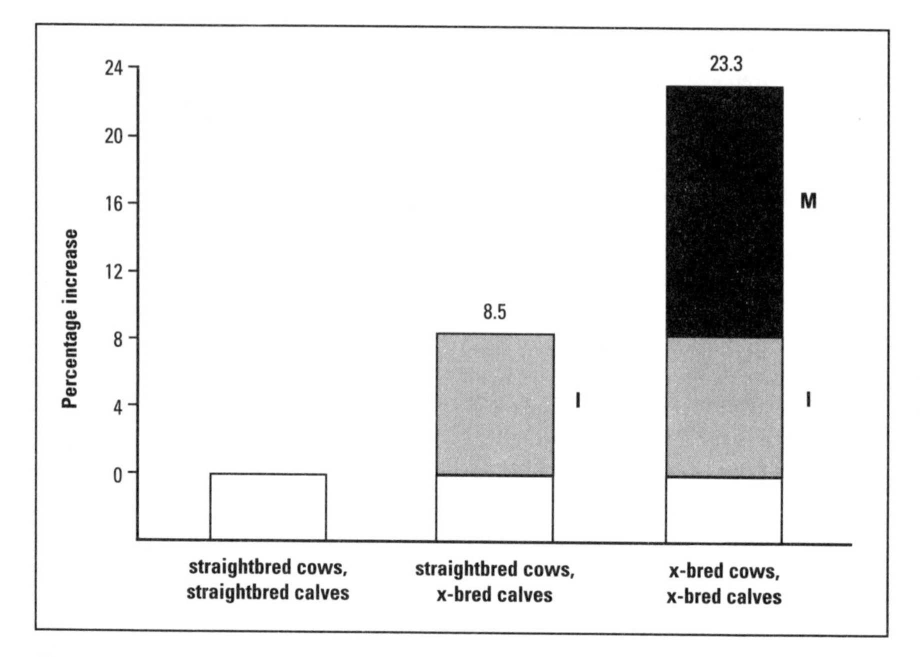

Figure 3.10 The cumulative benefits of individual (**I**) and maternal (**M**) heterosis on weight of calf weaned per beef cow exposed to breeding [5, 27].

stimulated by the availability of relatively cheap F1 females as a by-product of other enterprises (e.g. the use of beef x dairy suckler cows in the UK and Eire, the use of longwool x Merino ewes or longwool x hill ewes for lamb production in Australia and the UK respectively). Nonetheless, the benefits are still obtained.

The **stratified** system of sheep breeding in the UK provides a good example of the use of complementarity and both maternal and individual heterosis [26] (see Figure 3.9). Most hill sheep flocks in the UK are made up of purebred ewes of traditional hill breeds such as the Scottish Blackface, the Swaledale and the Welsh Mountain. Typically these ewes are bred pure in the hills for about four lamb crops. Then they are 'drafted' to better land on the same farm, or sold to other farms, for crossing – usually with a ram from a longwool or crossing breed, such as the Bluefaced Leicester or Border Leicester. The F1 females resulting from these matings are widely used in commercial flocks in the uplands and lowlands (see Plate 14). They, in turn, are mated to rams from the terminal sire breeds, such as the Suffolk and Texel, to produce lambs for slaughter (see Plate 15). The use of F1 longwool x hill females in commercial flocks makes full use of maternal heterosis. Crossing these to a third breed maximises the benefits of individual heterosis in the slaughter generation. Additionally, there are benefits from complementarity, because of the use of medium-sized F1 breeding females with good reproductive and maternal characteristics, but much larger terminal sires with good growth and carcass characteristics.

Similarly, the use of large terminal sire beef breeds in herds of cows made up of crosses between traditional British beef breeds, or crosses between beef and dairy breeds, makes full use of complementarity and both maternal and individual heterosis. Figure 3.10 illustrates the cumulative benefits of individual and maternal heterosis in beef cattle.

Systems of crossing

The simplest type of cross is that between two breeds. For obvious reasons this is called a **two-way cross**. The progeny resulting from a two-way cross are called F1 or first cross animals. In the previous section there were several examples of this type of cross, including longwool x hill breed commercial ewes, and beef x dairy suckler cows.

If animals from this two-way cross are then mated back to one of the parent breeds, this is termed a backcross, as mentioned earlier. When F1 crossbred animals are mated back to one of the parent breeds, individual heterosis of the resulting offspring is halved compared to that in the F1 generation. This is also true in all subsequent generations of backcrossing. Individual heterosis is at its maximum in the F1 generation, and is halved in every subsequent generation of backcrossing to the same parent breed. (This happens because each generation of backcrossing halves the number of loci which are heterozygous.) Similarly, when F1 animals are interbred to produce an F2 generation, heterosis is halved compared to that in the F1 generation. However, there is no further reduction in heterosis in the F3 and F4 generations, as long as there is no inbreeding.

Type of crossing system (breed of male is given first)	Fraction of heterosis relative to that in the F1		
	Individual heterosis (e.g. in survival of the crossbred animal itself)	Maternal heterosis (e.g. in fertility, or total weight of calf weaned by crossbred cows)	Paternal heterosis (e.g. in male fertility or libido)
Pure breed	0	0	0
Two-breed cross			
A x B	1	0	0
Backcross			
A or B with AB	$^1/_2$	1	0
AB with A or B	$^1/_2$	0	1
Three-breed cross			
C x AB	1	1	0
AB x C	1	0	1
Four-breed cross			
AB x CD	1	1	1
Rotational cross			
2 breed	$^2/_3$	$^2/_3$	0
3 breed	$^6/_7$	$^6/_7$	0

Table 3.9 Fractions of heterosis maintained in different crossing systems [12, 27].

Table 3.9 shows the fractions of individual, maternal and paternal heterosis which are maintained in different types of crossbreeding schemes, *relative to that in the F1*. For example, a value of 1 in the table indicates that a particular type of cross maintains the full amount of heterosis seen in the F1; a value of ½ indicates that heterosis is halved compared to that in the F1. As mentioned before, the value of a particular cross depends on its average merit compared to that of the best parent breed, i.e. it is a function of the additive genetic merit of both of the parent breeds, as well as heterosis.

One way of maintaining heterosis after producing a two-way cross is to mate to a third breed, to produce a **three-way cross**. The progeny of terminal sire sheep or beef breeds out of the type of F1 ewes and cows described above are three-way crosses. New breeds can be added to the mix indefinitely, in an attempt to maintain heterosis, but unless additional breeds have high additive genetic merit, the benefits of keeping heterosis high will soon be outweighed by the use of inferior breeds. Also, although three-way crosses maintain the *relative* level of heterosis compared to that in F1 generation, the *absolute* amount will depend on the specific breeds concerned – some combinations of breeds will lead to high absolute levels of heterosis, and other combinations will not.

Systems based on specific crosses can be very efficient, especially if specialised sire and dam breeds are used. However, as illustrated above, in cattle and sheep breeding these systems usually depend on a readily available external supply of replacement F1 females. Otherwise substantial numbers of at least one of the constituent breeds will need to be kept alongside the crossbred commercial herd or flock, just to breed sufficient replacement females.

An alternative to the use of specific crosses is **rotational crossing**. This involves the use of the same two or three (or more) breeds in rotation. In a two-breed rotational cross, breeds A and B would be mated to produce F1 offspring with 50% of the genes of each parent breed (abbreviated to AB here). These would be mated to sire breed A, to produce a second generation (abbreviated A(AB)) of offspring with an average of ¾ A genes and ¼ B genes. These in turn would be bred to sires from breed B, producing offspring with an average of ⅜ A genes and ⅝ B genes. This process continues until the proportions of genes from the two breeds stabilises at an average of about ⅓A, ⅔B and ⅔A, ⅓B in successive generations (see Table 3.10). In a three-breed rotational cross

Breed of parents (sire breed shown first)	Proportion of genes in offspring from	
	Breed A	Breed B
A x B	1/2 (0.5)	1/2 (0.5)
A x (AB)	3/4 (0.75)	1/4 (0.25)
B x [A(AB)]	3/8 (0.375)	5/8 (0.625)
A x (B[A(AB)])	11/16 (0.688)	5/16 (0.313)
B x [A(B[A(AB)])]	11/32 (0.344)	21/32 (0.656)
A x (B[A(B[A(AB)])])	43/64 (0.672)	21/64 (0.328)
B x [A(B[A(B[A(AB)])])]	43/128 (0.336)	85/128 (0.664)

Table 3.10 The average proportion of genes from two breeds in offspring after successive generations of rotational crossing [27].

the proportions of genes from the three breeds stabilises at an average of about $\frac{1}{7}$, $\frac{2}{7}$, $\frac{4}{7}$, with the highest proportion of genes coming from the sire breed used to produce the most recent generation, and *vice versa*. In practice, cattle or sheep live for varying lengths of time, and produce offspring at a range of ages, and so generations overlap rather than being discrete. This means that herds or flocks are soon made up of animals with different proportions of genes from the breeds involved, and several sire breeds need to be used each year. This can lead to additional difficulties in recording, arranging mating groups, and in managing and feeding animals of different genetic make-up if they differ markedly in size. Two and three breed rotational crossing maintains about $\frac{2}{3}$ and $\frac{6}{7}$, respectively, of the level of heterosis seen in the F1 generation. However, as mentioned already, it is the combined effect of additive genetic merit and heterosis in the particular crosses or crossbreeding systems concerned which needs to be evaluated.

Rotational crossing is used in the pastoral beef industries of several countries, particularly to breed replacement females which are two- or three-way crosses between the traditional British beef breeds (Aberdeen Angus, Hereford, Shorthorn). Also, there is growing interest in this type of system in the UK beef industry for two main reasons. Firstly, the combined effects of increasing yield per cow and the introduction of milk quotas in the dairy industry means that the number of dairy cows is declining, and so there are less beef x dairy heifers available as replacements for suckler herds. Secondly, Holsteins have poorer beef characteristics than the British Friesians they are replacing. So, the increasing proportion of Holstein genes in the UK dairy herd means that the beef x dairy heifers that are available are of declining beef merit (See Plates 16 and 17).

Conservation of genetic resources

Conservation of genetic resources has become a major issue of public interest in the last decade or so. Most of us associate this with saving pandas, tigers or tropical rainforests because of the media attention to wildlife. However, conservation of domestic animal and plant genetic resources is also potentially very important for the future welfare of the human race. All of the strategies for genetic improvement discussed in this chapter depend on genetic variation, so it is important for the success of future improvement programmes that genetic variation is used in a sustainable way. It is useful to consider two types of conservation of genetic variation in domestic animals, although they overlap. The first is the conservation of rare breeds or strains of livestock which are in danger of extinction. The second is conservation of genetic variation within breeds involved in active improvement programmes. The first of these is discussed below. Conservation of variation within breeds is discussed further in Chapter 4, in the context of designing improvement programmes.

Why conserve rare breeds?

In several Western countries interest in breed conservation was stimulated, a few decades ago, by farmers and breed enthusiasts who were concerned that the increasing use of a few specialised breeds was leading to a severe decline in the numbers of animals in many of the traditional breeds. Organisations like the Rare Breeds Survival Trust (RBST) in the UK, and the American Minor Breeds Conservancy (now the American Livestock Breeds Conservancy) were set up to promote conservation of those breeds thought to be at risk. In many third-world countries interest in conservation was stimulated at about the same time by the widespread crossing of indigenous breeds to imported breeds. Short-term productivity of the imported breeds and their crosses is often higher than that of the indigenous breeds when additional inputs are supplied. But imported breeds and their crosses are often less capable of surviving and producing on lower quality feeds, and often show less heat and disease tolerance than indigenous breeds. Hence, one argument put forward to support conservation, especially in Western countries, is aesthetic: that the wide variety of livestock breeds we have today is part of our cultural heritage, and deserves protection. Another is biological: that these breeds are reservoirs of genetic variation which are being overlooked or may become important in the future.

The arguments for conservation, particularly in third-world countries, are strengthened by predictions of the scale of human population growth. If the human population doubles in the next two generations as predicted, breeds which can prosper in harsh conditions may make a particularly important contribution to feeding those humans most in need. A further argument for conservation is that sustainable crossbreeding schemes require at least two viable purebred populations and often more. Trends towards fewer breeds will reduce the opportunities for improvement of productivity through crossing.

Scientific support for breed conservation has a rather chequered history. In the West, the subject has received relatively little scientific attention until recently, perhaps because support for conservation originated largely from grassroots enthusiasts and bypassed most mainstream livestock geneticists. However, there is also a counter-argument to conservation on biological grounds. This is that the markets for animal products do not usually change dramatically, and so there is scope for popular breeds to keep pace with new markets by changing the emphasis in selection, without the need to return to rare breeds for new genetic variation. There is already wide genetic variation in most breeds, which permits short- and medium-term responses to selection. In the longer term new variation is created by mutation, and so current breeds should be able to adapt to new markets indefinitely. While this argument probably holds for breeds that are not that much different from each other in the first place, it is probably not true globally. Many of the breeds or strains most at risk have evolved over a long time in very harsh environments. It would be difficult and time consuming to reinstate in 'improved' breeds some of the traits like tolerance to disease, to heat and to nutritional deprivation, which some rare breeds already possess.

New molecular techniques allow the transfer of genes across species, so some scientists argue that conservation of genetic resources for future animal production should be considered in the widest possible sense, and not confined to domestic animals [15]. While these techniques have yet to be applied to animals for agricultural purposes, there are analogies in plant breeding, where genes conferring resistance to viruses, resistance to insects and herbicide tolerance have been transferred into plants from viruses, bacteria and other plant species.

Which breeds should be conserved?

If the case for conserving breeds is accepted, the next difficult decision is which breeds to conserve. There are probably over 3,500 breeds or strains of domestic livestock [1], and limited resources to devote to conservation, so clearly priorities must be set. International cooperation is important if these resources are to be used efficiently. For instance, there is no point in two countries devoting their scarce resources to the conservation of the same or closely related breeds, while others get neglected. Hence, collecting information on the status of different breeds has been a priority for organisations involved in conservation (see Plates 18 and 19). The European Association for Animal Production (EAAP) and the Food and Agriculture Organization (FAO) of the United Nations have been collaborating since the late 1980s to produce a Global Animal Genetic Data Bank to collate information on:

- which breeds and strains exist in each country
- whether the same breeds exist in different places
- what the genetic characteristics of each breed are, and the similarity between breeds
- the productivity of breeds in particular environments
- the importance of each breed to the local human population
- whether the number of breeding animals in each population is changing
- whether conservation programmes already exist for breeds at risk [31].

This information helps in the identification of breeds or strains which are declining in numbers rapidly. Estimates of the number of animals required to maintain a breed in Western countries range from 150–1,500 breeding females, depending on the species and reproductive rate. However, populations in third-world countries which decline below 5,000 breeding females are considered at risk by FAO [19]. These are highlighted in a World Watch List produced regularly by FAO to encourage government or other agencies in the country concerned to take action. The information in this data bank helps in the appropriate *utilisation* of genetic resources, as well as conservation of these. For example, it can help in selecting appropriate breeds or strains for use in similar environments in other countries, and to identify distant or complementary breeds for crossbreeding programmes.

Until recently, the degree of similarity between breeds has been estimated from knowledge of the history of the breed, or from variation in the blood groups present. However, new molecular genetic tools mean that direct information on differences in the sequences of DNA present in different breeds can be used to measure the diversity or **genetic distance** between breeds more objectively, to ensure that the most diverse breeds are conserved and that the conserved population is based on individuals which are as dissimilar as possible. These techniques could also be used in future to target breeds or individuals carrying particularly rare or important alleles, if the existence of these is known. These approaches are already being used to assist in conservation of plant genetic resources.

Methods of conservation

The main methods of conservation which have been proposed or used are:

- **Maintaining breeds in their normal farm environment**. The advantages of this approach are that the genetic resources are still being utilised, they can be seen and enjoyed, the performance characteristics can be properly recorded, and the breeds have the opportunity to evolve, for example, to develop resistance to new disease challenges, or to adapt to changes in husbandry. The disadvantages are that selection and genetic drift (the chance change in gene frequency which occurs across generations due to small population size) may result in unfavourable genetic changes, there is a risk of increasing inbreeding and hence homozygosity, which is associated with reduced fitness, and the animals are at risk from disease or other natural disasters. Also, they are likely to be less productive and so more costly to maintain than more common breeds. This may increase the risk of individual breeders ceasing to keep rare breeds. Grants to farmers keeping rare breeds have been proposed in some countries as a way of promoting this type of conservation.

- **Maintaining breeds in farm parks or other collections**. Many of the pros and cons are similar to the method above. However, there is potentially more control over the management of the population. Also, there may be greater opportunities for education, and to recoup some of the costs of conservation by promoting public access. This approach, pioneered by the RBST, has become very popular in several Western countries.

- **Creating a gene pool**. This involves crossing several rare breeds together, then breeding them to maintain genetic variability [24]. It is probably an effective way of conserving genetic variation from two or three populations, but there is a greater risk of losing useful genes when more populations are combined. Although useful genes may be conserved with this approach, the ‘identity’ of different breeds is lost.

- **Frozen storage of semen or embryos (or DNA) from rare breeds**. These methods have the advantage that, after the initial investment, they are relatively inexpensive. Also, the population being conserved is free from unintended genetic change. During storage, frozen genetic material is at less risk from disease and natural disasters than live animals, but obviously at risk from technological failures. These risks can be minimised by splitting the material collected from a particular population and storing it in different locations. The disadvantages are that reproductive technologies are not uniformly successful for all individuals or all species, and the expertise is not always available in the places where it is needed most. Also, the breeds conserved in this way are not able to adapt to changes in the production environment or new disease challenges. If frozen semen is used as the only method of conservation, then several generations of backcrossing are needed to reinstate the breed concerned. In contrast, breeds can be reinstated rapidly from frozen embryos. Although preservation of DNA has been discussed as a method of conserving genetic variation, it is probably not practical at this stage to consider conserving parts of the genome of rare breeds in isolation from the whole.

When the costs of these different methods of conservation are compared, as well as their effectiveness in avoiding inbreeding, the combined use of live animals and

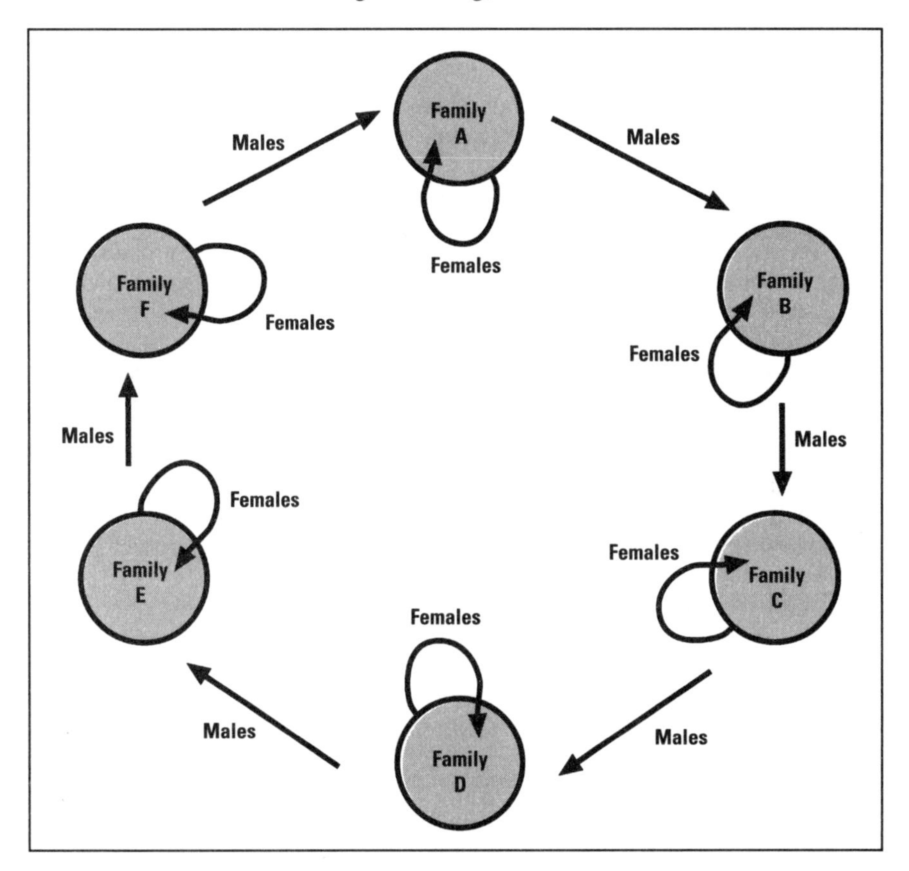

Figure 3.11 The use of within-family selection to limit inbreeding. (This method was used to limit inbreeding in the unselected 'control' line in a recent SAC selection experiment using Suffolk sheep. The control line was maintained to allow estimation of response to selection by comparing the performance of animals in the two lines each year. The control line comprised 75 breeding females divided into 6 families, and 6 males were used per annum. After 10 years the level of inbreeding in this flock was similar to that in the selection line which was over double the size of the control line, but did not use within-family selection.)

frozen semen appears to be the best strategy [23]. When conservation is based on live animal populations, or a combination of live animals and frozen semen, there are several guidelines which help to maintain genetic variation in the population [10, 15]:

- Start with as variable a population as possible.
- Start with as large a population as possible.
- Turn over generations as slowly as possible.
- Use enough parents (especially males) to keep inbreeding at acceptable levels. (What acceptable levels of inbreeding are, the effect of number of parents on rates of inbreeding, and mating designs to minimise inbreeding are discussed in more detail in Chapter 4.)
- Minimise the variation in family sizes to reduce inbreeding – ideally each male parent should be replaced by a son, and each female by a daughter.
- Subdivide the breeding population to reduce inbreeding and genetic drift. In small populations it may be most effective to organise the population into 'families' in a notional circle, and keep female replacements in the family in which they were born but move replacement males from the family in which they were born to the next family in the circle, in the same sequence each year (see Figure 3.11). In slightly larger populations it is more effective to mate males from one group to females of different groups in successive years.

Sometimes unselected control populations of animals are maintained as a benchmark against which to measure response to selection in an experimental or industry breeding scheme. The guidelines above are also helpful in establishing and maintaining such a population.

Summary

- Traditionally, three main strategies have been used for the genetic improvement of livestock. These are: (i) selection between breeds or strains; (ii) selection within breeds or strains and (iii) crossbreeding. In future new strategies, such as the ability to transfer genes within or between species, may become available as a result of developments in molecular biology.
- For any genetic improvement strategy to be effective, it is important to have a clear view of what the economically important traits are. Then it is logical to choose the most appropriate breed or cross, based on their performance in these traits. It is then sensible to consider whether this pure breed, or component breeds of the cross, can be improved further by within-breed selection.

- The structure of livestock breeding industries in most industrialised nations is often described schematically as a pyramid with elite or nucleus breeders at the top, one or more middle tiers of purebred or crossbred multipliers, and a final tier of commercial herds or flocks, or end users. Ideally, market signals from commercial herds or flocks influence breeding decisions in the tiers above, but there are many factors which can dilute these signals.

- Selection between breeds or strains can achieve dramatic and rapid genetic change when there are large genetic differences between populations in characteristics of economic importance. It is obviously costly to replace whole flocks or herds of breeding females at once. In practice, changes are often made more gradually by grading up to the new breed. It is important that choices among breeds are made on the basis of well-designed objective comparisons of performance in the relevant production environment.

- Selection within breeds involves comparing animals of the same breed and mating the preferred animals to produce the next generation. This process is usually repeated each generation and as long as there is genetic variation in the characters under selection this produces changes in each generation.

- Objective selection within breeds or strains is intended to increase the average level of additive genetic merit or breeding value of the population. Ideally, the steps involved are: (i) deciding on the breeding goal (it is sensible to identify the breed or cross which has the highest merit in this trait or combination of traits before embarking on within-breed selection in the best breed, or in component breeds of the best cross); (ii) deciding on the selection criterion; (iii) designing the breeding programme, e.g. numbers of males and females selected annually, ages at mating; (iv) implementing the programme, i.e. doing the routine recording, evaluation and mating of animals and (v) monitoring progress and redesigning the programme if necessary.

- Objective selection depends on having records of performance on the candidates for selection, or their relatives, or both. In most Western countries the majority of cattle and sheep breeders use recording schemes which are operated by regional or national agencies who specialise in recording and evaluation. Although these schemes are sometimes less flexible and more expensive than home-grown schemes, they encourage discipline and consistency in recording, they often allow access to more sophisticated methods of evaluation and they can improve communication with customers.

- Annual rates of genetic improvement in polygenic traits depend on four main factors: (i) the selection intensity achieved; (ii) the accuracy with which genetic merit in the trait of interest is predicted; (iii) the amount of additive genetic variation in the trait of interest and (iv) the generation interval. Generally speaking, the higher the selection intensity, accuracy and genetic variation, and the lower the generation interval, the higher the annual rate of genetic improvement.

- Profitability usually depends on several different animal characteristics rather than on any single trait. Hence, in genetic improvement programmes animals are usually selected on a combination of traits. This can be achieved by tandem selection, the use of independent culling levels or, most efficiently, by index selection.

- Crossbreeding involves mating animals of different breeds, lines or species. It is usually used for one or more of the following reasons: (i) to improve the overall efficiency of a production system by crossing breeds which each have high genetic merit in different traits (complementarity); (ii) to produce individual animals of intermediate performance between that of two more extreme parent breeds; (iii) for grading up to a new breed or strain; (iv) as an intermediate step in the creation of a new synthetic or composite breed; (v) to introduce new variation to numerically small breeds; (vi) to introduce a single gene for a favourable characteristic to an existing breed (introgression); or (vii) to exploit heterosis or hybrid vigour.

- Heterosis is defined as the advantage in performance above the mid-parent mean. It is most useful when it leads to the average performance of crossbred animals exceeding that of the best parent breed. Individual heterosis directly influences the performance of crossbred animals themselves. Maternal and paternal heterosis arise when dams or sires are crossbred, and the effects are often measured in terms of improved reproductive efficiency or improved performance of offspring.

- Different systems of crossing lead to different proportions of individual, maternal and paternal heterosis being maintained. However, the most appropriate system of crossing depends not only on this, but also on the additive merit of the breeds available, and the absolute level of performance of crossbreds.

- Each of the strategies for genetic improvement depends on genetic variation, so it is important for the success of future improvement programmes that genetic variation is used in a sustainable way. However, there are aesthetic as well as biological reasons for conservation of genetic resources.

- It is difficult to decide which breeds should be conserved. International organisations, such as FAO, are attempting to identify those breeds most in need of conservation programmes and are helping to implement these. In future, molecular genetic tools may help to ensure that the most diverse breeds are conserved.

- The main methods of conservation which have been proposed or used are: (i) maintaining breeds in their normal farm environment; (ii) maintaining breeds in farm parks or other collections; (iii) creating a gene pool and (iv) frozen storage of semen, embryos or DNA from rare breeds. When the costs of these different methods of conservation are compared, as well as their effectiveness in avoiding inbreeding, the combined use of live animals and frozen semen appears to be the best strategy.

- When conservation is based on live animal populations, it helps in maintaining genetic variation to: (i) start with as variable a population as possible; (ii) start with as large a population as possible; (iii) turn over generations as slowly as possible; (iv) use enough parents (especially males) to keep inbreeding at acceptable levels; (v) minimise the variation in family sizes to reduce inbreeding and (vi) subdivide the breeding population to reduce inbreeding and genetic drift.

References

1. Barker, J.S.F. 1994. 'A global protocol for determining genetic distances among domestic livestock breeds.' *Proceedings of the 5th World Congress on Genetics Applied to Livestock Production*, Vol. 21 pp. 501–8.
2. Beatson, P.R. 1993. 'Coopworths: making a concept happen.' *Proceedings of the A.L. Rae Symposium on Animal Breeding and Genetics*, H.T. Blair and S.N. McCutcheon (eds), Department of Animal Science, Massey University, Palmerston North, New Zealand, pp. 154–9.
3. Cameron, N.D., Smith, C. and Deeble, F.K. 1983. 'Comparative performance of crossbred ewes from three crossing sire breeds.' *Animal Production*, 37:415–21.
4. Cartwright, T.C. 1982. 'Objectives in beef cattle improvement.' *Proceedings of the World Congress on Sheep and Beef Cattle Breeding*, R.A. Barton and W.C. Smith (eds), Vol. II, pp. 19–27.
5. Cundiff, L.V. and Gregory, K.E. 1977. *Beef cattle breeding*. United States Department of Agriculture, Agriculture Information Bulletin No. 286. USDA, Washington DC.
6. Cundiff, L.V., Gregory, K.E. and Koch, R.M. 1982. 'Effects of heterosis in Hereford, Angus and Shorthorn rotational crosses.' *Beef Research Program Progress Report No. 1*. (ARM-NC-21) pp. 3–5. Roman L. Hruska US Meat Animal Research Center, Nebraska.
7. Cundiff, L.V., Gregory, K.E., Koch, R.M. and Dickerson, G.E. 1986. 'Genetic diversity among cattle breeds and its use to increase beef production efficiency in a temperate environment.' *Proceedings of the 3rd World Congress on Genetics Applied to Livestock Production*, Vol. IX, pp. 271–82.

8. Cunningham, E.P. and Syrstad, O. 1987. *Crossbreeding* ***Bos indicus*** *and* ***Bos taurus*** *for milk production in the tropics*. FAO Animal Production and Health Paper No. 68. Food and Agriculture Organisation of the United Nations, Rome.

9. Davis, G.P. and Arthur, P.F. 1994. 'Crossbreeding large ruminants in the tropics: current knowledge and future directions.' *Proceedings of the 5th World Congress on Genetics Applied to Livestock Production*, Vol. 20, pp. 332–9.

10. de Rochambeau, H. and Chevalet, C. 1990. 'Genetic principles of conservation.' *Proceedings of the 4th World Congress on Genetics Applied to Livestock Production*, Vol. XIV, pp. 434–42.

11. de Vaccaro, L.P. 1990. 'Survival of European dairy breeds and their crosses with Zebus in the tropics.' *Animal Breeding Abstracts*, 58:475–94.

12. Dickerson, G.E. 1974. 'Evaluation and utilisation of breed differences.' *Proceedings of Working Symposium on Breed Evaluation and Crossing Experiments in Farm Animals*. Instituut voor Veeteeltkundig Onderzoek, Netherlands, pp. 7–23.

13. Dickerson, G.E. 1978. 'Animal size and efficiency: basic concepts.' *Animal Production*, 27:367–79.

14. Fell, H.R. 1979. *Intensive Sheep Management*. Farming Press, Ipswich.

15. Frankham, R. 1994. 'Conservation of genetic diversity for animal improvement.' *Proceedings of the 5th World Congress on Genetics Applied to Livestock Production*, Vol. 21, pp. 385–92.

16. Gregory, K.E., Cundiff, L.V. and Koch, R.M. 1982. 'Comparison in crossbreeding systems and breeding stocks used in suckling herds of continental and temperate areas.' *Proceedings of the 2nd World Congress on Genetics Applied to Livestock Production*, Vol V, pp. 482–503.

17. Harris, B.L., Clark, J.M. and Jackson, R.G. 1996. 'Across breed evaluation of dairy cattle.' *Proceedings of the New Zealand Society of Animal Production*, 56:12–15.

18. Harris, D.L., Stewart, T.S. and Arboleda, C.R. 1984. *Animal breeding programs: a systematic approach to their design*. Advances in Agricultural Technology, Publication No. AAT-NC-8, Agricultural Research Service, North Central Region, US Department of Agriculture, Illinois.

19. Hodges, J. 1990. 'The global organisation of animal genetic resources.' *Proceedings of the 4th World Congress on Genetics Applied to Livestock Production*, Vol. XIV, pp. 466–72.

20. Kempster, A.J., Cook, G.L. and Southgate, J.R. 1982. 'A comparison of different breeds and crosses from the suckler herd. 2. Carcass characteristics' *Animal Production* 35:99–111.

21. Kempster, A.J., Croston, D., Guy, D.R. and Jones, D.W. 1987. 'Growth and carcass characteristics of crossbred lambs by ten sire breeds, compared at the same estimated subcutaneous fat proportion.' *Animal Production*, 44:83–98.

22. Livestock Improvement. 1996. *Dairy Statistics 1995–1996.* Livestock Improvement Corporation Ltd, New Zealand Dairy Board, Hamilton, New Zealand.

23. Lömker, R. and Simon, D.L. 1994. 'Costs of and inbreeding in conservation strategies for endangered breeds of cattle.' *Proceedings of the 5th World Congress on Genetics Applied to Livestock Production*, Vol. 21, pp. 393–6.

24. Maijala, K. 1970. 'Need and methods of gene conservation in animal breeding.' *Annales de Génétique et de Sélection Animale*, 2:403–15.

25. Mayala, K. 1974. 'Breed evaluation and crossbreeding in sheep. A summarising report.' *Proceedings of Working Symposium on Breed Evaluation and Crossing Experiments in Farm Animals.* Instituut voor Veeteeltkundig Onderzoek, Netherlands, pp. 389–405.

26. Meat and Livestock Commission, 1988. *Sheep in Britain.* MLC, Bletchley.

27. Nicholas, F.W. 1987. *Veterinary Genetics.* Oxford University Press, Oxford.

28. Owen, J.B. 1976. *Sheep Production.* Baillière Tindall, London.

29. Parker, A.G.H. and Rae, A.L. 1982. 'Underlying principles of co-operative group breeding schemes.' *Proceedings of the World Congress on Sheep and Beef Cattle Breeding*, R.A. Barton and W.C. Smith (eds.),Vol. II, pp. 95–101.

30. Simm, G., Conington, J. and Bishop, S.C. 1994. 'Opportunities for genetic improvement of sheep and cattle in the hills and uplands.' In T.L.J. Lawrence, D.S. Parker and P. Rowlinson (eds), *Livestock Production and Land Use in Hills and Uplands.* Occasional publication No. 18, British Society of Animal Production, pp. 51–66.

31. Simon, D. 1990. 'Data banks and the conservation policy.' *Proceedings of the 4th World Congress on Genetics Applied to Livestock Production*, Vol. XIV, pp. 423–6.

32. Smith, C. 1964. 'The use of specialised sire and dam lines in selection for meat production.' *Animal Production*, 6:337–44.

33. Smith, C. 1984. 'Rates of genetic change in farm livestock.' *Research and Development in Agriculture*, 1:79–85.

34. Snedecor, G.W. and Cochran, W.G. 1989. *Statistical Methods.* 8th edn. Iowa State University Press, Ames, Iowa.

35. Southgate, J.R., Cook, G.L. and Kempster, A.J. 1982. 'A comparison of different breeds and crosses from the suckler herd. 1. Liveweight growth and efficiency of food utilization.' *Animal Production*, 35:87–98.

36. Thiessen, R.B., Hnizdo, E., Maxwell, D.A.G. et al. 1984. 'Multibreed comparisons of British cattle. Variation in body weight, growth rate and food intake.' *Animal Production*, 38:323–40.

37. Timon, V.M. 1974. 'The evaluation of sheep breeds and breeding strategies.' *Proceedings of Working Symposium on Breed Evaluation and Crossing Experiments in Farm Animals.* Instituut voor Veeteeltkundig Onderzoek, Netherlands, pp. 367–87.

38. Wiener, G. 1994. *Animal Breeding*. The Tropical Agriculturalist series. Macmillan, London, and CTA, Wageningen, Netherlands.

Further reading

Amer, P.R., Kemp, R.A. and Smith, C. 1992. 'Genetic differences among the predominant beef cattle breeds in Canada: An analysis of published results.' *Canadian Journal of Animal Science*, 72:759–71.

Ark, The (Quarterly journal of the Rare Breeds Survival Trust. Includes a list of UK breeds at risk of extinction).

Baker, R.L. and Rege, J.E.O. 1994. 'Genetic resistance to diseases and other stresses in improvement of ruminant livestock in the tropics.' *Proceedings of the 5th World Congress on Genetics Applied to Livestock Production*, Vol. 20, pp. 405–12.

Barker, J.D., Smith, C. and Bateman, N. 1984. 'Crossbred ewes for hill conditions.' In T.J. Maxwell and R.G. Gunn (eds), *Hill and Upland Livestock Production*. Occasional Publication No. 10, British Society of Animal Production, p. 164.

Bichard, M. 1971. 'Dissemination of genetic improvement through a livestock industry.' *Animal Production*, 13:401–11.

Falconer, D.S. and Mackay, T.F.C. 1996. *Introduction to Quantitative Genetics*. 4 edn. Longman, Harlow.

Food and Agriculture Organization. 1993. *World Watch List for Domestic Animal Diversity*. R. Loftus and B Scherf (eds), 1st edn. FAO, Rome.

Gregory, K.E., Cundiff, L.V. and Koch, R.M. 1992. *Composite breeds to use heterosis and breed differences to improve efficiency of beef production*. U S Department of Agriculture, Agricultural Research Service, Roman L. Hruska US Meat Animal Research Center, Nebraska.

Gregory, K.E., Cundiff, L.V. and Koch, R.M. 1994. Germplasm utilisation in beef cattle. *Proceedings of the 5th World Congress on Genetics Applied to Livestock Production*, Vol. 17, pp. 261–8.

Guy, D.R. and Smith, C. 1981. 'Derivation of improvement lags in a livestock industry.' *Animal Production*, 32:333–6.

Hazel, L.N. 1943. 'The genetic basis for constructing selection indexes.' *Genetics*, 28:476–90.

Lush, J.L. 1945. *Animal Breeding Plans*. 3rd edn. Iowa State College Press, Ames, Iowa.

Proceedings of Working Symposium on Breed Evaluation and Crossing Experiments in Farm Animals. 1974. Instituut voor Veeteeltkundig Onderzoek, Netherlands.

Van Vleck, L.D., Pollak, E.J. and Oltenacu, E.A.B. 1987. *Genetics for the Animal Sciences*. W.H. Freeman and Company, New York.

CHAPTER 4

What affects response to selection within breeds?

Introduction

Over the last few decades objective selection within breeds has been practised widely in pig and poultry breeding and, to a slightly lesser extent, in dairy cattle breeding. However, it has been used much less widely in beef cattle and sheep breeding. There are several reasons for this. Pig and poultry breeding in industrialised countries has become concentrated in a small number of companies which sell breeding stock into a large and very competitive sector, which sends clear market signals. So setting breeding goals has been relatively simple. In contrast, market signals have often been less clear in beef cattle and sheep production, and commercial breeding objectives have often been obscured by fashions in the show ring. Those responsible for breeding decisions in pig and poultry breeding companies have been committed to objective methods and less influenced by traditional breeding practices than cattle and sheep breeders. In contrast, beef and sheep breeding is in the hands of individual breeders, many of whom have a long family history in the profession. They are often sceptical about new approaches, which is understandable if traditional methods have served them well. This feeling may be particularly strong in Britain, where so many breeds of livestock were developed and exported worldwide – though history shows the dangers of resting on your laurels in livestock breeding. Many breeds which were very popular even a few decades ago are minority breeds today.

Dairy cattle breeding has fallen between these two extremes. Perhaps this is because market signals have been clearer than in beef cattle and sheep production, and because breeders, and their customers, are confronted each milking time with visible evidence of the link between their breeding decisions and profitability. For commercial beef and sheep producers selection between breeds and crossbreeding have been easier strategies for improvement, because there are often more obvious differences between breeds and crosses than between individual bulls or rams of the same breed at a sale. Also, there are often easily identified 'markers', like coat colour, for particular breeds or crosses of proven merit.

It would be too simplistic to set up pig and poultry breeding as a model to which ruminant breeders should aspire without qualification. There is no doubt that pig and poultry breeders have employed objective breeding techniques very successfully to improve major performance traits. However, it is probably fair to say that they have had rather narrow breeding goals in the past. In some cases this has contributed to problems of functional fitness, such as leg weakness in pigs and poultry, and the inability of the males of some strains of turkey to mate naturally because of excessive breast muscle development. There is now a growing awareness that there are both ethical and economic arguments against the pursuit of very narrow breeding goals. While this increases the importance of choosing appropriate breeding goals, and paying attention to the functional fitness of selected animals, it should not detract from the value of objective selection. The success of objective selection methods in pigs and poultry is sometimes attributed to the intensive systems in which they are usually kept. Although performance recording may be easier in these systems, intensive production systems are not a prerequisite for successful selection programmes. If cattle and sheep breeders are going to meet the more exacting demands of their local customers, and if animal agriculture is going to help to meet the demands of a growing global population, then it is vital that there is wider use of relevant objective methods of within-breed selection. The key to this will be a better understanding of the procedures involved, their value and their limitations. Hence, the purpose of the next two chapters is to expand on the introduction to within-breed improvement given in Chapter 3. This chapter deals with predicting response to selection, and designing breeding programmes to achieve the maximum response to selection. Chapter 5 deals with predicting the breeding values of individual animals – a vital component in achieving responses once the overall design of a breeding scheme has been decided.

Factors affecting rates of genetic gain

The factors affecting rates of genetic gain were introduced in Chapter 3. These will be easier to understand if we consider the special case where animals are selected on a single measurement of their own performance in one trait, ignoring information from relatives. It is also simplest to consider the process of selection in a single closed flock or herd, that is, a flock or herd which is producing all of its own replacement breeding stock rather than buying these in from other flocks or herds. Each year calves or lambs are born, their performance is recorded, and the best performing animals are selected and retained. Once they reach normal breeding age, they replace those adult males or females in the breeding herd or flock which have died of natural causes, or which the breeder wishes to **cull** and replace by superior animals. Although this is an annual process, it is easiest to follow if we concentrate initially on a single group of recorded animals and their offspring.

For example, let us consider a group of lambs of both sexes, each with a record of 20-week weight (generation 1). From this group the lambs of each sex with the highest 20-week weight are selected for breeding to each other. The same performance trait, 20-week weight, is recorded eventually in their offspring (generation 2). The **response** is measured by comparing the average performance of all the lambs in generation 1 with the average performance of all the lambs in generation 2 (i.e. the offspring of those animals selected for breeding from generation 1). In this case, the response to selection *per generation* depends on just two things: the **selection differential** achieved and the **heritability** of the trait selected on.

Selection differential

The selection differential (abbreviated to **S** here) is the difference between the mean performance of selected animals (i.e. those identified to become parents of the next generation) and the overall mean of the group of animals from which they were selected. (To continue the example above, it is the difference between the mean performance of selected animals in generation 1, and the overall mean of generation 1 animals.) This is illustrated in Figure 4.1. Because there are usually

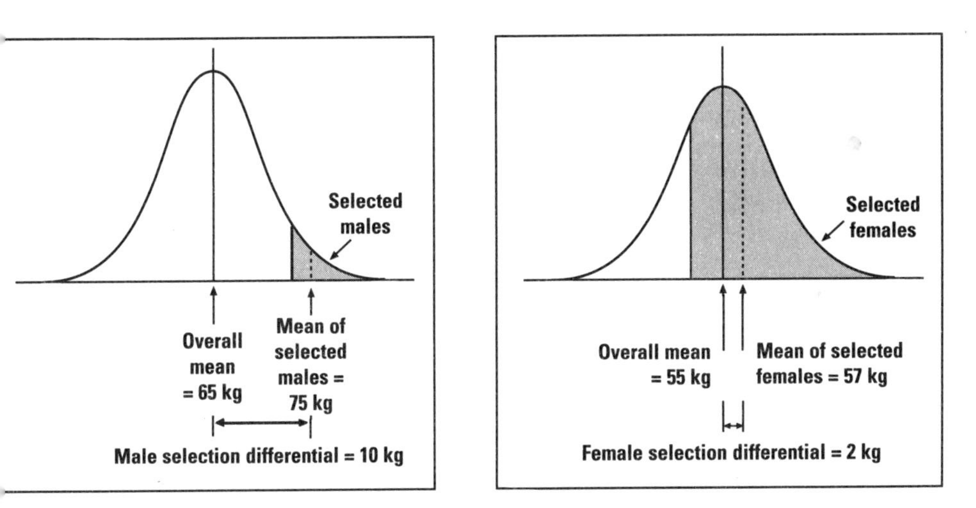

Figure 4.1 (a) The selection differential achieved in males by selecting the heaviest ram lambs from a performance-recorded group at 20 weeks of age. The selection differential among males is the average weight of selected males minus the average of the group of males as a whole. (b) The selection differential achieved by selecting a larger proportion of ewe lambs from a recorded group, at the same age.

unequal numbers of male and female parents selected, but each sex eventually contributes half their genes to the offspring, it is important to calculate selection differentials separately for males and females and then average them to get an overall selection differential. In this example, the selection differential achieved in males is +10 kg, while that achieved in females is +2 kg, so the average selection differential achieved is +6 kg ([10 + 2]/2). Figure 4.1 illustrates that the fewer the number of animals selected for high performance, the further above the mean they will be, and so the higher the selection differential will be. A further point to note is that the wider the variation in performance of the group, the higher the selection differential which can be achieved. (Although, as we will see later, this is only of value if there is a lot of genetic rather than environmental variation).

Heritability

The second factor affecting the response to selection per generation is the heritability of the trait concerned (abbreviated $\mathbf{h^2}$). There are several ways of defining heritability. The simplest definition is that the heritability is the proportion of superiority of parents in a trait (i.e. the proportion of the selection differential) which, on average, is passed on to offspring. So, if the heritability of a trait is high, we can expect much of the superiority of parents to be passed on to offspring. If the heritability of a trait is low, this implies that only a little of the superiority of parents is a result of the genes they carry, and so only a little of this superiority will get passed on to offspring. Heritabilities are expressed as proportions from 0 to 1, or as percentages from 0% to 100%.

It will be easiest to understand selection differentials and heritabilities, and how they act together to influence response, by returning to the sheep example. In the case above, the average selection differential achieved across both sexes was +6 kg.

If the heritability of live weight at 20 weeks of age is 0.3 or 30%, then we expect on average 30% of this 6 kg superiority to be passed on to offspring. So, in this example we expect a 1.8 kg improvement in 20-week weight of the next generation (0.3 x 6 kg). This is illustrated in Figure 4.2. It is important to note that selection achieves an increase in the *average* performance of the next generation – it does *not* add 1.8 kg to the weight of every lamb.

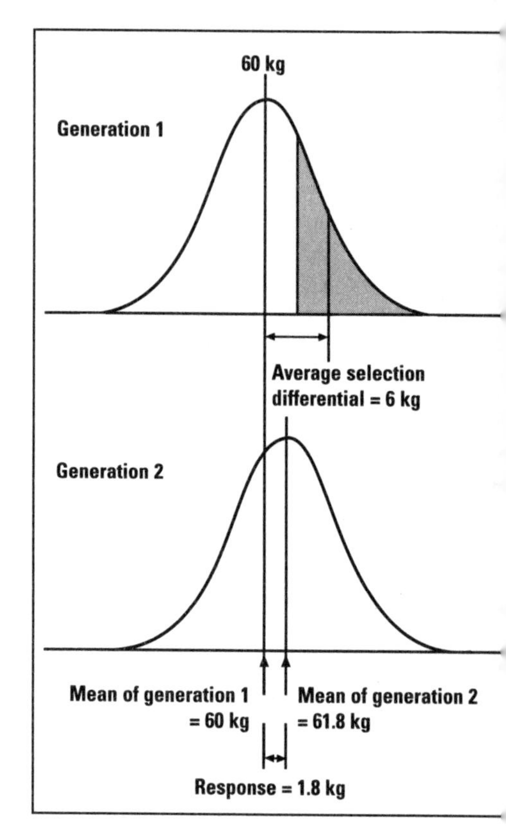

Figure 4.2 Expected response to selection on 20-week weight in sheep, with an average selection differential of +6 kg, and a heritability of 0.3.

So how do we know what the heritability of a particular trait is in the first place? It may sound circular, but the simplest way of estimating the heritability for a particular trait is to select parents on this trait, measure the selection differential achieved, mate them, measure the same trait in their offspring and then calculate, using the regression techniques outlined in Chapter 2, how much of the superiority of parents was passed on to offspring. There are other more sophisticated methods which involve measuring the degree of resemblance in performance between different classes of relatives, but these will be easier to understand later.

In addition, there is an alternative definition of heritability, which follows from the discussion of different types of variation in Chapter 2. The heritability of a trait is the additive genetic variation in that trait (or the variation in breeding values), expressed as a proportion of the total phenotypic variation in that trait. (The fact that the heritability is the ratio of two variances explains why it is abbreviated to $\mathbf{h^2}$. The square root of the heritability, abbreviated to **h**, is the ratio of the additive genetic standard deviation of a trait to the phenotypic standard deviation.) As explained in Chapter 2, it is only the effects of selection on the additive genetic variation which is reasonably predictable. Using the abbreviations introduced in Chapter 2 (pages 49-50) [4]:

$$\textbf{Heritability } (\mathbf{h^2}) = \frac{\mathbf{V_A}}{\mathbf{V_P}}$$

Distributions of $\mathbf{V_A}$ and $\mathbf{V_P}$ for two traits with different heritabilities are shown in Figure 4.3. The inner distributions show the proportion of the total phenotypic variation which is a result of additive genetic variation, or variation in breeding values. In one case, additive genetic variation accounts for 10% of the total

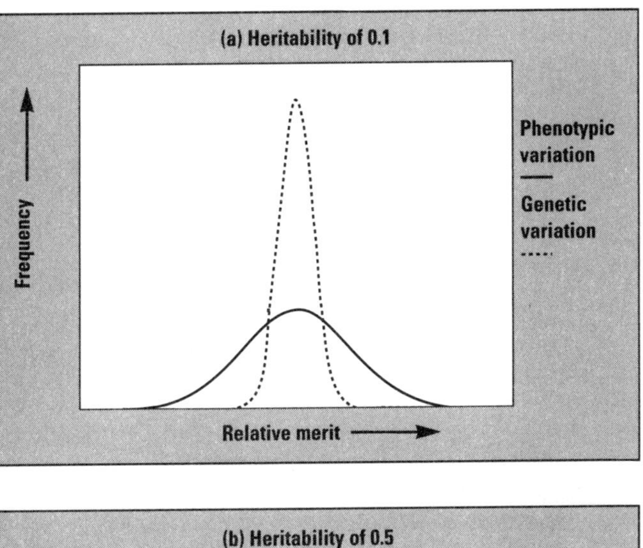

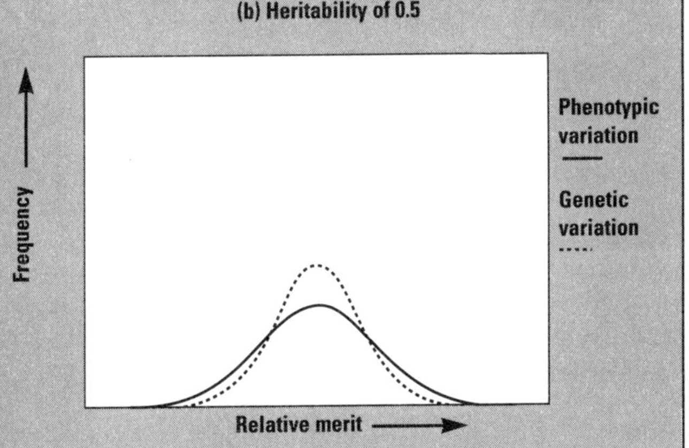

Figure 4.3 Distributions of additive genetic variation (**V_A**) and phenotypic variation (**V_P**) for two traits with different proportions of additive genetic variation, and hence different heritabilities. (a) Heritability of 0.1 or 10%. (b) Heritability of 0.5 or 50%.

phenotypic variation so, as we have seen, the heritability is 10%. In the other case a higher proportion of the total variation, 50%, is additive genetic variation, so the heritability is 50%. The graphs illustrate that when the heritability of a trait is low, the distribution of breeding values will be narrower, relative to phenotypic variation, than when the heritability of the trait is high. Table 4.1 shows typical examples of heritabilities for some traits of interest in cattle and sheep.

The examples above have illustrated that we can predict the response to selection (**R**) per generation from a knowledge of the selection differential (**S**) and the heritability of the trait under selection (**h^2**) [4]:

$$\mathbf{R = S \times h^2}$$

The ability to predict response to selection is important in planning breeding programmes, and especially in comparing alternative schemes. The problem with the approach described so far is that it predicts response per generation, but generations can be of variable length. For instance, in the sheep example, if the

Species/breed	Trait	Typical heritability values
Dairy cattle	Milk yield	25 – 40 %
	Milk fat %	50 – 60 %
	Milk protein %	50 – 60 %
	Various measures of fertility	1 – 5 %
	Incidence of various production diseases	0 – 25 %
Beef cattle	Birthweight	25 – 35 %
	Weaning weight	20 – 30 %
	Yearling weight	30 – 40 %
	Carcass fat depth	35 – 50 %
	Calving ease	5 – 15 %
Sheep	Birthweight	10 – 25 %
	Weaning weight	15 – 25 %
	Yearling weight	20 – 35 %
	Carcass fat depth	25 – 35 %
	Various combined measures of fertility and lamb output	5 – 15 %
	Fleece weight	30 – 40 %
	Fibre diameter	45 – 55 %

Table 4.1 Typical heritabilities for some traits of economic importance in cattle and sheep breeding. See Chapters 6 to 8 for average values for these and many other traits from comprehensive literature searches.

selected males and females had been mated at seven months of age, the earliest age possible in most breeds, then their offspring would be born when they were about one year old. If, on the other hand, the selected animals were mated at nineteen months of age, the offspring would be born a year later than before, when the parents were two years old. So, two breeders may be achieving equal selection differentials and equal responses per generation, but achieving these responses in very different lengths of time. The solution to this problem is to express predicted responses per annum rather than per generation. To do this, we need to know the **generation interval** in the flock or herd concerned.

Generation interval

The generation interval (abbreviated **L**) is the average age of parents when their offspring are born. Because there are unequal numbers of male and female parents, and they often breed at different ages, generation intervals have to be calculated separately for each sex of parent, and then averaged. An additional complication is that in most flocks or herds there are different proportions of animals in each age group. Also, if the fertility of parents of different age groups differs, then this too must be taken into account. This means that we cannot simply average the parents' age classes, we must weight them according to the number of offspring produced by each age class. For example, if we had a flock of ewes comprising 50% two-year-olds and 50% three-year-olds, the average age would simply be two and a half years. If, as is usually the case, we had a higher proportion of young animals in the flock to allow for natural wastage, then we could not simply average the age classes. If we had 60% two-year-olds and 40% three-year-olds, the average age of the ewes would be (0.6 x 2) + (0.4 x 3) = 2.4 years. If these two age groups produce equal numbers of offspring per ewe, then the female generation interval would simply be 2.4 years. If younger ewes are less prolific than older ones, so that 55% of lambs come from two-year-old ewes, and 45% from three-year-old ewes, then the female generation interval would be 2.45 years ([0.55 x 2] + [0.45 x 3] = 2.45). A more realistic example is given in Table 4.2.

To predict responses to selection per annum, rather than per generation, we simply divide the formula shown above for response per generation by the average generation interval (**L**):

$$\mathbf{R}\ \textbf{(per annum)} = \frac{\mathbf{S \times h^2}}{\mathbf{L}}$$

So, if we selected on 20-week weight each year in a flock with the age structure shown in Table 4.2, and achieved the selection differentials shown in Figure 4.1 each year, we would have;

$$\mathbf{S} = \textbf{6 kg}$$
$$\mathbf{h^2} = \textbf{0.3 and}$$
$$\mathbf{L} = \textbf{2.67 years}$$

So, we would expect to get a response per annum of:

$$\mathbf{R} = \frac{\mathbf{6 \times 0.3}}{\mathbf{2.67}}$$
$$= \textbf{0.67 kg per annum (approx.)}$$

A more useful formula for predicting response

As mentioned already it is very useful to be able to predict rates of response to selection in order to compare breeding programmes with different designs – for example, to choose the design likely to maximise annual response. At first sight it appears that we need to know quite a lot of information in advance in order to predict response. We need to know the heritability of the trait under selection, but for most traits there are already estimates available which we can use. We also need to know the generation interval, but usually this can be predicted accurately from the expected age distribution of parents, together with typical levels of reproductive performance for parents of each age group. A greater limitation of the formula used so far is that we need to know the selection differential in advance in order to predict response. It is easy to calculate a selection differential once animals have been performance recorded and parents have been selected, but how can we predict what selection differential is likely to be achieved before embarking on a breeding programme?

1. Age of ewes at lambing (years)	2. Number of ewes in this age group	3. Number of lambs from ewes in this age group	4. Proportion of lambs from ewes in this age group (approx.)	5. Contribution of ewes in this age group to weighted average age (col. 1 x col. 4)
2	25	35	0.22	0.44
3	22	37	0.23	0.69
4	20	34	0.21	0.84
5	18	30	0.19	0.95
6	15	24	0.15	0.90
Total	100	160	1.00	3.82
Weighted average age of ewes = 3.82 years (approx.)				
Female generation interval = 3.82 years (approx.)				

Table 4.2 An example of calculating the average generation interval in a flock of sheep of mixed ages. (a) Calculating the female generation interval. In this example, the 100-ewe flock has a typical age distribution, with progressively fewer ewes in successive age groups, as a result of natural wastage. The average litter size is 1.6 lambs, but prolificacy is lowest in the youngest ewes, it is highest in ewes of intermediate age, and then declines slightly in the oldest ewes. Calculating the female generation interval involves recording the number of ewes in each age group at lambing (column 2), calculating the number of lambs (column 3) and then the proportion of lambs (column 4) from ewes in each age group. Each ewe age is then multiplied by the relevant proportion of lambs (column 1 x column 4 = column 5), to get the contribution of this age group to the average weighted age. These values are then summed over all age groups to get a weighted average age, which is the female generation interval.

1. Age of rams at birth of their offspring (years)	2. Number of rams in this age group	3. Number of lambs from rams in this age group	4. Proportion of lambs from rams in this age group (approx.)	5. Contribution of rams in this age group to weighted average age (col. 1 x col. 4)
1	2	78	0.49	0.49
2	2	82	0.51	1.02
Total	4	160	1.00	1.51
Weighted average age of rams = 1.51 years (approx.)				
Male generation interval = 1.51 years (approx.)				

(b) Calculating the male generation interval. In this example four rams are used in the flock, and they are mated to groups of ewes of mixed ages. Hence the two age groups of rams have roughly equal numbers of lambs born. Weighted average ages are calculated as described above for ewes.

Female generation interval = 3.82 years (approx.)
Male generation interval = 1.51 years (approx.)
Average generation interval = 2.67 years (approx.)

(c) Calculating the average generation interval.

We saw in Chapter 2 that one of the features of normal distributions is that we can predict remarkably accurately what proportion of animals will fall within different ranges of performance around the mean. For example, about 68% of animals will have performance in the range from 1 standard deviation below the mean to 1 standard deviation above the mean, 95% of animals fall in the range –2 to +2 standard deviations, and so on. So, if we know how many standard deviations better than average our selected animals are, and we know how many kg of live weight or litres of milk each standard deviation is worth, we can predict the selection differentials achieved.

If we know the total number of animals recorded, and the number selected from this group, there are standard statistical tables which allow us to express the merit of selected animals in standard deviation units above the mean. A full table of these values is shown at the end of the book. Table 4.3 shows part of it. This table shows, for example, that if we select the best 5 rams out of a total of 75 which are performance recorded, on average they are expected to be 1.893 standard deviations above the mean of all 75 rams. If the standard deviation of 20-week weight is about 5 kg, then we expect the average weight of the 5 selected rams to be about 9.5 kg heavier than the average of all 75 recorded rams (1.893 x 5 kg = 9.465 kg).

The superiority of animals selected, expressed in standard deviation units, is called the standardised selection differential or **selection intensity** (abbreviated **i**). Table 4.3 shows that the selection intensity is largely a function of the proportion of animals selected – the lower the proportion selected, the higher the selection intensity. This is also illustrated in Figure 4.4. However, with a fixed proportion of animals selected, selection intensities increase slightly as the total number of animals tested rises. For instance, compare the selection intensity when 10 animals are selected from 50 tested (1.372) with that when 20 animals are selected from 100 tested (1.386). As for selection differentials, selection intensities have to be calculated separately for each sex, then averaged.

Number of animals selected	Total number of animals from which they were selected			
	25	50	75	100
	Selection intensity (i)			
1	1.965	2.249	2.403	2.508
5	1.345	1.705	1.893	2.018
10	0.936	1.372	1.588	1.730
15	0.624	1.139	1.381	1.536
20	0.336	0.951	1.217	1.386
25	0	0.786	1.079	1.259
35	–	0.488	0.843	1.050
50	–	0	0.539	0.792
75	–	–	0	0.420
100	–	–	–	0

Table 4.3 Examples of the selection intensity achieved, in standard deviation units, by selecting different numbers of animals from groups of 25, 50, 75 or 100 animals in total. See the table at the end of this book for a fuller set of selection intensities from the same source [2].

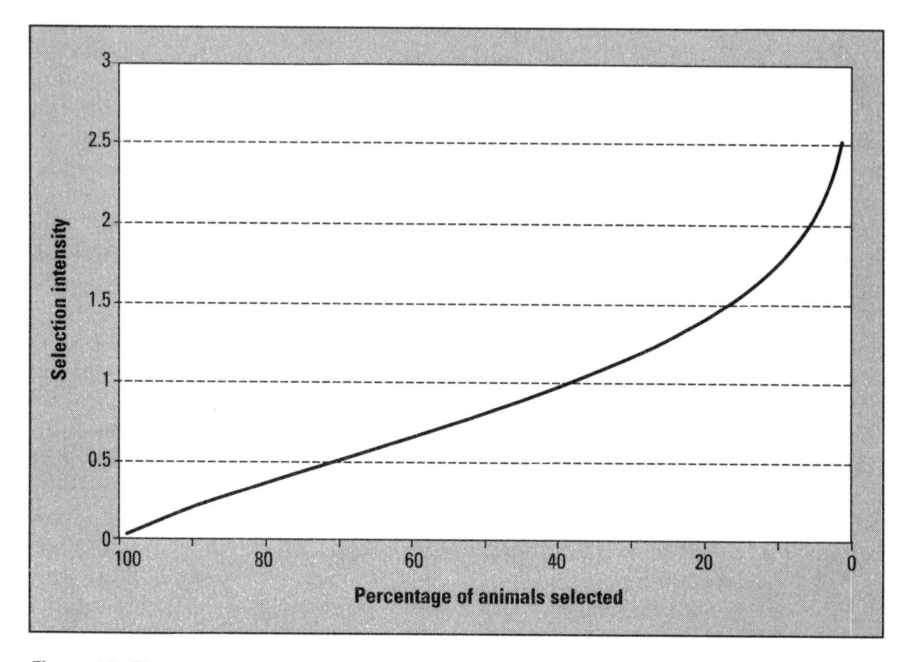

Figure 4.4 The relationship between proportion of animals selected and selection intensity – the selection differential measured in standard deviation units. The values shown assume that animals are selected from a very large total number of animals tested.

We can calculate the phenotypic variance (**V_P**) or phenotypic standard deviation (**sd_P**) of the trait under selection in the initial population of animals, as explained in Chapter 2. Alternatively, we could use published estimates of variation from similar animals to those we are interested in, kept in a similar production system (usually variances for the same trait are relatively constant within a given breed and level of production). Putting this together, we can predict selection differentials by multiplying the selection intensity, which tells us how many sd units better than average our selected animals are, by the phenotypic sd, which tells us how much each sd unit is worth in kg of live weight, litres of milk etc. So [4]:

$$\mathbf{S} = \mathbf{i} \times \mathbf{sd_P}$$

Substituting **i** x **sd_P** instead of **S** in the formula used before gives a new formula for predicting response per annum [4]:

$$\mathbf{R} = \frac{\mathbf{i} \times \mathbf{sd_P} \times \mathbf{h^2}}{\mathbf{L}}$$

It is easiest to see the connection with the earlier formula when the terms are presented as above, but most commonly the terms are re-ordered and the formula is shown as [4]:

$$\mathbf{R} = \frac{\mathbf{i} \times \mathbf{h^2} \times \mathbf{sd_P}}{\mathbf{L}}$$

To illustrate the use of this formula, let us consider a beef cattle breeding programme to improve calf weaning weight (200-day weight) in a herd of 120 Simmental cows. Let us assume that there are 100 calves reared per annum, so on average there are 50 calves of each sex available for selection each year. From performance records collected over many years in Britain by MLC, we estimate that the phenotypic standard deviation of weaning weight in the Simmental breed is 35 kg. If we pick the top five bull calves each year, based on their own weaning weight, then we predict a male selection intensity of 1.705 standard deviations (see Table 4.3). If we select the top 35 heifers each year from 50 recorded, then the female selection intensity will be 0.488 standard deviations (see Table 4.3). This gives an average selection intensity across both sexes of 1.097. If the bulls are mated once only, at fifteen months of age, then their progeny will be born when they are two years old, and the male generation interval will be two years. If the herd is made up of 35 heifers (three years old), 30 four-year-old cows, 25 five-year-old cows, 20 six-year-old cows and 10 seven-year-old cows, and they have equal calving rates, then the weighted average age of dams, and hence the female generation interval, is 4.5 years. The male and female generation intervals together give an average generation interval of 3.25 years. The heritability of weaning weight in beef cattle is usually about 25%. Putting these values into the new formula gives:

$$\mathbf{i} = \mathbf{1.097 \text{ sd units}}$$
$$\mathbf{sd_P} = \mathbf{35 \text{ kg}}$$
$$\mathbf{h^2} = \mathbf{0.25}$$
$$\mathbf{L} = \mathbf{3.25 \text{ years}}$$

and so:

$$\mathbf{R} = \frac{\mathbf{1.097 \times 35 \times 0.25}}{\mathbf{3.25}} \mathbf{\text{ kg per annum}}$$

$$\mathbf{R} = \mathbf{2.95 \text{ kg per annum (approx.)}}$$

The link between selection intensity and generation interval

The formula above shows that the annual response to selection will be highest when the selection intensity is high, the heritability is high, and the generation interval is low. There is little that breeders can do about the heritability of a trait – this is largely a biological characteristic of the trait concerned. But, within biological limits, breeders can increase selection intensities and decrease generation intervals, although these two measures are linked. The strength of that link varies between sexes, and species. For any breeding herd or flock to remain the same size the breeding animals have to be replaced at least when they die of natural causes, or are culled for unavoidable reasons. Because relatively few males are required for breeding, there is a plentiful supply of potential replacements in all farmed species. This means that it is usually possible to keep male generation intervals short and, at the same time, keep male selection

Ewe age at lambing (years)	Number of ewes in each age group					
	Flock 1	Flock 2	Flock 3	Flock 4	Flock 5	Flock 6
2	25	35	45	55	65	75
3	22	31	40	45	35	25
4	20	27	15	–	–	–
5	18	7	–	–	–	–
6	15	–	–	–	–	–
Female generation interval (weighted average age)	3.76 years	3.06 years	2.70 years	2.45 years	2.35 years	2.25 years
Female selection intensity achieved in 20-week weight (in standard deviation units)	1.079	0.843	0.637	0.443	0.244	0

Table 4.4 An example of the female generation intervals and selection intensities achieved in six flocks of 100 ewes when different proportions of available female replacements are brought into each flock. The extreme cases are in flock 1, where females are retained for as long as possible (up to 5 lamb crops), and in flock 6, where all available females are brought into the flock, displacing older ewes. In all cases natural wastage of about 10% is assumed, and ewes are assumed to lamb first at 2 years of age, and to rear an average of 1.5 lambs, of which half are males and half females (for simplicity, it is assumed that ewes of different ages have the same reproductive rate). Selection intensities can be converted to selection differentials by multiplying them by the standard deviation of the trait concerned. For example, if selection is for 20-week weight, the selection intensities would be multiplied by the sd of 20-week weight – assumed to be about 5 kg. So, the female selection differential achieved in flock 1 would be about 5.4 kg (1.079 x 5 = 5.395 kg); that in flock 2 would be about 4.2 kg, and so on.

intensities high. But this is not the case for females. The females of most breeds of cattle and sheep rear less than two offspring per annum, and so less than one female offspring per annum on average. This automatically sets an upper limit to the number of breeding females which can be replaced each year.

It is easiest to see the link between selection intensity and generation interval by considering two extreme examples of a herd or flock replacement policy. At one extreme, all the available young replacement females could be brought into a herd or flock to replace older females. In this case, the average age of the herd or flock would be low, and female generation intervals would be short. However, there would be no selection at all among female replacements (i.e. the female selection intensity would be zero). At the other extreme, if breeding females were only replaced when they died of natural causes, then the average age of the herd or flock would be high and the female generation interval would be long. However, in this case few replacement females would be needed, so they could be highly selected, and selection intensities would be high. These two extreme situations, and some of the intermediate ones, are illustrated in Table 4.4.

The optimum flock or herd age structure to maximise response is often intermediate to the two extremes described. This is illustrated in Table 4.5 and Figure 4.5. The table and graph show the predicted annual rate of response in 20-week weight for the different female replacement rates shown in Table 4.4. In

	Flock 1	Flock 2	Flock 3	Flock 4	Flock 5	Flock 6
Female generation interval (years)	3.76	3.06	2.70	2.45	2.35	2.25
Male generation interval (years)	1.00	1.00	1.00	1.00	1.00	1.00
Average generation interval (years)	2.38	2.03	1.85	1.725	1.675	1.625
Female selection intensity	1.079	0.843	0.637	0.443	0.244	0
Male selection intensity	1.893	1.893	1.893	1.893	1.893	1.893
Average selection differential (kg)	1.486	1.368	1.265	1.168	1.069	0.947
Predicted annual response to selection (kg per annum)	**0.94**	**1.01**	**1.03**	**1.02**	**0.96**	**0.88**

Table 4.5 Predicted annual response to selection on 20-week weight in sheep, with different female replacement policies. Female generation intervals and selection intensities are those shown in Table 4.4. The heritability of 20-week weight is assumed to be 0.3 and the phenotypic standard deviation of 20-week weight is assumed to be 5 kg. Responses are calculated from the formula shown in the text.

all cases it is assumed that 5 ram lambs are selected from 75 available in each flock each year, and so a male selection intensity of 1.893 standard deviations is achieved. Each ram is used for one year only, so the male generation interval is one year in all cases. In this example the maximum rate of gain is achieved by replacing 45 females per annum, though the difference between alternative schemes is fairly small. The difference between schemes would be more marked if more males were selected, so reducing male selection intensity, or if male generation intervals were longer.

In the past, optimising generation intervals and selection intensities was central to the design of breeding programmes. Deciding on the optimum proportion of animals to select from each age group was important because animals were usually only evaluated once, and the results of these evaluations could only be compared within age groups. The modern methods of genetic evaluation described in Chapter 5 allow comparison of animals of different age groups. Also, advances in the methodology for evaluation and in the power of computers means that it is feasible to do evaluations much more often than in the past. When these modern evaluation methods are used, genetic progress will be maximised by selecting animals with the highest predicted breeding value, regardless of age, rather than aiming for fixed proportions of animals in each age group. The age structure of a flock or herd will still end up close to the optimum calculated as described above. However, there will be some additional gains as a result of keeping animals which are better than expected for their particular age group longer, and culling animals which are poorer than expected for their particular age group sooner.

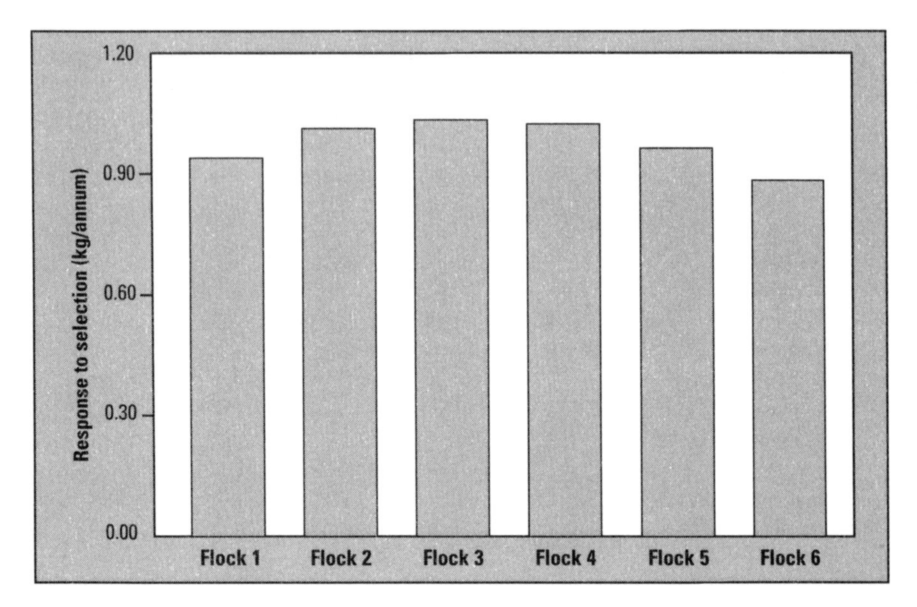

Figure 4.5 Predicted annual response to selection on 20-week weight in sheep in flocks with different female replacement policies, as shown in Tables 4.4 and 4.5.

Using information from relatives

So far we have considered only a single measurement of performance on the animal itself as the selection criterion. However, this is only one of a number of sources of information which are commonly used in animal breeding. These include records of performance from:

- **The animal's ancestors**. Selection on ancestors' information is called **pedigree selection**. An index combining information on ancestors' performance is usually called a **pedigree index**, and selection on pedigree indexes is common in young animals until they get a record of performance themselves.

- **The animal itself**. Selection on the animal's own performance is the simplest method of selection. Historically, most beef cattle and sheep performance recording schemes have been based on this source of information alone. This is usually termed a **performance test**. More recent performance recording schemes use information from all available relatives to predict breeding values.

- **The animal's full or half siblings** (sibs for short – these are the animal's brothers and sisters). Full sibs are animals which have both parents in common; half sibs are animals which have only one parent in common. A selection programme based on information from siblings is called **sib selection** or a **sib test**. They are particularly useful if the trait of interest is only measurable on one sex. For instance, milk records from their full sisters are being used increasingly to help select young dairy bulls. Similarly, many pig breeding programmes to improve carcass composition were based on sib selection, following carcass dissection of a full sib. (This approach is less common now, because of the widespread use of ultrasonic scanning to assess carcass characteristics in live animals.)

● **The animal's progeny**. A breeding programme based on the performance of an animal's offspring is termed a **progeny test**. Again, information on the performance of progeny is most useful when the trait of interest can be measured in only one sex. In most countries, selection for milk production in dairy cattle is based on progeny testing. Bulls undergoing test are used in many milk-recording herds through artificial insemination. They are then selected for wider use, or

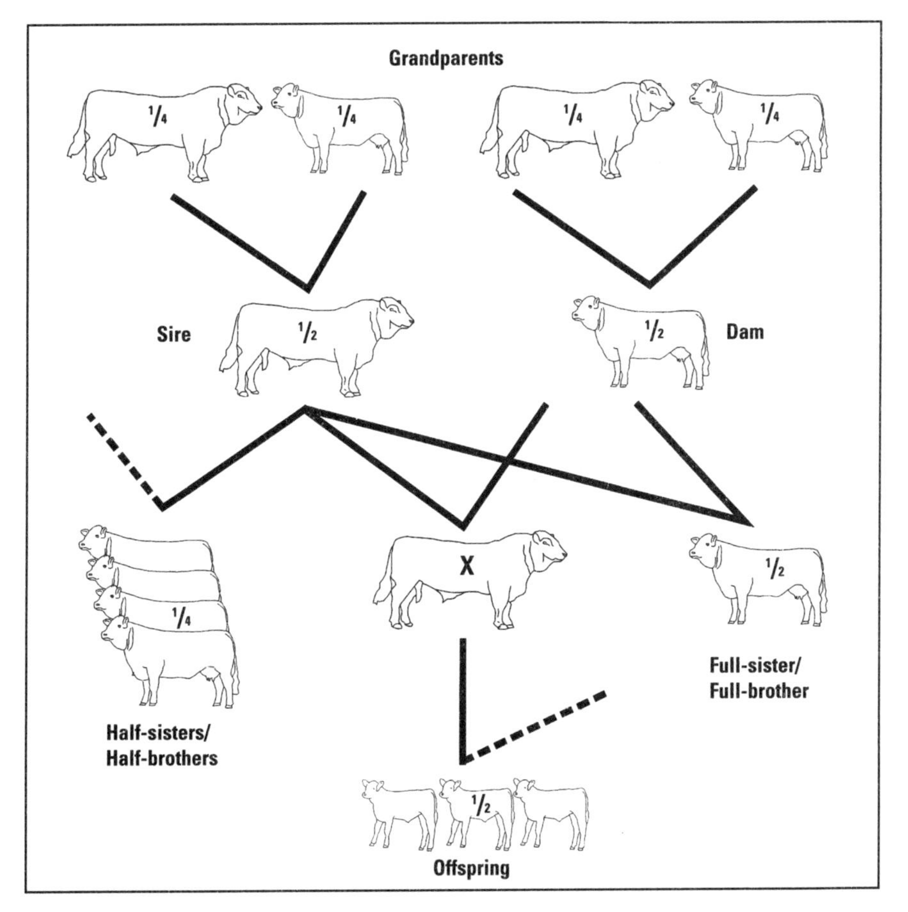

Figure 4.6 The proportion of genes in common between the bull marked X, and various classes of relatives. Parents and offspring have exactly half their genes in common. The proportions of genes in common between other classes of relatives are expected average proportions. Dotted lines indicate relationships to dams which are not shown in the pedigree [17].

culled, on the basis of the milk production records of their daughters, compared with those of contemporary animals.

- **Any other relatives of the animal.**

- **Combinations of the classes of relatives listed above.** (An outline of how this is achieved is given in Chapter 5).

Records of performance from relatives are useful in selection because related animals have genes in common, and so the performance of relatives can provide clues to the genetic merit or breeding value of the candidates for selection. Figure 4.6 shows the proportion of genes in common between the bull marked X, and various classes of relatives. These same proportions are true for all animals, including humans (but take a close look at your relatives before you advertise this fact!).

Offspring always have exactly one half of their genes in common with each parent (apart from the slight inequalities caused by the different size of the X and Y chromosomes). This is because every animal develops from an embryo which got one member of each pair of its chromosomes from the father, via the sperm, and one member from the mother, via the egg. However, for other classes of relatives the proportion of genes in common are averages. This is because segregation and recombination lead to chance variation in the proportion of genes from ancestors (this was explained in Chapter 2). Each sperm and each egg carry copies of half the genes of the animals that produced them. Which sample of genes make up this half is determined entirely by chance. So, some pairs of full sibs have more than half of their genes in common, by chance, and others have less. Each generation which separates two relatives leads on average to a halving of the genes they have in common.

In order to predict response to selection using these new sources of information, we need a modified version of the formula used so far:

$$\mathbf{R} = \frac{\mathbf{i} \times \mathbf{h}^2 \times \mathbf{sd_P}}{\mathbf{L}}$$

For reasons which are explained in the appendix at the end of this chapter, this can be rewritten as [4]:

$$\mathbf{R} = \frac{\mathbf{i} \times \mathbf{h} \times \mathbf{sd_A}}{\mathbf{L}}$$

In this formula the **R**, **i** and **L** have exactly the same meaning as before. The new terms are **h**, which is the square root of the heritability, also referred to as the **accuracy of selection on a single record of the animal's own performance**, and $\mathbf{sd_A}$, which is an abbreviation for the **additive genetic standard deviation**, or **standard deviation of true breeding values**.

This formula still refers to the special case when selection is on the animal's own performance. There is a more general version in which **h** is substituted by **r**, which is the accuracy of selection on any combination of records from the animal and its relatives [4]:

$$\mathbf{R} = \frac{\mathbf{i} \times \mathbf{r} \times \mathbf{sd_A}}{\mathbf{L}}$$

As mentioned in Chapter 2, the symbol **r** is usually used to denote a correlation. In this case the accuracy is the correlation between the animals' true breeding values for the trait(s) under selection, and the measurement(s) on which selection is based.

Accuracy of selection

The accuracy of selection itself depends mainly on three things: (i) the heritability of the trait concerned – the higher the heritability the higher the accuracy; (ii) the source of information on which selection is based, e.g. the class of relative – the closer the relatives on whose records selection is based, the higher the accuracy of selection and (iii) the amount of information available from relatives – the more relatives of a given class recorded (and the more often they are recorded if the trait can be measured repeatedly), the higher the accuracy, although there are diminishing returns as the numbers of relatives increase. Table 4.6 shows the relationship between accuracy and the square root of the heritability, for different classes of relative, assuming that selection is based on a single record of performance from that type of relative. To predict response to selection on such a record, we simply use the appropriate value of **r** in the formula above. So, for example, if selection is on a single record on the animal itself, the accuracy of selection is equal to the square root of the heritability of the trait concerned (**r** = **h**). To continue an example used earlier, the heritability of weaning weight in beef cattle is about 0.25, so its square root, **h**, is 0.5. Hence the accuracy of selection for weaning weight, when selection is based on a single record of the animal's own performance, is 0.5, or 50%. If selection is based on a single measurement from one parent, or one offspring, then the accuracy is only half that when selection is based on a record on the animal itself. In the beef cattle example, selection on a single parent or offspring record would give an accuracy of 0.25 or 25%.

Table 4.6 shows that the accuracy of selection on a single record of performance for different types of relative is directly related to the proportion of genes they have in common. This is a very important principle in objective animal breeding.

Single measurement of performance on	Accuracy (r)
Animal itself	h
One parent	$^1/_2$ h
One grandparent	$^1/_4$ h
One great-grandparent	$^1/_8$ h
An identical twin	h
One full sib	$^1/_2$ h
One half sib	$^1/_4$ h
One progeny	$^1/_2$ h

Table 4.6 A comparison of the accuracy of selection on a single record of performance on the candidate animal itself, or on a single record from a relative. Accuracies are presented as proportions of **h**, the square root of the heritability of the trait under selection. So, for example, if $\mathbf{h^2}$ = 0.25, h = 0.5, so the accuracy of selection on the animal's own performance is 0.5, the accuracy of selection on one parent's record, or one full sib's record, or one progeny's record, is 0.25 ($^1/_2$ x 0.5 = 0.25), the accuracy of selection on one grandparent's record, or one half sib's record, is 0.125 ($^1/_4$ x 0.5 = 0.125), and so on [12].

In subjective selection it is easy to overplay the importance of a particular favourite ancestor, or other relative, in making selection decisions (e.g. if it has won prizes in major shows, or was sold for a very high price). In objective selection, the aim is to give each performance record from a relative its proper emphasis, and this emphasis depends on the expected proportion of genes in common between the relative and the candidate for selection.

Calculating the accuracy of selection with records from greater numbers of relatives is slightly more complicated, although the results are easy to understand. Figure 4.7 shows the accuracy of selection on records from varying numbers of half or full sibs and progeny.

These graphs illustrate that:

- The closer the relative, the more valuable the record. For example, a single record from an offspring is twice as valuable as a single record from a half sib, other things being equal (i.e. the accuracy of selection is twice as high).
- As the number of progeny increases, the accuracy of selection tends towards 1. As the number of full sibs increases, the accuracy tends towards 0.7. With half sibs, the accuracy tends towards 0.5 as numbers increase.
- A single record from a full sib is of equal value to a single progeny record, but additional records from progeny are more valuable in terms of increasing the accuracy of selection.
- Initially, the more relatives of a given class, the higher the accuracy, although there are diminishing returns.
- Initially, the higher the heritability, the higher the accuracy.
- Initially, the lower the heritability, the greater the proportional contribution which records from relatives make (e.g. compared to the accuracy of selection on a single record on the animal itself, selection on records from 50 progeny increases the accuracy of selection by a factor of about 5, when the heritability is 0.1, but only by a factor of about 1.6 when the heritability is 0.3).

Additive genetic standard deviation

The value of the additive genetic standard deviation, or standard deviation of breeding values can be derived from the heritability of the trait concerned and its phenotypic standard deviation. We have already seen that:

$$\textbf{Heritability } (\mathbf{h^2}) = \frac{\mathbf{V_A}}{\mathbf{V_P}}$$

Standard deviations are the square root of variances, and so we can take the square root of all terms in this equation to get:

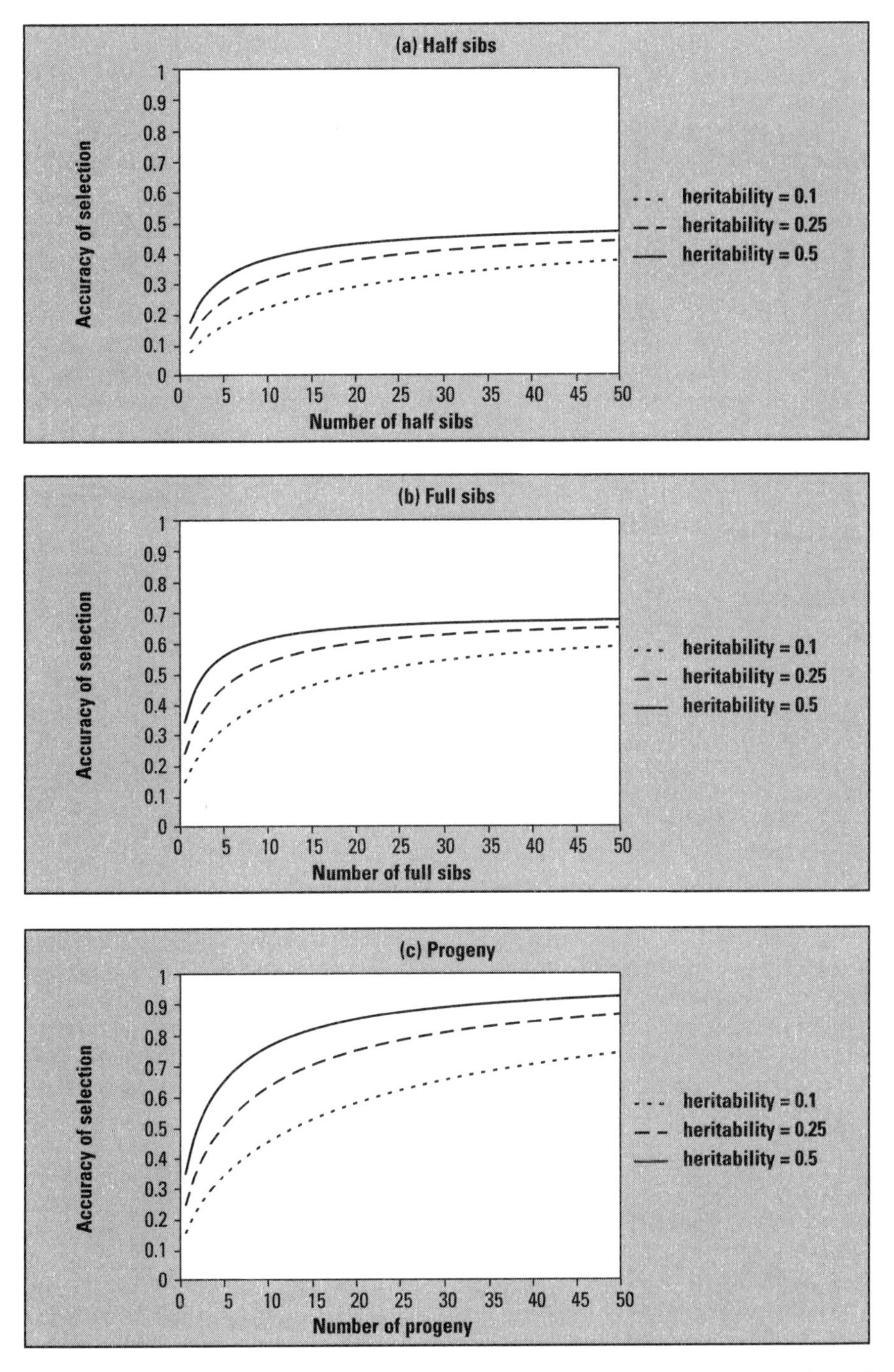

Figure 4.7 The accuracy of selection on performance records from relatives with varying numbers of records, and traits of different heritability. In graph (a) the records are from half sibs, in graph (b) they are from full sibs, and in graph (c) they are from progeny [12].

$$h = \frac{sd_A}{sd_P}$$

We can rearrange the terms in this equation (by multiplying both sides by sd_P, then cancelling out the two sd_P's which appear on the right hand side of the equation) to give:

$$sd_A = h \times sd_P$$

So, the additive genetic standard deviation is simply the phenotypic standard deviation multiplied by the square root of the heritability.

A practical example

Although using records of performance from relatives can increase the accuracy of selection, it is important to recognise that this benefit can be outweighed by increases in the generation interval. For example, if we want to select for increased 400-day weight in a 120-cow herd of beef cattle, we might consider two main options: performance testing and progeny testing. For simplicity, let us assume that selection is based on males only, and that all available replacement females enter the herd giving a female generation interval of 3.85 years and a female selection intensity of 0. Also, let us assume that progeny testing takes place in several separate crossbred herds. In each case, if we select 5 males from a total of 50 which are either performance recorded in the pure herd, or progeny tested in the crossbred herds, then the selection intensity among males is 1.705. The average selection intensity is then 0.8525 ([1.705 + 0]/2).

The heritability of 400-day weight is about 0.4, so if we select bulls on their own performance alone, the accuracy of selection is about 0.632 (the square root of the heritability). Subject to them reaching sexual maturity, we should be able to mate bulls at about 15 months of age, giving a male generation interval of two years. This means that in the herd using performance testing the average generation interval is 2.925 years ([3.85 + 2]/2). If we select bulls on the records from 20 progeny, the accuracy is higher – about 0.830 – but the crossbred progeny do not reach 400 days of age until the bulls are over three years old. So the minimum male generation interval possible in the purebred herd is approaching four years (it is safest to assume four years exactly, if we want to maintain a

	Performance testing	**Progeny testing**
Average selection intensity (**i**)	0.8525	0.8525
Approx. accuracy of selection (**r**)	0.632	0.830
Additive genetic standard deviation ($\mathbf{sd_A}$ in kg)	28.4	28.4
Average generation interval (**L** in years)	2.925	3.925
Predicted response (kg per annum, approx.)	5.23	5.12

Table 4.7 Characteristics of the two beef cattle breeding schemes described in the text – one based on performance testing, the other on progeny testing. In both cases selection is to increase 400-day weight.

seasonal pattern of calving). This means that in the herd using progeny testing the average generation interval is 3.925 years ([3.85 + 4]/2).

To predict annual response to selection in the two schemes, we also need to know the additive genetic standard deviation of 400-day weight. This is about 28.4 kg for the larger beef breeds. (This is calculated from the phenotypic standard deviation of 400-day weight (about 45 kg) multiplied by the square root of the heritability of 400-day weight (square root of 0.4 = 0.632), so 45 x 0.632 = 28.4 kg). Table 4.7 summarises the characteristics of the two breeding schemes.

The predicted annual response to selection can then be calculated from these characteristics, using the formula introduced in the last section. This is illustrated below for the scheme based on performance testing:

$$\mathbf{i} = \mathbf{0.8525}$$
$$\mathbf{r} = \mathbf{0.632}$$
$$\mathbf{sd_A} = \mathbf{28.4\ kg}$$
$$\mathbf{L} = \mathbf{2.925}$$

and so:

$$\mathbf{R} = \frac{\mathbf{0.8525 \times 0.632 \times 28.4}}{\mathbf{2.925}}$$

$$\mathbf{R} = \mathbf{5.23\ kg\ per\ annum\ (approx.)}$$

The equivalent response for the scheme based on progeny testing is 5.12 kg per annum. In this example, the gains in accuracy from progeny testing are outweighed by the longer generation interval among males which is needed in order to obtain records of performance on progeny. However, with a lower heritability, or more progeny records, the reverse would be true. So it pays to check the expected responses from different types of selection, using the relevant values for the heritability and number of relatives to derive the accuracy, before embarking on selection. Usually records from ancestors and at least some sibs (or other **collateral** relatives – those from the same generation as the animal itself) can be used to increase accuracy of selection without lengthening the generation interval. Using records from descendants usually increases the generation interval, and so there is a trade-off between increasing accuracy and minimising generation interval. The methods of combining information from different types of relative are discussed in later sections.

Comparisons of different breeding schemes also need to consider the resources needed, as well as predicted responses. Even when it appears justified by higher predicted responses, it would be very expensive to maintain crossbred herds solely for progeny testing. In some cases equivalent responses could be achieved at lower cost from performance testing by expanding the purebred herd to increase selection intensity. Also, in most cases it would be quite wasteful to progeny test all available bulls. If the number tested is reduced arbitrarily, then this will lead to lower selection intensity than that possible from performance testing. In practice, it makes sense to use all available information in selection – in the example above, it would have been more efficient to progeny test only those bulls which themselves had high 400-day weight (a process called **two-stage selection**). Also, records from half sibs are generated as a by-product of most performance testing schemes, so these can often increase accuracy at no extra cost in money or time. (See the book by Weller in the Further Reading section at the end of this chapter for methods of comparing the cost-effectiveness of different breeding schemes.)

Using repeated records of performance

So far, we have considered only single records of performance on candidates for selection, or their relatives. However, many traits of interest can be measured more than once over an animal's lifetime. For example, growing meat animals can be weighed on several occasions, most dairy cows have about four lactations during their lifetime, most beef cows would wean six or seven calves during their lifetime, and sheep would have four or five litters during their lifetime and be shorn annually for four or five years.

Often there are very strong genetic associations between performance at different stages in an animal's life. In these cases it is safe to assume that performance at these different stages is being influenced by the same genes. Hence, repeated records of performance can give additional clues to the breeding value of animals, and enhance the accuracy of selection. If repeated measurements are influenced by exactly the same genes, then any differences in performance from contemporaries, for example in yield in different lactations, is due to differences in the environment or management experienced by the animal. We have already seen that the performance of an animal can be split into genetic and environmental components [4]:

$$\mathbf{P = G + E}$$

Splitting the genetic part down further into additive genetic merit (breeding value), plus non-additive genetic effects, as explained in Chapter 2, gives:

$$\mathbf{P = A + NA + E}$$

However, the environmental component of performance can also be split further into **permanent (E_p)** and **temporary (E_t) environmental effects**. This gives:

$$\mathbf{P = A + NA + E_p + E_t}$$

Like the genes, permanent environmental effects stay with the animal for life but, unlike the genes, they do not get passed on to offspring. Temporary environmental effects have a much shorter impact, for example affecting only one lactation, or one weight record. If a dairy heifer suffered a severe mastitis infection in one quarter of her udder and, as a result of damage to the secretory tissue in that quarter, never produced milk from it again, that would be a permanent environmental effect. Yield in the other quarters may rise slightly to compensate, but the animal would produce less milk in total in each successive lactation than if it had not been affected by mastitis. In a single lactation, the yield of the same animal may be temporarily depressed because it happened to be housed in a group of cows which was fed very poor quality silage. However, when the animal was fed on better silage in subsequent lactations this effect would disappear. So, this would be classified as a temporary environmental effect.

The value of repeated records in selection depends on a measure called the **repeatability**. This is the correlation between repeated records from the same animal. It is defined as the proportion of the total phenotypic variation which is explained by the combined effects of the genes and the permanent environmental variation [4]:

$$\textbf{Repeatability} = \frac{V_G + V_{E_p}}{V_P}$$

or, in full:

$$\textbf{Repeatability} = \frac{V_G + V_{NA} + V_{E_p}}{V_P}$$

This looks similar to the equation used earlier to calculate the heritability, except that there are a few extra terms on the top of the equation. The heritability is concerned only with the 'predictable' additive genetic part of the equation, whereas the repeatability also includes the variation due to non-additive genetic and permanent environmental effects, which are specific to individual animals, and hence difficult to predict. As a result, the repeatability sets an upper limit to the heritability. Like heritabilities, repeatabilities range from 0 to 1, or from 0% to 100%. Some examples of the repeatability of different traits are shown in Table 4.8.

It may seem ironic at first sight, but the higher the repeatability of a trait, the lower the value of repeated records in selection for that trait. In other words, if the repeatability is high, the first record of performance gives a good indication of the animal's genetic merit, and adding subsequent records only improves the accuracy slightly. If the repeatability is low, the first record of performance gives a poor indication of subsequent performance. Here repeated records help to build up a more accurate estimate of the animal's genetic merit, but the law of diminishing returns applies. This is illustrated in Figure 4.8, which shows the accuracy of selection for traits with different repeatability, depending on the

Breed/species	Trait	Repeatability	Source
Dairy cattle	Milk yield	0.55	[1]
	Days to first service	0.08	[6]
	Days open	0.10	"
	No. services per conception	0.07	"
Beef cattle	Cows calving per cow pregnant	0.06	[11]
	Calves weaned per cow calving	0.05	"
	Cow weight pre-mating	0.57	"
	Calf weaning weight (age adjusted)	0.37	"
Sheep	Greasy fleece weight	0.58	[5]
	Wool fibre diameter	0.70	"
	Ovulation rate across years	0.31	"
	No. lambs weaned per ewe mated	0.08	"

Table 4.8 Estimates of the repeatabilities of some traits of economic importance in cattle and sheep.

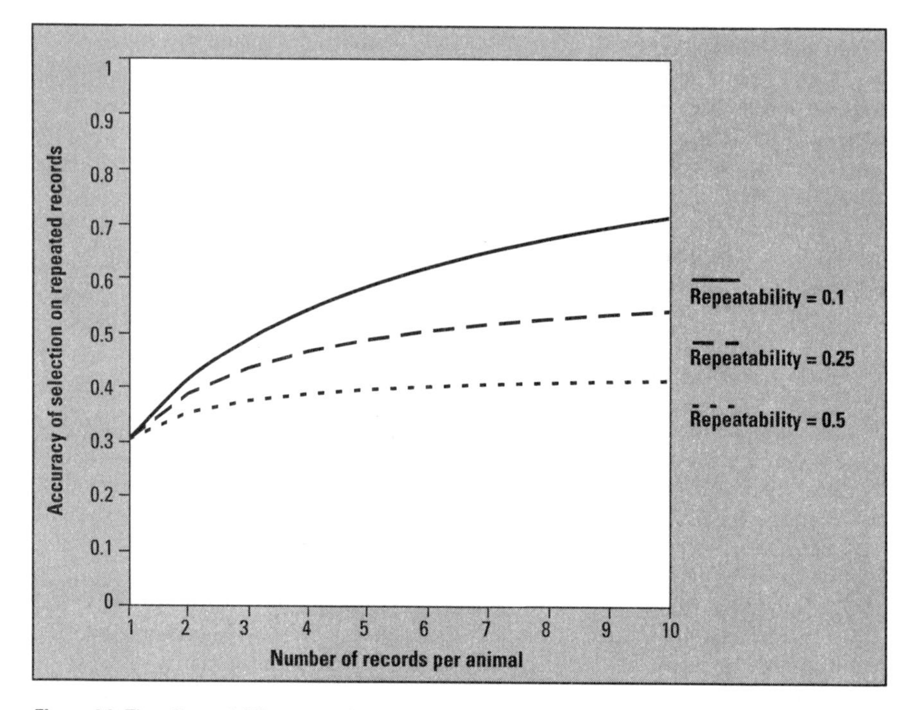

Figure 4.8 The effect of different numbers of repeated records on the accuracy of selection, for traits with different repeatabilities. In this example the heritability of the trait is 0.1 [12].

number of repeated records available. As with the use of information from relatives, it is important to take into account any time lag in obtaining repeated records. In principle, the benefits of higher accuracy are often outweighed by the penalty from increases in generation interval. However, in practice, selection is usually based on a single record initially, with subsequent records being used to refine later estimates of genetic merit.

Repeatabilities measure the similarity between repeated records of the same trait, assuming that performance measured in different years (or at other intervals) is controlled by exactly the same genes. However, there are often associations between different traits which are not likely to be influenced by exactly the same genes. For instance, animals that are heavier than their contemporaries at a given age tend to be fatter too; and higher yielding cows generally have lower milk protein percentage. Even when the same trait is measured at different times, different genes may influence the trait at different times. For example, animals which are heavier than average at an early age are often heavier at later ages, but not all the genes influencing early weight also influence mature weight. Similarly, heifers which have a high milk yield compared to contemporaries often have comparatively high yields in later lactations and *vice versa*, but not all the genes which influence early lactations also influence later lactations. The degree of association between different characteristics is measured by the correlation coefficient, as explained in Chapter 2.

Predicting correlated responses to selection

Before embarking on a selection programme it is useful to predict what the consequences will be, not only in the trait under selection, but also in other associated traits. Providing that the heritabilities of the trait under selection and other traits of interest are known, their phenotypic variances (or standard deviations) are known, and the genetic correlation between them is known, the **correlated response** to selection can be predicted. It is easiest to think of this as a three-step procedure:

1. Calculate the predicted annual response in the trait under selection (trait X). If we want to predict correlated responses it is simplest if we first predict the direct response in the trait under selection in *additive genetic standard deviation units*, rather than units of measurement. If selection is based on a single record of the animal's own performance, this becomes:

$$\mathbf{R_X} = \frac{\mathbf{i \times h_X}}{\mathbf{L}} \quad \text{(in additive genetic sd units of X per annum)}$$

(Note that this formula is identical to the one presented earlier, except that the term $\mathbf{sd_A}$ is omitted in order to express the response in additive genetic sd units rather than units of measurement.)

2. Multiply the predicted annual response from (1) by the genetic correlation between the traits under selection and the second trait of interest (trait Y), to get the predicted correlated response in the second trait, expressed in additive genetic standard deviation units.

$$\mathbf{CR_Y = R_X \times r_{AXY}} \quad \text{(in additive genetic sd units of Y per annum)}$$

3. Multiply the predicted correlated response in standard deviation units from (2) by the additive genetic standard deviation of the second trait ($\mathbf{sd_{AY}}$) to express the predicted correlated response in the units of measurement per annum.

$$\mathbf{CR_Y = R_X \times r_{AXY} \times sd_{AY}} \quad \text{(in units of measurement per annum)}$$

Writing this out in full gives [4]:

$$\mathbf{CR_Y} = \frac{\mathbf{i \times h_X \times r_{AXY} \times sd_{AY}}}{\mathbf{L}}$$

To extend an example used before, if we select for 20-week weight in sheep we expect an increase in fat depth as a result of a positive genetic correlation between these two traits. We can predict the direct response in weight (in units of measurement) from one of the formulae presented earlier in this chapter, and the correlated response in fat depth using the formula above. If we assume that:

$$\mathbf{i} = \mathbf{1.4}$$
$$\mathbf{h^2_X} = \mathbf{0.3}$$

$h_X = 0.55$

$h^2_Y = 0.3$

$h_Y = 0.55$

$r_{AXY} = 0.4$

$sd_{AX} = 2.75$

$sd_{AY} = 0.715$

$L = 2.0$

then the predicted response in 20-week weight is:

$$R_X = \frac{i \times h_X \times sd_{AX}}{L} \quad \text{(in kg per annum)}$$

$$R_X = \frac{1.4 \times 0.55 \times 2.75}{2.0}$$

$$R_X = 1.06 \text{ kg per annum (approx.)}$$

and the correlated response in fat depth is:

$$CR_Y = \frac{1.4 \times 0.55 \times 0.4 \times 0.715}{2.0}$$

$$CR_Y = 0.11 \text{ mm per annum.}$$

Limitations of these predictions

There are several limitations to the methods of predicting response described above which need to be mentioned. The first concerns the amount of variation in the trait under selection. Selection itself causes a reduction in the amount of genetic variation in the first few generations. In other words, choosing parents with performance at one end of the distribution results in a narrower distribution in the performance and breeding values of their offspring. The reduction in variation depends on the heritability of the trait concerned and the intensity of selection. The higher the heritability, and the higher the selection intensity, the greater the reduction in variation from one generation to the next. As a result of this, predictions of long term responses to selection based on the initial variation in the population may be too high. For example, with intense selection for a trait with a high heritability there may be a loss of about 20% in response to the second generation of selection, compared to that obtained in the first [4]. This reduction in genetic variation happens in the first few generations of selection, and then the amount of variation stabilises, or approaches **equilibrium**. Typically the loss in response to selection compared to that in the initial generation, stabilises at about 25%. There are two ways to deal with this problem. The first is to re-estimate the amount of variation and the heritability after the first few generations of selection, and then recalculate predicted responses from these equilibrium values. The second, which is less costly, is to use more complex formulae to predict long-term response to selection, which account for the fact that the heritability and genetic variation were estimated from populations in which selection had not been practised [3, 13].

After many generations of selection the variation in the trait of interest may become exhausted – for example, if each animal in the population has copies of only those alleles which have a favourable effect on performance. As a result, no further response to selection can be achieved. While these **selection limits** or **plateaux** have been reached in selection experiments with fruit flies and other laboratory species, they are rarely of practical concern in farm animals. This is because they occur after many generations of selection, which implies many decades of selection for the same trait in farm livestock. Over this time span it is unusual for market requirements to remain the same, and so it is unlikely that selection will be for exactly the same animal characteristics. Also, livestock populations are rarely closed over this length of time – the introduction of new animals introduces new variation and hence delays the approach to the selection limit. Even in a closed population, with selection for the same trait, new mutations prolong the period over which responses to selection are achieved. So, in theory, predictions of response do not hold true indefinitely. But, in practice, they are usually satisfactory over the lifetime of a particular breeding programme.

The final limitation which needs to be mentioned concerns the variation in response which may be achieved in different selection programmes. The formulae presented in this chapter predict the average response to selection. However, if several identical breeding programmes were started at the same time, they would not achieve identical responses. Some schemes would achieve lower than predicted responses, others would achieve higher responses, but on average the response should match that predicted. It is usually too expensive to demonstrate this variation in response by running identical experimental breeding programmes in farm livestock, but there are many good examples of it in laboratory animal selection experiments. However, for fairly simple breeding programmes, it is possible to predict the variation in response.

The variation in the response achieved occurs because there is an element of chance as to which genes are present in the initial group of animals. There is also an element of chance in which genes are lost and which ones are retained each time parents are selected. (These elements of chance are fundamental to the study of genetic improvement of livestock, and help to make it such a fascinating subject for scientists and practical breeders alike.) The main factors affecting the variation in response from that predicted are: (i) the size of the population – the smaller the population, the greater the risk of a difference between predicted and achieved responses, and (ii) the number of parents selected each generation – the smaller the number of parents selected, particularly males as there are usually fewer of these needed in the first place, the greater the risk of a difference between predicted and achieved responses.

Controlling inbreeding

As explained earlier, inbreeding is the mating of two related animals. Sometimes breeders pursue a deliberate policy of mating related animals, with the aim of increasing the frequency of favourable genes. However, even when steps are taken to avoid it, inbreeding is inevitable in closed populations. Sooner or later, matings will occur between related animals. In smaller populations this will tend to occur sooner, and in larger populations it will tend to occur later. The chances of related animals mating are increased when selection is practised in a closed population. This is because related animals have genes in common and hence their performance and breeding values are more alike than those of unrelated animals. As a result, selection for a particular characteristic increases the frequency of matings between related animals and hence increases inbreeding compared to that in a random mating population.

Animals	Breed	Trait	Inbreeding depression expressed as change in trait per 1% increase in inbreeding	Source
Dairy cattle	Holstein	Milk yield	–29.6 kg	[16]
		Fat yield	–1.08 kg	
		Protein yield	–0.97 kg	
	Holstein	Milk yield	–25 kg (–0.37%)[1]	[10]
		Fat yield	–0.9 kg (–0.35%)	
		Protein yield	–0.8 kg (–0.36%)	
		Fat %	+0.05% (+1.32%)	
		Protein %	+0.05% (+1.55%)	
		Somatic cell score	+0.012 (+0.51%)	[9]
Beef cattle	Hereford (cows)	Pregnancy rate (2-year-old cows)	–0.23% (–0.33%)[2]	[8]
		Prenatal survival (2 y.o.)	–1.67% (–1.73%)	
		Weaning rate (2 y.o.)	–1.24% (–2.26%)	
		Weaning weight (as trait of cow; all ages)	–0.47 kg (–0.24%)	
Sheep	S. Blackface, Cheviot, W. Mountain	Gross lifetime income per ewe	–£1.27 (–1.16%)[2]	[14]
		Greasy fleece weight	–0.01 kg (–0.53%)	[15]
	Many breeds	Greasy fleece weight	–0.017 kg	[7]
		Weaning weight	–0.111 kg	"
		Ewe fertility (ewes lambing/ewe joined)	–0.014	"
		Lamb survival	–0.028[3]	"
		(lambs weaned/lambs born)	–0.012[4]	"

[1] Figures in brackets express inbreeding depression as a percentage of the mean of the trait concerned when the mean was presented; [2] figures in brackets express inbreeding depression as a percentage of the mean performance of crosses between inbred and control lines; [3] per 1 % increase in inbreeding of lamb; [4] per 1 % increase in inbreeding of dam.

Table 4.9 Some examples of inbreeding depression in traits of economic importance in cattle and sheep.

There are two main reasons for wishing to limit inbreeding. The first is that inbreeding reduces the amount of genetic variation in a population, and so can reduce response to selection (and increase variation in response to selection). The second is that inbreeding can lead to a decline in performance in traits associated with fitness, such as reproductive rate and disease resistance. This decline in performance is known as **inbreeding depression**. Table 4.9 gives some examples of inbreeding depression in a range of characteristics in cattle and sheep. Inbreeding depression is thought to be the result of an increase in the frequency of recessive genes which adversely affect those characters associated with survival and overall 'fitness'. These are generally the same sort of traits which show heterosis as a result of crossbreeding. So, it is helpful to think of inbreeding and inbreeding depression as being the opposite of crossbreeding and heterosis respectively.

The decline in genetic variation following inbreeding, and the amount of inbreeding depression both depend on the amount of inbreeding. The closer the relationship between two animals which are mated, the greater the amount of

Figure 4.9 How an animal can inherit two alleles which are identical by descent, i.e. they are copies of the same allele from a single ancestor. In this example Bull 1 with genotype Bb is mated to Cows 1 and 2, both with genotype bb. A son of Cow 1 (Bull 2) with genotype Bb is then mated to his half sister (Cow 3) also of genotype Bb. The resulting calf has the genotype BB. This calf must have inherited both copies of the B allele from Bull 1, as this bull is the only animal in the grandparent generation to carry a copy of this allele.

inbreeding in the resulting offspring. Hence it is important to be able to measure the amount of inbreeding of existing animals, or the animals which could be produced from a particular mating. It is also important to be able to predict the rate at which inbreeding will accumulate over time in a herd, flock or breed involved in a particular breeding programme.

The amount of inbreeding is measured by the **inbreeding coefficient** (abbreviated **F**). The inbreeding coefficient is formally defined as the probability that two alleles at any locus are identical by descent. This is not as complicated as it sounds at first. Consider two calves, both of which inherit two genes for black coat colour (BB). The first calf is the result of a mating between an unrelated bull and cow. It has two alleles which are **identical in kind**, i.e. they are both B alleles, and they have the same effect, but they originated in unrelated ancestors. The second calf is the result of a mating between two heterozygous black half sibs (Bb). The pedigree of the second calf is shown in Figure 4.9. In this example the bull and cow that are mated are paternal half sibs, i.e. they have the same sire (the calf's grandsire). If this is a heterozygous black bull (Bb), and the calf's granddams are both homozygous red cows (bb), then the calf must have obtained both copies of his B allele from the common grandsire. In this case the two B alleles are **identical by descent**, i.e. they are copies of a single allele in a common ancestor.

Figure 4.10 shows an abbreviated version of a pedigree of an animal which is the result of a mating between two half sibs (such as the calf in the previous example). The inbreeding coefficient of the animal at the bottom of the pedigree (animal D) can be calculated by counting the number of individuals in each path through the pedigree which leads from the animal whose inbreeding coefficient is being calculated, through a common ancestor and back to the original animal. This is easiest to follow by referring to Figure 4.10. If we start with animal D, we can trace a path through B to the common ancestor A and back through C to D again. In other words there are three animals in the common path excluding D itself – these are B, A and C. If there is more than one common ancestor then this process is repeated for each one. The inbreeding coefficient can be calculated from the formula [4]:

$$\mathbf{F_D = \Sigma(\tfrac{1}{2})^n\ )\ )\ (1+F_A)}$$

where:

the symbol **Σ** indicates that the results are summed for each path through a common ancestor

n is the number of individuals in each path (the ½ appears in the formula to account for the fact that there is a 50% chance of a copy of any particular allele passing from a parent to an offspring). The formula shows ½ 'raised to the power of n' – so if n=2, the ½ is squared (½ x ½), if n=3, the ½ is cubed (½ x ½ x ½) and so on.

$\mathbf{F_D}$ is the inbreeding coefficient of animal D

$\mathbf{F_A}$ is the inbreeding coefficient of animal A

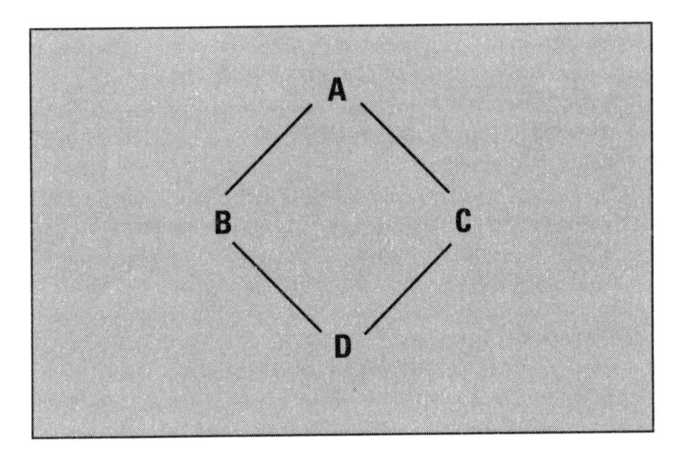

Figure 4.10 An abbreviated and more general version of the pedigree shown in Figure 4.9, where two half sibs are mated together. Animal A is the common ancestor i.e. the father of both animals B and C if they are paternal half sibs, or their mother if they are maternal half sibs. Animals B and C are of opposite sex and they are mated to produce animal D.

If the inbreeding coefficient of animal A is unknown, then it is assumed to be 0, and the formula simplifies to:

$$F_D = \Sigma(\tfrac{1}{2})^n$$

So, in the example above where there is only a single common ancestor, and three individuals in the path, the inbreeding coefficient of animal D is calculated from:

$$F_D = (\tfrac{1}{2})^3$$

$$F_D = 0.125 \text{ or } 12.5\%$$

In other words there is a 12.5% chance that the two alleles at any locus are identical by descent in an animal which results from a mating between half sibs. This example also illustrates the fact that, generally, the inbreeding coefficient of an animal is one half of the additive relationship between its parents. In this case the parents were half sibs, which are expected to have 25% of their genes in common, so they are said to have an additive relationship of 25%. So, when half sibs are mated to each other we expect their offspring to have inbreeding coefficients of 12.5%. Table 4.10 shows pedigrees and inbreeding coefficients for some other types of mating between close relatives.

If pedigrees could be traced back far enough, then all animals in a species would be related to each other, though the relationship would be very distant in most cases. It is not practical, or necessary, to go this far back in calculating inbreeding coefficients in livestock breeding, but it is important to define the reference population from which coefficients are calculated. For example, we might define the first generation of animals recorded in a herd or flock book, or animals born in some other significant year, as the initial reference population. Then, in calculating inbreeding coefficients, we assume that none of the animals in this initial population are inbred themselves. If the reference population chosen

Type of mating	Number of individuals in pathway through common ancestor	Inbreeding coefficient (F)
Father–daughter 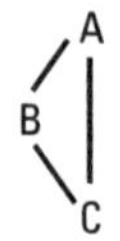	2 (B, A)	$F_C = (^1/_2)^2$ $F_C = 0.25$ or 25%
Full brother–full sister 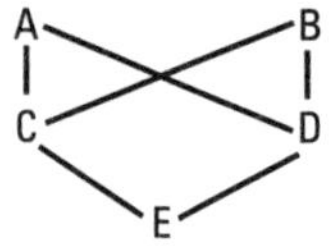	3 (C, A, D) + 3 (C, B, D)	$F_E = (^1/_2)^3 + (^1/_2)^3$ $F_E = 0.125 + 0.125$ $F_E = 0.25$ or 25%
Half brother–half sister 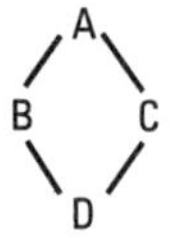	3 (B, A, C)	$F_D = (^1/_2)^3$ $F_D = 0.125$ or 12.5%
Full cousins 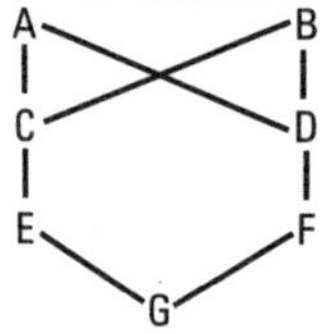	5 (E, C, A, D, F) + 5 (E, C, B, D, F)	$F_G = (^1/_2)^5 + (^1/_2)^5$ $F_G = 0.03125 + 0.03125$ $F_G = 0.0625$ or 6.25%
Half uncle–niece 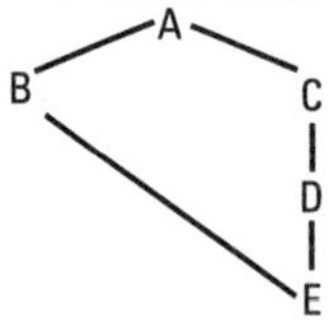	4 (B, A, C, D)	$F_E = (^1/_2)^4$ $F_E = 0.0625$ or 6.25%
Half cousins 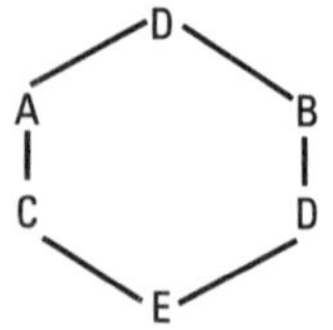	5 (D, B, A, C, E)	$F_F = (^1/_2)^5$ $F_F = 0.03125$ or 3.125%

Table 4.10 Some examples of calculating the inbreeding coefficient for animals resulting from matings between different types of close relative. In each case it is assumed that the common ancestor is not inbred itself, so the simpler version of the formula is used.

is too recent, or if there is a lot of missing information in pedigrees, then inbreeding coefficients may be severely underestimated.

Matings between close relatives are relatively uncommon in cattle and sheep breeding today because of the recognition of the high risk of undesirable side effects from high levels of inbreeding. However, high levels of inbreeding can occur also as a result of matings between less closely related animals for several generations. It is important to recognise this in designing breeding programmes. In this case the average inbreeding coefficient across a herd, flock or breed, and the rate at which this changes, are useful measures. Inbreeding coefficients can be calculated for large numbers of animals by a number of methods which are more amenable to computer programming than the method outlined above [12].

It is also useful to be able to predict the rate of inbreeding in advance, for example when comparing alternative designs for a breeding programme. Although it is relatively easy to predict rates of inbreeding for unselected populations, it is very difficult to do so accurately for populations under selection. Similarly, it is easier to predict rates of inbreeding when family sizes are equal than when they are not. Many of the simpler formulae underpredict rates of inbreeding, and so can give misleading results. However, the more accurate formulae are too complex to describe here. (For more information on this problem see the papers by Caballero and by Villanueva and colleagues, listed under Further Reading at the end of this chapter.) The formulae below are reasonably simple and illustrate some important points about rates of inbreeding, but they should be used with caution because they depend on assumptions which are rarely met in livestock breeding (equal family sizes, no selection, discrete generations).

The rate of inbreeding (usually abbreviated to **ΔF**) *per generation* can be calculated from [4]:

$$\Delta\mathbf{F}\ \textbf{per generation} = \frac{\mathbf{1}}{\mathbf{8M}} + \frac{\mathbf{1}}{\mathbf{8F}}\quad \textbf{(approx.)}$$

where:

M = the total number of males entering the population *each generation*
F = the total number of females entering the population *each generation*

Since the number of females used is usually much larger than the number of males used in most livestock breeding programmes, it is the number of males that influences the rate of inbreeding most.

The rate of inbreeding *per annum* can be calculated from a modified version of this formula:

$$\Delta\mathbf{F}\ \textbf{per annum} = \frac{\mathbf{1}}{\mathbf{8mL^2}} + \frac{\mathbf{1}}{\mathbf{8fL^2}}\quad \text{(approx.)}$$

where:

m = the total number of males entering the population *each year*
f = the total number of females entering the population *each year*
$\mathbf{L^2}$ = the average generation interval in years, squared (i.e. multiplied by itself)

Number of males used per annum	Number of females used per annum	Predicted rate of inbreeding (% per annum)
1	35	3.21
5	25	0.53
5	35	0.71
5	75	1.01
10	35	0.40
20	35	0.25

Table 4.11 Predicted rates of inbreeding with different numbers of males and females selected. The generation intervals assumed are rounded versions of those shown in Table 4.5.

So, for example, in a closed flock of 100 sheep with 5 different rams and 35 new females entering the flock each year, and an average generation interval of 2 years:

$$\mathbf{m} = 5$$
$$\mathbf{f} = 35$$
$$\mathbf{L} = 2$$
$$\mathbf{L}^2 = 4$$

and

$$\Delta F = \frac{1}{8 \times 5 \times 4} + \frac{1}{8 \times 35 \times 4}$$

$$= \frac{1}{160} + \frac{1}{1120}$$

$$= 0.00625 + 0.00089$$

= approximately 0.007 per annum, or 0.7% per annum (of which 0.625% is due to males)

Table 4.11 shows the expected annual rates of inbreeding for several other breeding schemes calculated using this formula. These results show that the number of parents selected per annum, and especially the number of sires selected, has a major impact on rates of inbreeding, as mentioned above. The values shown assume that selected animals are unrelated. However, in practice, related animals are often selected together because their breeding values are more alike than those of unrelated animals. This is particularly true when modern methods of predicting breeding values are used, which maximise the use of performance information from relatives. (These methods are explained in the next chapter.) This co-selection of relatives increases the rate of inbreeding further. It is important in designing breeding programmes to achieve a balance between selecting fewer parents to achieve high selection intensities, which maximise response, and using enough parents to keep inbreeding at acceptable levels.

The results in Table 4.11 also show that shorter generation intervals increase the annual rate of inbreeding. This explains why turning over generations slowly is a useful method of minimising inbreeding in conservation programmes, as explained in Chapter 3. Although there is little hard evidence available, it is generally accepted that absolute levels of inbreeding below 10%, or annual rates of inbreeding below 1%, are unlikely to result in serious inbreeding depression. There is evidence that achieving a given level of inbreeding quickly through close matings is likely to be more harmful than achieving the same level of inbreeding more slowly through successive matings between less closely related animals.

Summary

- It is important to be able to predict response to selection in order to compare alternative breeding programmes and so choose those which will achieve a high response to selection in a cost-effective way.

- When animals are selected on a single measurement of their own performance in one trait, the response to selection (**R**) per generation can be predicted by multiplying the selection differential (**S**) achieved by the heritability (**h**^{2}) of the trait selected on:

$$\mathbf{R = S \times h^2}$$

- The selection differential is the difference between the mean performance of selected animals (i.e. those identified to become parents of the next generation) and the overall mean of the group of animals from which they were selected.

- The heritability is the proportion of superiority of parents in a trait (i.e. the proportion of the selection differential) which, on average, is passed on to offspring. Alternatively, the heritability of a trait can be defined as the additive genetic variation in that trait (or the variation in breeding values) expressed as a proportion of the total phenotypic variation in that trait. Heritabilities are expressed as proportions from 0 to 1, or as percentages from 0% to 100%.

- The ability to predict response to selection is important in planning breeding programmes, and especially in comparing alternative schemes. However, it is usually more useful to predict responses per annum rather than per generation. To do this, we have to divide the response per generation by the generation interval (the weighted average age of parents when their offspring are born: **L**) in the flock or herd concerned:

$$\mathbf{R\ (per\ annum)} = \frac{\mathbf{S \times h^2}}{\mathbf{L}}$$

- From properties of the normal distribution, if we know what proportion of animals are selected as parents we can predict their superiority in standard deviation units. This is called the standardised selection differential or selection intensity (**i**). This allows us to rewrite the formula to predict annual response to selection as:

$$R = \frac{i \times sd_P \times h^2}{L}$$

- The annual response to selection will be highest when the selection intensity is high, the heritability is high, and the generation interval is low. There is little that breeders can do about the heritability of a trait – this is largely a biological characteristic of the trait concerned. Within biological limits, breeders can increase selection intensities and decrease generation intervals.

- Records of performance from relatives are useful in selection. This is because related animals have genes in common, and so the performance of relatives can provide clues to the genetic merit or breeding value of the candidates for selection. Records of performance are often used from the animal's ancestors, the animal itself, the animal's full or half sibs, the animal's progeny, other more distant relatives of the animal, or combinations of these classes of relatives.

- To predict the response to selection when using information from relatives, a more general version of the formula above is used. This involves **r**, the accuracy of selection on any combination of records from the animal and its relatives. The accuracy is the correlation between animals' true breeding values for the trait(s) under selection, and the measurement(s) on which selection is based:

$$R = \frac{i \times r \times sd_A}{L}$$

- The accuracy of selection on a single record of performance from a relative is calculated as the product of the proportion of genes they have in common and the square root of the heritability of the trait under selection. With records from greater numbers of relatives, in general: (i) the closer the relative the more valuable the record; (ii) initially, the more relatives of a given class, the higher the accuracy, although there are diminishing returns; (iii) initially, the higher the heritability, the higher the accuracy and (iv) initially, the lower the heritability, the greater the proportional contribution which records from relatives make.

- Using records of performance from relatives can increase the accuracy of selection, but this benefit can be outweighed by increases in the generation interval. Usually records from ancestors and collateral relatives can be used to increase accuracy of selection without lengthening the generation interval. Using records from descendants usually increases the generation interval, and so there is a trade-off between increasing accuracy and increasing generation interval. Comparisons of different breeding schemes also need to consider the resources needed, as well as predicted responses.

- Many traits of interest can be measured more than once over an animal's lifetime. Repeated records of performance can give additional clues to the breeding value of animals, and enhance the accuracy of selection. The value of repeated records in selection depends on the repeatability. This is the correlation between repeated records from the same animal. It is defined as the proportion of the total phenotypic variation which is explained by the combined effects of the genes and the permanent environmental variation:

$$\textbf{Repeatability} = \frac{V_G + V_{E_P}}{V_P}$$

● The repeatability sets an upper limit to the heritability. Like heritabilities, repeatabilities range from 0 to 1, or from 0% to 100%. The higher the repeatability of a trait, the lower the value of repeated records in selection for that trait. As with the use of information from relatives, it is important to take into account any time lag in obtaining repeated records.

● Repeatabilities measure the similarity between repeated records of the same trait, assuming that performance measured in different years (or at other intervals) is controlled by exactly the same genes. However, there are often associations between different traits which are not likely to be influenced by exactly the same genes. The degree of association between these characteristics is measured by the correlation coefficient.

● Correlated response to selection can be predicted from the direct response if the genetic correlation between the two traits and the additive genetic standard deviation of the second trait are known:

$$CR_Y = R_X \times r_{AXY} \times sd_{AY}$$
(in units of measurement per annum)

● There are several limitations to the methods of predicting response outlined. The first is that selection itself causes a reduction in the amount of genetic variation in the first few generations of selection, so predictions of long-term responses to selection based on the initial variation in the population may be too high. However, there are more complex formulae to predict long-term response to selection. The second limitation is that after many generations of selection the variation in the trait of interest may become exhausted. But, predictions of response are usually satisfactory over the lifetime of most livestock breeding programmes. The final limitation is that the formulae presented predict the average response to selection. Individual breeding programmes may achieve higher or lower responses than this as a result of chance.

● Inbreeding is the mating of two related animals. Sometimes breeders pursue a deliberate policy of mating related animals. However, even when steps are taken to avoid it, inbreeding is inevitable in closed populations. There are two main reasons for wishing to limit inbreeding:

(i) it reduces the amount of genetic variation in a population, and so can reduce response to selection, and (ii) it can lead to a decline in performance in traits associated with fitness, known as inbreeding depression. Effectively, inbreeding and inbreeding depression are the opposite of crossbreeding and heterosis respectively.

- The decline in genetic variation following inbreeding, and the amount of inbreeding depression both depend on the amount of inbreeding. The closer the relationship between two animals which are mated, the greater the amount of inbreeding in the resulting offspring. Hence it is important to be able to measure the amount of inbreeding of existing animals, or the animals which could be produced from a particular mating. The amount of inbreeding is measured by the inbreeding coefficient (abbreviated **F**). The inbreeding coefficient is formally defined as the probability that two alleles at any locus are 'identical by descent'.

- The inbreeding coefficient of an animal can be calculated from its pedigree by counting the number of individuals in each path through a common ancestor and back to the original animal. If there is more than one common ancestor then this process is repeated for each one. The inbreeding coefficient of the animal concerned ($\mathbf{F_D}$) is then calculated from the formula:

$$\mathbf{F_D} = \Sigma(\tfrac{1}{2})^n\,(1+\mathbf{F_A})$$

where n is the number of individuals in each path through the common ancestor, and $\mathbf{F_A}$ is the inbreeding coefficient of the common ancestor A.

- It is also useful to be able to predict the rate of inbreeding in advance, for example when comparing alternative designs for a breeding programme. Many of the simpler formulae underpredict rates of inbreeding. However, the more accurate formulae are complex. The formulae below are reasonably simple but they should be used with caution because they depend on assumptions which are rarely met in livestock breeding. The rate of inbreeding (ΔF) *per generation* can be calculated from:

$$\Delta\mathbf{F}\ \textbf{per generation} = \frac{1}{8M} + \frac{1}{8F}\quad\textbf{(approx.)}$$

where **M** and **F** are the total number of males and females entering the population each generation. The rate of inbreeding per annum can be calculated from:

$$\Delta\mathbf{F}\ \textbf{per annum} = \frac{1}{8mL^2} + \frac{1}{8fL^2}\quad\textbf{(approx.)}$$

where m and f are the total number of males and females entering the population *each year*.

- The number of parents selected per annum, and especially the number of sires selected, has a major impact on rates of inbreeding. It is important in designing breeding programmes to achieve a balance between selecting fewer parents to achieve high selection intensities, which maximise response, and using enough parents to keep inbreeding at acceptable levels. Related animals are often selected together because their breeding values are more alike than those of unrelated animals. This co-selection of relatives increases the rate of inbreeding further. Also, shorter generation intervals increase the annual rate of inbreeding. Absolute levels of inbreeding below 10%, or annual rates of inbreeding below 1% are unlikely to result in serious inbreeding depression.

Appendix

The connection between the two formulae presented to predict response to selection

The original formula presented on page p.116 is:

$$\mathbf{R} = \frac{\mathbf{i} \times \mathbf{sd_P} \times \mathbf{h^2}}{\mathbf{L}}$$

This can be converted to the second more general formula in several steps. Firstly, we can expand the heritability to become:

$$\mathbf{h^2} = \frac{\mathbf{sd_A} \times \mathbf{sd_A}}{\mathbf{sd_P} \times \mathbf{sd_P}}$$

Substituting this in the formula above, instead of $\mathbf{h^2}$ gives:

$$\mathbf{R} = \frac{\mathbf{i} \times \mathbf{sd_P} \times \mathbf{sd_A} \times \mathbf{sd_A}}{\mathbf{L} \times \mathbf{sd_P} \times \mathbf{sd_P}}$$

The $\mathbf{sd_P}$ on the top of the equation cancels out with one of the $\mathbf{sd_P}$'s on the bottom (i.e. both can be removed, to simplify the formula). Also, one of the $\mathbf{sd_A}$'s on the top of the equation, and the remaining $\mathbf{sd_P}$ on the bottom can be removed, and rewritten as **h**, giving [4]:

$$\mathbf{R} = \frac{\mathbf{i} \times \mathbf{h} \times \mathbf{sd_A}}{\mathbf{L}}$$

This formula is appropriate if selection is on a single record of performance of the animal itself (i.e. $\mathbf{r} = \mathbf{h}$). For other types of selection **h** is replaced by **r**, the more general value of accuracy. If you do not follow the algebra involved in deriving these formulae, but want to prove that they really do mean the same thing, then recalculate the response to selection for weaning weight in Simmental cattle, as in the example on page 117 but using the new formula, with $\mathbf{h} = 0.5$ (from $\mathbf{h^2} = 0.25$) and $\mathbf{sd_A} = 17.5$ kg (from $\mathbf{h} \times \mathbf{sd_P} = 35 \times 0.5$).

References

1. Animal Data Centre. 1997. *UK Statistics for Genetic Evaluations (Dairy). February 1997.* ADC, Rickmansworth, Herts.

2. Becker, W.A.. 1984. *Manual of Quantitative Genetics.* 4th edn. Academic Enterprises, Pullman, Washington.

3. Dekkers, J.C.M. 1992. 'Asymptotic response to selection on best linear unbiased predictors of breeding values.' *Animal Production*, 54:351–60.

4. Falconer, D.S. and Mackay, T.F.C. 1996. *Introduction to Quantitative Genetics.* 4th edn. Longman, Harlow.

5. Fogarty, N.M. 1995. 'Genetic parameters for live weight, fat and muscle measurements, wool production and reproduction in sheep: a review.' *Animal Breeding Abstracts*, 63:101–43. (Also, see erratum in Vol. 63, No. 12.)

6. Hayes, J.F., Cue, R.I. and Monardes, H.G. 1992. 'Estimates of repeatability of reproductive measures in Canadian Holsteins.' *Journal of Dairy Science*, 75:1701–6.

7. Lamberson, W.R. and Thomas, D.L. 1984. 'Effects of inbreeding in sheep: a review.' *Animal Breeding Abstracts*, 52:287–97.

8. MacNeil, M.D., Dearborn, D.D., Cundiff, L.V. et al. 1989. 'Effects of inbreeding and heterosis in Hereford females on fertility, calf survival and preweaning growth.' *Journal of Animal Science*, 67:895–901.

9. Miglior, F., Burnside, E.B. and Dekkers, J.C.M. 1995. 'Nonadditive genetic effects and inbreeding depression for somatic cell counts of Holstein cattle.' *Journal of Dairy Science*, 78:1168–73.

10. Miglior, F., Burnside, E.B. and Kennedy, B.W. 1995. 'Production traits of Holstein cattle: estimation of nonadditive genetic variance components and inbreeding depression.' *Journal of Dairy Science*, 78:1174–80.

11. Morris, C.A., Baker, R.L., Hickey, S.M. et al. 1993. 'Evidence of genotype by environment interaction for reproductive and maternal traits in beef cattle.'*Animal Production*, 56:69–83.

12. Nicholas, F.W. 1987. *Veterinary Genetics.* Oxford University Press, Oxford.

13. Villanueva, B., Wray, N.R. and Thompson, R. 1993. 'Prediction of asymptotic rates of response from selection on multiple traits using univariate and multivariate best linear unbiased predictors.' *Animal Production*, 57:1–13.

14. Wiener, G., Lee, G.J. and Woolliams, J.A. 1994. 'Consequences of inbreeding for financial returns from sheep.' *Animal Production*, 59: 245–9.

15. Wiener, G., Lee, G.J. and Woolliams, J.A. 1994. 'Effects of breed, rapid inbreeding, crossbreeding and environmental factors on fleece weight and fleece shedding in sheep.' *Animal Production*, 59:61–70.

COLOUR SECTION

Plate 1 Domestication has brought about many changes in animals, including changes in size and fatness. This plate shows a Suffolk ram – this breed has undergone many generations of selection for growth and meat production, compared to a Soay ram – this breed has undergone very little artificial selection and retains many of the characteristics of its wild predecessors, including a leaner carcass than many 'improved' breeds. (Courtesy of Dr John Robinson.)

Plate 2 'Robert Bakewell at Dishley Grange' by John Boultbee. (Courtesy of the Royal Agricultural Society of England.)

Plate 3 'Prize Livestock at Dishley Grange' by Thomas Weaver, 1802, showing Longhorn cattle and New Leicester sheep. (Courtesy of Iona Antiques, London.)

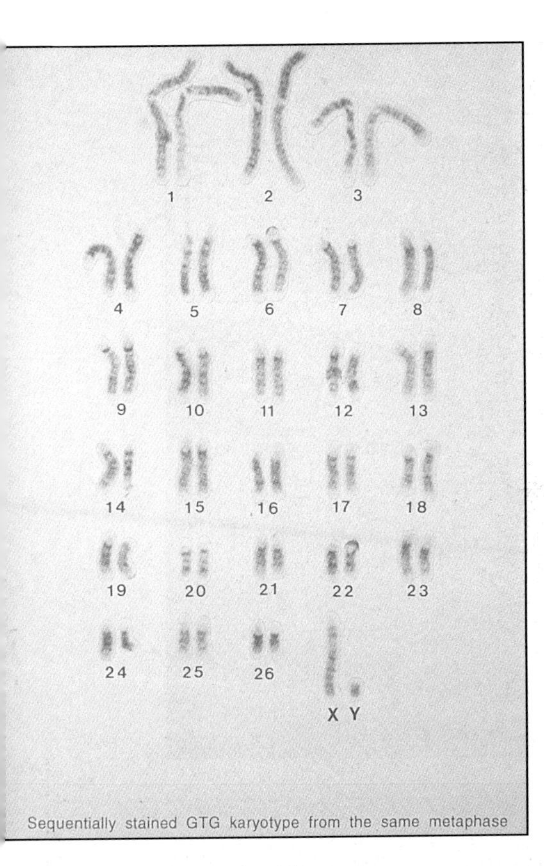

Plate 5 G-banded karyotype of the sheep – a photograph of the chromosomes at a particular stage of cell division, which has been cut up and rearranged to show all pairs of chromosomes in order. The numbering of the chromosomes is consistent with the 1995 (Texas) standardisation. (From [2]; courtesy of Dr HA Ansari, AgResearch Grasslands, Palmerston North, New Zealand.)

Plate 6 Red, White and Roan Shorthorn cattle. (Courtesy of Carey Coombs.)

Plates 7 & 8 Icelandic cattle showing some of the many coat colours still present in this breed.

Plate 9 Highland cattle still show wide variation in coat colour, despite many generations of selection. This variation has been reduced by chance or by selection in many other breeds. (Courtesy of Mrs Ùna F Cochrane.)

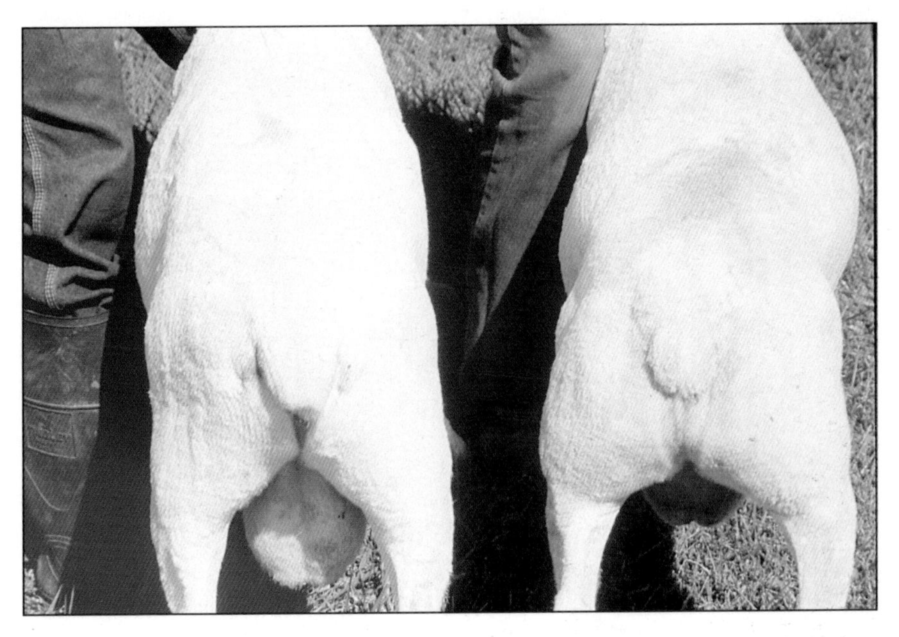

Plate 10 Dorset-Romanov F1 lambs – the lamb on the left is not carrying a callipyge gene, the lamb on the right is heterozygous for the callipyge gene. (Courtesy of Dr K A Leymaster.)

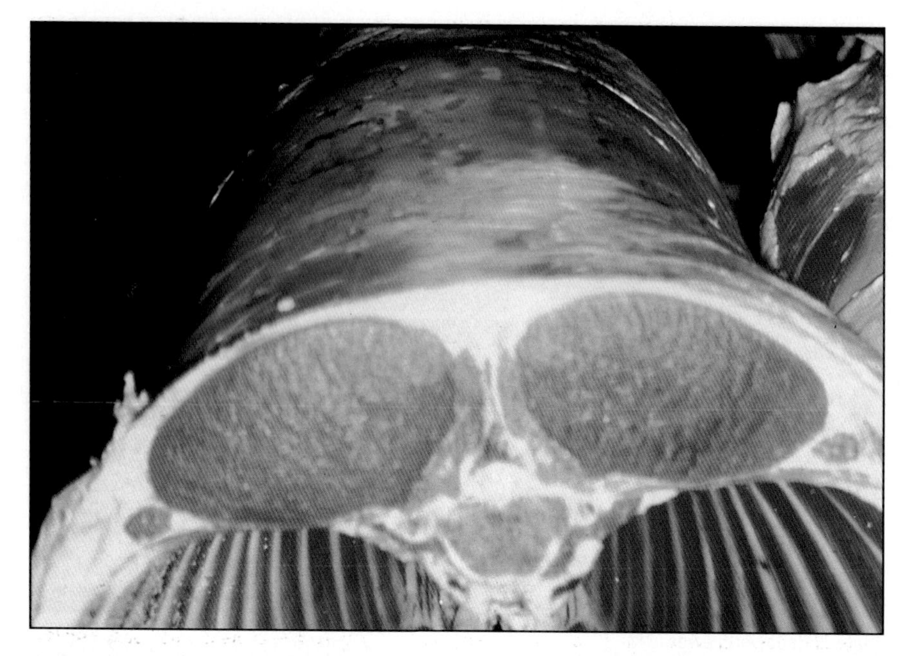

Plate 11 Cross-section of a carcass from a lamb carrying the callipyge gene. (Courtesy of Dr GD Snowder.)

Plate 12 North American Holstein Friesians have had a dramatic impact on the populations of dairy cattle in most temperate countries over the last few decades.

Plate 13 Coopworth ewes in Mr John Wilkie's flock, near Wanganui, New Zealand. The Coopworth is a synthetic breed, the development of which has been based on objective performance recording.

Plate 14 Crossbred ewes such as these Scottish Mules (Bluefaced Leicester x Scottish Blackface) form the backbone of the commercial sheep industry in Britain.

Plate 15 Suffolk ewes and lambs from the Stewart family's flock at Sandyknowe, Kelso, Scotland. Rams from breeds such as the Suffolk, Texel and Charollais are widely used as terminal sires in Britain to produce lambs for slaughter from F1 crossbred ewes.

Plate 16 Beef x dairy suckler cows have been very important historically in the UK and Eire. However increasing specialisation in the dairy herd, and reduced dairy cow numbers, mean that more beef producers in these countries are breeding their own beef x beef replacements.

Plate 17 First cross Simmental suckler cows with second cross Simmental calves at foot. (Courtesy of the British Simmental Cattle Society Ltd.)

Plate 18 Gloucester cattle are classified as an endangered breed in the Rare Breeds Survival Trust priority list. The trust aims to promote conservation of endangered breeds of farm livestock of all species in Britain.

Plate 19 Belted Galloways are classified as a minority breed in the RBST priority list.

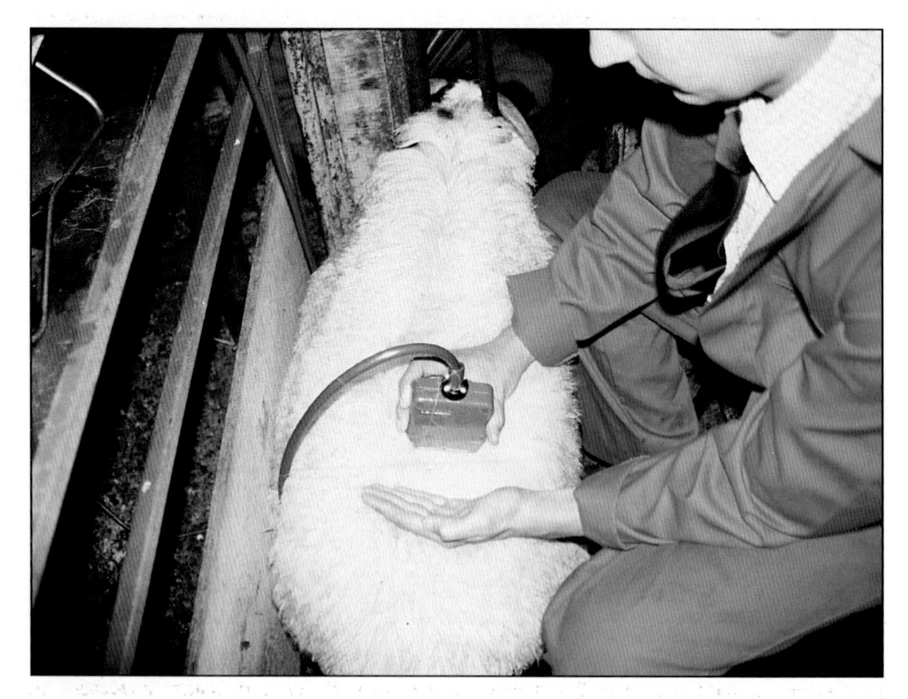

Plate 20 Lamb being ultrasonically scanned.

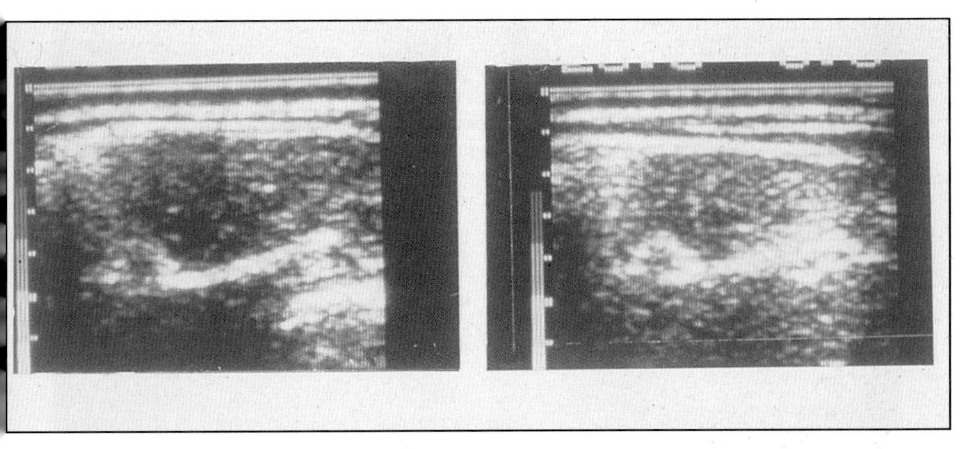

Plate 21 Ultrasonic scans from the loin area of two Suffolk ram lambs. The scan on the right is from an animal with a relatively high fat depth and relatively low muscle depth, that on the left is from an animal with a lower fat depth and a higher muscle depth.

Plate 22 With sufficient data, BLUP can help disentangle direct and maternal influences on the performance of offspring. South Devon cow and calf from Mr W H D Scott's Grove herd in Gloucestershire, England.

Plate 23 Bulls at the biannual Perth Bull Sales in Scotland – one of the largest sales of bulls in Europe. An increasing number of buyers of bulls and rams use objective information to assist their selection decisions. One of the major benefits of BLUP is that, providing that there are strong enough genetic links for effective across-herd or across-flock evaluations, it permits the fair comparison of animals from different herds or flocks.

Plate 24 Holstein Friesian cows grazing in Northland, New Zealand. The NZ climate allows low-cost production of milk – entirely from grass in the majority of herds.

Plate 25 Jersey and Holstein Friesian and Holstein Friesian x Jersey cows at the Ruakura Agricultural Research Centre, New Zealand. The New Zealand dairy industry is one of the few in temperate areas to make use of systematic crossbreeding.

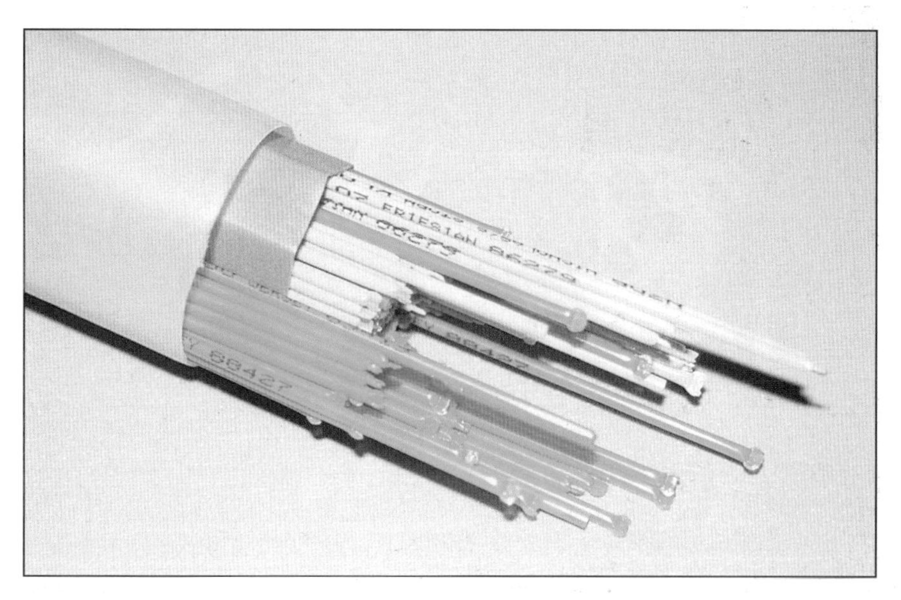

Plate 26 Straws of frozen semen. Artificial insemination has had a dramatic effect on rates of genetic improvement in dairy cattle in many countries. It permits higher selection intensities, shorter generation intervals and more accurate prediction of breeding values than the use of natural mating.

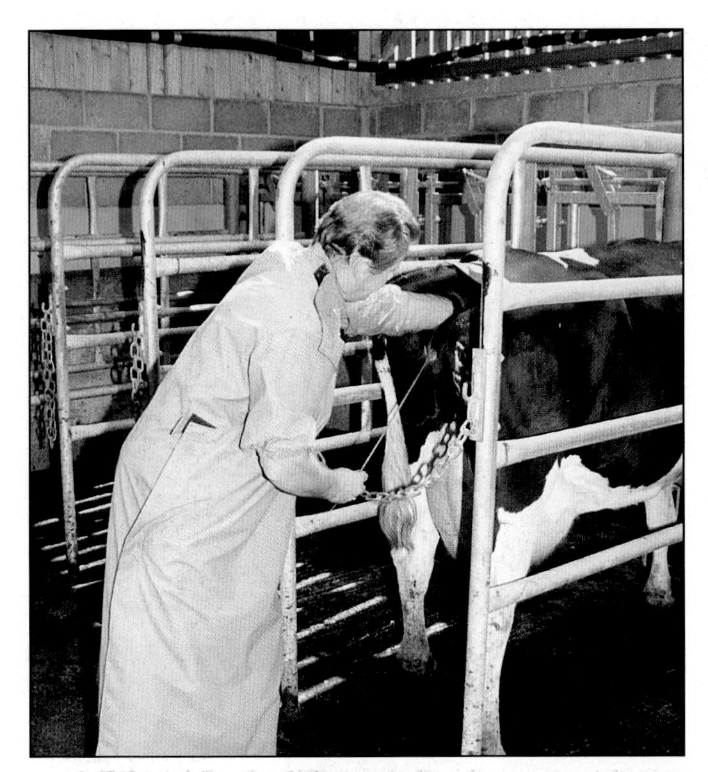

Plate 27 Artificial insemination in progress.

Plate 28 Skalsumer Sunny Boy, one of the most popular AI sires ever. In 1992 and 1993 more than 600 calves sired by him were born each day around the world – an average of one calf every $2^1/_2$ minutes. (Courtesy of Genus.)

Plate 29 Cow at the Langhill Dairy Cattle Research Centre in Scotland feeding through a Calan gate, which permits monitoring of individual cow intakes.

Plate 30 Hereford cattle in Northland, New Zealand. Traditional British beef breeds such as the Hereford, Aberdeen Angus and Shorthorn, or crosses between them, are widely used as maternal lines in extensive pastoral beef industries in temperate countries.

Plate 31 Angus cattle on Mr John Wilkie's farm near Wanganui, New Zealand.

Plate 32 Beef cattle in a feedlot in Alberta, Canada.

16. Wiggans, G.R., VanRaden, P.M. and Zuurbier, J. 1995. 'Calculation and use of inbreeding coefficients for genetic evaluation of United States dairy cattle'. *Journal of Dairy Science*, 78:1584–90.

17. Wray, N.R. and Simm, G. 1991. *BLUP for pedigree beef and sheep breeders*. SAC Technical Note T265. Scottish Agricultural College, Perth.

Further reading

Becker, W.A.. 1984. *Manual of Quantitative Genetics*. 4th edn. Academic Enterprises, Pullman, Washington.

Caballero, A. 1994. 'Developments in the prediction of effective population size.' *Heredity*, 73:657–79.

Cameron, N.D. 1997. *Selection Indices and Prediction of Genetic Merit in Animal Breeding*. CAB International, Wallingford.

Dickerson, G.E. and Hazel, L.N. 1944. 'Effectiveness of selection on progeny performance as a supplement to earlier culling of livestock.' *Journal of Agricultural Research*, 69:459–76.

Falconer, D.S. and Mackay, T.F.C. 1996. *Introduction to Quantitative Genetics*. 4th edn. Longman, Harlow.

Goddard, M.G. and Smith, C. 1990. 'Optimum number of bull sires in dairy cattle breeding.' *Journal of Dairy Science*, 73:1113–22.

Lush, J.L. 1945. *Animal Breeding Plans*. 3rd edn. Iowa State College Press, Ames, Iowa.

Lush, J.L. 1994. *The Genetics of Populations*. Special Report 94. Iowa Agriculture and Home Economics Experiment Station, College of Agriculture, Iowa State University, Ames, Iowa.

Meuwissen, T.H.E. and Woolliams, J.A. 1994. 'Response versus risk in breeding schemes.' *Proceedings of the 5th World Congress on Genetics Applied to Livestock Production*, Vol. 18, pp. 236–43.

Nicholas, F.W. 1987. *Veterinary Genetics*. Oxford University Press, Oxford.

Rendel, J.M. and Robertson, A. 1950. 'Estimation of genetic gain in milk yield by selection in a closed herd of dairy cattle.' *Journal of Genetics*, 50:1–8.

Van Vleck, L.D., Pollak, E.J. and Oltenacu, E.A.B. 1987. *Genetics for the Animal Sciences*. W.H. Freeman and Company, New York.

Villanueva, B., Woolliams, J.A. and Gjerde, B. 1996. 'Optimum designs for breeding programmes under mass selection with an application in fish breeding.' *Animal Science*, 63:563–76.

Weller, J.I. 1994. *Economic Aspects of Animal Breeding*. Chapman and Hall, London

CHAPTER 5

Predicting breeding values

Introduction

The last chapter concentrated on how to *predict* responses to selection in simple breeding programmes. This is useful for comparing alternative breeding programmes, and helping to make sensible investment decisions. When it comes to actually doing the selection there are a number of steps that can be taken to improve the chances of *achieving* the responses predicted. These include: (i) managing the animals in a way which will make it easier to disentangle genetics and environment, (ii) adjusting records of performance for known environmental effects, and then (iii) predicting the breeding values of individual animals. Modern methods of genetic evaluation achieve the second and third of these steps simultaneously, but it will be easier to understand why this is beneficial by considering the traditional approach first.

Management of candidates for selection

In most breeding programmes we attempt to disentangle the effects of genes and environment, in order to select animals that have high genetic merit, and not those that perform well simply because they are well fed and managed. The environmental factors that obscure true genetic merit can be divided into two types – those which it is difficult to attribute to individual animals, and those which we can identify as affecting particular animals and can do something about.

An example of the first type would be a subclinical disease affecting the performance of some animals in a herd or flock, but not others. In general we know that this sort of thing happens, but it is difficult to identify with certainty all the animals that have been affected. Probably the best that we can do in this situation is to discard the performance records from animals that clearly have been affected. Although it is difficult to avoid influences such as that of subclinical disease, as far as possible animals which are going to be compared should be managed to give them equal opportunity to express their genetic merit. For example, comparisons of post-weaning growth of animals will be fairest if they are all weaned at the same age and have equal access to supplementary feed.

The second type of environmental influences are those that we can attempt to adjust for. These include factors such as age of dam, birth rank (single, twin, triplet etc.), lactation number, season or date of birth or age at measurement. Although we can never know the *exact* effect that any of these factors has on an individual animal's performance, we can estimate the *average* effect that any of these factors has in a group of animals. Adjusting for this is usually better than doing nothing about it. Methods of adjusting records for this type of environmental effect are discussed in the following section.

It is the environmental effects of the first type, and those of the second type which have not been removed fully by adjustment, which contribute to the environmental and phenotypic sources of variation in animal performance which were described in earlier chapters.

Objective methods of genetic improvement rest heavily on comparisons of the performance of animals which have been treated in a similar way, e.g. born over a relatively short period of time, on the same farm, and fed and managed similarly. These animals are often called **contemporaries**, and the groups they belong to are called **contemporary groups**. The accuracy of selection will be improved by ensuring that animals in a contemporary group get as similar treatment as possible. Ideally, contemporary groups should be as large as possible to allow the best possible separation of genetic and environmental effects on performance. In practice, a compromise has to be reached between making the groups large but including animals which are not really contemporaries and accepting smaller groups of animals which really do share a similar environment. For instance, in pedigree beef herds there is often quite a wide spread in calving date. When records from these herds are analysed the size of contemporary groups can be increased by expanding the range of birth dates over which calves are eligible to join a group. But at some stage this defeats the object, as ignoring the increasing seasonal effects on calf performance outweighs any advantage from making the contemporary groups larger. There are various statistical methods of deciding on the optimal group size in these situations, and references to some of these are listed at the end of the chapter.

Adjusting records of performance

Additive correction factors

There are a number of methods of adjusting records of performance to help to deal with the second type of environmental influence described above. The simplest approach is to use **additive correction factors**, so called because they involve adding amounts to (or subtracting amounts from) the performance records of animals which belong to particular classes such as singles or twins. For example, we could weigh lambs at weaning and note which animals were born and reared as singles or twins. If we averaged the weights for the singles and twins separately, we might find that singles were 3 kg heavier than twins, on average. If we wanted to select for growth rate alone, regardless of litter size, the simplest way to compare animals fairly would be to add 3 kg to the weaning weight of all twin lambs (or subtract 3 kg from the weight of all single lambs) to create a set of adjusted records. Then all the lambs could be compared as a single group and those with the heaviest adjusted weights selected for breeding.

Additive correction factors have been very widely used in the past in all species. The advantages of this type of adjustment are that:

- They are easy to use.

- They make no assumptions about the average genetic merit of animals in different contemporary groups.

The disadvantages of additive correction factors are that:

- Many records of performance are needed to estimate correction factors accurately.

- They have to be estimated from data prior to the selection of animals.

- They may be fairly specific to a particular herd, flock or management system.

- In some circumstances they can over-correct records of performance and remove some genetic differences. For example, if correction factors were estimated for a flock in which two rams had been used, and by chance the genetically largest ram had more single-born offspring, the effect of being born as a single lamb could be overestimated because part of the advantage to singles was in fact due to having a genetically heavier sire. When the records of performance were adjusted, some genetic differences due to the sires of the lambs would be removed inadvertently.

- In other circumstances they may under-correct records of performance – for example, other things being equal, cows of intermediate ages produce calves with the highest weaning weights. However, if the worst cows in each age group in a herd are culled each year on the basis of the weaning weight of their calves, those cows that are left in each age group are progressively better than their former contemporaries. However, a simple comparison of the weaning weights of calves from the remaining dams of different ages would not recognise this. As a result the true effect of age of dam on offspring weaning weight would be underestimated for older cows.

Additive correction factors were used to adjust the performance of Suffolk sheep involved in the SAC selection experiment mentioned in Chapter 2. The sheep were selected on an index made up of live weight together with fat and muscle depths measured with an ultrasonic scanner (see Plates 20 and 21). Before calculating index scores, the weight, fat depth and muscle depth of individual sheep were adjusted for their birth rank (i.e. whether they were born as a single, twin or triplet), the age of their dam (two-year-old versus older) and for their age at scanning (animals were weighed at exactly 150 days of age, but were scanned in groups two or three times a week at about 150 days of age, so their ages at scanning differed by a few days). Table 5.1 shows the adjustment factors used for ram lambs in the SAC flock. Lambs which were born as twins from ewes which were older than two years of age were the most numerous category, and so they were treated as the norm, and their records were not adjusted. Records from other types of animal were then adjusted to this level. So, for example, the weights of single ram lambs had 4.25 kg subtracted, and those of triplets had 2.51 kg added, to bring them to the level expected had they been born as twins. Similarly, records from ram lambs born to two-year-old ewes had 1.75 kg added to them.

Trait	Correction factors for						
	Birth rank			Dam age (years)		Age at scanning (days)	
	Single	Twin	Triplet	2	3 and over	under 150	over 150
Live weight	–4.25 kg	0	+2.51 kg	+1.75 kg	0	—	—
Fat depth	–0.52 mm	0	+0.21 mm	—	0	+0.07 mm/day	–0.07 mm/day
Muscle depth	–0.69 mm	0	+0.63 mm	—	0	+0.10 mm/day	–0.10 mm/day

Table 5.1 Additive correction factors used to adjust live weights, ultrasonic fat and muscle depths in SAC Suffolk ram lambs, prior to calculating index scores. A separate set of correction factors was used for ewe lambs. The values shown were subtracted from or added to the records from the class of lambs shown, to make them equivalent to records from twin born lambs from ewes of 3 years of age and older. The values shown for age at scanning were subtracted for *each* day of age in excess of 150 days at scanning, or added for *each* day of age under 150 days on the day of scanning.

Multiplicative correction factors

Multiplicative correction factors are similar to additive factors but, as the name suggests, the records of performance are adjusted by multiplying by the correction factors rather than adding the correction factors. For example, rather than expressing the correction factor for 20-week weight of triplet-born lambs as +2.51 kg, as in Table 5.1, we could calculate that triplets are 4% lighter than twins, and so multiply all weights from triplets by 1.04 to bring them to the level expected for twins. Most of the advantages and disadvantages of multiplicative correction factors are similar to those for additive correction factors. However, multiplicative correction factors are more appropriate when the scale of the correction depends on the mean level of performance in the herd or flock. For example, in flocks with a much lower average 20-week weight than that of the SAC flock, the difference between the weights of twins and triplets is likely to be less than 2.51 kg. In flocks with a much higher average 20-week weight than that of the SAC flock, the difference between the weights of twins and triplets is likely to be more than 2.51 kg. The percentage difference in weights is likely to be more similar across flocks. So, when correction factors are going to be used across herds or flocks which have very different levels of performance, multiplicative correction factors may be more appropriate than additive adjustments. Similar arguments apply to the effectiveness of multiplicative correction factors for dealing with animals with different levels of performance within flocks or herds.

Standardising to adjust records

A third method of adjusting records involves assigning records from animals born in a specified time period to a contemporary group, based on the factors to be adjusted for. Within each of these groups each record is then expressed as a deviation from the mean of the group, in standard deviation units. For example, records from lambs born over a period of a few weeks (a season) could be assigned to four groups: single reared from two-year-old dams, single reared from older dams, multiple reared from two-year-old dams and multiple reared from older dams. The mean and standard deviation of the trait concerned are then calculated separately for each of the four groups. Finally, the performance record of each animal is expressed as a deviation from the mean of its own group, and then divided by the sd for that group. This gives records expressed in sd units (i.e. typically ranging from about –3 to about +3), rather than the units in which the trait was measured – though they can easily be converted to units of measurement again. These standardised measurements can be compared directly across contemporary groups within a flock or herd.

The advantages of this more complicated method of adjustment are that:

- It needs no prior information on the effects of the different factors, e.g. the effect of being born as a single. So it can be used in a wide range of herds or flocks, even if the variation in performance differs widely among them.

- It reduces possible bias from preferential treatment of some animals – sometimes particular groups of animals have been given special treatment, knowingly or unknowingly. This may be detected in the performance records if some contemporary groups have a lot wider variation (higher sd) than others. Expressing records in sd units removes this effect, and effectively equalises the variation in each contemporary group.

The disadvantages are that:

- Standardising records in each contemporary group assumes that the different groups are of equal genetic merit, which may not be true. For example, in an active selection programme the youngest dams in the herd or flock will have the highest genetic merit, on average. As a result, their progeny are expected to be of higher merit than those from older dams. This method of correction ignores this potential advantage to the progeny of younger dams.

- The results for a whole group can be greatly influenced by an animal in that group with extremely high or low performance.

- Corrections may be unreliable when groups are small.

- Of all three methods described, it is most likely to over-correct and remove some genetic differences.

This method of adjustment was used widely by the Meat and Livestock Commission (MLC) in beef and sheep performance recording in Britain until recently. Similar or slightly more sophisticated versions of standardising records are used in dairy cattle and beef evaluations in several countries now [2, 3, 15].

Predicting breeding values

Clues to an animal's breeding value

Having treated animals as equally as possible, and then adjusted for any known environmental differences, the next step is to attempt to rank them on additive genetic merit or breeding value. We never know the *true* breeding value of an animal, though we can come close to it by recording very large numbers of offspring. Usually this is impractical and very expensive, and so we have to rely on **predicted** or **estimated breeding values** (**PBVs** or **EBVs**) of the candidates for selection. (Strictly speaking, breeding values are predicted, because they relate to future breeding performance, but the terms predicted and estimated get used interchangeably. The term predicted breeding values is used mainly in this chapter but in later chapters when talking about industry practice, both terms are used depending on the convention in the species and country concerned.)

The clues we can use to predict breeding values have already been mentioned. They include records of performance from:

- the animal itself
- the animal's ancestors
- the animal's full or half sibs
- the animal's progeny
- any other relatives of the animal
- combinations of the classes of relatives listed above

Predicting an animal's breeding value is a bit like completing a large, complicated jigsaw puzzle, where each piece of the puzzle is a record of performance from the animal itself or one of its relatives. The more pieces of the puzzle we have, the easier it is to see the true picture. However, some pieces of the puzzle are more informative than others. Generally, the higher the proportion of genes in common between the animal and a given relative, the more useful the record of performance from that relative. But, as we saw in the last chapter, records from progeny are of most value. As the number of records on progeny increases, the correlation between predicted and true breeding values (the accuracy of selection) approaches 1. So widespread progeny tests produce predicted breeding values which are very close to true breeding values. With other classes of relative the accuracy of prediction never reaches 1, and for all classes of relative there are diminishing returns in accuracy as the number of records increases.

Calculating PBVs

Using the animal's own performance

In the simplest case, when we have a single record of performance on the animal itself, the predicted or estimated breeding value is the deviation in performance from contemporaries, multiplied by the heritability of the trait concerned. The deviation in performance is calculated after adjusting the performance records for the type of environmental effects discussed in the last section:

PBV or EBV = h^2 x deviation in performance from contemporaries

This is equivalent to the formula used at the beginning of the last chapter to predict response to selection, except that the PBV refers to a single animal, whereas the response refers to all the progeny born from selected parents. Table 5.2 illustrates how PBVs are calculated for two groups of animals in separate herds when each animal has a single record of performance. When PBVs are calculated in this way, the animals rank in exactly the same order within a herd as they rank on their performance record, or on their deviation from the mean of contemporaries. However, the PBV predicts how much of the superiority or inferiority in the animal's performance is due to its (additive) genes. The table also illustrates several other features of PBVs:

- They can have positive or negative values, or be equal to zero.

- The sign indicates whether they are expected to be genetically above (+) or below (–) the mean of the group of animals on which the calculations were performed, or some other defined group of animals whose PBVs are set to average zero, e.g. those animals born in a particular year. It is important to know which these animals are. The group of animals from which deviations in breeding value are expressed is often called the **genetic base**.

Herd 1				Herd 2			
Animal	Adjusted 400-day weight (kg)	Deviation from mean (kg)	PBV (kg)	Animal	Adjusted 400-day weight (kg)	Deviation from mean (kg)	PBV (kg)
1	450	–70	–28	11	470	–70	–28
2	600	+80	+32	12	530	–10	–4
3	440	–80	–32	13	550	+10	+4
4	480	–40	–16	14	490	–50	–20
5	530	+10	+4	15	640	+100	+40
6	570	+50	+20	16	480	–60	–24
7	500	–20	–8	17	520	–20	–8
8	520	0	0	18	570	+30	+12
9	550	+30	+12	19	610	+70	+28
10	560	+40	+16	20	540	0	0
Mean	520	0	0	Mean	540	0	0

Table 5.2 An example of the calculation of PBVs for 400-day weight in beef cattle when each animal has a single record of performance. There are groups of 10 contemporary animals in each of two herds. The performance records have been adjusted for the type of environmental effects discussed in the last section e.g. age of dam. The heritability of 400-day weight is assumed to be 0.4. In this example we have no way of knowing whether the 20 kg difference in the average weight of animals in the two herds is due to breeding, feeding and management, or a combination of these. So, the best we can do is to calculate and use the PBVs within herd only.

- They span a narrower range than the deviations in performance – this regression or 'shrinking' reflects the fact that part of the variation in performance is environmental (as shown in Figure 2.19). For an animal with a single record of performance the regression coefficient, which determines the extent of this shrinking, is simply the heritability of the trait concerned. The higher the heritability, the lower the proportion of environmental variation, and so the less severe the shrinking.

- They are expressed (at least initially) in the same units as the record of performance (e.g. kg of live weight, litres of milk, mm of fat).

- In the example in Table 5.2 we have no way of knowing whether the 20 kg difference in the average weight of animals in the two herds is due to breeding or to feeding and management, or a combination of these. In this case, the best we can do is to calculate and use the PBVs within herd only.

As outlined in Chapter 4, using repeated records of performance from the same animals can increase the response to selection. Similarly, the use of repeated records of performance can increase the accuracy of predicting breeding values for individual animals. In this case:

PBV = b x average deviation of individual's performance records from mean of contemporaries

where **b** is a regression coefficient which depends on the number of repeated records, and on the heritability and repeatability of the trait concerned (see [14] for details of the calculation of this regression coefficient).

Using information from relatives

Calculating PBVs from the performance of relatives can be a bit more difficult. The simplest case is when the only records available are from the parents. If we first calculate PBVs for each parent, then the PBV of their offspring is simply:

PBV of offspring = ½PBV of sire + ½PBV of dam

So, if we mated a bull with a PBV for 400-day weight of +30 kg to a cow with a PBV of +10 kg, the PBV of the offspring for 400-day weight would be +20 kg. This is illustrated in Figure 5.1(a). This formula should come as no surprise, because we have already seen that offspring get exactly half their genes from each parent. If we only have a PBV for one parent, then the best we can do is to assume that the other parent is of average genetic merit. If PBVs are expressed relative to the average merit of animals born in the recent past then a parent of unknown PBV is usually given a PBV of zero. So, for example, if we mated the bull mentioned above with a PBV of +30 kg for 400-day weight to a cow of unknown PBV, the PBV of the offspring would be +15 kg ([30+0]/2; see Figure 5.1(b)). Or more directly, when the PBV of one parent is unknown, we simply halve the PBV of the other parent to get the offspring's PBV.

In some cases we have a PBV for the sire, and for the dam's sire, but not for

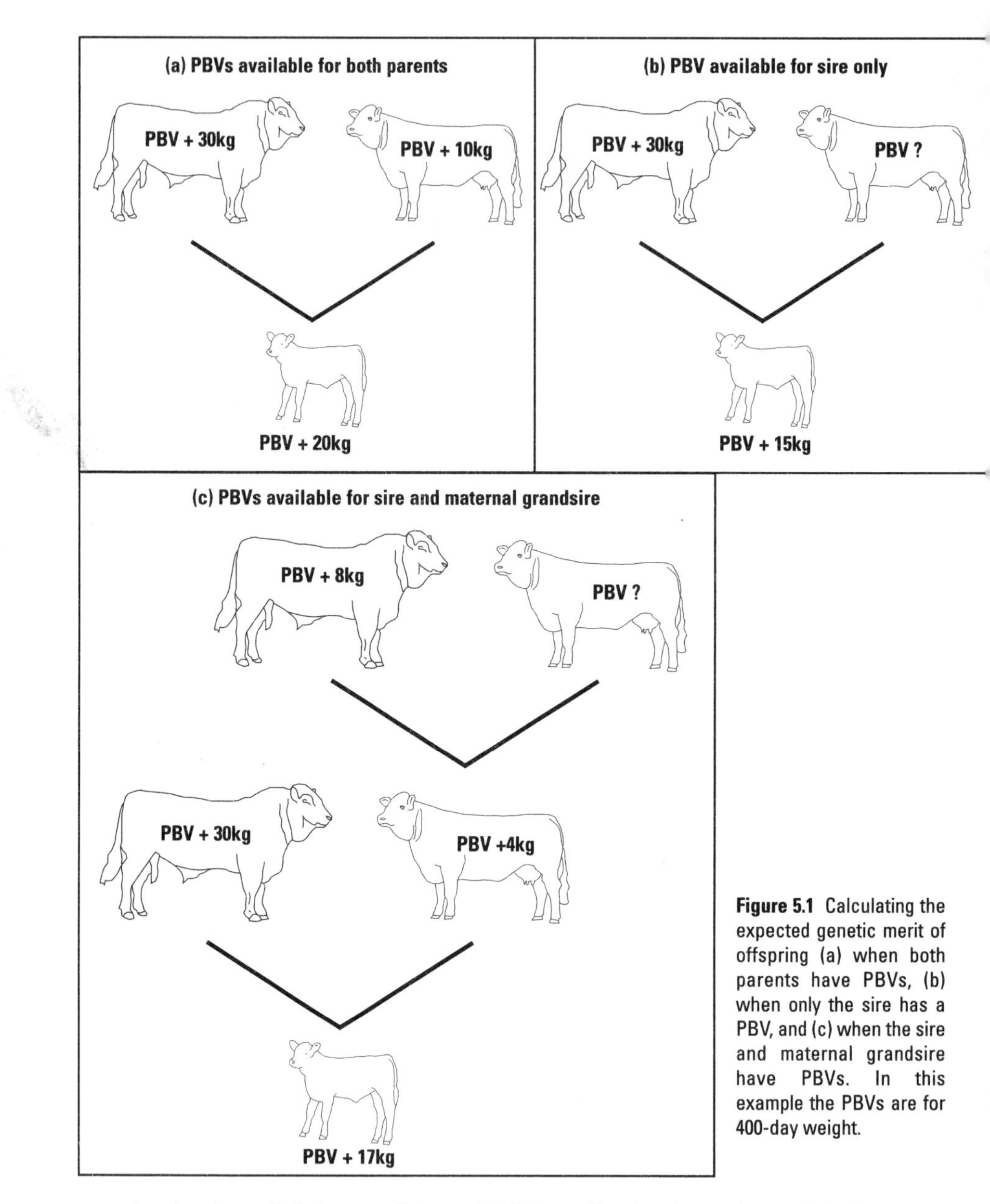

Figure 5.1 Calculating the expected genetic merit of offspring (a) when both parents have PBVs, (b) when only the sire has a PBV, and (c) when the sire and maternal grandsire have PBVs. In this example the PBVs are for 400-day weight.

the dam herself. We can still get a PBV for offspring in two steps. Firstly we get a PBV for the dam by halving her sire's PBV. Then we average the sire's and the dam's PBVs as before. For example, if we stick with the bull mentioned above, but mate him to a cow of unknown PBV whose sire has a PBV of +8 kg, then our prediction of the cow's BV is +4 kg (8/2) and so the offspring PBV is +17 kg ([30 + 4]/2; see Figure 5.1(c)). Alternatively, and more directly, we can add half of the sire's PBV to one quarter of the maternal grandsire's PBV. PBVs calculated in this way from ancestor's PBVs are known as pedigree indexes. They are very valuable in providing an early prediction of an animal's genetic merit, before it has performance records of its own or records from collateral relatives.

Similarly, if we only had a PBV based on a *single* record of performance from any relative, our best prediction of the BV of a related animal would be:

PBV = proportion of genes in common with the relative x PBV of the relative

So, for example, if we had PBVs of a full or a half sib based on a single measurement of their own performance we would multiply their PBV by 0.5 or 0.25 respectively to get a PBV for the unrecorded relative. (See Figure 4.6 for the proportions of genes in common between an animal of interest and some other relatives.)

We saw earlier that if we have PBVs for both parents we can average these to get PBVs for offspring. This is because the two parents contribute independent samples of genes to their offspring (unless the parents are related). However, if we have more than one of any other type of relative we cannot simply average their PBVs, because progeny or sibs, for example, have genes in common with each other as well as the animal whose PBV is being derived, and so there is 'overlap' in the information they provide – this was illustrated in the last chapter by the diminishing returns in accuracy with increasing numbers of relatives.

As an example let us consider the prediction of a bull's BV from single records on his progeny, which are not related to each other except through him. In this case:

PBV = b x deviation of progeny records from overall mean

where **b** is a regression coefficient which depends on the number of progeny (**n**), and on the heritability (**h^2**) of the trait concerned [14]:

$$\mathbf{b} = \frac{\mathbf{2n}}{\mathbf{n + (4 - h^2)/h^2}}$$

This looks frightening at first sight, but all we need to note here is that (i) the higher the number of progeny, the greater the value of b and (ii) the higher the heritability, the higher the value of b. Table 5.3 shows the values of b for traits with different heritabilities, and for sires with different numbers of progeny. The values in the table show that b approaches a maximum value of 2 when there are records on many progeny. In other words, with many progeny a bull's PBV is simply the superiority or inferiority of his progeny multiplied by 2. Again, this value of 2 should come as no surprise – progeny have half their genes in common with each parent, so the progeny performance is a measure of half the bull's breeding value.

To illustrate this further, let us consider two dairy bulls, one whose daughters yield 500 kg more milk than average, and one whose daughters yield 500 kg of milk less than average. The PBVs of the two bulls depend on the number of daughters on which these yield deviations are based. Table 5.4 shows PBVs for the two bulls, assuming that these daughter yield deviations are produced from 1, 5, 10, 100, 1,000 or 10,000 daughters of each bull. The heritability of milk yield is about 0.35, so with 1 to 10,000 daughters we get values of b between 0.175 and 1.998, as shown in Table 5.3. With a record from only a single daughter of each bull, we have

	Value of 'b'			
No. of progeny	$h^2 = 0.1$	$h^2 = 0.25$	$h^2 = 0.35$	$h^2 = 0.5$
1	0.050	0.125	0.175	0.250
5	0.227	0.500	0.648	0.833
10	0.408	0.800	0.979	1.176
100	1.439	1.739	1.811	1.869
1,000	1.925	1.970	1.979	1.986
10,000	1.992	1.997	1.998	1.999

Table 5.3 Values of the multiplier or regression coefficient (b in the formula presented in the text) used to derive PBVs from progeny records for traits with different heritabilities, and for sires with different numbers of progeny.

Average daughter yield deviation (kg milk)	Bull A	Bull B	Bull C
	+500	–500	+300
PBV (in kg milk) if yield deviation is based on single records from the following number of daughters			
1	+87.5	–87.5	+52.5
5	+324.0	–324.0	+194.4
10	+489.5	–489.5	+293.7
100	+905.5	–905.5	+543.3
1,000	+989.5	–989.5	+593.7
10,000	+999.0	–999.0	+599.4

Table 5.4 PBVs of three bulls with average daughter yield deviations of +500 kg, –500 kg and +300 kg milk, assuming that these deviations are produced from 1, 5, 10, 100, 1,000 or 10,000 daughters of each bull, and that the heritability of milk yield is 0.35.

very little information to base our prediction on. This is reflected by the fact that the value of b is low, and so the PBVs are severely 'shrunk' compared to those derived from the same daughter yield deviation based on large numbers of daughters.

The table also shows results for a third bull, Bull C, which has an average daughter yield deviation of +300 kg milk. Comparing results for this bull with those for Bull A illustrates that, even though the average daughter yield deviation is lower, the PBV of Bull C may be higher than that for Bull A when Bull C has more daughters recorded than Bull A (e.g. compare the PBV for Bull A with 10 daughters with that for Bull C with 100 daughters).

Combining information from different types of relatives can become complex. A common method is to produce an index which weights the contributions from different types of relatives according to the class of relative and the number of records available from each class. This approach is described in the next section.

Selection indexes

Selection indexes are used to combine adjusted records of performance in (i) *a single trait* measured on the animal itself, and one or more classes of relative; (ii) *several traits* measured on the animal itself; or (iii) *several traits* measured on the animal itself and on one or more classes of relative. Indexes with more than one trait are termed **multi-trait indexes**.

Combining information on a single trait from relatives

In the first case listed above, the object is to produce a single score for selecting animals, based on information from different types of relative. In this case we can define the index as:

$$\mathbf{I = b_1P_1 + b_2P_2 + b_3P_3 \ldots}$$

where:

I = the index score for an individual animal.

$\mathbf{b_1, b_2, b_3}$ = the weighting factors or **index coefficients** by which the phenotypic measurements $\mathbf{P_1}$, $\mathbf{P_2}$, and $\mathbf{P_3}$ respectively are multiplied. (These **b's** are equivalent to the regression coefficients used earlier to calculate PBVs from records of performance from an individual or averages from groups of progeny.)

$\mathbf{P_1, P_2, P_3}$ = the phenotypic measurement on the animal itself, or the average measurement from different groups of relatives, for the single trait under selection, after adjustment for known environmental effects. For example, if we were selecting for weaning weight in beef cattle or sheep, and had a herd or flock with 100 breeding females and five sires used per annum, we would have three main sources of information: a record of weaning weight for each of the candidates for selection which we call $\mathbf{P_1}$, plus an average weaning weight of the two parents which we call $\mathbf{P_2}$, and an average weaning weight of around 20 paternal half sibs which we call $\mathbf{P_3}$. In this example there are three sources of information, but in practice there may be more or less than this.

The problem is then to derive a set of index coefficients which give appropriate emphasis to each of the three sources of information in order to get the highest possible correlation between the index score for weaning weight (which is a prediction of breeding value) and the animals true breeding value for weaning weight. The problem can be solved using algebra, but we will not go into that here. The relative size of the resulting index coefficients, and thus the emphasis on the different sources of information, depends mainly on: (i) the heritability of the trait under selection: the higher the heritability the greater the emphasis which goes on the animal's own record of performance, and the lower the emphasis on records from relatives; (ii) the class of relative concerned: the closer the relationship, the more emphasis that these records receive; as mentioned before, progeny records are of most value, the value of other records is proportional to the expected proportion of genes in common with the animal being scored and (iii) the number of records available for each class of relative: the more records, the greater the emphasis which is put on the average measurement from this group. Table 5.5 shows some examples of the relative importance of different sources of information when indexes are derived for traits with different heritabilities.

Although this type of index makes optimal use of information from different classes of relatives, it does have drawbacks. The main one is that it does not deal adequately with records from animals reared in different environments. Also, in practice, the animals being evaluated have records available from different combinations of relatives, and they have different numbers of relatives of each type. This means that many different index coefficients have to be calculated and used. More modern (BLUP) methods, which overcome these problems, are discussed in a later section.

Combining information on different traits

In most livestock production systems profitability depends on several different animal characteristics rather than on any single trait. It is important to reflect this in genetic improvement programmes, and animals are usually selected (whether objectively or subjectively) on a combination of traits. This can be achieved in a number of ways, as described in Chapter 3. However, multi-trait index selection is widely agreed to be the most efficient method of improving several traits at once. In this context, an **economic selection index** is used. (This is an extension of the type of index already discussed.) As well as combining information on different traits, this type of index still allows records from the animal itself and from different classes of relatives to be combined into a single score. To derive an economic selection index we need to know:

- Which traits we want to improve – collectively called the **breeding goal** or **breeding objective**.

- The **economic values** of each of the traits in the breeding goal. The economic value of a trait is often defined as the **marginal profit** resulting from a genetic change of one unit in that trait. That is, for example, the increase or decrease in profit resulting from a change of 1 kg in live weight, 1 litre of milk, or 1 mm of backfat compared to the current average value in the herd or flock, with no change in other traits in the breeding goal. Economic values can be calculated from several different perspectives, e.g. with the aim of maximising the profitability of an enterprise for an individual producer, or with the aim of improving the efficiency of a national livestock industry. Often there are heated arguments over which method is most appropriate. (See the references by Amer and Weller in the Further Reading section, which discuss the pros and cons of the different approaches, and the attempts to unify them.) Despite these differences of approach, economic values are usually fairly robust. Also, in practice, it is the *relative* economic values of goal traits (i.e. the value of each goal trait relative to the others) rather than their *absolute* values which affects response most. Relative economic values tend to be particularly robust.

- The set of traits for which measurements will be available for all candidates for selection – called the **index measurements** or **selection criteria**. These may be the same as the traits in the breeding goal (e.g. if the goal traits can all be measured in both sexes, and on the live animal) or they may be different from the goal traits (e.g. if they are only available on one sex, or are measured after slaughter, in which case measurements from relatives or indirect measurements will often be used).

- The additive genetic and phenotypic variances for traits in the breeding goal and the index measurements, and the additive genetic and phenotypic covariances among them.

With this information it is possible to calculate a set of weighting factors or index coefficients which, when used in selection, will maximise genetic progress in overall economic merit. These index coefficients are applied to the measurements

No. of records of erformance on the nimal itself	No. of paternal half sibs with a performance record	No. of progeny with a performance record	% emphasis on record from animal's sire	% emphasis on record from animal's dam	% emphasis on record from animal itself	% emphasis on records from half sibs	% emphasis on records from progeny	Accuracy of PBV
0	0	0	50.00	50.00	0	0	0	0.22
1	0	0	24.32	24.32	51.35	0	0	0.38
1	5	0	14.88	16.60	34.18	34.34	0	0.40
1	10	0	10.72	13.19	26.61	49.47	0	0.42
1	50	0	3.31	7.13	13.15	76.41	0	0.49
1	10	5	5.67	3.14	7.89	32.54	50.76	0.64
1	10	10	4.79	2.66	6.68	27.53	58.34	0.67
1	10	50	2.15	1.19	2.99	12.33	81.34	0.81
1	10	100	1.27	0.70	1.77	7.30	88.96	0.87
1	10	500	0.30	0.17	0.41	1.71	97.41	0.97

(a) low heritability **(h^2 = 0.1)**

No. of records of erformance on the nimal itself	No. of paternal half sibs with a performance record	No. of progeny with a performance record	% emphasis on record from animal's sire	% emphasis on record from animal's dam	% emphasis on record from animal itself	% emphasis on records from half sibs	% emphasis on records from progeny	Accuracy of PBV
0	0	0	50.00	50.00	0	0	0	0.50
1	0	0	20.00	20.00	60.00	0	0	0.76
1	5	0	13.02	17.67	50.70	18.60	0	0.76
1	10	0	9.66	16.55	46.21	27.59	0	0.77
1	50	0	3.15	14.38	37.53	44.94	0	0.78
1	10	5	5.27	3.69	11.85	31.13	48.06	0.88
1	10	10	4.51	3.15	10.14	26.63	55.58	0.90
1	10	50	2.09	1.46	4.70	12.34	79.42	0.95
1	10	100	1.25	0.87	2.81	7.38	87.68	0.97
1	10	500	0.30	0.21	0.67	1.75	97.08	0.99

(b) high heritability **(h^2 = 0.5)**

Table 5.5 Some examples of the approximate emphasis given to different sources of information on an animal and its relatives by selection indexes for a trait with (a) low heritability (h^2 = 0.1), and (b) high heritability (h^2 = 0.5). Accuracies of the resulting PBVs are also shown – see later sections in this chapter for definition and discussion of accuracy. The emphasis shown for half sibs and progeny is the total emphasis on records from this source – the emphasis on each record from a half sib or a progeny can be obtained by dividing this total emphasis by the number of animals of the type concerned. The percentage emphasis in each row does not always add up to 100 because of rounding. (Dr R E Crump; after [7].)

from candidate animals, or their relatives, in order to calculate a single index score for each candidate.

The aim of multi-trait index selection is to maximise the change in breeding value for total economic merit. The breeding value for total economic merit can be expressed as the sum of the breeding values for all the traits in the breeding goal, each weighted by its economic value:

$$\mathbf{BV_{TEM} = v_1BV_1 + v_2BV_2 + v_3BV_3 \ldots}$$

where:

$\mathbf{BV_{TEM}}$ = true breeding value for total economic merit (TEM)

$\mathbf{v_1, v_2, v_3}$ = economic values for breeding goal traits 1, 2 and 3

$\mathbf{BV_1, BV_2, BV_3}$ = true breeding values for breeding goal traits 1, 2 and 3. There are only three goal traits in this case, e.g. carcass weight, carcass fat class and carcass conformation class, but there may be more in comprehensive indexes.

In theory the breeding goal for total economic merit should include all heritable characters which influence profitability. In practice there is often insufficient information on the genetic and phenotypic variances and covariances of all of these to do so, and so only the main characters are included.

The index on which selection is based is the sum of the phenotypic measurements on index traits from the animal itself, or the average measurements from groups of relatives, each weighted by the appropriate index coefficient:

$$\mathbf{I = b_1P_1 + b_2P_2 + b_3P_3 + b_4P_4 + b_5P_5 + b_6P_6 \ldots}$$

where:

I = the index score for an individual animal; unless the index has been rescaled, the index scores are the predicted breeding values for total economic merit, expressed in £, $ or whatever currency the economic values were measured in. However, indexes are often rescaled to make the numbers more manageable, or to make the mean index score equal in different flocks or herds – this reduces the temptation to compare absolute values across flocks if the indexes were not calculated in a way which makes this valid.

$\mathbf{b_1}$ **to** $\mathbf{b_6}$ = the index coefficients which are applied to the phenotypic measurements $\mathbf{P_1}$ to $\mathbf{P_6}$ respectively. These are calculated to make maximum genetic gain in the breeding value for total economic merit, as defined in the equation above. (This equation defining the breeding goal is used solely to allow optimum **b** values to be calculated.)

P_1 to P_6 = the phenotypic measurements on the animal itself, or groups of relatives, for the index traits or selection criteria, adjusted for known environmental effects. In this case there are six different criteria which contribute to the overall index score: e.g. the animal's own live weight (**P_1**), ultrasonic fat depth (**P_2**) and ultrasonic muscle depth (**P_3**), and the average weight (**P_4**), fat depth (**P_5**) and muscle depth (**P_6**) of its half sibs. There may be more or less measurements than this in practice.

As before, the problem in constructing a multi-trait selection index is to find the values of **b_1**, **b_2**, **b_3** and so on, which will maximise the change in breeding value for total economic merit. The mathematical solution to the problem is to set up a series of simultaneous equations and solve them to get the values of the index coefficients. (These equations contain the economic values and the genetic and phenotypic variances and covariances mentioned above). If there are only a few goal traits and a few index measurements, it is feasible to calculate the b values with a pocket calculator. Examples of the type of calculations involved can be found in the references listed at the end of this chapter. However, with more traits it becomes cumbersome to do the calculations this way and it is easiest to solve the large number of equations using a mathematical technique called matrix algebra on a computer. This can be done using software packages such as Mathcad, Matlab, Genstat or SAS (see also the *Proceedings of the World Congresses on Genetics Applied to Livestock Production* for details of more specialised animal breeding software.) Although calculating index coefficients may be difficult, the principles of index selection, actually using the index coefficients and understanding the outcome are all much more straightforward.

A practical example

A practical example of the steps involved in deriving a multi-trait index, and the results obtained, may be useful. In the mid-1980s an index was derived at Lincoln College, New Zealand, to help select for leaner sheep [13].

The first step was to decide on the breeding goal. It was decided that two traits would be included: carcass lean weight and carcass fat weight, both measured at a constant age. The aim was for selection on the index to increase carcass lean weight at this age and to reduce carcass fat weight, or at least limit any further increase in it. (The fairly strong positive genetic correlation between carcass lean and fat weights makes it difficult to increase lean weight without also increasing fat weight at a constant age. However, if lean weight is increasing faster than fat weight, the proportion of fat in the carcass can still be reduced.)

The next step was to calculate economic values for carcass lean and fat weights. At the time, payment for lamb carcasses destined for export from NZ was based on carcass weight and the tissue depth (mainly fat) over the twelfth rib, at a point 11 cm from the midline (the so-called GR site). Payments on this scale were first converted to payments per kg lean and per kg fat by using published information on the carcass lean and fat weights of carcasses of different weight and GR depth. The marginal costs of production of lean and fat, which were mainly feed costs, were then estimated and subtracted from the respective marginal returns. This gave **marginal profits** of NZ$ +5.65 per kg carcass lean, and NZ$ –4.12 per kg fat. In other words, increasing carcass lean weight by 1 kg, compared to the current average, at the same fat weight, was worth $5.65. The minus sign on the marginal profit from increasing fat weight indicates that there is a **marginal loss** from increasing fat weight. Hence, increasing carcass fat weight by 1 kg compared to the current average and with lean weight remaining the same would incur a penalty of $4.12. These marginal profits were used as the economic values in the index. So we can write down the breeding goal as:

$$\mathbf{BV_{TEM} = [5.65 \times BV_{lean\ wt}] + [-4.12 \times BV_{fat\ wt}]}$$

The third step was to decide on index measurements. Since neither of the traits in the breeding goal could be measured directly on the live animal, three index measurements were used as indirect predictors of carcass merit. These were live weight (LW), ultrasonic fat depth (UFD) and ultrasonic muscle depth (UMD), all measured on candidate animals for selection, at a constant age. So the index to be constructed can be written:

$$\mathbf{I = b_1LW + b_2UFD + b_3UMD}$$

The next task was to find appropriate values for $\mathbf{b_1}$, $\mathbf{b_2}$ and $\mathbf{b_3}$ to maximise change in the breeding value for total economic merit. As described above, this requires values of the phenotypic and genetic variances and covariances, as well as the economic values. In this example, average values of phenotypic variances, heritabilities, phenotypic and genetic correlations for traits in the breeding goal and index were obtained from a comprehensive review of the scientific literature, and the variances and covariances required for the index calculations were derived from these. Ideally the variances and covariances should be estimated from the population of animals in which the index will be used. However, in many circumstances this is too expensive and time consuming and so values taken from existing literature have to be used initially. Once sufficient data have been collected on the animals under selection, it makes sense to estimate the required genetic parameters, and to update the index if the new parameters differ from the ones used originally. In this case, the index coefficients calculated to maximise response were +0.10 per kg LW, –0.45 per mm UFD and +0.30 per mm UMD, so we can write the formula to calculate each animal's own index score as:

$$\mathbf{I = [+0.10 \times LW] + [-0.45 \times UFD] + [+0.30 \times UMD]}$$

Although, we have not gone through the calculations in detail here, intuitively the signs on the index coefficients look sensible. We would expect to favour heavier animals and those with larger muscle depths if we want to increase lean weight, so positive weightings for LW and UMD look reasonable. Conversely, we would expect to penalise animals with high fat depths, so the negative weighting on UFD also looks sensible. If there are more traits than this in an index, especially with complex associations between them, it is not always this easy to check that the size and direction of index weights make sense.

Table 5.6 shows the index scores calculated from this formula for several animals with different live weight and ultrasonic measurements. The table illustrates

Animal	Live weight (kg)	Ultrasonic fat depth (mm)	Ultrasonic muscle depth (mm)	Index score (NZ $)
1	55	4	25	11.20
2	60	4	25	11.70
3	65	4	25	12.20
4	65	3	25	12.65
5	65	5	25	11.75
6	65	4	20	10.70
7	65	4	30	13.70
8	65	5	30	13.25

Table 5.6 Index measurements and index scores for eight sheep. Index scores were calculated using the coefficients of +0.10 per kg LW, –0.45 per mm UFD and +0.30 per mm UMD, as described in the text.

that animals with the more favourable combinations of measurements get the highest index scores. It also illustrates that animals which have poor performance in one trait can still get comparatively high index scores if they have excellent performance in other traits. In this particular index there is a lot of emphasis on reducing fat, so it is harder for animals to compensate for high fat than it is to compensate for low weight or muscle depth. The fact that animals can still get a high score without excelling in all components of the index makes it difficult for some breeders to accept index selection as they are looking for *individual animals* which excel in all characteristics. However, index selection is still the most efficient method to genetically improve total economic merit *of the whole herd or flock.*

Further calculations for this particular index showed that including measurements from groups of 10, 20 or 30 half sibs in the index was expected to improve the accuracy of selection by 11.5%, 17.3% and 20.8% respectively, compared to selection on an index with measurements from the individual only. Hence, selection on indexes with these numbers of half sibs should lead to corresponding increases in response in total economic merit, compared to that from selection on the original index. Other examples of selection indexes are given in Chapters 6, 7 and 8.

BLUP

Some shortcomings of the traditional methods of adjusting records of performance and predicting breeding values have already been mentioned. These include:

- PBVs can only be compared fairly for animals which are managed and fed similarly, e.g. within herds or flocks.

- The results from some of the methods for adjusting performance are specific to the herds or flocks for which they were derived, and it may be unwise to apply them more generally.

- Most of the methods of adjusting records run the risk of removing some true genetic differences (e.g. between the progeny of dams of different ages).

- Several methods of adjusting performance make the assumption that animals in different contemporary groups are of equal genetic merit.

- Although conventional indexes combine information from relatives in an appropriate way, they are not very flexible to use, for example, different weighting factors are needed whenever there are different numbers of records from a particular class of relatives. With large numbers of animals and relatives the task of calculating all of the index coefficients becomes very difficult or impossible.

Much of the impetus to produce better methods of evaluating animals arose following the commercial uptake of artificial insemination (AI) in dairy cattle. In Britain, the first cattle AI station was opened at Cambridge in 1942, following research on the technique by Sir John Hammond and his colleagues in the then Cambridge School of Agriculture. Although the introduction of AI was intended initially to reduce the spread of venereal diseases in cattle, and reduce the risks from handling bulls on farms, its potential for assisting the genetic evaluation of dairy bulls and accelerating genetic improvement was soon recognised.

Progeny testing schemes for dairy bulls were introduced in the 1950s, and bulls were evaluated by a method known as the **contemporary comparison (CC)**. Much of the work on statistics and methodology was undertaken by Professor Alan Robertson and his co-workers at the then Institute of Animal Genetics in Edinburgh. The CC system involved comparing the average production of the first lactation daughters of a bull undergoing progeny testing with the average production of the other heifers milked in the same herd, in the same year and in the same season [10]. Deviations in production for daughters of the bull being tested were then combined across herds, after weighting to take into account the number of daughters of the bull undergoing test, and the number of contemporaries against which these were compared. Comparing cows on a within-herd basis recognised the fact that milk production is not only affected by genetic merit but also by management and feeding (environment). However, CCs were based on the assumption that all herds were of equal genetic merit and that, apart from sires and their daughters, other animals were unrelated – two assumptions which were increasingly violated by the wide uptake of AI.

Much of the research to overcome these problems and produce a fairer system of predicting breeding values was initiated by Professor CR Henderson at Cornell University in the United States [4,5]. He first proposed a statistical procedure known as **best linear unbiased prediction**, or **BLUP**, in 1949. Although the workings are complex, and are not described in detail here, BLUP is basically a statistical technique which disentangles genetics from management and feeding in the best possible way, and so produces more accurate predictions of breeding value. It achieves this by:

- Estimating environmental effects (like dam age, season of calving, birth rank) and predicting breeding values simultaneously.

- 'Recognising' that some performance records are from related animals, and so they are expected to be more alike than those from unrelated animals. Related animals in different contemporary groups provide **genetic links** between the groups (see Figure 5.2). These links are necessary in order for BLUP to estimate environmental effects and predict breeding values simultaneously.

BLUP is essentially an extension of the methods used in selection indexes. However, there are several steps involved in conventional index selection: deriving index coefficients, adjusting the records of performance for environmental effects, and then applying the index coefficients to the adjusted records to get PBVs for individual animals. BLUP can achieve these simultaneously, and so it produces PBVs in one step.

How BLUP works

We have already seen that one of the main tasks in deriving a selection index is to calculate the weightings to apply to performance records from different classes of relatives, or to records on different traits. With BLUP there are similar problems in calculating the emphasis that ought to be given to different sources of information when predicting animals' breeding values and in estimating environmental effects.

In both selection indexes and BLUP (as well as in many other applications of statistics), the problem of finding the appropriate emphasis to give to different pieces of information is solved by setting up a series of simultaneous equations. This sounds scary at first, but most of us have solved simultaneous equations at one time or another. If it was at school, then it was probably finding the price of two commodities, given the total price of different quantities of each. For example, if we bought 10 apples and 5 oranges on day one for a total cost of £3.00, we can write this as:

10 apples + 5 oranges = £3.00 [equation 1]

If, on the following day, we bought 5 apples and 15 oranges for £5.25, we can write this as:

5 apples + 15 oranges = £5.25 [equation 2]

One way to find the price of apples and oranges is to multiply both sides of equation 2 by 2, so that it has the same number of apples as equation 1:

10 apples + 30 oranges = £10.50 [equation 2 multiplied by 2]

We can then subtract equation 1 from the new equation, to get a price for oranges alone:

(10 apples + 30 oranges) – (10 apples + 5 oranges) = £10.50 – £3.00

so:

25 oranges = £7.50

and:

1 orange = £0.30

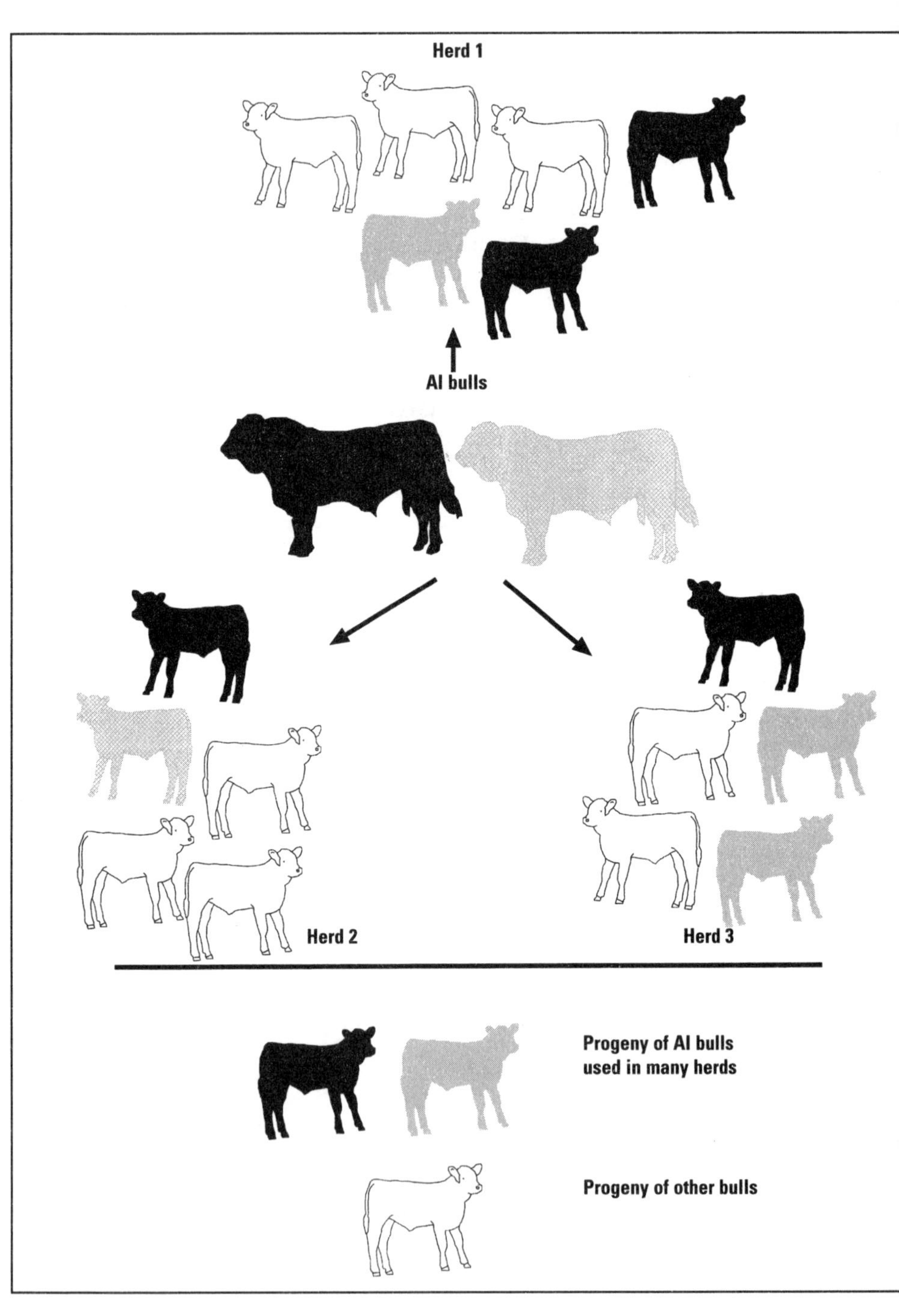

Figure 5.2 Genetic links between contemporary groups are necessary for BLUP to estimate environmental effects and predict breeding values simultaneously. In cattle these links occur most commonly through the use of AI bulls in many different herds [17].

We can then substitute the price of oranges in one of the original equations to get the price of apples. For example, in equation 1 we now know that the 5 oranges cost £1.50, and so:

10 apples = £3.00 – £1.50 = £1.50 and therefore: **1 apple = £0.15**

In this simple example we were able to solve the equations easily just by rearranging them. However, it is possible to solve this sort of equation directly using matrix algebra. In BLUP, the equations to be solved link the performance records to the environmental effects and to the animals whose BVs we are predicting. Instead of calculating prices for apples and oranges from a total amount of money spent, we are estimating environmental effects (e.g. herd, dam age, birth rank) and predicting breeding values from a total amount of live weight or milk produced. For example, Table 5.7 summarises the milk production records from the first lactation daughters of two unrelated dairy bulls. Each bull has some daughters in two different herds, but within each herd the heifers are managed as a single group.

The BLUP equations to solve in order get PBVs for the two sires would be as follows. The abbreviations **S1** and **S2** are used for sires 1 and 2, and **H1** and **H2** are used for herds 1 and 2:

15H1	**+ 0H2**	**+ 5S1**	**+ 10S2**	**= 85,500**
0H1	**+ 18H2**	**+ 12S1**	**+ 6S2**	**= 113,400**
5H1	**+ 12H2**	**+ (17+w)S1**	**+ 0S2**	**= 101,900**
10H1	**+ 6H2**	**+ 0S1**	**+ (16+w)S2**	**= 97,000**

Just as the earlier equations explained the total amount of money spent in terms of the number of apples and oranges bought, these BLUP equations explain the total amount of milk produced in terms of the numbers of daughters' lactations recorded in each herd and from each sire. It is the addition of the term w in the last two equations which distinguishes them from the equations in the fruit example. BLUP requires a slightly modified approach to solve equations, since we are interested in getting predictions of breeding values and not just fixed prices of commodities as in the fruit example. The term **w** is a weighting factor which accounts for the heritability of the trait concerned and for the relationship between the animals who produced the records and those whose BVs are being predicted. In this case, the records are single lactation records from daughters of two unrelated sires, and the weighting factor is $(4–h^2)/h^2$. Note that this term also appeared in the formula on page 157 for predicting the breeding value of a sire based on progeny records. (See [8,14] for the equivalent weighting factors for records from other relatives.) The heritability of milk production is about 0.35, and so the weighting factor in this case becomes approximately 10.43.

Solving these four equations looks like hard work compared with those in the fruit example. However, modern powerful computers can solve vast numbers of these equations very rapidly indeed. Solving these equations and scaling the results appropriately (see later section on scale of presentation) gives predicted breeding values for the two bulls of:

S1 = –175.3 kg milk

	Sire 1	Sire 2	Total
Herd 1	5	10	15
Herd 2	12	6	18
Total	17	16	33

Table 5.7 Summary of milk production records available from the daughters of two bulls, milked in two different herds. (a) Numbers of daughters in each herd.

	Number of heifers	Total milk production (kg)	Average milk production (kg)
Herd 1	15	85,500	5,700
Herd 2	18	113,400	6,300
Sire 1	17	101,900	5,994
Sire 2	16	97,000	6,063

(b) Total and average milk production for each herd and each sire.

and

S2 = +175.3 kg milk.

In this example the two sires were unrelated. However, in practice the animals being evaluated are often related to each other. These relationships are accounted for in BLUP evaluations by creating what is known as a **relationship matrix** – essentially a table showing the expected proportion of genes in common between all the animals being evaluated. A simple example is shown in Table 5.8. In this case there are four bulls being evaluated, A, B, C and D. Bulls A and B are full brothers, so they are expected to have 0.5 of their genes in common, but they are unrelated to the other two bulls. Bulls C and D are half-brothers, so they are expected to have 0.25 of their genes in common. The relationship matrix is used to derive the appropriate weightings to use in the BLUP equations; these weightings are equivalent to the value of **w** used in the example above when the sires were unrelated. The relationship matrix can also be modified to account for inbreeding – related inbred animals have more genes in common than related outbred animals. (See [8, 9, 14] for more details on the mechanics of BLUP.)

BLUP PBVs have to be expressed relative to a base. When relationships between animals are included in BLUP evaluations, the pedigrees are usually traced right back until a group of animals is reached whose parents are unknown. Animals with unknown parents are termed **base population** animals. These base population animals are assumed to have an average PBV of zero, and the PBVs of all other animals are expressed relative to the base population. So, for example, cows with higher genetic merit for yield than the average of base population animals would have positive PBVs, and those with lower merit for yield would have negative PBVs. Often, base populations are made up of groups of animals of different average genetic merit, for example if they originate from different countries. This can cause bias in predicting breeding values, but it can be accounted for by assigning base animals to different **genetic groups** [8].

The problem with expressing PBVs relative to the merit of distant ancestors is that, if genetic progress is being made, the average genetic merit of base population

Bull	A	B	C	D
A	1.0	0.5	0	0
B	0.5	1.0	0	0
C	0	0	1.0	0.25
D	0	0	0.25	1.0

Table 5.8 An example of a simple relationship matrix, showing the degree of relationship between four bulls. An animal's relationship to itself is 1.0 (i.e. it has all its genes in common with itself), the relationship between full sibs (bulls A and B) is 0.5, and that between half sibs (bulls C and D) is 0.25. (In this example, none of the animals are inbred).

animals could be a lot lower than that of the *current* breeding population. This may make the current animals appear much better than they really are. There are two main solutions to this. One is to express PBVs relative to a more recent group of animals – for example those born in a recent year – and to set their average merit equal to zero. This is called a **fixed base**, if the definition of base animals remained unchanged for a few years. Fixed bases are usually updated every few years, as they too can become outdated if genetic progress is high. An out-of-date fixed base can be quite misleading because many animals which have positive PBVs are, in fact, well below the average genetic merit of the current breeding population. Fixed bases are used in beef, sheep and dairy cattle evaluations in many countries. The alternative is a **rolling base**. This changes each year or at each new evaluation so, for example, the current calf crop may be set to have PBVs averaging zero. A rolling base avoids the problem of highlighting animals of low merit, but the frequent changes in absolute values of PBVs can lead to confusion. Rolling bases are used, for example, in dairy cattle evaluations in Canada and France.

BLUP models

BLUP can be applied under different sets of assumptions called **models**, which differ in sophistication. The most common BLUP models are: (i) **sire models**; (ii) **sire-maternal grandsire models** and (iii) **individual animal models**. The name of the model indicates the animals for which BVs are predicted, and the relationships used to predict them. So, sire models predict BVs for sires from their progeny records, sire maternal grandsire models predict BVs for sires using records from both their progeny and maternal grand-progeny, and individual animal models predict BVs for all animals included in the evaluation, using all relationships among them.

To use a motoring analogy, most cars serve the same function, but they do so with different levels of performance. In this analogy the performance of the different BLUP models is the accuracy with which they predict the animal's true breeding value. The more sophisticated BLUP models recognise more relationships between animals, and hence predict BVs more accurately. The sire model is equivalent to a basic model of family car, the sire-maternal grandsire models are middle of the range models, and the individual animal model is the top of the range, fuel-injected, 16 valve model. However, as with cars, there is a cost for higher performance. An equation has to be solved for each animal which gets a PBV. In national genetic evaluations there may be hundreds or thousands of sires, and so there are hundreds or thousands of equations to solve in a sire model BLUP evaluation. But there may be millions of individual animals, and so millions of equations to solve in an individual animal model BLUP evaluation. As a result, one of the main factors governing the uptake of more sophisticated models has been the cost and availability of computing power.

As mentioned above, the impetus for new methods of evaluation came primarily from dairy cattle breeding. Sire-model BLUP evaluations were first used in dairy cattle in the US in the early 1970s, and in several other countries soon afterwards. For example, BLUP evaluations were introduced in the UK dairy industry in 1979. Advances in computing allowed a sire-maternal grandsire model to be used in the UK from the start. This improved the accuracy of evaluations compared with those from sire-model BLUP by recognising that not all cows to which test bulls were mated were of equal merit. However, the sire-maternal grandsire model only accounted for differences in the merit of cows via their sire (the maternal grandsire of heifers whose records were being evaluated, hence the name of this model). The emphasis was still on predicting BVs for sires – indexes were produced for cows by combining their sire's and maternal grandsire's PBVs with their own production records. These difficulties were overcome with the introduction of individual animal-model BLUP evaluations in the UK in 1992. Many other countries introduced animal-model BLUP evaluations for dairy cattle around this time.

Although BLUP evaluations were first used in dairy cattle, the benefits are also relevant to other species. Over the last couple of decades BLUP methods have become the methods of choice for evaluating most farm livestock species. Because performance-recorded populations of beef cattle and sheep are usually smaller than those of dairy cattle, and because BLUP methods have been adopted later in these groups, it has often been possible to skip several generations of BLUP model and adopt animal model evaluations from the start. For example, in Britain, BLUP evaluations were used in some sheep breeding schemes for the first time in 1990, and they were used in beef cattle for the first time in 1991. In both cases, individual animal models were used from the start. More details are given in the following chapters.

Direct and maternal genetic effects

There are some traits of interest in livestock breeding where the performance of offspring is affected not only by their own genes and environment but also by their mother's genes and environment. For example, the weaning weight of a suckled beef calf is affected not only by the genes for growth which the calf inherits, and the environment it gets, but also by its mother's genes for maternal characteristics such as uterine capacity and milk production, and the environmental influences on her performance in these traits (see Figure 5.3). So, it is often useful to think of direct and maternal breeding values for traits such as weaning weight which are influenced by both direct and maternal genetic merit. Both males and females carry genes for milk production and other aspects of maternal performance, but only females get the chance to express them. However, with sufficient data, BLUP can predict both direct and maternal breeding values for males and females, from the relationships between animals with performance records (e.g. cows with calf weaning weights) and those without (e.g. bulls) (see Plate 22).

The concept of separating direct and maternal genetic effects appears rather complicated at first sight, but it is really just an extension of the principles we have

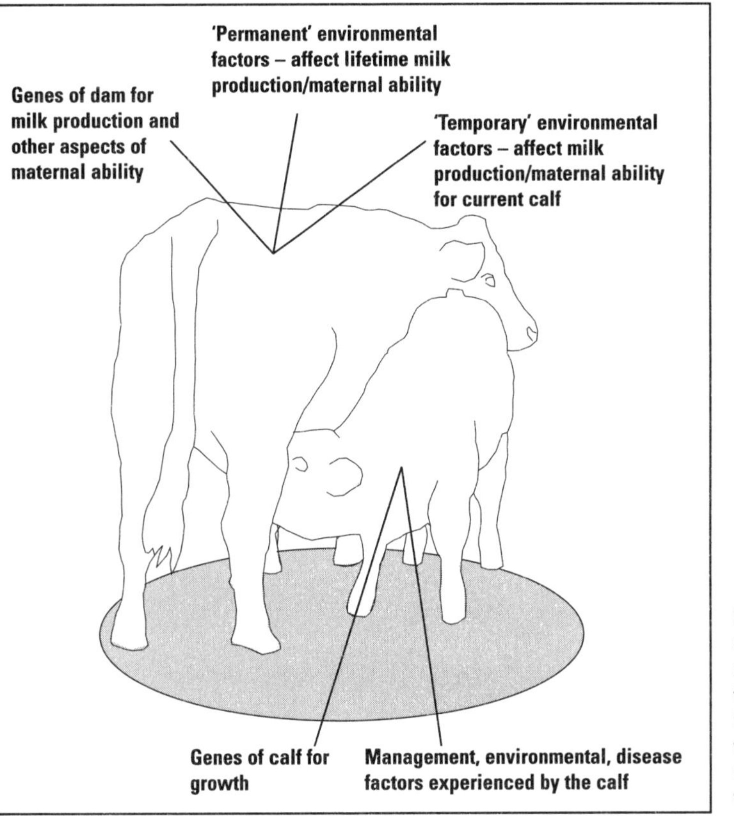

Figure 5.3 Some of the genetic and environmental factors which influence calf weaning weights, either directly or via the dam [18].

already considered. We can predict the breeding value of dairy bulls for milk traits based on their daughters' milk production records but, because we cannot easily record the milk yield of beef cows or non-dairy sheep, and because we are interested in other aspects of maternal performance as well as milk yield, we have to rely on the early growth or weaning weights of offspring to assess maternal genetic merit. Many different types of relative can provide clues to an animal's genes for maternal performance, but the principle is easiest to understand by considering the performance of grandoffspring of a bull, produced by his daughters, as a means to assess his maternal genetic merit. These calves get, on average, a quarter of their genes for growth, directly from their grandsire. But their performance is also influenced by the fact that their dams received half of the grandsire's genes for milk production and maternal performance. In contrast, calves sired by sons of the bull we are interested in are only influenced by the genes they inherited for growth (again, a quarter of their genes on average come from their grandsire). If, on average, grandoffspring from daughters of the original bull have heavier weaning weights than grandoffspring from sons of the original bull, then this is likely to be a consequence of that bull carrying genes for high maternal performance. Conversely, if grandoffspring from daughters of the original bull have lower weaning weights than grandoffspring from his sons, then this is likely to be a result of that bull carrying genes for low maternal performance.

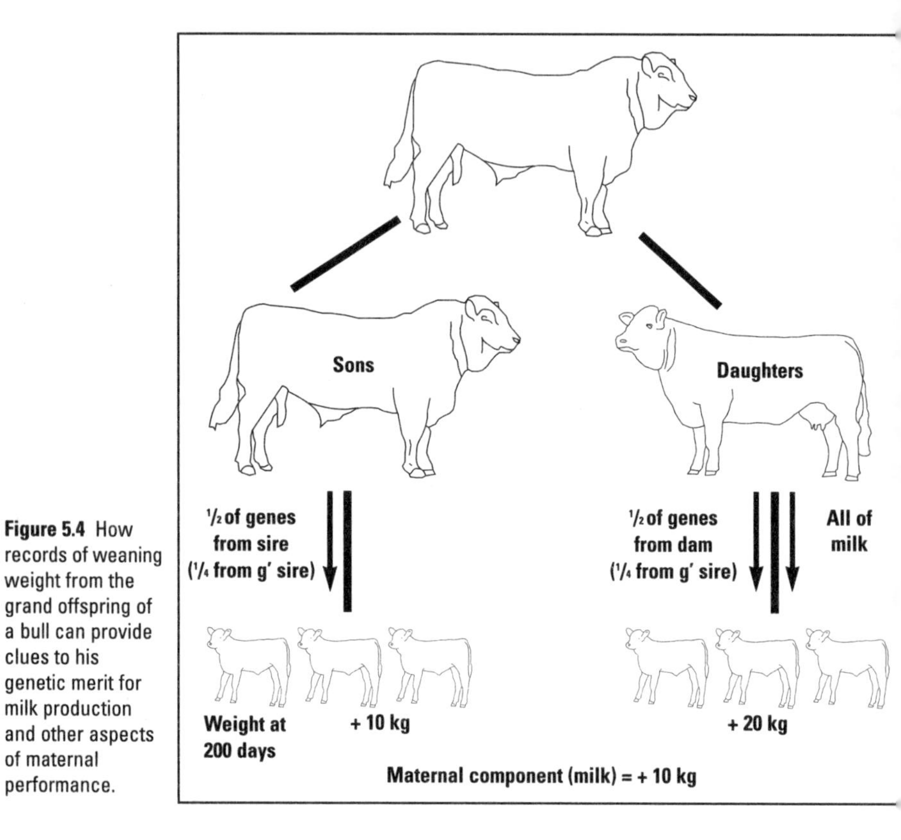

Figure 5.4 How records of weaning weight from the grand offspring of a bull can provide clues to his genetic merit for milk production and other aspects of maternal performance.

This is illustrated in Figure 5.4. By comparing these two types of descendants of a sire – those which were influenced by the maternal performance of his female relatives and those which were not – BLUP can predict breeding values for both direct and maternal components of weaning weight or other traits influenced by both direct and maternal genetic merit. (See Chapter 7 for a practical example.)

Single and multi-trait BLUP

Initially most BLUP evaluations were for one trait at a time (**single-trait evaluations**). However, as mentioned earlier in this chapter, when there is a correlation between two or more different traits a record of performance in one trait can help to predict an animal's breeding value in other traits. Hence the increasing use of **multi-trait BLUP evaluations**. However, because these evaluations are even more demanding on computing resources than single-trait BLUP, they have been used more widely to date in smaller beef cattle and sheep evaluations than in larger national dairy cattle evaluations. As well as the usual information on relationships between animals, and estimates of the heritabilities (or variances) of the traits to be evaluated, multi-trait BLUP evaluations require estimates of the correlations (or covariances) between all of the traits included.

Benefits of BLUP

As mentioned earlier, there must be genetic links between contemporary groups for BLUP to be able to estimate environmental effects and predict breeding values simultaneously. Providing these links exist:

- The effects of environment and breeding will be separated more effectively by BLUP than with traditional methods which first adjust records of performance for environmental effects and then predict breeding values. With traditional methods genetic merit and environment can easily be confounded, e.g. corrections for age of dam may ignore the fact that genetic improvement is being made. As a result, younger animals of higher merit may be undervalued. Similarly, the use of sires of different genetic merit in different contemporary groups would be overlooked with traditional methods. Also, the use of sires of different merit at different times during the mating season could result in some true genetic differences being treated inadvertently as seasonal effects. In BLUP evaluations, the more relationships between animals that are recognised, the greater the chance of overcoming these problems.

- The resulting PBVs can be compared directly across different contemporary groups. This means that PBVs can be compared across age groups, so allowing more accurate sequential selection or culling, and removing the need to decide in advance on the optimum age structure in the herd or flock.

- Similarly, PBVs can be compared across herds and flocks, as long as these are genetically linked via related animals. This has a major effect on the selection intensities which can be applied (see Plate 23). The benefits of selection across herds or flocks are particularly high when AI is used, as in most dairy cattle industries, because very high male selection intensities can be achieved. However, across-herd or across-flock evaluations are still valuable in industries with less widespread use of AI. This is particularly true in countries like Britain where pedigree herds and flocks are often small, and so within-herd or within-flock selection intensities are low. Similarly, PBVs can be compared across years, as long as there are genetic links between years – that is, related animals which are recorded in successive years (e.g. progeny from the same sires or dams). Plots of PBVs across time show the estimated genetic trend which is occurring in a particular herd, flock or national population, for the traits evaluated. These trends allow pedigree breeders to monitor the success of their breeding programmes, and they allow commercial buyers of rams or bulls to select from breeds or breeding schemes which are making genetic progress. Several examples of estimated genetic trends are given in the following chapters.

The relationships between some animals have to be identified and included in BLUP evaluations in order to achieve the benefits above. However, BLUP evaluations increasingly involve all relationships, i.e. they are animal model evaluations.

There are additional benefits here:

- Including performance information for all relatives in predicting breeding values increases the accuracy of selection, and hence response to selection. An example of this is shown in Table 5.9. This particular example is from pigs, and the benefits vary from one species to the next depending on the typical numbers of relatives of different classes with performance records. However, these results give an indication of the increase in accuracy which can be obtained. Note that this example shows only the benefits of including information from all relatives. These benefits could also be achieved by using a very comprehensive selection index, incorporating information from many different classes of relatives, but it is more practical to incorporate this information using BLUP. Further benefits, not measured here, would result from the better separation of genetic and environmental effects which BLUP achieves.

- When all relationships between animals are included, and all the traits under selection are being evaluated, BLUP can account for the fact that selection reduces the amount of variation in a trait. It can also account for non-random matings. For example, depending on the perceived merit of a young sire, he may be mated to females of lower or higher than average merit in a herd or flock, which will clearly influence the performance of his progeny compared to the progeny of sires mated to average females. Animal model BLUP can account for this, and so produce unbiased PBVs.

Using economic values with BLUP

Both multi-trait selection indexes and multi-trait BLUP use information on the associations between traits to produce more accurate predictions of breeding value. Multi-trait economic selection indexes then go one stage further, combining

Information used, and method of predicting breeding value	Heritability			
	0.1	0.2	0.4	0.6
Record from individual only	100 (0.30)	192 (0.43)	356 (0.59)	511 (0.72)
Records from individual and full brothers and sisters in optimum index	113 (0.35)	208 (0.47)	365 (0.62)	517 (0.73)
Records from individual and all relatives in BLUP	158 (0.50)	244 (0.56)	390 (0.60)	526 (0.75)

Table 5.9 Predicted rates of genetic improvement with various sources of information and methods of predicting breeding values. Rates of improvement are expressed in percentage units relative to that expected from selection on a record of performance on individual animals, for a trait with a heritability of 0.1 (= 100). Accuracies of selection are shown in brackets. The results are from a simulated pig breeding scheme with 10 males and 100 females, with records from three offspring of each sex per annum, and a generation interval of one year. ([16] and Dr N R Wray.)

the PBVs for breeding goal traits into a single score based on their economic values. We have already seen that BLUP has several advantages over conventional indexes in predicting breeding values. How do we capitalise on these and yet still get a PBV for total economic merit? There are basically two approaches, both based on the multi-trait selection index methods outlined earlier. In the first case, if we have multi-trait BLUP PBVs for all the breeding goal traits we are interested in, we can simply multiply these by their economic values and add them together to get a PBV for total economic merit [12]:

$$\mathbf{PBV_{TEM} = v_1PBV_1 + v_2PBV_2 + v_3PBV_3 \ldots}$$

This formula is very similar to that presented on page 162 as the breeding goal for multi-trait indexes. If we do not have multi-trait BLUP PBVs for the breeding goal traits (e.g. carcass traits), but we do have them for other correlated traits (e.g. ultrasonic measurements), then an intermediate step is required. This involves deriving PBVs for goal traits from the PBVs for indicator traits, using the genetic correlation between the two traits. The newly calculated PBVs for breeding goal traits can then be weighted by their economic values as above.

Getting estimates of heritabilities and correlations

All the methods for predicting breeding values discussed in this chapter require estimates of the heritabilities (or genetic and phenotypic variances) of the traits being evaluated, and some require estimates of the correlations (or genetic and phenotypic covariances) among traits. It is particularly important that accurate and relevant estimates of these parameters are used in indexes and BLUP evaluations, so that the emphasis is apportioned properly among traits, and among records of performance from the animal itself and various classes of relative. Accurate estimates of heritabilities and correlations are also important if the estimates of genetic trends from BLUP are going to be meaningful.

Estimation of variances, covariances, heritabilities and correlations is based on measuring the degree of similarity in performance between relatives. For example, heritabilities can be estimated from regression of offspring on parent performance, or from the covariance in performance among half sibs. In the past, parameters were usually estimated from designed experiments involving these particular classes of relatives. However, there are now more sophisticated methods of estimating variances and covariances, and hence heritabilities and genetic correlations. These use similar methodology to that outlined earlier for BLUP to account for the similarity in performance between many different classes of relatives and to separate genetic and environmental influences on performance in the best possible way. As a result, they can be used to estimate parameters from field records, as well as from designed experiments. (The former usually contain records from many different types of relatives and these records are often spread unevenly across many herds or flocks, seasons etc. which complicates analysis.) For more details on parameter estimation, and some of the computer packages available to do this, see references listed in the Further Reading section, especially Hill and Mackay, Cameron and the *Proceedings of the World Congresses on Genetics Applied to Livestock Production*.

Scale of presentation of genetic merit

In all the examples discussed so far the PBVs referred to the genetic merit of the animals being evaluated. The beef cattle and sheep genetic evaluation schemes in many countries express genetic merit on this scale. This can cause confusion among bull or ram buyers, because when they breed from the sire they buy they can expect to see only half the PBV in terms of extra progeny performance, as only half the genes of the progeny come from the sire. However, dairy cattle producers are used to thinking about the merit of a bull in terms of the extra milk produced by his daughters, i.e. genetic merit expressed in terms of the merit of offspring, rather than the merit of the animal being evaluated. This difference in the way genetic merit is visualised is understandable, as dairy bulls do not produce milk themselves, whereas many of the traits of interest in beef cattle and sheep can be recorded on both sexes. Hence, the results of dairy cattle evaluations in some countries (and sheep and beef evaluations in a few) are expressed as **predicted** or **estimated transmitting abilities** (**PTAs** or **ETAs**) or **expected progeny differences** (**EPDs**). These three terms mean exactly the same thing, but the preferred name varies between countries. PTAs, ETAs or EPDs are simply one half of the PBV (or EBV) for the trait concerned, reflecting the fact that offspring receive half their genes from each parent. So, it is easy to move from one scale to the other, but it is vitally important to know which scale you are working on! (Unless they are rescaled, the results of sire-model BLUP evaluations are expressed as PTAs, while those from animal model evaluations are expressed as PBVs. Hence, the solutions to the equations in the example on pages 169-70 of the two dairy sires used in two herds were multiplied by 2 to put them on the PBV scale.)

To summarise:

PTA (or ETA, or EPD) = ½PBV (or ½EBV)

The expected genetic merit of offspring can be calculated from the PTAs or ETAs of parents in exactly the same way as it can from PBVs or EBVs, as shown in Figure 5.1. That is:

PTA of offspring = ½PTA of sire + ½PTA of dam

However, there is a very important difference in the way the two scales of measuring genetic merit are used to predict the expected difference in *performance* of offspring, as opposed to the *genetic merit* of offspring. If genetic merit is expressed on the BV scale, the expected difference in performance of offspring, compared to using parents of average merit (i.e. PBVs of 0) is:

expected difference in performance of offspring
= ½PBV of sire + ½PBV of dam

If the PBV of the dam is unknown, for example when a bull or ram is being used on crossbred commercial females, the dams are assumed to be of average genetic merit, and so have a PBV of 0. Since half of 0 is 0, the contribution from the dam is ignored and:

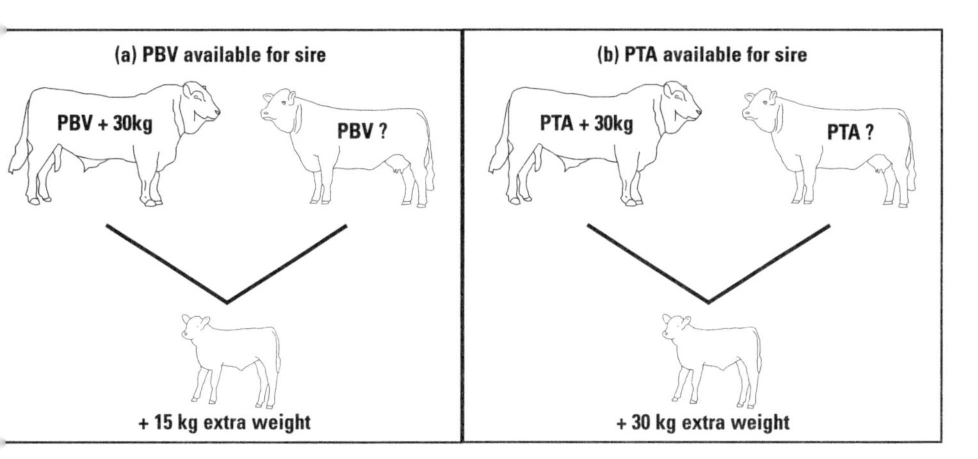

Figure 5.5 (a) Calculating the expected difference in offspring performance when PBVs (or EBVs) are available on sires. **(b)** Calculating the expected difference in offspring performance when PTAs (or ETAs or EPDs) are available on sires. In this example the PBVs or PTAs are for 400-day weight, and the difference in weight of offspring is recorded at 400 days of age.

expected difference in performance of offspring = ½PBV of sire

For example, if a bull with a PBV of +30 kg for 400-day weight is used across a herd of commercial suckler cows, his calves are expected to weigh 15 kg more at 400 days than those sired by an average bull i.e. a bull with a PBV of 0 kg (see Figure 5.5(a)).

In contrast, if genetic merit is expressed on the transmitting ability scale, the merit of parents is already expressed in terms of the expected difference in performance of offspring, and so there is no need to halve the PTAs of parents:

expected difference in performance of offspring
= PTA of sire + PTA of dam

If PTAs are only available for the sire (the most common case) then the dams are assumed to have PTAs of zero, and so:

expected difference in performance of offspring = PTA of sire

This is illustrated in Figure 5.5(b). Similarly, if semen from a dairy bull with a PTA of +20 kg protein is used in a herd of cows without PTAs, our best guess is that his daughters will yield 20 kg more protein per lactation than daughters of a bull with a PTA of 0 kg protein.

Accuracies and reliabilities

Whatever method of genetic evaluation is used, it is useful to know just how accurate individual PBVs are. In statistical terms, accuracies are correlations between predicted and true breeding values, and they range from 0 to 1. However, they are often expressed as percentages by multiplying the correlation by 100. High accuracies indicate that the PBVs are expected to be close to the true BVs, low accuracies indicate that the association between PBVs and true BVs is weaker.

There are several factors that affect the accuracies of PBVs:

- the heritability of the trait concerned – the higher the heritability of the trait, the higher the accuracy

- the amount of information on the trait concerned on the animal itself – the more information, the higher the accuracy

- the amount of information on the trait concerned on relatives of the animal – the more information, and the more useful the class of relative in predicting breeding values, the higher the accuracy (as before, progeny are the most valuable class of relatives; others contribute in proportion to the genes they have in common with the animal concerned)

- if it is a multi-trait evaluation, the amount of information on related traits on the animal itself and its relatives – the more information, and the higher the correlation with the trait of interest, the higher the accuracy

- the number of contemporaries recorded – the more contemporaries recorded, the higher the accuracy

Generally the accuracies of PBVs for an individual animal start off quite low when it is young. For example, the PBV of a young dairy bull or maiden heifer might be based exclusively on its parents' PBVs. Young beef calves or lambs might have an additional measure themselves, e.g. a birthweight, but their PBV would still be based mainly on ancestors' performance. The accuracies increase as records of performance from the animal itself and from its collateral relatives are included in the evaluation. For instance, milk records from sisters would contribute to the PBVs of young dairy bulls; heifers would have information from their sisters plus their own milk record. Likewise beef cattle and sheep would have records of their own plus those from sibs. Accuracies increase further once performance records are available from progeny and other descendants. However, as indicated earlier, there is a trade-off between accuracy and generation interval. In most cases waiting for many progeny to be recorded in order to get very accurate PBVs increases generation intervals, and can reduce annual responses to selection.

The principle behind genetic evaluation in general, and multi-trait animal model BLUP evaluation in particular, is to use all available information on the animal and its relatives to give the most accurate prediction of breeding value possible at that time. Repeating genetic evaluations quite frequently allows new performance information to be included, producing more accurate PBVs. Accuracies also give an indication of the chance that future PBVs will differ from current ones. Low accuracies indicate that there is a high chance that the PBV will change, e.g. when extra performance records become available on the animal itself or its relatives. High accuracies indicate that the PBV is already based on a substantial amount of performance information, and that new information is unlikely to lead to big changes in PBVs. This is illustrated in Figure 5.6, which shows the possible range of true BVs for animals with PBVs of different accuracies. Typical levels of accuracy achieved in dairy and beef cattle evaluations are given in Chapters 6 and 7.

An important feature of BLUP PBVs is that they are already scaled to account for the amount of performance information on which they are based. PBVs based on very little information get adjusted or 'shrunk' back (regressed) towards the mean PBV. This shrinking applies both to high and low PBVs. In other words, it is difficult to get either a very high, or a very low PBV, on the basis of little information. The more information available on an animal and its relatives, the less the PBVs are shrunk, and the easier it is to get very high or very low PBVs. This feature of BLUP is a very valuable way of accounting for the risk involved in breeding decisions. It helps to avoid selecting animals which initially appear to have high genetic merit on the basis of flimsy evidence, and which might turn out later to be much poorer than expected. The chance of an individual animal's BLUP PBV going down is exactly the same as the chance of it going up, but obviously breeders are more concerned about selecting an animal which turns out to be worse than expected than about selecting an animal which turns out to be better than expected.

Because BLUP PBVs already account for accuracy, many scientists argue that it is irrelevant to look at the accuracy any further; rates of genetic progress in the trait concerned will be highest if selection is based solely on BLUP PBVs. These PBVs have already been adjusted to balance the risk against the potential gain in

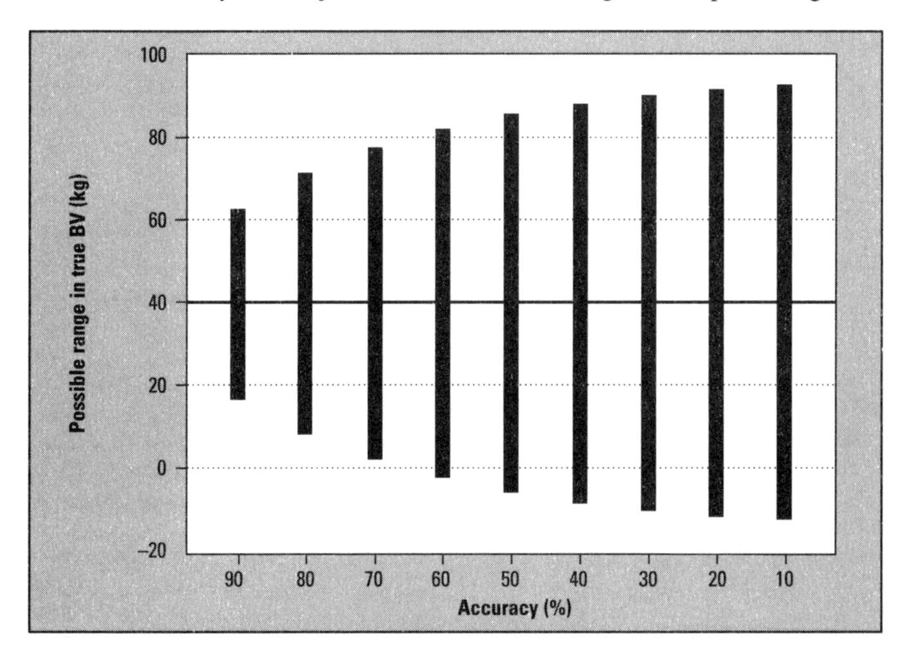

Figure 5.6 Possible range of true breeding values for a bull with a PBV for 400-day weight of +40 kg with various levels of accuracy. In this example, there is a 90% chance that the true breeding value lies in the range shown. (Dr RE Crump, see [14] for more details on methodology.)

order to identify animals which are the 'best bet' overall. Breeders who take any further account of accuracies may reduce the risk but, on average, would also reduce the genetic gain. While this is true, accuracies are still valuable in allowing breeders to assess the risk involved in the purchase of individual animals, and in interpreting changes in an individual animal's PBVs over time.

Accuracies are often presented alongside PBVs from beef cattle and sheep evaluations. However, most dairy cattle PBVs or PTAs are accompanied by **reliabilities** or **repeatabilities**. Reliabilities and repeatabilities of PBVs or PTAs are simply squared accuracies. In other words a PBV with an accuracy of 0.5 would have a reliability of 0.25 (0.5 x 0.5), and a PBV with an accuracy of 0.9 would have a reliability of 0.81 (0.9 x 0.9). The relationship between accuracy and reliability is shown in Figure 5.7.

The fact that two different measures are used with the same aim is a bit confusing, and it is difficult to say with certainty why these different conventions have been adopted. Because of the widespread use of AI and progeny testing in dairy cattle, most bulls end up with a very accurate PBV or PTA. The risk of PBVs changing is identical for a bull with an accuracy of 0.9 or a reliability of 0.81. But the fact that reliabilities give lower numbers than accuracies may deter people from using all but the most reliably tested animals. So the convention of using reliabilities for dairy cattle evaluations probably arose to give particular emphasis to widely proven bulls and to reduce the emphasis on young bulls and other bulls with a limited proof. Generally beef cattle and sheep evaluations are based on fewer performance records and so they are less accurate than those for dairy bulls. Adopting reliabilities, rather than accuracies, could well be self-defeating, as beef and sheep breeders may be deterred from using the majority of animals. Also, accuracies make it a bit easier to discriminate among animals when they all have relatively little information available, simply because the scale of accuracies is wider than the scale of reliabilities 'at the bottom end' (i.e. when animals' PBVs are based on relatively little information).

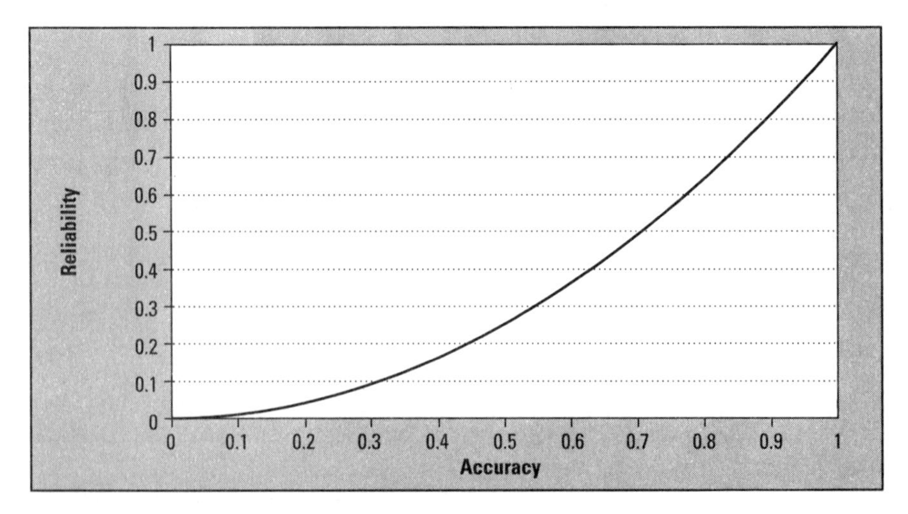

Figure 5.7 The relationship between accuracy and reliability.

International conversions and evaluations

Cattle breeding and sheep breeding have become increasingly international businesses over the past few decades – aided largely by the development of techniques for artificial insemination and semen freezing and, more recently, by the development of equivalent procedures for embryos. At the same time there has been an increase in awareness of the value of objective performance recording and associated genetic evaluation of animals. This is particularly true in dairy cattle breeding, and much effort has gone into developing procedures to allow breeders in one country to make use of genetic evaluations of bulls in other countries. Much of the groundwork in this area has been done under the guidance of an organisation called INTERBULL – a subcommittee of the International Committee for Animal Recording, based at the Swedish University of Agricultural Sciences in Uppsala.

Where there is likely to be an active international trade in genetic material, it is logical to consider comparing evaluations across national boundaries. In the past this has usually been achieved by deriving formulae to convert PBVs or PTAs calculated in the exporting country to equivalent PBVs or PTAs in the importing country. However, there is now growing interest in combining records from several countries into single international evaluations. Some of the issues involved in the two approaches are discussed below.

International conversions

International conversion formulae are established using the regression methods outlined in Chapter 2. Essentially this involves plotting the PBVs or PTAs of a group of bulls, derived from progeny tests in the exporting country, against their PBVs or PTAs derived from progeny tests in the importing country. The slope of the line passing through these points on the graph, and the point at which the line crosses an axis (the intercept) are then used to derive converted PBVs or PTAs for bulls from the exporting country which have yet to get a progeny test in the importing country. Although conversion formulae are relatively simple to derive and use in principle, in practice there are many factors which affect their feasibility and reliability. The most important of these are discussed in the next section. An example of a conversion formula is given in the section after that.

Factors affecting feasibility and reliability

Perhaps the most fundamental issue affecting the feasibility and reliability of international conversions is the extent to which there are genetic links between any two countries. In practice this usually means: are there enough sires which have been in recent widespread use in each of the two countries? INTERBULL guidelines for dairy cattle suggest that a minimum of 20 sires born within a 10-year period, and with very reliable progeny tests (reliability greater than 75%) in both countries are required in order to derive conversion factors [6]. Clearly, accurate and unique identification of animals is very important if reliable international conversion factors are to be derived or evaluations performed.

Where there are insufficient sires used in common between any two countries to obtain 'direct' conversion factors, it may be possible to obtain 'indirect' conversions via a third country. For example, when US Holstein dairy bulls were used in the UK in the mid-1980s there were too few sires used in both of these countries to derive reliable conversions. But there were already good links between the UK and Canada, and between Canada and the US. So, US to UK conversions were obtained in two steps by first converting US to Canadian PTAs, and then converting Canadian to UK PTAs. While this might be a useful temporary measure, indirect conversions are inferior to direct conversions. Planned exchanges of semen may be needed in order to establish these.

Another important issue to be addressed is whether or not the same traits are being recorded and evaluated in the two countries of interest. In dairy cattle there is a great deal of similarity in the milk traits recorded in different countries, and this has greatly assisted international conversions (i.e. most countries now record milk, fat and protein yields, and fat and protein proportions). However, there is also a great deal of interest in the shape, size and other conformation measures of a bull's daughters. Over the last few decades these have been scored in many countries using a system known as **linear type classification**. Basically, this involves inspecting cows, and scoring their appearance in a number of characteristics in relation to the biological extremes for the trait concerned. For example, in most countries there are several udder traits which are scored, udder depth, the strength of attachment of the udder, and teat length and placement (see Chapter 6 for more details). In the past there was greater disparity in the traits involved in linear type classification, but this is changing as a result of moves towards global harmonisation of classification. As a result the reliability of international conversions for these traits is increasing. Although live weight recording is a fairly standard feature of most beef and sheep recording schemes, the wide array of other traits recorded creates difficulties for international onversions. A pragmatic approach is to start with the traits recorded in common across countries, or those which are highly correlated to the traits of most interest, and work towards standardisation, at least in those traits of international importance.

The issue of which traits are recorded may be more complex than it first seems. This is because what is apparently the same characteristic on paper may be affected by different genes if the animals are reared in very different environments. For example, a beef bull may have very good genetic merit for 400-day weight in a country with fairly intensive management, but might have much poorer genetic merit for 400-day weight under extensive range conditions. Similarly dairy bulls with high PBVs for milk production in intensive environments do not always rank highly when tested in extensive environments. This is referred to as a genotype x environment interaction, as explained in Chapter 3. This needs to be checked before embarking on conversions between two countries, by comparing the evaluations of a team of bulls with reliable progeny test results (**proofs**) in the two countries, for what is apparently the same trait. The correlation between evaluations in the two countries will give a measure of the value of international conversions to the importing country in identifying bulls likely to rank highly in a domestic evaluation. The fact that genotype x environment interactions are absent or relatively small for milk production traits among most temperate countries has allowed the widespread and effective use of international conversion formulae to date.

Different countries use different procedures for evaluation of animals, and this can complicate the process of converting PBVs. Generally speaking, the greater the similarity in the procedures used, the more accurate the conversion of breeding values from the two countries will be. Also, the more accurate the evaluations are in the home country, the more accurate the converted PBVs will be. There is widespread agreement that animal model BLUP is the most accurate procedure available for estimating breeding values. This is already widely used in dairy cattle evaluations, and many major beef-producing countries and several major sheep-producing countries have already adopted, or plan to adopt this. The heritability of the trait concerned, the size of the contemporary group in which the animal was recorded, and the amount of information available on other traits, or from relatives, also influence the accuracy of evaluations. INTERBULL recommendations are that the reliability should exceed 50% before converted dairy proofs are published.

Some countries present PBVs or PTAs in the units in which the trait was measured, while in others results are scaled in some way, e.g. by expressing them relative to 100, or in standard deviation units. The first of these methods is the simplest to deal with if conversions are to be made. As discussed earlier, there is variation between countries and between species in whether results are presented on the breeding value or transmitting ability scale. Clearly it is important to know which scale is being used, and to account for any differences. There is also variation in the type and date of the genetic base used, and this must also be accounted for.

In addition to those listed above, several other factors affect the accuracy of conversions. For example, the bulls used to derive conversions may have been used on the best cows in the country (e.g. as a result of high semen price) or their offspring may have received preferential treatment. Also, the non-genetic factors accounted for in the evaluations may differ between countries. Some of these factors are difficult to account for, and they may explain some of the anomalies in the use of conversions.

An example

Table 5.10 gives some examples of formulae used recently to convert foreign Holstein Friesian proofs to UK PTAs. Foreign proofs are converted to UK proofs by multiplying them by the b value shown in the table and then adding the a value. So, a US bull with a US evaluation of +1500 lbs milk, +55 lbs fat and +45 lbs

Country	Trait	a	b
USA	milk (lbs)	450	0.30
	fat (lbs)	10.3	0.26
	protein (lbs)	10.5	0.27
	fat %	–0.16	0.99
	protein %	–0.08	0.98
Netherlands	milk (kg)	319	0.44
	fat (kg)	14.4	0.40
	protein (kg)	11.0	0.39
	fat %	0.02	0.46
	protein %	0.00	0.43

Table 5.10 Some examples of the 'a' and 'b' values for formulae used recently to convert foreign Holstein Friesian proofs to UK PTAs. The formulae are applied by multiplying the foreign proof by 'b', and then adding 'a' to get a UK PTA95 [1].

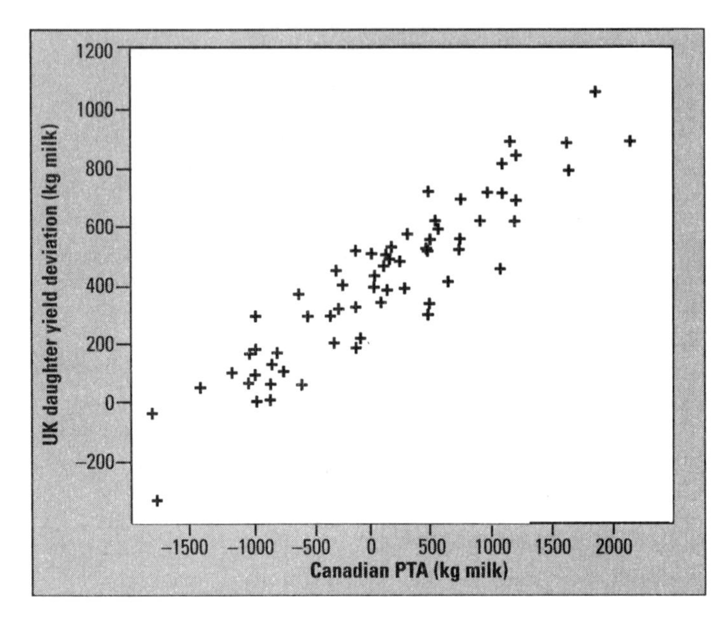

Figure 5.8 Plot of Canadian daughter yield deviations for milk against UK PTAs for milk (PTA95) for the 64 Holstein Friesian bulls with proofs in both countries which were used to derive the Canadian to UK conversions in February 1997. The a and b values produced from these data were 390 and 0.27 kg milk respectively (ADC).

protein would have the following UK converted proof:

(**+1500** x **0.30**)	**+ 450**	=	**+900 kg milk**
(**+55** x **0.26**)	**+ 10.3**	=	**+24.6 kg fat**
(**+45** x **0.27**)	**+ 10.5**	=	**+22.65 kg protein**

Figure 5.8 shows the good association between Canadian daughter yield deviations for milk and UK PTAs for milk for the Holstein Friesian bulls with proofs in both countries which were used to derive the February 1997 Canadian to UK conversions.

International evaluations

Conversion formulae such as those described above are based on sires' proofs in two countries – the foreign and the importing country. As a result, they have to be derived separately for each foreign country of interest. Also, when conversion formulae are used as shown above, the ranking of sires will be identical in the foreign and the importing countries. In practice many sires have proofs in several countries, and their rankings may differ between countries.

To surmount these problems, and make better use of the international information available on a sire, a procedure has been developed recently to allow **international evaluations**. This is called **multi-trait across country evaluation**, or **MACE**, and it is based on the multi-trait BLUP method described earlier [11]. However, rather than evaluating many different traits at once, MACE treats records on the same trait from different countries as if they were different traits. This allows for differences among countries in the heritability of the trait of interest (e.g. milk production), and allows for different genetic correlations between this trait recorded in different countries. MACE uses average daughter yield deviations from sires in different countries (these can be derived approximately from the sires' local PTAs, the numbers of daughters contributing to these PTAs and the heritability used locally), rather than individual daughter records, to reduce the computing demands and ensure that the records coming from different countries have been adjusted for the non-genetic effects known to be important in those countries. (When data sets are smaller, or if computing power allows it, international evaluations could be based on records from individual animals.) Evaluations are produced for each sire, for each country. The emphasis on information from different countries and the reliability of the MACE evaluation of any sire in any participating country depends on the number of daughters he has in each of the participating countries and the correlation between the trait being evaluated in the local country and other countries in which he has daughters.

The feasibility and reliability of international evaluations is affected by many of the factors which affect international conversions. In particular, good genetic links are needed between participating countries. Additionally, MACE requires estimates of the genetic correlations between the trait being evaluated in the participating countries. Examples of the genetic correlations between production in the UK and some other countries are shown in Table 5.11. The correlations are all fairly high. However, the fact that there are particularly high correlations between the production traits in the UK and the corresponding traits in Ireland, the Netherlands, France and Denmark shows that daughter information in these countries is of proportionally greatest value in producing UK MACE proofs. This is probably largely a result of the similarity of climate and production systems in these countries. Conversely, the correlations between production traits in the UK and the corresponding traits in New Zealand and Australia were the lowest, probably reflecting differences in climate and production systems.

There are several potential benefits from the use of MACE. Firstly it produces separate PTAs for sires for each country involved in the evaluation; these are expressed on the local scale and are ready to use. Secondly, because they use information from daughters in all the countries in which they are present, MACE evaluations are more accurate than the conversions they replace. Thirdly, because MACE allows different values for genetic correlations between a trait in different countries, the rankings of sires evaluated can differ among countries. This is particularly useful where there is a genotype x environment interaction, e.g. for the same milk production trait in countries with very different management systems. (The correlations in Table 5.11 indicate that the greatest likelihood of re-ranking of sire proofs between the UK and the other countries shown will be with New Zealand.) The possibility for different rankings among countries is also useful where the traits being recorded in different countries are similar, but not identical (e.g. some linear type traits). These benefits are also seen in cows when MACE evaluations on male relatives are used in calculating cow PTAs.

The INTERBULL Centre now produces MACE evaluations routinely for milk, fat and protein yields. At the time of writing, the evaluations were based on data from nineteen countries. Evaluations were produced for Ayrshire, Brown Swiss, Guernsey, Holstein and Jersey sires. Some of the countries involved use the INTERBULL international evaluations in place of domestic evaluations, others use them until a sire has a reliable local proof, and others do not use them at all. However, it is clear that as MACE evolves further, and the results are more widely used, it will improve the accuracy of selection of dairy bulls (and

	Genetic correlation with corresponding trait in UK		
Country	Milk (kg)	Fat (kg)	Protein (kg)
Australia	0.84	0.84	0.84
Austria	0.89	0.89	0.88
Belgium	0.89	0.89	0.89
Canada	0.94	0.93	0.93
Czech Republic	0.90	0.90	0.90
Denmark	0.93	0.93	0.93
Finland	0.91	0.91	0.91
France	0.93	0.93	0.93
Germany	0.90	0.90	0.90
Ireland	0.95	0.95	0.95
Italy	0.91	0.91	0.91
Netherlands	0.94	0.93	0.93
New Zealand	0.82	0.82	0.81
Spain	0.87	0.87	0.87
Sweden	0.91	0.91	0.91
Switzerland	0.89	0.89	0.89
USA	0.91	0.91	0.91

Table 5.11 Estimates of the genetic correlations between milk, fat and protein production in the UK and the same traits in several other countries. These estimates were produced by INTERBULL for the February 1997 MACE evaluations of Holstein Friesians. (INTERBULL; ADC.)

potentially other breeds/species) internationally, for a wide range of traits.

Conversion formulae are also produced by INTERBULL from the data used for international evaluations. These are more accurate than the conversion formulae produced solely from data from two countries. They are useful for producing converted proofs for animals which have not been included in the international evaluation, or as an interim measure when a foreign country produces a local evaluation between INTERBULL evaluations.

Summary

- There are a number of steps involved in selection of animals to maximise the response achieved. These include: (i) managing the animals as equally as possible so as to make it easier to disentangle genetics and environment; (ii) adjusting records of performance for known environmental effects and then (iii) predicting the breeding values of individual animals by the most appropriate method. Modern (BLUP) methods of genetic evaluation achieve the second and third of these steps simultaneously,

- Records of performance may be adjusted using additive or multiplicative correction factors, or by standardising. Additive and multiplicative correction factors are the simplest to use, but both have to be derived from prior analysis of large data sets. Standardising records is more complex, but requires no prior information. Using multiplicative correction factors or standardising records may be most appropriate when the size of the effect being adjusted for is related to the level of performance of animals either within or between herds or flocks. Each of the methods can separate genetic and environmental effects poorly in some circumstances, e.g. when sires have been used unequally across different groups of animals. This type of problem can be reduced or overcome by simultaneous estimation of environmental effects and prediction of breeding values using BLUP methods.

- Predicting an animal's breeding value is a bit like completing a large, complicated jigsaw puzzle, where each piece of the puzzle is a record of performance from the animal itself or one of its relatives. Generally, the higher the proportion of genes in common between the animal and a given relative, the more useful the record of performance from that relative. However, records from progeny are of most value. As the number of records on progeny increases, the correlation between predicted and true breeding values approaches 1. With other classes of relatives the accuracy of prediction never reaches 1, and for all classes of relatives there are diminishing returns in accuracy as the number of records increases.

- Predicted breeding values (PBVs) can have positive or negative values, or be equal to zero. The sign indicates whether they are expected to be genetically above (+) or below (–) the mean of the group of animals on which the calculations were performed, or some other defined group of animals whose PBVs are set to average zero (the base). PBVs are expressed (at least initially) in the same units as the record of performance (e.g. kg of live weight, litres of milk, mm of fat). The PBVs of animals can only be compared within contemporary groups, herds or flocks, unless there are genetic links between these groups and the PBVs were from across-herd or across-flock BLUP evaluations.

- In the simplest case, when we have a single record of performance on the animal itself, the predicted or estimated breeding value (PBV or EBV) is the deviation in performance from contemporaries, multiplied by the heritability of the trait concerned. The deviation in performance is calculated after adjusting the performance records for environmental effects. PBVs calculated from a single record of performance span a narrower range than the deviations in performance – the higher the heritability, the lower the proportion of non-genetic variation and so the less severe the shrinking.

- Calculating PBVs from the performance of relatives is more difficult. The simplest case is when the only records available are from the parents. If we first calculate PBVs for each parent, then the PBV of their offspring is calculated by simply adding half the PBV of the sire to half the PBV of the dam. PBVs calculated from ancestor's PBVs are known as pedigree indexes. They are very valuable in providing an early prediction of an animal's genetic merit.

● PBVs can be derived from a *single record of performance* on any relative, by multiplying the PBV of the relative by the proportion of genes in common with the animal without a PBV.

● If we have more than one of any other type of relative apart from parents, we cannot simply average their PBVs, because progeny or sibs, for example, have genes in common with each other as well as with as the animal whose PBV is being derived, and so there is overlap in the information they provide. In these cases PBVs are calculated from the average deviation in performance of the progeny or sibs from their contemporaries, multiplied by a factor which depends on the class of relative concerned, the number of records available, and the heritability of the trait concerned.

● Combining information from different types of relatives is more complex. The most common method is to produce an index which, again, weights the contributions from different types of relatives according to the class of relative, the number of records available from each class, and the heritability of the trait concerned.

● In most livestock production systems profitability depends on several different animal characteristics rather than on any single trait. Index selection is also used to combine adjusted records of performance in several traits measured on the animal itself, or several traits measured on the animal itself and on one or more classes of relatives.

● To derive a multi-trait economic selection index we need to know: (i) which traits we want to improve (the breeding goal or objective); (ii) the economic values of each of the traits in the breeding goal; (iii) the set of traits for which measurements will be available for all candidates for selection (the index measurements or selection criteria); (iv) the additive genetic and phenotypic variances for traits in the breeding goal and the index measurements, and the additive genetic and phenotypic covariances among them. With this information it is possible to calculate a set of index coefficients which, when used in selection, will maximise genetic progress in overall economic merit.

● Adjusting records of performance and then predicting breeding values in two separate stages has several shortcomings: (i) PBVs can only be compared fairly for animals which are managed and fed similarly e.g. within herds or flocks; (ii) the results from some of the methods for adjusting performance are specific to the herds or flocks for which they were derived; (iii) most of the methods run the risk of removing some true genetic differences; (iv) several methods of adjusting performance make the assumption that animals in different contemporary groups are of equal genetic merit and (v) although conventional indexes combine information from relatives in an appropriate way, they are not very flexible to use, e.g. different weighting factors are needed whenever there are different numbers of records from a particular class of relatives.

- BLUP is a statistical technique which disentangles genetics from management and feeding in the best possible way, and so produces more accurate predictions of breeding value. It achieves this by estimating environmental effects and predicting breeding values simultaneously. BLUP 'recognises' that some performance records are from related animals. Related animals in different contemporary groups provide genetic links between the groups. These links are necessary in order for BLUP to estimate environmental effects and predict breeding values simultaneously.

- In both selection indexes and BLUP the problem of finding the appropriate emphasis to give to different pieces of information is solved by setting up a series of simultaneous equations. In BLUP, the equations to be solved link the performance records to the environmental effects, and to the animals whose BVs we are predicting. BLUP equations also include weighting factors which account for the heritability of the trait concerned and for the relationship between the animals who produced the records and those whose BVs are being predicted.

- PBVs are expressed relative to a group of animals whose average merit is set equal to zero, e.g. ancestors with unknown parents, or animals born in a specific year. The latter is called a fixed base. Fixed bases are usually updated every few years, as they can become outdated if genetic progress is high. The alternative is a rolling base which changes each year, or at each new evaluation, so that animals born in a recent year are set to have PBVs averaging zero. A rolling base avoids the problem of highlighting animals of low merit, but the frequent changes in absolute values of PBVs can lead to confusion.

- BLUP can be applied under different sets of assumptions – called models – which differ in sophistication. The most common BLUP models are: (i) sire models; (ii) sire-maternal grandsire models and (iii) individual animal models. The name of the model indicates the animals for which BVs are predicted, and the relationships used to predict them. The more sophisticated BLUP models recognise more relationships between animals, and hence predict BVs more accurately. An equation has to be solved for each animal which gets a PBV. As a result, one of the main factors governing the uptake of more sophisticated models has been the cost and availability of computing power.

- It is often useful to think of direct and maternal breeding values for traits such as weaning weight which are influenced by both direct and maternal genetic merit. Both males and females carry genes for milk production and other aspects of maternal performance, but only females get the chance to express them. However, with sufficient data, BLUP can predict both direct and maternal breeding values for males and females from the relationships between animals with performance records (e.g. cows with calf weaning weights) and those without (e.g. bulls).

- BLUP evaluations can be performed for one trait (single-trait evaluations) or several traits (multi-trait evaluations) at a time. When traits are correlated, multi-trait evaluations produce more accurate PBVs than single-trait evaluations. However, multi-trait evaluations are more demanding on computing resources than single-trait BLUP, and require accurate estimates of covariances among traits.

- There are several benefits from using BLUP. As long as there are genetic links among contemporary groups in different herds or flocks, and across years, BLUP separates genetic and environmental effects more effectively than traditional methods, and the resulting PBVs can be compared directly across different contemporary groups, and hence across age groups, herds, flocks, or years. Plots of PBVs across time show the estimated genetic trend which is occurring. The relationships between some animals have to be identified and included in BLUP evaluations in order to achieve these benefits. However, when BLUP evaluations involve all relationships there are additional benefits. Including performance information for all relatives in predicting breeding values increases the accuracy of selection, and, when all the traits under selection are being evaluated, BLUP can account for the fact that selection reduces the amount of variation in a trait, and it can also account for non-random matings.

- If we have multi-trait BLUP PBVs for all the breeding goal traits we are interested in, we can simply multiply these by their economic values and add them together to get a PBV for total economic merit. If we only have PBVs for other correlated (indicator) traits, we have to derive PBVs for goal traits from the PBVs for indicator traits, using the genetic correlation between the two traits.

- The predicted genetic merit of animals is expressed on one of two scales. PBVs refer to the genetic merit of the animals being evaluated. But predicted or estimated transmitting abilities (PTAs or ETAs) or expected progeny differences (EPDs) express genetic merit in terms of the expected superiority or inferiority in the performance of progeny. PTAs, ETAs or EPDs are one half of the PBV (or EBV) for the trait concerned. The predicted genetic merit of offspring is simply the average of the PBVs or PTAs of parents. However, the method of predicting differences in offspring performance varies depending on the scale on which genetic merit is expressed. If parents' genetic merit is expressed as PBVs, we *average* these to get the expected difference in offspring performance, but if parent genetic merit is expressed as PTAs we *sum* these to get the expected difference in offspring performance.

● Accuracies are correlations between predicted and true breeding values. High accuracies indicate that the PBVs are expected to be close to the true BVs, low accuracies indicate that the association between PBVs and true BVs is weaker. The accuracies of PBVs depend on the heritability of the trait concerned, the amount of information on the trait concerned on the animal itself and its relatives, the amount of information on related traits on the animal itself and its relatives (if it is a multi-trait evaluation), and the number of contemporaries recorded. Accuracies also give an indication of the chance that future PBVs will differ from current ones. Low accuracies indicate that there is a high chance that the PBV will change when new information becomes available, and vice versa. Reliabilities or repeatabilities of PBVs or PTAs are squared accuracies.

● International trade in genetic material has created much interest in comparing evaluations across national boundaries. In the past this has usually been achieved by deriving formulae to convert PBVs or PTAs calculated in the foreign country to equivalent PBVs or PTAs in the importing country. There are several disadvantages to this approach. Conversion formulae have to be derived separately for each foreign country of interest. Also, when conversions are used, the ranking of sires will be identical in the foreign and the importing countries. In practice many sires have proofs in several countries, and their rankings may differ between countries.

● To surmount these problems, and make better use of the international information available on a sire, a procedure has been developed recently to allow international evaluations. This is called multi-trait across country evaluation (MACE). MACE treats records on the same trait from different countries as if they were different traits. This allows for differences among countries in the heritability of the trait of interest, and allows for different genetic correlations between this trait recorded in different countries.

● MACE has several potential advantages: (i) it produces separate PTAs for sires for each country involved in the evaluation, and these are expressed on the local scale; (ii) because they use information from daughters in all the countries in which they are present MACE evaluations are more accurate than conversions and (iii) because MACE allows different values for genetic correlations between a trait in different countries, the rankings of sires evaluated can differ between countries (this is particularly useful where there is a genotype x environment interaction, or where the traits being recorded in different countries are not identical).

References

1. Animal Data Centre. 1997. *UK Statistics for Genetic Evaluations (Dairy). February 1997.* ADC, Rickmansworth, Herts.
2. Brotherstone, S. and Hill, W.G. 1986. 'Heterogeneity of variance amongst herds for milk production'. *Animal Production*, 42:297–303.
3. Crump, R.E., Simm, G., Nicholson, D. et al. 1997. 'Results of multivariate individual animal model genetic evaluations of British pedigree beef cattle.' *Animal Science*, 65:199–207.
4. Henderson, C.R. 1973. 'Sire evaluation and genetic trends'. *Proceedings of the Animal Breeding and Genetics Symposium in Honor of J.L. Lush.* American Society for Animal Science, Blackburgh, Champaign, Illinois, pp. 10–41.
5. Henderson, C.R. 1975. 'Best linear unbiased estimation and prediction under a selection model.' *Biometrics*, 31:423–47.
6. INTERBULL. 1990. *Recommended procedures for international use of sire proofs.* INTERBULL Bulletin No. 4, International Bull Evaluation Service, Department of Animal Breeding and Genetics, Uppsala, Sweden.
7. Johansson, I. and Rendel, J. 1968. *Genetics and Animal Breeding.* Oliver and Boyd, Edinburgh.
8. Mrode, R.A. 1996. *Linear Models for the Prediction of Animal Breeding Values.* CAB International, Wallingford.
9. Nicholas, F.W. 1987. *Veterinary Genetics.* Oxford University Press, Oxford.
10. Robertson, A., Stewart, A. and Ashton. E.D. 1956. 'The progeny assessment of dairy sires for milk: The use of contemporary comparison.' *Proceedings of the British Society of Animal Production 1956*, pp. 43–50.
11. Schaeffer, L.R. 1994. 'Multiple-country comparison of dairy sires.' *Journal of Dairy Science*, 77:2671–8.
12. Schneeberger, M., Barwick, S.A., Crow, G.H. and Hammond, K. 1992. 'Economic indices using breeding values predicted by BLUP.' *Journal of Animal Breeding and Genetics*, 109:180–7.
13. Simm, G., Young, M.J. and Beatson, P.R. 1987. 'An economic selection index for lean meat production in New Zealand sheep.' *Animal Production*, 45:465–75.
14. Van Vleck, L.D., Pollak, E.J. and Oltenacu, E.A.B. 1987. *Genetics for the Animal Sciences.* W.H. Freeman and Company, New York.
15. Wiggans, G.R. and VanRaden, P.M. 1991. 'Method and effect of adjustment for heterogeneous variance.' *Journal of Dairy Science*, 74:4350–7.
16. Wray, N.R. 1991. 'The estimation of genetic merit.' *Summaries of Papers, British Society of Animal Production, Winter Meeting*, March 1991, paper No. 32.
17. Wray, N.R. and Simm, G. 1991. *BLUP for pedigree beef and sheep breeders.* SAC Technical Note T265. The Scottish Agricultural College, Perth.
18. Wray, N.R., Simm, G., Thompson, R. and Bryan, J. 1991. 'Application of BLUP in beef breeding programmes.' *British Cattle Breeders' Club Digest No. 46*, pp. 29–39.

Further reading

Amer, P.R. 1994. 'Economic theory and breeding objectives.' *Proceedings of the 5th World Congress on Genetics Applied to Livestock Production*, Vol. 18, pp. 197–204.

Banos, G. 1994. 'International genetic evaluation of dairy cattle.' *Proceedings of the 5th World Congress on Genetics Applied to Livestock Production*, Vol. 17, pp. 3–10

Cameron, N.D. 1997. *Selection Indices and Prediction of Genetic Merit in Animal Breeding*. CAB International, Wallingford.

Falconer, D.S. and Mackay, T.F.C. 1996. *Introduction to Quantitative Genetics*. 4th edn. Longman, Harlow.

Harris, D.L. and Newman, S. 1994. 'Breeding for profit: synergism between genetic improvement and livestock production (a review).' *Journal of Animal Science*, 72: 2178–200.

Hazel, L.N. 1943. 'The genetic basis for constructing selection indexes.' *Genetics*, 28:476–90.

Hill, W.G. and Mackay, T.F.C. (eds). 1989. *Evolution and Animal Breeding. Reviews on Molecular and Quantitative Approaches in Honour of Alan Robertson*. CAB International, Wallingford.

INTERBULL Bulletins. International Bull Evaluation Service, Department of Animal Breeding and Genetics, Uppsala, Sweden.

Mrode, R.A. 1996. *Linear Models for the Prediction of Animal Breeding Values*. CAB International, Wallingford.

Nicholas, F.W. 1987. *Veterinary Genetics*. Oxford University Press, Oxford.

Proceedings of the World Congresses on Genetics Applied to Livestock Production. (Especially volumes covering prediction of breeding values, estimation of genetic parameters, computing strategies and software, e.g. Vols. 18 and 22 in the 1994 *Proceedings of the 5th World Congress*.)

VanRaden, P.M. and Wiggans, G.R. 1991. 'Derivation, calculation and use of national animal model information.' *Journal of Dairy Science*, 74:2737–46.

Van Vleck, L.D., Pollak, E.J. and Oltenacu, E.A.B. 1987. *Genetics for the Animal Sciences*. W.H. Freeman and Company, New York.

Weller, J.I. 1994. *Economic Aspects of Animal Breeding*. Chapman and Hall, London.

CHAPTER 6

Dairy cattle breeding

Introduction

The first five chapters in this book concentrated on the scientific principles of genetic improvement. The aim of the next four chapters is to give some more details of the current and possible future applications of these principles in practical cattle and sheep breeding. Chapters 6 to 8 follow a similar format. They begin by examining breeding goals in a general way to identify animal characteristics likely to be important in genetic improvement programmes. Next they examine breeds and crosses in widespread use and methods of selection within breeds, including systems of testing, traits recorded and methods of genetic evaluation. Finally, each chapter has a section on the evidence for genetic improvement and its value and a list of practical guidelines on selection.

Most of the examples of dairy cattle breeding discussed in this chapter are from the UK, elsewhere in Europe, New Zealand, Canada and the US. However, they are intended to illustrate issues of wide relevance in dairy cattle breeding, at least in temperate areas. To show the context into which production in these countries fits, Figure 6.1 shows the world production of cow milk by continent in 1995. Europe had the highest production, accounting for about 40% of global milk production. Figure 6.2 shows the amount of milk produced in 1995 by the fifteen highest producing countries in the world. The United States was by far the largest producer, but the Russian Federation, India, several European countries, New Zealand and Australia featured in the top fifteen.

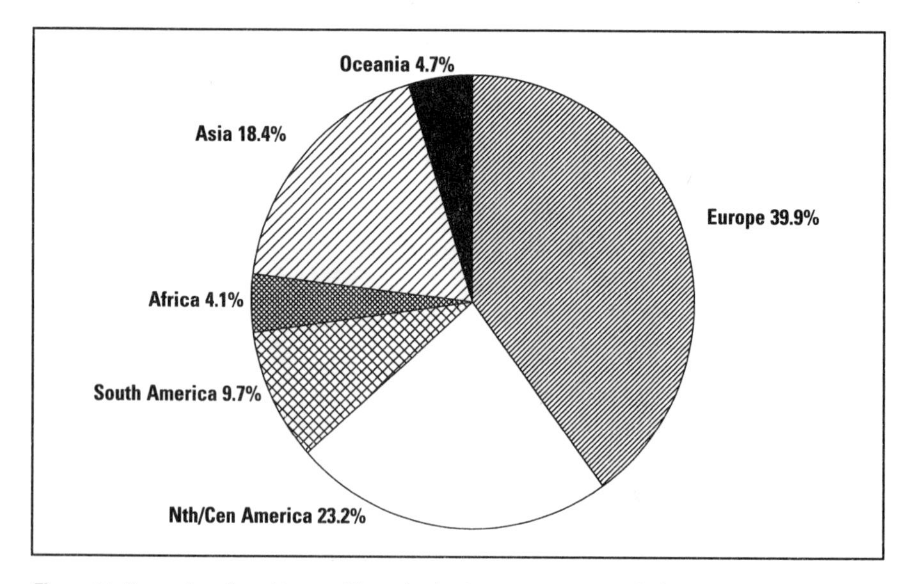

Figure 6.1 Proportion of world cow milk production by continent in 1995 [14].

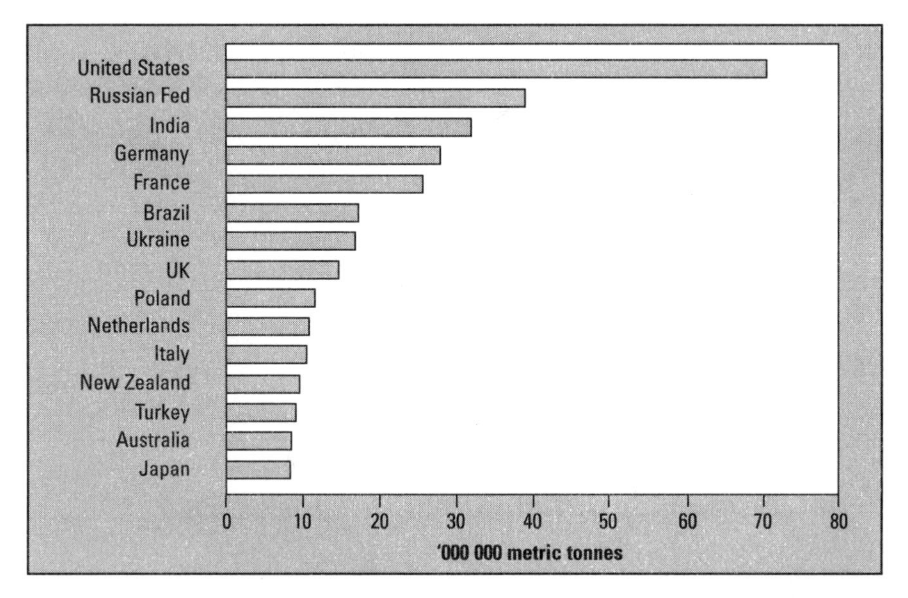

Figure 6.2 Production of cow milk by the fifteen highest producing countries in the world in 1995 [14].

Breeding goals

In broad terms, the breeding goal of most livestock producers is to increase the profitability of their animals. Many breeders and producers would add that this should be achieved without detriment to the animals' health and welfare. While there may be broad agreement on this aim, there is far less agreement on what the main components of profitability are, and how to improve them most effectively. Figure 6.3 shows the main sources of returns and the main variable costs in relatively high input and relatively low input dairying systems in Britain. While these figures do not tell the whole story as far as profitability is concerned, they do highlight some of the key animal characteristics that influence it. Also, although the absolute values of the different inputs and outputs shown varies between countries, there is greater similarity in their relative importance. For example, both milk prices and production costs will be substantially lower in countries such as New Zealand selling at world market prices and with predominantly pastoral production (see Plate 24). The results show that:

- returns from milk production are by far the most important returns and are of similar relative importance in both high and low input systems

- returns from the sale of calves and cull cows are relatively low in both systems, but slightly more important in low than in high input systems

- feed costs are by far the most important variable costs in both high and low input systems, although the relative importance of concentrate versus forage costs differs substantially between systems

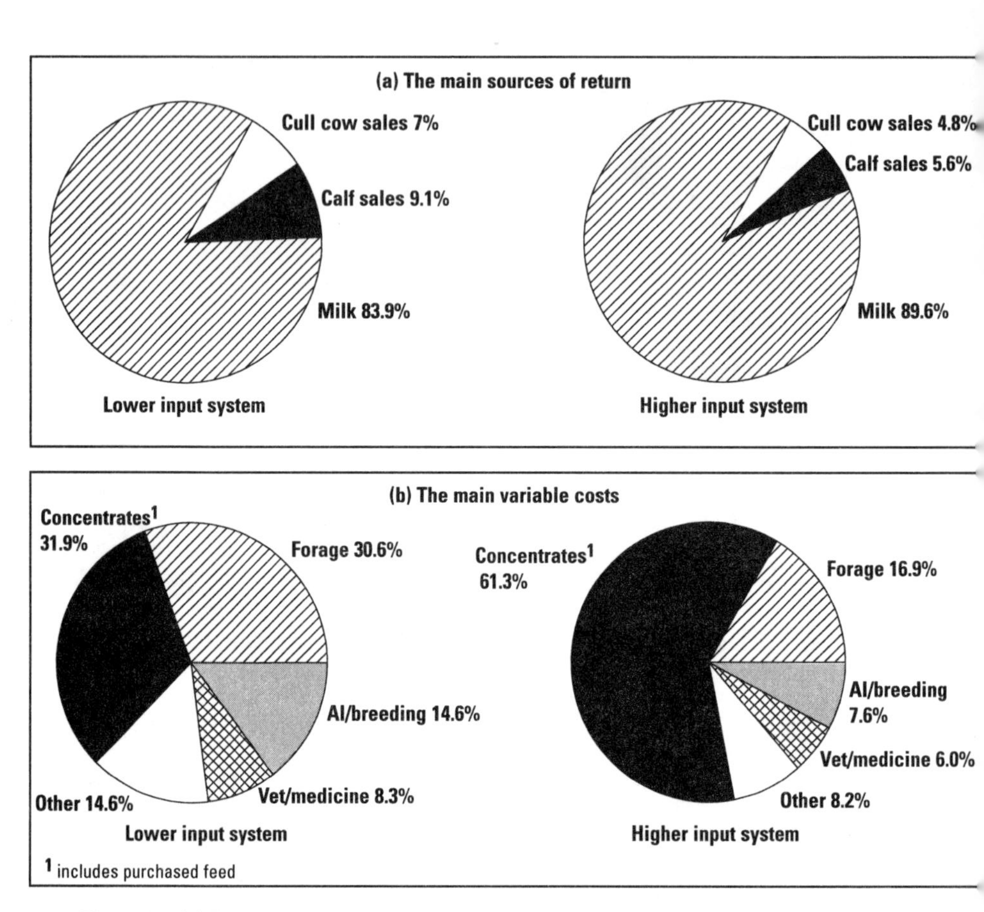

Figure 6.3 (a) The main sources of return in relatively low input and relatively high input dairying systems in Britain and (b) the main variable costs in relatively low input and relatively high input dairying systems in Britain. The returns and costs are based on a lower input system with spring calving, an average yield of 5,000 litres milk per cow per annum and 500 kg concentrate feed per cow per annum, and a higher input system with autumn calving, an average yield of 7,800 litres milk per cow per annum and 2,300 kg concentrate feed per cow per annum. An annual share of heifer replacement costs (about £250 in this case) would normally be deducted from the returns [54, 56].

- veterinary costs account for a relatively low proportion of the variable costs, although there are other hidden associated costs, such as lost production due to disease itself, or due to milk withdrawal e.g. following antibiotic treatment

For most of the last few decades yield of milk has been the main objective criterion for selecting between and within dairy breeds in most temperate countries. Figure 6.4 shows the approximate change in yield per cow in Britain over the last 70 years or so. By the 1920s and 1930s yield per cow was about double that produced by a cow suckling a calf. The current average yield is about five- to sixfold higher than that required by a calf. These changes have been brought about by a combination of selection, both between and within breeds, and improved feeding, health and management. Separating the contributions of breeding and management to changes in production is difficult, especially over this time span. Some examples of attempts to do so over shorter periods of time are given later in the chapter.

Although yield is clearly a major component of profitability, the emphasis it has received is also due to the ease of measurement compared to some of the other components of profitability. Over the last decade or so the relevance of continued selection for higher yields has been questioned on at least three counts. The first is the increased emphasis in payment schemes in many countries on the composition of the milk. The second is the introduction of milk quotas in some countries as a means of controlling national production. (For example, quotas were introduced throughout the European Union (EU) in 1984, and there have been several subsequent reductions in quota allocations.) These topics are discussed further below. The third is the actual or perceived deleterious effect of selection for yield on the health, fertility and welfare of cows. The evidence for this, and some possible solutions, are discussed later in the chapter.

The trend for payment schemes to attach more value to milk composition began with schemes which paid premia or deducted penalties for high or low fat or total solids content. Later premia and penalties were attached to solids-non-fat content, and then directly to protein content. More recently there has been a steady increase in the importance of milk protein relative to milk fat in most countries. Figure 6.5 shows how the payment for protein, relative to that for fat, has increased in a sample of herds in England and Wales over a seven-year period.

The progressive increase in emphasis on milk solids, especially protein, in milk payment schemes has fuelled the debate on the merits of selection for breeds and individuals with high yields of milk. Although payment schemes are often couched in terms of a base price for liquid milk with deviations from this depending on the percentage of fat and protein, overall profitability is primarily determined by yield of milk and milk solids (kg), and milk solids content (%) is

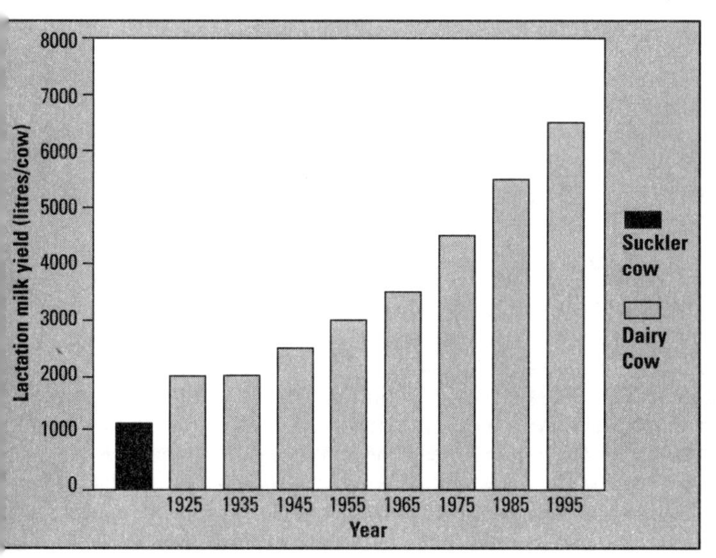

Figure 6.4 Changes in approximate average yield per cow in Britain since the 1920s, compared to the estimated yield of a cow suckling a calf. These changes have been brought about by a combination of selection, both between- and within-breeds, and improved feeding, health and management [2, 51, 53].

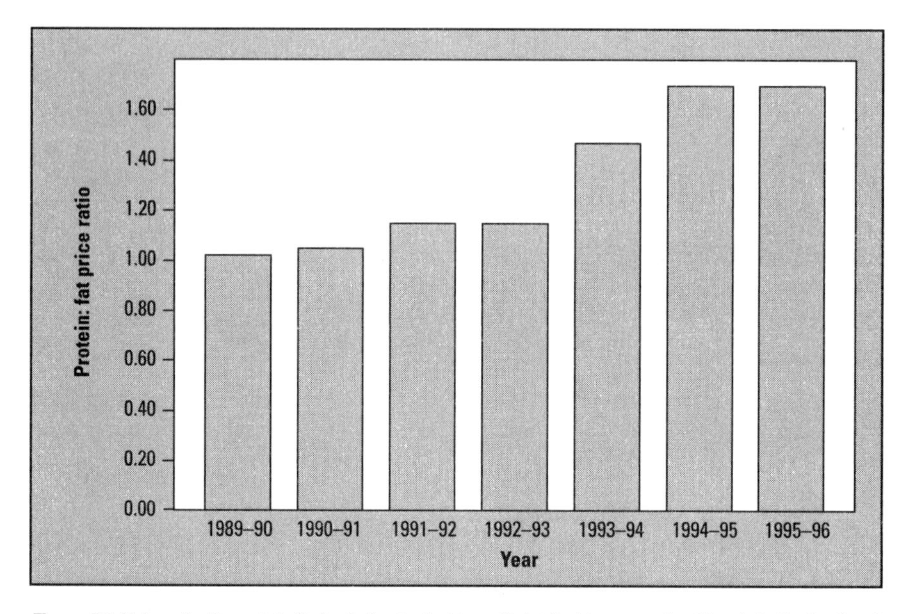

Figure 6.5 Value of milk protein (kg) relative to that for milk fat (kg) in a sample of herds in England and Wales over a seven-year period [16].

usually of secondary importance. Generally, buyers specialising in liquid milk sales tend to put less emphasis on milk composition than those specialising in processed dairy products. In theory, this sort of divergence in payment scheme may justify separate breeding goals for producers selling into different markets. However, in practice, the strong genetic association between total milk yield and yield of protein or fat makes separate goals unnecessary in most countries.

In the EU, and some other countries, milk quotas effectively limit total output per herd. This has led many farmers in these countries to question the wisdom of selecting between and within breeds for higher yields of milk and milk solids *per se*. However, several studies have shown that the highest yielding breeds and individuals are generally the most efficient converters of feed energy to milk energy [9, 27, 38, 39, 57]. The results from one of these studies are shown in Table 6.1. The Jersey breed appears to be an exception to this rule in several studies, having a higher efficiency than expected for its yield. This may be due to a higher yield and intake per unit of body size than in other breeds [4, 40].

The higher gross efficiency of higher yielding breeds is thought to be largely due to spreading the costs of maintenance (i.e. feed used to maintain body functions) over a higher total output of milk, rather than to any intrinsic difference in the net efficiency of food use [63]. Also, higher yields per cow can lead to lower production costs per litre, because of the spreading of some fixed costs associated with the number of cows (e.g. labour). So, selection for higher yield is likely to increase efficiency and profitability in most circumstances, even if total output is restricted. This, together with the option to lease or purchase milk quota in some countries, and the view that quotas are only a temporary method of regulating supply, have meant that yields per cow have continued to rise in most countries with quotas. Consequently, the number of dairy cows in these countries has declined.

Characteristics	Breed		
	Holstein Friesian	Dutch Friesian	Dutch Red and White
Number of cows	20	20	23
Milk (kg)	5331	4660	4562
Fat (%)	3.96	4.22	4.20
Protein (%)	3.25	3.40	3.49
Dry matter intake (kg)	5088	4822	4769
Energy intake (MJ net energy (NE))	32601	30884	30581
Average body weight (kg)	524	511	538
Gross efficiency (milk energy (MJ NE)/energy intake (MJ NE))	0.508	0.488	0.484

Table 6.1 Differences in milk production, feed intake and gross efficiency among Dutch Friesian, Holstein Friesian and Dutch Red and White dairy cattle over 52 weeks of lactation [39].

For example, the number of dairy cows in the UK declined from around 3.1 million to 2.6 million between 1985 and 1996 [35]. Similarly, between 1987 and 1993 the numbers of dairy and dual-purpose cows in the then twelve member states of the EC declined from about 26 million to about 21.5 million (including adjustment for the addition of the former East Germany to the EU; [50]).

Traditionally, the sale of calves for beef production has been seen as an important by-product for many European dairy industries. In several European countries dairy bulls entering progeny tests for milk production are performance tested first for beef characteristics. In Britain, Ireland and a few other countries there is a further integration of the dairy and beef industries because of the widespread use of beef x dairy suckler cows. This has led to much debate on the merits of breeding for dual-purpose versus more specialised dairy cows. This debate intensified following the influx of specialised North American Holstein strains of black and white cattle, with comparatively poor beef characteristics, to most European countries in the 1970s. The results in Table 6.1 and those from experimental studies comparing the profitability of dual-purpose versus specialised dairy strains of black and white cattle show that, in most circumstances, specialised dairy strains are the most profitable. As a result, the debate appears to have subsided now, and most European and other temperate dairy industries are following a path of increased specialisation. In this case, the returns from surplus calves will usually be maximised by mating those cows not required to breed replacements to beef bulls. Good fertility and, where appropriate, a compact calving period both help to maximise the number of cows available for mating to a beef bull, and so help to maximise returns from the sale of beef-cross calves. Similarly, the choice of beef breed for mating to surplus dairy cows, and the choice of sire within this breed, both have an important influence on profitability. These issues, together with selection for beef traits in dairy or dual-purpose breeds, are discussed further in Chapter 7. In future, if cost-effective methods of sexing semen and embryos are developed, even fewer cows will be needed to breed replacement dairy heifers. So there will be further scope to increase returns from sale of beef-cross calves, or pure beef calves if these are derived from embryos.

Figure 6.3(b) shows the importance of feed costs, or at least feed costs relative to milk value, in determining profitability. Because of the difficulty of measuring intake, and the evidence that higher yielding breeds and individuals are more

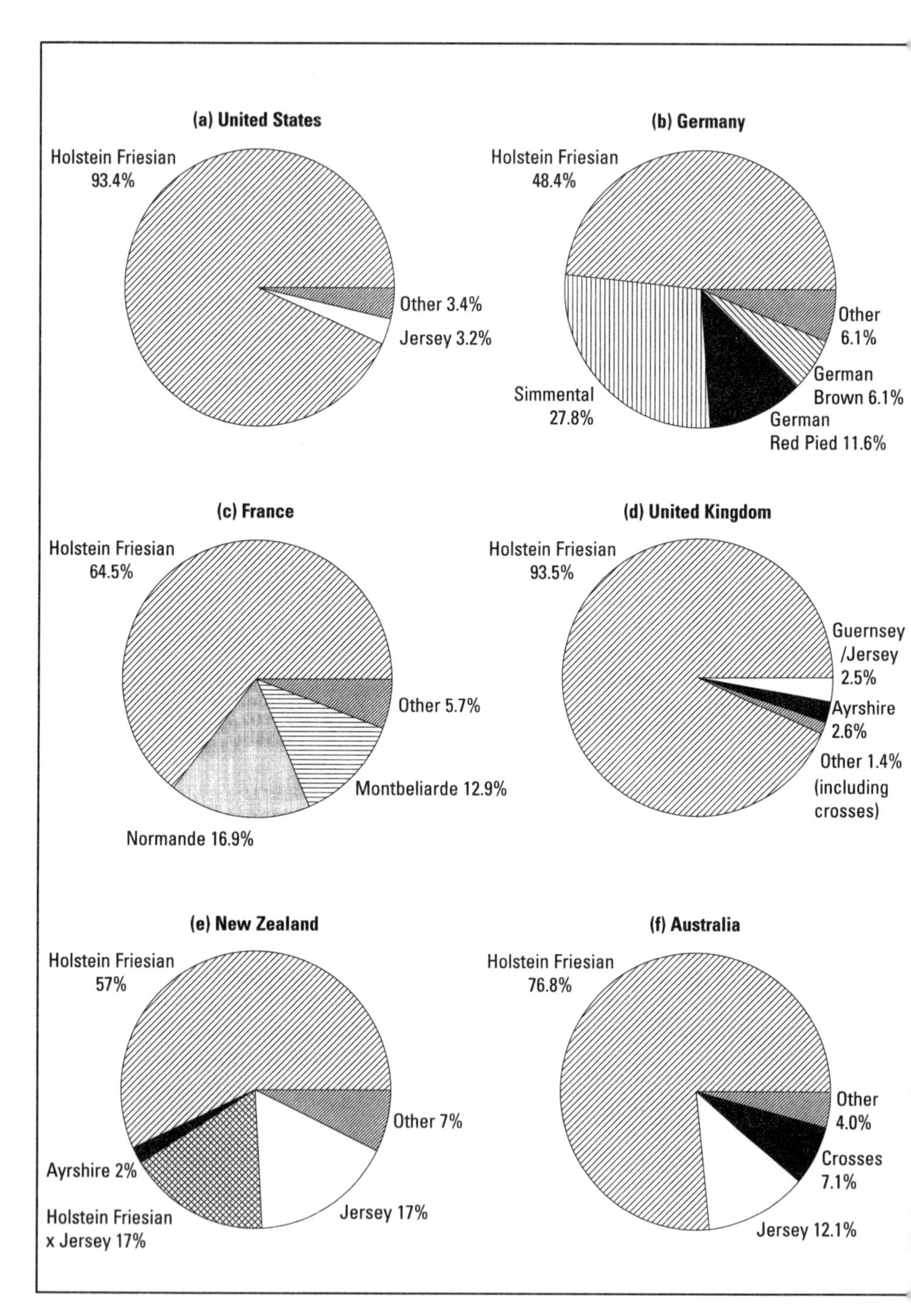

Figure 6.6 The proportion of different breeds and crosses which make up the dairy cattle populations of some major dairying countries [28, 50, 51].

efficient converters of feed energy to milk energy, most dairy cattle breeding schemes have not considered food intake or efficiency directly as part of the breeding goal. Also, philosophical (but important) arguments over whether yield drives intake, or intake drives yield, and on the importance of maintaining the capacity for high roughage intake in ruminants, make it difficult to know whether breeding programmes should aim to increase or decrease feed intake! Some of these issues are beginning to be addressed in dairy cattle breeding, and they are discussed in more detail later in this chapter.

Direct health costs appear to be relatively minor components of profitability from Figure 6.3(b). However, the indirect costs through lost production, and the fact that there are animal welfare implications of disease, suggest that genetically improving health deserves greater emphasis than a superficial economic analysis might suggest. Similarly, the direct costs associated with reproduction appear relatively low, but there are indirect costs here too. There is evidence of a genetic decline in some aspects of health and reproduction as a result of selection for yield [13, 46]. Hence, there is a need for greater consideration of the long-term, cumulative effects of ignoring small, unfavourable changes in health and reproduction.

While the evidence of the higher gross efficiency of higher yielding breeds and higher yielding animals within breeds is strong, there is a growing awareness of the need to develop more comprehensive breeding goals for dairy cattle. In future it is likely that many more of the animal characteristics discussed above will feature, either directly or indirectly, in dairy cattle breeding programmes. Some of the steps towards this, such as the introduction of national genetic evaluations for a wider range of traits, and the development of multi-trait selection indexes for dairy cattle, are discussed in more detail later in this chapter.

Breeds and crosses used in dairying

Figures 6.6(a)–(f) show the proportion of different breeds and crosses which make up the national dairy cattle populations in several major dairying countries. In some cases the proportions are for all cows, in others they refer only to milk recorded cows. Charts for France and Germany include dual-purpose breeds. These charts illustrate that: (i) black and white strains predominate and (ii) there is little use of systematic crossbreeding in most temperate systems, except in New Zealand and to a lesser extent Australia (see Plate 25). In most other cases, crossing is only a step towards breed or strain substitution.

Designed breed comparisons in many countries, together with the results of commercial semen importations, have demonstrated the higher total yield of milk and milk solids from North American black and white strains, compared to local black and white strains or other breeds [25, 27]. The average milk yields of different breeds from milk recording schemes in England and Wales show the same trend (see Table 6.2). Although they do not conform to some of the rules recommended for proper breed comparisons in Chapter 3, such as comparing different breeds in the same environment, results from these recording schemes have the advantages of including more breeds, being based on more animals and being more up to date than most of the designed comparisons. The results in Table 6.2 also show that although Holstein Friesians have higher total milk and milk solids yield, they have a lower compositional quality than several other breeds.

Some of the tools developed to put the appropriate emphasis on solids yield versus content, on protein versus fat, and on other components of profitability, are discussed later.

Figure 6.7 shows the increasing proportion of Holstein blood in the population of black and white cows registered by the Holstein Friesian Society (HFS) in the UK and Eire. The fact that breeding companies, breeders and producers are 'voting with their feet' by continuing to breed, test and select animals with an increasing proportion of Holstein blood probably provides further evidence of the validity of the results on the profitability of this strain compared to others, at least in relatively high input temperate systems.

The most appropriate measure of physical and economic performance of dairy herds depends on the most limiting resource in the herd concerned. While production per cow is a major component of profitability in higher input dairying systems, production per hectare is usually more important in lower input pastoral systems. The ranking of breeds on production per hectare is less clear than that on production per cow. There is evidence that some strains of Jersey have a higher or equal output per hectare to black and white strains, but choosing equivalent

Breed	Milk yield (kg)	Fat %	Protein %
Holstein Friesian	6366	4.02	3.19
Ayrshire	5682	4.04	3.26
Jersey	4339	5.54	3.80
Guernsey	4556	4.79	3.49
Dairy Shorthorn	5470	3.84	3.22
All breeds	6269	4.05	3.20

Table 6.2 Average annual milk yields of different breeds in recorded herds in England and Wales in 1994/95. The breeds are ranked by numerical importance. The absolute yields of the main breeds are similar in Scotland and Northern Ireland, and their rankings are identical [35].

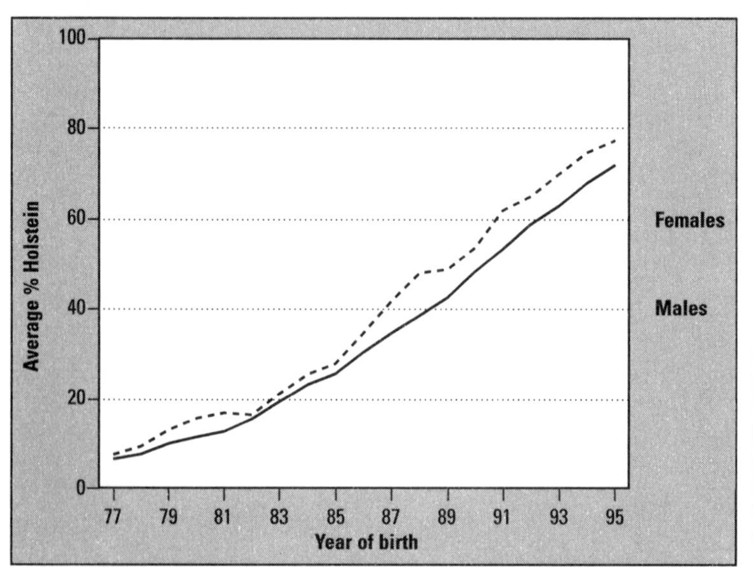

Figure 6.7 Change in the proportion of Holstein blood in bulls and cows registered by the Holstein Friesian Society in the UK and Eire. (HFS)

stocking rates to make fair comparisons is difficult (see Table 6.3). Table 6.4 shows various measures of the physical and economic performance of herds comprised of Holstein Friesian or Channel Island breeds recorded by Genus, the largest UK dairy cattle breeding company. These data do not allow unbiased comparison of breeds, since some breeds may be more prevalent in some management systems, but they do illustrate how rankings of breeds change, depending on the type of financial margin used to compare them (i.e. margin over purchased feed expressed per litre, per kg fat, per cow or per hectare). The greater numerical importance of the Jersey breed in countries with mainly pastoral production is probably a result of greater emphasis on production per hectare than on production per cow. For example, in New Zealand, although the proportion of Jersey cows has declined, they still account for about a fifth of the dairy cattle population.

There is a similar explanation for the low use of crossbreeding in countries with higher input systems. In these systems there appear to be no other breeds which consistently produce first crosses (F1s) which have higher yields than pure black and white animals. This is partly due to the advantage in additive genetic merit for milk and milk solids yield per cow seen in the black and white breeds, and partly due to the typically low level of heterosis for production. In contrast, because of the greater similarity among breeds in production per hectare,

Breed	Jersey		Friesian	
Stocking rate (cows/ha)	3.57	4.53	3.02	3.98
Fat plus protein yield:				
kg per cow	334	272	372	268
kg per ha	1192	1231	1121	1064
Days in milk	260	223	261	226

Table 6.3 Production of Jersey and Friesian cows grazed at low or high stocking rates in New Zealand [8, 22].

Breed	Holstein Friesian	Channel Island
Yield per cow (litres)	6253	4536
Yield from forage (litres per cow)	2497	1442
Milk price (pence per litre; ppl)	24.79	29.41
Concentrate used per litre (kg)	0.27	0.34
Total concentrate per cow (kg)	1680	1529
Margin over purchased feed (MOPF) per litre (ppl)	20.77	23.87
MOPF per kg fat (£)	4.94	4.49
MOPF per cow (£)	1298	1078
MOPF per hectare (£)	2829	2894

Table 6.4 Various measures of economic performance of Genus Milkminder herds in England and Wales, comprised of Holstein Friesian or Channel Island (Jersey and Guernsey) breeds, in 1995/96. These data do not allow unbiased comparison of these breeds, since some breeds may be more prevalent in some management systems, but they do illustrate how rankings of breeds change, depending on the type of margin used to compare them [16].

crossbred dairy cows do appear to be competitive in pastoral systems. This advantage may be augmented by the fact that heterosis for live weight appears to be lower than that for yield (see Table 3.8). So crossbred cows benefit from higher yield due to heterosis, but show a proportionally lower increase in live weight, and hence in maintenance costs, due to heterosis. The growing awareness of the importance of health and fertility traits in dairy cattle breeding may lead to an increased interest in crossbreeding in future. Also, some of the new breeding technologies outlined below and in Chapter 9 may create opportunities for other breeds to compete more effectively with the Holstein Friesian in terms of additive genetic merit. This could increase the value of crossbreeding in future, if the resulting F1s outperform both parent breeds.

Selection within breeds

Systems of testing

Progeny testing schemes

One of the fundamental problems in dairy cattle breeding is that most of the traits of interest are expressed only in females. However, because fewer males than females are needed in livestock breeding, the scope for genetic improvement is greatest through selection of males, especially when AI is used. As outlined in Chapter 5, it was the uptake of AI forty to fifty years ago which provided both the stimulus and the means for improved methods of genetic improvement of dairy cattle (Plates 26 and 27). The progeny testing schemes which evolved then have become central to genetic improvement of dairy cattle in most countries.

How progeny testing works. The basis of progeny testing schemes is the comparison of the milk production of the daughters of young bulls being tested with the daughters of other bulls recorded in the same herd, year and season (see Figure 6.8).

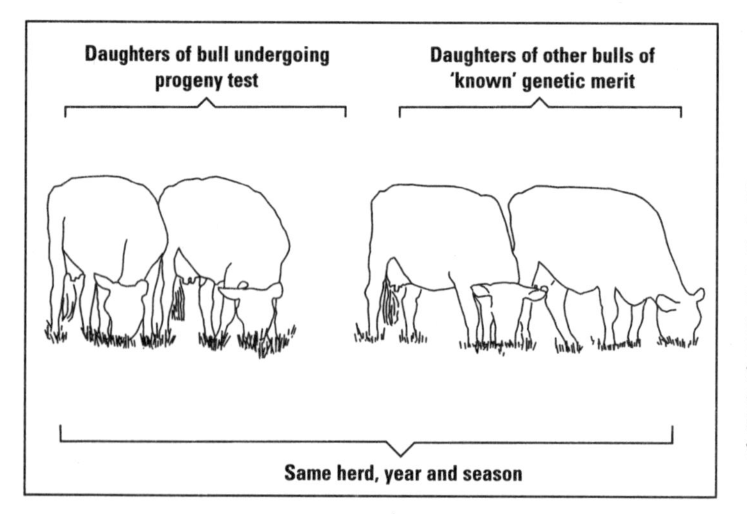

Figure 6.8 The basis of progeny testing schemes is the comparison of the milk production of the daughters of young bulls being tested, with the daughters of other bulls recorded in the same herd, year and season.

Results are then combined across herds, years and seasons and used to predict the genetic merit of the bulls being tested, and of other animals in the recorded population. The procedures involved were introduced in Chapter 5, and they are expanded on later in this section.

Dairy cattle progeny testing schemes in most countries share the following features, which are illustrated in Figure 6.9.

- The identification of young bulls of high predicted genetic merit for progeny testing. Often young bulls are bred specifically for progeny testing by **contract matings** between the top cows available and the best available AI bulls. These matings are arranged by the breeding organisations (e.g. milk boards, farmer cooperatives and breeding companies) or individual breeders engaged in progeny testing. The increased global trade in semen, embryos and animals has highlighted differences between countries in the genetic merit of dairy cattle. As a result, the importance of contract matings to homebred cows has declined in countries which are not currently near the top of the international league table of genetic merit. In these countries, many of the agencies involved in progeny testing obtain young bulls either by arranging contract matings abroad and importing the resulting embryos or offspring, or by identifying high predicted merit young bulls or embryos already available. However, the principle is exactly the same. Pedigree indexes are widely used to identify young bulls of interest, and to monitor their predicted merit at various stages throughout testing.

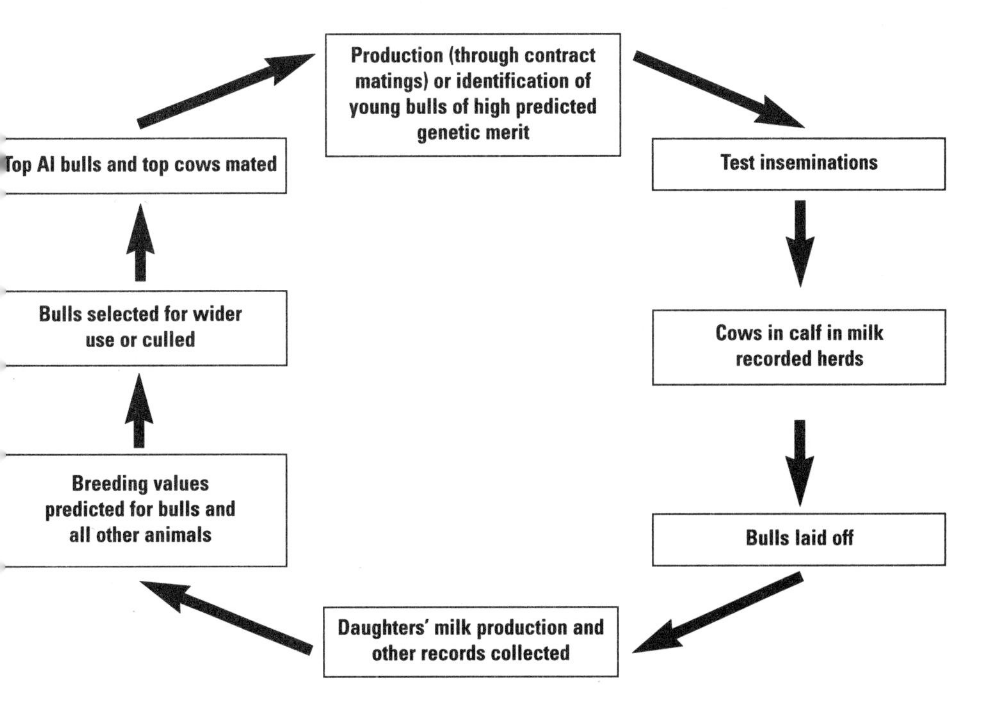

Figure 6.9 The main features of progeny testing programmes for dairy bulls.

The widespread use of imported animals and semen, and improvements in the local mechanisms for genetic improvement, are leading to a rapid increase in the genetic merit of the dairy cows in several countries, such as the UK, which were previously well down the international league table. This, together with a trend towards broader breeding goals which put less emphasis on high yields, should lead to an increase in local contract matings in these countries in future.

- Cows in milk recorded herds are inseminated with semen from young bulls being progeny tested. For example, in the UK, Genus – the largest breeding company engaged in progeny testing – aims to have about 60 daughters of a bull with completed first lactations. To achieve this about 400 to 500 cows are inseminated, over a period of a few months, per young bull being tested. This number allows for about 50% conception rate, 50% of calves born being female, and for subsequent losses during rearing and lactation [32]. The herds used for progeny testing by most of the larger agencies are commercial dairy herds, but they are expected to have high standards of recording, and to treat their animals uniformly. They are contracted to rear, milk and record female calves resulting from these test inseminations alongside the daughters of other bulls used in their herds. One of the criticisms of some smaller progeny testing schemes, or syndicates, is that they are open to bias through the preferential treatment of the daughters of particular bulls. Even with the most modern methods of evaluation these problems are difficult to overcome, although any initial bias is usually diluted later when records from more daughters in other herds get included in the evaluation of the bull. If the information is available, checking the number of herds in which the bull was tested, and the distribution of his daughters across these herds, helps to identify bulls at risk from this type of problem. (Picking bulls with a high reliability alone does not necessarily guard against this problem, though the use of more sophisticated statistical methods in evaluation, which are mentioned later, helps to reduce it.)

- At this point the young bulls are laid off, or temporarily retired, until their daughters' milk production and other records have been collected and their breeding values have been predicted. (In some countries larger quantities of semen are collected and frozen at the start of testing, and bulls are slaughtered before the results of the progeny test are known.)

- Milk production and other performance records on daughters are collated and the bulls' breeding values are predicted. In most countries breeding values are predicted for kg milk, kg fat, kg protein, % fat and % protein. Additionally, in most countries, the appearance, conformation or type of a bull's daughters is scored for a range of characteristics including body size, the size, shape and placement of udders and teats, and the shape and angle of feet and legs. Type classification is done either by the progeny testing agency itself, or more commonly by a

> breed society. (Type traits are discussed in more detail in a later section.) Most testing agencies also score additional traits such as the fertility of the bull, the calving ease when his offspring are born, the presence of any genetic abnormalities among calves and, once daughters are milking, their temperament, speed of milking and other 'workability' traits. In some countries the incidence of common diseases is also recorded. Breeding values for production, type and some of the additional characteristics recorded are usually predicted by a national agency established for that purpose (e.g. the Animal Data Centre in the UK, which is funded by a levy on all milk produced), or by a breed society, university or government department. Bulls with high predicted breeding values for some combination of these recorded traits are retained for semen production, while those which fail to meet the desired standard are culled. Typically less than 10% of bulls which enter progeny testing are retained as AI sires for wider use.

The number of bulls' progeny tested annually in some major dairying nations is shown in Table 6.5. Numbers have risen recently, but the relatively low number of bulls tested annually in the UK in the 1970s and 1980s probably contributed to the low rates of genetic improvement made. (Limited semen imports and comparatively late access to the US market were also important factors.) However, testing more bulls is only helpful if they are of high predicted genetic merit and there are sufficient recorded cows available to ensure an accurate test. The increasing international trade in dairy cattle, semen and embryos has stimulated restructuring in many dairy cattle breeding industries worldwide, directed towards improving the merit of bulls entering tests and increasing the speed and accuracy with which the tests are conducted. (In the UK examples include the revamping of the Genus Sire Improvement Programme, several intranational and international agreements on cooperative progeny testing (e.g. between the Supersires group of breeding companies, and between Genus and Holland Genetics respectively), new initiatives to encourage more progeny testing such as financial incentives to register daughters of test bulls by the HFS, and the launch of a new cooperative company, Cogent, which combines the formation of an elite nucleus herd with progeny testing in members' herds.)

It is worth noting that the number of bulls tested, and their merit, are not the only factors affecting rates of gain in national dairy cattle populations. The level of use of top bulls is another criterion which has a major effect on overall genetic gain. A study in the early 1980s clearly showed that most European countries, except the

Country	Average number of Holstein Friesian AI bulls progeny tested annually
Canada	364
Denmark	345
Finland	41
France	591
Germany	489
Italy	267
Netherlands	396
New Zealand	160
Sweden	93
UK	147
USA	1300

Table 6.5 Average number of Holstein Friesian AI bulls progeny tested annually in some major dairying nations. Only bulls born in 1987-89 are shown. (Dr G Banos, INTERBULL, for all countries except New Zealand; NZ figures from [28].)

UK and Eire, tested a high number of bulls relative to the size of their national dairy herds, but used the best ones on a fairly limited scale. In contrast, in the US and New Zealand proportionally fewer bulls were tested, but the best ones were much more widely used. (In New Zealand this widespread use of top bulls is facilitated by a compact, seasonal calving pattern together with the common use of fresh semen, which can be used at lower concentration than frozen semen.) This gave similar or even higher potential rates of gain than those possible in countries testing more bulls [11]. The figures in Table 6.5 show a similar pattern to that reported in the early 1980s in terms of number of bulls tested relative to the size of national dairy herds. However, top bulls are more widely used in Europe now than they were then (see Plate 28).

Pathways of genetic improvement with progeny testing. In dairy cattle populations which depend on progeny testing for genetic improvement, progress can be attributed to four 'pathways' [49, 52]. These are the selection of: (i) bulls to breed bulls – that is, the selection of top AI bulls to sire the next generation of young bulls for progeny testing; (ii) cows to breed bulls – the selection of top cows to breed young bulls for testing; (iii) bulls to breed cows – the choice of bulls to breed cows in herds which contribute bulls for testing and (iv) cows to breed cows – the choice of cows to breed replacement heifers in herds which contribute bulls for testing.

The contributions of these four pathways to overall genetic improvement in a dairy cattle population are potentially about 30%, 39%, 28% and 3% respectively [71]. In other words, about 70% of the progress made in a population is down to the choice of parents to breed the next generation of bulls for testing – choices which are in the hands of the breeding companies rather than individual breeders. Of the breeding options available to individual breeders, the choice of AI bull has the greatest impact by far. To allow for losses due to involuntary culling, and to allow for fluctuations in the sex ratio of calves born and losses during rearing, often 60% of cows, or more, have to be selected as potential dams of replacement heifers. This gives little scope for selection among them on genetic merit. One of the few opportunities to practice selection among females arises if extra female calves are reared as potential replacements. In this case it is worth calculating a pedigree index for all female calves prior to selection and retaining those with high index scores.

Breeding practices in herds which do not contribute bulls for progeny testing (herds in the multiplier or commercial tiers described in Chapter 3) do not affect rates of genetic gain in national breeding schemes. However, widespread use of the top AI bulls in these herds, and efficient selection of replacement heifers, will reduce the genetic lag, or gap in genetic merit, between them and the elite or nucleus tier of herds contributing animals for progeny testing.

Pros and cons of progeny testing. The main advantage of progeny testing is that, providing sufficient daughters are recorded, it produces accurate predictions of a bull's genetic merit. Also, the results are based on the performance of daughters in commercial herds under management systems which are typical for the country concerned, and so they are of direct relevance to many other herds in that country (i.e. the risk of genotype x environment interactions is minimised). The disadvantages are that progeny testing schemes are costly to run, and the results take many years to materialise. For instance, most bulls will be about 1½ to 2 years old when the test inseminations are completed, 2½ to 3 years old when their daughters are born, and about 5½ to 6 years old by the time these daughters have completed their first lactations and evaluations have been produced.

Progeny testing also depends on access to a large population of milk recorded cows. While most temperate dairying countries have sufficiently large populations of recorded black and white cows to run effective progeny testing schemes, it is difficult to sustain these in numerically small breeds. As a result these breeds often have too few bulls tested, progeny test results of low accuracy, or both of these problems. These problems in numerically small breeds could probably be reduced by greater international cooperation on the identification and testing of young bulls of high predicted merit, if the breed concerned occurs in several countries. The use of alternative breeding schemes such as the nucleus schemes described below, may also be helpful in these breeds.

Nucleus breeding schemes

Theoretical benefits. Before the advent of progeny testing, one approach to the genetic improvement of dairy cows was the use of a central cow performance test station (e.g. in Denmark). Comparing the performance of animals under a single management and feeding system has many attractions, and various modifications to this approach have been proposed since. One proposal was the creation of a large nucleus herd for recording the daughters of high predicted merit young bulls [20]. The advantages suggested for this type of scheme, compared to progeny testing in commercial herds, include: (i) greater operational control – e.g. it is easier to decide on aims, and to control the timing and precision of measurements, in a single herd than in a national programme; (ii) removing the risk of preferential treatment; (iii) higher accuracy of recording and selection; (iv) the opportunity to record and select on a wider range of traits than those recorded in progeny testing schemes based on commercial herds and (v) smaller scale of operation resulting in reduced cost, and opportunities for use in situations where the infrastructure for recording or the population size prevent effective progeny testing schemes in commercial herds [20, 37]. Wider interest in this type of scheme was generated by investigations on the use of **multiple ovulation and embryo transfer (MOET)** in dairy cattle breeding. MOET is a series of reproductive techniques including superovulation of a donor female (hormone treatment leading to higher numbers of ova being shed than usual), mating, recovery of the resulting embryos, and transfer of fresh or frozen embryos to recipient females. This permits elite females to leave many more progeny than they would be able to do by natural means.

One of the sources of inefficiency in traditional progeny testing schemes is that a lot of extra contract matings have to be made to account for low conception rates and the fact that, on average, only half the matings will produce bull calves for testing. This erodes the selection intensity among 'cows to breed bulls'. Some of the earliest studies on MOET in dairy cattle breeding looked at the possible gains from using it on contract mated cows to ensure that virtually all of them produced a bull calf for progeny testing. This increases the selection intensity achieved and produces a relatively small but worthwhile increase in response. Most agencies involved in progeny testing now use MOET routinely for this purpose.

In the late 1970s and the early 1980s Drs Frank Nicholas and Charles Smith proposed a new role for MOET in nucleus breeding schemes, as an alternative to progeny testing [36, 37]. Their proposed new schemes used MOET to create families of full sibs for testing and selecting both young bulls and future embryo donors. In these schemes (referred to from now on as MOET schemes) the selection of males was based primarily on the performance of their sisters, rather than the performance of their daughters as in conventional progeny testing schemes. In other words, breeding companies operating MOET nucleus breeding schemes would market semen from young bulls on the basis of a sib test rather than a progeny test. Information on sisters is available much sooner than that on offspring, and so these proposed schemes reduced generation intervals substantially compared to those possible in progeny testing, but they also reduced the accuracy of selection on males. However, on balance, the gains from shorter generation intervals outweighed the losses from lower accuracy, and the proposed schemes were predicted to give higher rates of genetic gain.

Nicholas and Smith examined both 'adult' and 'juvenile' MOET nucleus schemes. In the adult schemes, selection of embryo donors was based on one lactation of the candidate, plus information from other relatives. In the juvenile schemes selection of embryo donors was based solely on performance records from ancestors. A timetable of events for these two types of scheme is shown in Table 6.6. They estimated that generation intervals of 1.8 and 3.7 years could be achieved for the juvenile and adult MOET schemes compared to 6.3 years for conventional progeny testing schemes. As a result, adult and juvenile schemes were predicted to give up to about 80% and 90% extra response respectively, compared to conventional progeny testing schemes. However, in most MOET schemes, especially juvenile schemes, very large increases in the rates of inbreeding were expected. Also, schemes predicted to give the highest gains required higher success rates for MOET than were common at the time. An adult MOET scheme using more males and donor females than the scheme with the

Month	Juvenile MOET scheme	Adult MOET scheme
1	Born	Born
–		
13	Select on pedigree, MOET	
14		
15	Mate	Mate
–		
22	MOET progeny born	
23		
24	Calve	Calve
–		
34	Complete lactation, select	Complete lactation, select
35	MOET progeny for MOET	and MOET
36		Mate for further lactations
–		
44	MOET progeny born	MOET progeny born
Generation interval	22 months = 1.83 years	44 months = 3.67 years

Table 6.6 A comparison of the timetable of events in juvenile and adult MOET schemes as originally proposed by Nicholas and Smith [37].

highest predicted gains, was thought to be technically feasible at the time, and to produce acceptable levels of inbreeding. This scheme was predicted to give around 30% more response than a progeny testing scheme.

These initial results attracted a great deal of interest from scientists, breeders, breeding companies and investors in the potential value of MOET in genetic improvement programmes for dairy cattle. This alone acted as a catalyst, causing many breeding companies to review and improve their operations. The initial results also highlighted the difficulties in predicting rates of genetic gain and rates of inbreeding accurately, especially in small populations such as those in MOET nucleus herds. Most of the early work made assumptions which led to overestimation of response and underestimation of inbreeding in MOET schemes (e.g. ignoring reductions in genetic variation due to selection and inbreeding, use of optimistic success rates for MOET, ignoring variation in MOET family sizes). Since then here has been a great deal of work refining some of these assumptions, and optimising the design of MOET nucleus schemes. These studies have investigated variations in the size of schemes, the mating ratios and mating designs used, the sources of information used in selection, including the use of progeny tested rather than sib-tested bulls ('mixed' MOET schemes), whether schemes should be open or closed to the importation of males and females, and whether nucleus herds should be centralised or dispersed [67]. As well as producing many results of direct relevance to MOET schemes, these studies have produced many results of more general value in animal breeding.

On balance, current opinion is that with appropriate modifications to their design, MOET nucleus schemes can still offer increased rates of response compared to conventional schemes, although the extra gains are likely to be much lower than first predicted. These modifications include keeping nucleus herds open to importation, especially of high merit progeny tested bulls. (In this case MOET schemes become an adjunct, rather than an alternative to progeny testing schemes.) This reduces the risk of achieving lower rates of response than predicted due to the small population size and it reduces the rate of inbreeding. Rates of inbreeding can be reduced further by mating donors to different bulls for successive embryo collections and by reducing the emphasis on information from relatives in selection [67].

In future, the advantage to MOET nucleus schemes may be augmented through the application of so-called juvenile predictors of genetic merit (e.g. blood characteristics which can be measured at a young age to give a prediction of genetic merit for milk yield; [71]), the use of molecular genetic markers, as discussed in Chapter 9, and the use of sexing and other new reproductive techniques.

The potential benefits of nucleus schemes have been outlined, but there are also potential disadvantages. These include higher risks from a disease outbreak in the nucleus herd. Also, if management and feeding policy in the nucleus herd is markedly different from those in commercial herds which will use bulls bred in the nucleus, then there is a greater risk of genotype x environment interactions with nucleus breeding schemes than with progeny testing in commercial herds. However, steps can be taken to reduce these risks. For instance, different age groups of nucleus animals can be reared in different locations, and frozen embryos can be used as an insurance against disease outbreak. Also, management and feeding policies in the nucleus can be geared towards those in target commercial herds to reduce the risk of interactions.

MOET schemes in practice. Since the mid-1980s dairy cattle MOET breeding schemes have been established in Britain, Denmark, the Netherlands, Germany, France, Italy, Finland, Poland and Canada, though not all schemes still operate. The Genus MOET scheme in Britain was designed to be close to the original concept of Nicholas and Smith, involving a central nucleus herd and sib testing. However, the nucleus herd has remained open to importation of new genetic material, and in practice it has depended largely on the use of progeny tested rather than sib-tested bulls [31]. The Genus nucleus herd, in common with most others, was founded on embryos collected from high genetic merit North American cows. Female calves from these embryos have been used to build up a 250-cow nucleus herd. In each generation, 32 cows are selected to become embryo donors on the basis of their own performance and that of their sisters and other relatives. Selection is based on a combination of breeding values for kg milk, kg protein and kg fat, together with breeding values or phenotypic assessment of feed conversion efficiency, type classification results, fertility, health, milking speed, temperament etc. In the first few generations embryo donors were mated exclusively to top progeny tested AI bulls. However, some young bulls born in the MOET herd are now being used as sires. They are selected on breeding values derived mainly from their ancestors' and sisters' performance. On average, each young bull has information available on four full sisters and twelve half sisters.

Several of the other MOET programmes established worldwide have opted for dispersed rather than central nucleus herds, and few of these involve feed intake recording. For example, the TEAM (Total Evaluation of Animals with MOET) programme in Canada began in the Holstein and Ayrshire breeds in 1988. The aim was to select 60 females for MOET from these breeds to produce full sib families with at least fifteen progeny. Six Canadian AI centres cooperated in selecting donors, collecting embryos from them and transferring embryos into recipients on commercial farms. This gave more progeny per bull dam, and dispersed these more widely than in conventional schemes. Because of shorter generation intervals, the TEAM programme also gave the opportunity for earlier evaluations of young bulls. The benefits achieved in the initial phase of the TEAM project were lower than those theoretically possible. However, the project provided useful information on field-scale operation of a dispersed nucleus scheme, and highlighted key areas in which technical performance needed to be improved to maximise gains [29, 30].

In the Netherlands, the Delta programme combines some of the features of MOET nucleus herds as described above with features of the central performance testing stations mentioned earlier. The scheme produces elite heifer calves by embryo transfer. About 250 of these are used annually as embryo donors prior to lactating (i.e. analogous to the juvenile MOET scheme described above). Their subsequent first lactation milk production is compared in a central test. But additional elite heifers born outside the nucleus are also tested alongside nucleus-born heifers. That is, the scheme has an open nucleus, which allows contributions from non-nucleus females as well as males. Further embryo collections are made from about 60 donors on the basis of these central test lactation results [24]. The Cogent MOET nucleus scheme initiated recently in the UK has a similar design.

Essentially most MOET schemes have evolved into 'fast track' testing schemes which have become integrated into progeny testing schemes, rather than replacing them as originally proposed (see Figure 6.10). There are two main routes through which benefits feed back to progeny testing schemes. Firstly, MOET nucleus herds can provide unbiased and more accurate assessment of potential dams of young bulls for progeny testing. Selection of bull dams is often a weak link in conventional progeny testing because of preferential treatment of some elite cows when these are located in many individual herds. In several countries, large numbers of heifers or young cows from industry herds are introduced annually to nucleus herds specifically to allow fairer comparisons among them under uniform conditions. The accuracy of selection among bull dams is also increased because of the greater number of close relatives created through MOET. Secondly, MOET nucleus herds provide a valuable source of young bulls for progeny testing. Compared to young bulls sourced conventionally, those coming from MOET schemes have additional information because they have larger groups of half and full sisters. This information increases the reliability of early proofs, although eventually this is diluted by subsequent progeny information.

Because predicted transmitting abilities (PTAs) of animals in MOET schemes are usually based on much smaller numbers of records than those for progeny tested bulls their reliability is lower. The PTA of a bull is still the best guide to its true genetic merit, whether it was sib tested or progeny tested. But, if the reliability of PTAs for a group of bulls is low, then we expect a wider range in true transmitting abilities among them compared to that in a group of reliably proven bulls of equivalent average merit. The risks of breeding replacement females from bulls with low reliability proofs, whether these are sib tested bulls or young bulls being progeny tested, can be reduced by using several of them at once (i.e. it is

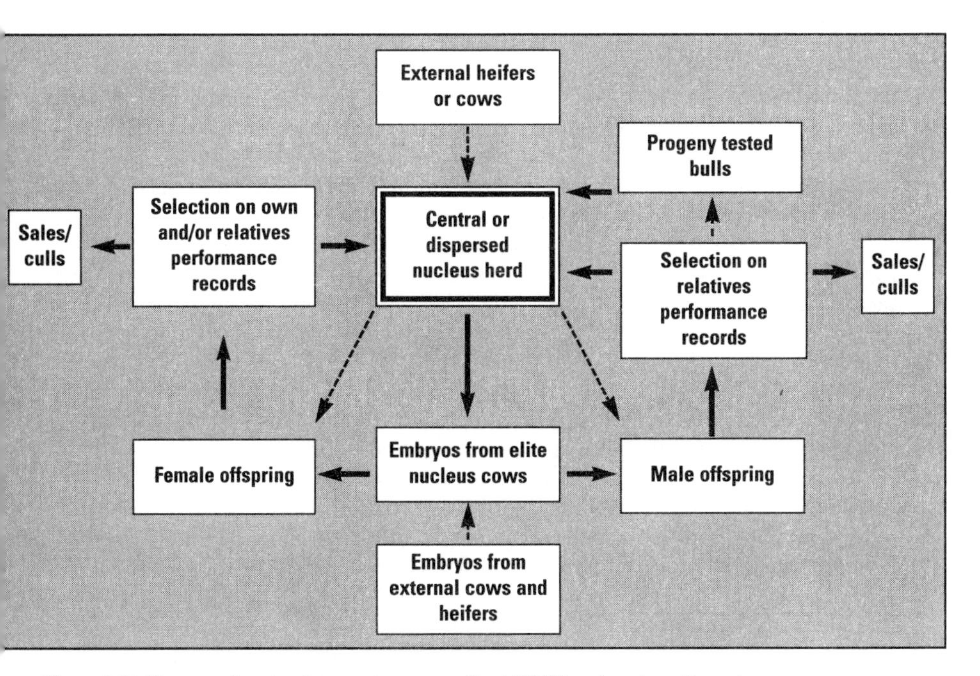

Figure 6.10 The annual cycle of events in a generalised MOET nucleus breeding scheme.

Reliability of PTA (%)	Category of reliability	Example of animals falling into this category
10–29%	Extremely low	A few young animals with pedigree indices only, and few recorded relatives contributing to these.
30–39%	Very low	Most animals with pedigree indices, with reliable information available on relatives. Individual MOET sib-tested bull with small full-sib family.
40–49%	Low	A few animals with pedigree indices from very reliably-tested ancestors. Cow with one or two lactations. Individual MOET sib-tested bull with larger full-sib family.
50 –59%	Low to moderate	Bulls with officially published PTAs (50% is minimum reliability in UK). Cows with 3 or 4 lactations.
60–75%	Moderate	Cows with 5 lactations.
76–89%	Moderate to high	Bulls with initial AI progeny test result. Cows with many ET daughters. Team of 4 MOET sib-tested bulls.
90–98 %	High	Proven AI bulls with second crop of daughters
99%	Very high	Widely used proven AI bulls.

Table 6.7 Approximate levels of reliability of transmitting abilities for milk production for individual MOET bulls and teams of MOET bulls compared to those for cows and progeny tested bulls. ([60]; Dr B J McGuirk.)

advisable not to put all of your eggs in one basket). Hence, semen from MOET schemes is often marketed from teams comprising several sib-tested bulls, rather than from individual bulls. Table 6.7 shows the approximate levels of reliability of transmitting abilities for individual sib-tested bulls and teams of sib-tested bulls compared to those for cows and progeny tested bulls.

Do MOET nucleus schemes work? There has been much debate since the early 1980s among dairy cattle breeding theorists and practitioners on the value of MOET schemes. This alone has been a valuable catalyst in generating renewed interest in the design and operation of all types of dairy cattle breeding schemes. However, evidence on the value of MOET schemes has become available over the last few years, as progeny test results of the first few teams of bulls from these schemes have appeared. There are two relevant questions to ask here.

The first is: do the bulls rank similarly on MOET and progeny test proofs? This gives an indication of whether breeders can rely on future MOET proofs as a guide to the true genetic merit of bulls. Figure 6.11 show that there is close agreement between the initial proofs of five groups of sib-tested bulls from the Genus MOET scheme and their subsequent national progeny test proofs, so the answer to the first question is yes, at least for this scheme.

The second question is: are the PTAs of MOET tested bulls as good as those from competing progeny testing schemes? This question can be answered by comparing

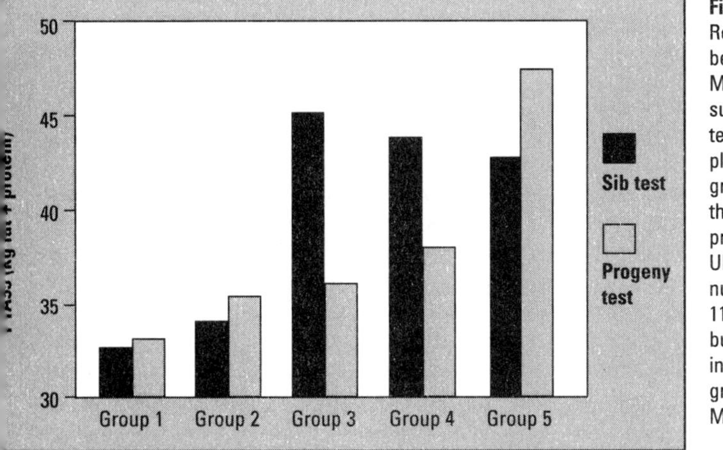

Figure 6.11 Relationship between the initial MOET sib proofs and subsequent progeny test proofs for kg fat plus protein for five groups of bulls from the Genus MOET programme in the UK. Each group numbered from 7 to 11 bulls, but some bulls were included in more than one group. (Dr B J McGuirk.)

the average PTAs of MOET bulls with the best available progeny tested bulls. At the time of writing, 12 of the top 50 Holstein Friesian AI bulls available in the UK, ranked on an index of overall economic merit, are from MOET breeding schemes in Britain or the Netherlands [1]. This suggests that these schemes are beginning to play a useful role in genetic improvement of dairy cattle, even if this is as an adjunct to progeny testing, rather than as an alternative to it.

Traits recorded

Many of the traits recorded and used in dairy cattle breeding programmes have been mentioned already. The purpose of this section is to provide a bit more detail on the manner of recording, on the heritabilities of these traits and the associations among them.

Milk recording

Progeny testing schemes for dairy bulls in most countries are based on national or regional milk recording schemes. For example, in the UK there are currently three large milk recording organisations which evolved in conjunction with the former milk marketing boards (National Milk Records, which operates mainly in England and Wales, the Scottish Milk Records Association and United Dairy Farmers, which operates in Northern Ireland). About half the dairy cows in the UK are officially milk recorded, a lower proportion than that in most other major dairying nations in Europe and elsewhere [35, 50, 51]. In most of these countries recording services must operate to standards set by the International Committee for Animal Recording before records are used in genetic evaluation.

The basis of most milk recording schemes in Western countries is that a recorder visits the herd at monthly (or longer) intervals, and records the milk yield of cows at the two or three milkings in a 24-hour period. Samples of milk are also obtained from individual cows at each milking for analysis of protein and fat contents. Increasingly, somatic cell counts (SCC) are obtained from these samples too. (Somatic cells in milk are those produced by the cow in response to infection. High somatic cell counts in milk are usually associated with mastitis. Also, high cell counts in the bulk milk tank result in substantial penalties in milk price in many countries. Thus, individual cow cell counts provide a management tool for culling cows with abnormally high counts and they provide the basis for genetic evaluations of bulls and cows for resistance to mastitis.) Farmers participating in milk recording schemes receive statements listing yields and other details of their cows at monthly or other regular intervals for management purposes. For the purposes of genetic evaluation, records are collated at regular intervals by the milk recording organisations and sent to the organisation responsible for evaluation.

Traditionally, milk records of cows have only been used for genetic evaluation once lactations are complete. The normal length of lactation for recording purposes in many countries with semi-intensive production systems is 305 days, but some shorter lactations (e.g. those in excess of 200 days in the UK) are considered to be complete for genetic evaluations if the cow has dried off or been culled. (It is often difficult to achieve 365-day calving intervals in high yielding herds, so inevitably many lactations are longer than 305 days. There is a view that planning lactations longer than 305 days may be more biologically and economically efficient for high yielding cows in some non-seasonal production systems.) All the available milk records (called **test day records**) for an individual cow would be used to predict that cow's 305-day yield of milk, fat and protein. Typically, there would be eight or nine of them at monthly intervals. However, there are strong genetic associations between the yields of milk, fat and protein at successive sampling occasions and so the total lactation yield can be predicted quite well from a smaller number of test day records (see Table 6.8). Consequently, there is a growing use of early lactation test day records (**records in progress**) to predict total lactation yields for genetic evaluations. (For example, records in progress from heifers have been used in UK genetic evaluations since 1995.) The use of records in progress can reduce substantially the time lag between making test inseminations from a young bull and getting an initial prediction of breeding value, and so it can shorten male generation intervals and improve the efficiency of progeny testing.

	Test day (approximately monthly intervals)									
Lactation record	**1**	**2**	**3**	**4**	**5**	**6**	**7**	**8**	**9**	**10**
Milk yield	0.87	0.89	0.97	0.98	0.99	0.97	0.98	0.97	0.95	0.88
Fat yield	0.77	0.85	0.95	0.97	0.97	0.93	0.97	0.96	0.97	0.98
Protein yield	0.84	0.90	0.97	0.97	0.98	0.95	0.98	0.98	0.95	0.91
Fat %	0.81	0.91	0.98	0.98	0.99	0.98	0.96	0.98	0.94	0.91
Protein %	0.76	0.87	0.95	0.97	0.96	0.98	0.92	0.94	0.89	0.77

Table 6.8 Estimates of genetic correlations between individual test day records for several milk production traits, obtained at approximately monthly intervals throughout lactation, and predicted full lactation records for the same traits. The records were from British Holstein Friesian heifers; all animals had at least 8 test-day records, but only a subset had 10 test-day records. Except for the first one or two test-day records, most show a high genetic correlation with total lactation production; heritabilities of test-day records follow a similar pattern [41].

Trait	Milk (kg)	Fat (kg)	Protein (kg)	Fat (%)	Protein (%)
Milk (kg)	**0.39**	0.84	0.95	–0.31	–0.42
Fat (kg)	0.75	**0.36**	0.87	0.26	–0.13
Protein (kg)	0.91	0.82	**0.36**	–0.16	–0.10
Fat (%)	–0.42	0.28	–0.20	**0.58**	0.51
Protein (%)	–0.50	–0.09	–0.10	0.60	**0.62**

Table 6.9 Estimates of the heritabilities of the most commonly recorded milk traits, and the correlations among them, for the first lactation of Holstein Friesian cows. Heritability estimates for yield of milk, fat and protein in later lactations tend to be slightly lower. Those for fat and protein content are similar to the first lactation values. Heritabilities are shown on the diagonal in bold, phenotypic correlations are shown above the diagonal and genetic correlations below the diagonal [68].

	First and second lactation		First and third lactation		Second and third lactation	
Trait	Phenotypic	Genetic	Phenotypic	Genetic	Phenotypic	Genetic
Milk (kg)	0.58	0.87	0.54	0.84	0.60	0.98
Fat (kg)	0.56	0.86	0.51	0.85	0.57	0.97
Protein (kg)	0.58	0.86	0.52	0.82	0.61	0.98
Fat (%)	0.73	0.85	0.71	0.84	0.69	0.74
Protein (%)	0.70	0.78	0.69	0.76	0.71	0.78

Table 6.10 Estimates of the phenotypic and genetic correlations among milk production traits recorded in the first, second and third lactations of Holstein Friesian cows [68].

Table 6.9 shows estimates of the heritabilities of the most commonly recorded milk traits and the correlations among them for the first lactation. Yields of milk, fat and protein are all moderately highly heritable, while fat and protein content are both highly heritable. There are high genetic correlations among milk, fat and protein yields, indicating that selection for any one of these three traits is expected to lead to increases in the other two. However, there are negative genetic correlations between all three yield traits and protein %, and between yields of milk or protein and fat %, indicating that selection for yield is likely to lead to a reduction in protein or fat concentration, or both. Table 6.10 shows the correlations between production in successive lactations. The reasonably high genetic correlations indicate that there is a reasonably close association between genetic merit for milk production in early and later lactations.

Type traits

As mentioned earlier, assessments of type or conformation are widely used in dairy cattle breeding. Type classification is usually performed by breed society staff, or staff from progeny testing agencies. The records are used both to provide a description of the cow concerned, and to enable predictions of breeding values for cows and bulls. The methods of type classification, and the emphasis that ought to be placed on type versus production have been a major source of disagreement between scientists and breeders for decades. Type classification schemes evolved as a method of describing the physical attributes of cows,

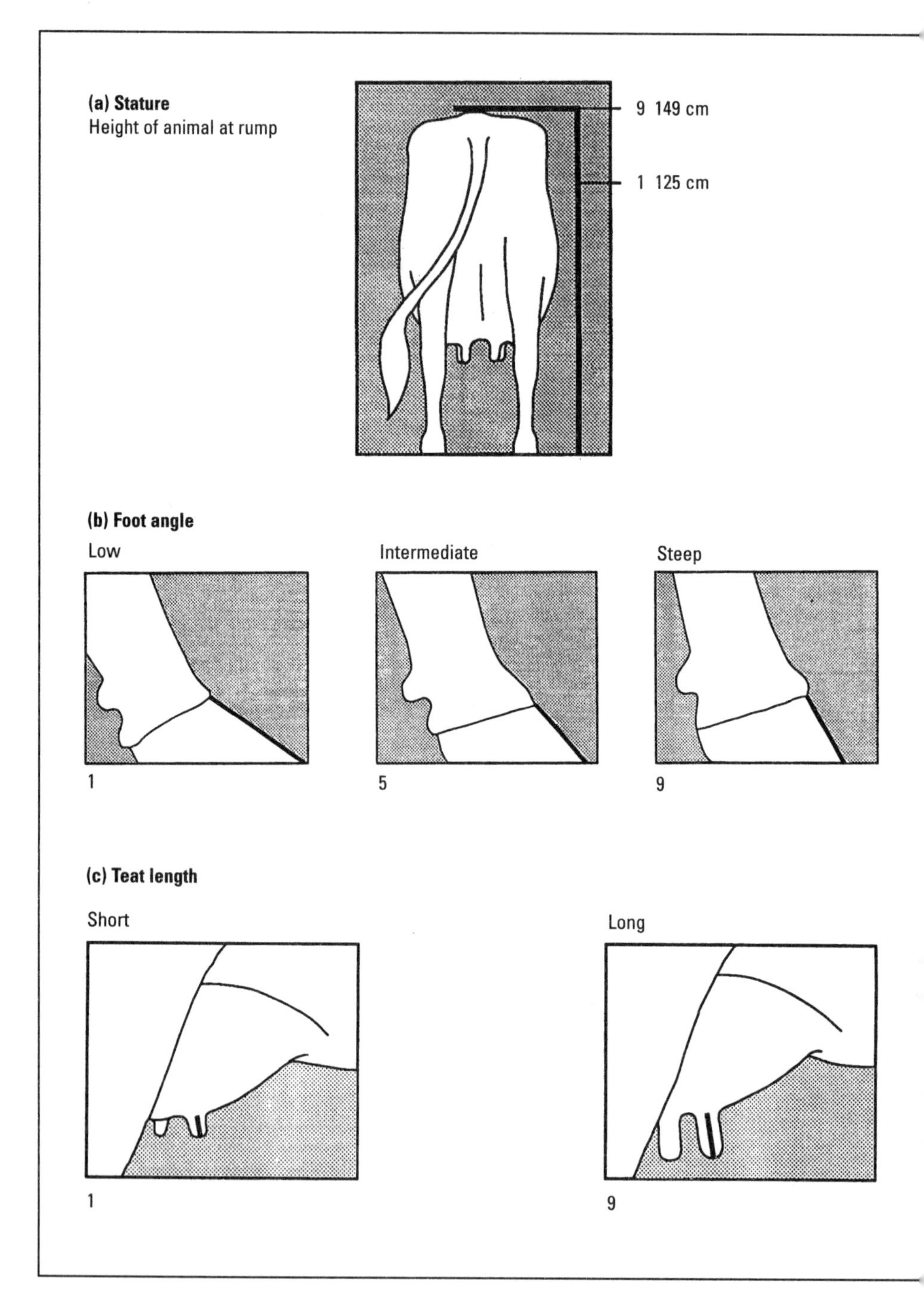

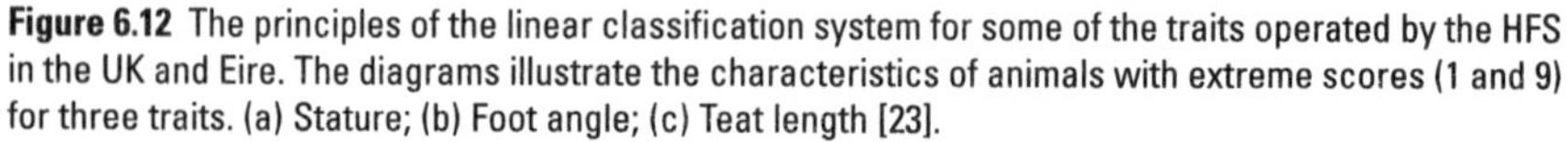
Figure 6.12 The principles of the linear classification system for some of the traits operated by the HFS in the UK and Eire. The diagrams illustrate the characteristics of animals with extreme scores (1 and 9) for three traits. (a) Stature; (b) Foot angle; (c) Teat length [23].

Trait	Interpretation of extreme scores: 1	9	Heritability
Stature	125 cm	149 cm	0.48
Chest width	Narrow	Wide	0.27
Body depth	Shallow	Deep	0.35
Angularity	Thick and coarse	Angular	0.26
Rump angle	High pins	Low pins	0.29
Rump width	Narrow pins	Wide pins	0.22
Rear legs, side view	Straight	Sickled	0.19
Foot angle	Low	Steep	0.27
Fore udder attachment	Loose	Strong	0.27
Rear udder height	Low	High	0.20
Udder depth	Below hocks	22 cm above hocks	0.39
Central ligament[1]	Flat	Strong	0.16
Front teat placement, rear view	Wide	Close	0.43
Teat placement, side view	Close	Far apart	0.41
Teat length	Short	Long	0.44

[1] Formerly called udder support

Table 6.11 The interpretation and heritabilities of linear type traits recorded by the Holstein Friesian Society in Britain and Ireland. ([7, 23]; Dr S Brotherstone.)

particularly those thought to be associated with functional fitness, health and longevity, or success in the show ring. However, the early classification systems were based on subjective opinions of individual classifiers, and so they were not very repeatable or easy to interpret. Many breed societies retain classification schemes which include some subjective judgement, but increasingly these schemes are using more objective measurements. In particular, in the late 1970s and early 1980s a new system called linear assessment was developed in North America. This has since been adopted in most major dairying countries.

For most traits, linear assessment still involves visual appraisal of the cow. However, unlike the original classification systems, linear assessment scores the characteristics of the cow in relation to the biological extremes, rather than labelling her as good, bad or indifferent. This means that the results are more repeatable, and more amenable to scientific analysis. Figure 6.12 illustrates the principles of the linear assessment system for some of the traits scored by HFS in the UK and Eire. Table 6.11 lists the fifteen linear type traits recorded by HFS, and shows the interpretation of high and low scores for each of these traits and their heritabilities. In addition, HFS operates a type classification scheme which allocates scores for four composite traits: body conformation, dairy character, legs and feet and mammary. These scores are assigned subjectively, but they are closely related to the relevant linear type scores. A Total Score, ranging from 40 to 97 points, is calculated from the cows score for these four composite traits, with 20% of total emphasis given to body conformation, 20% to dairy character, 20% to legs and feet and 40% to mammary. Based on the Total Score, cows are assigned to Final Classes of Excellent (90–97 points), Very Good (85–95 points), Good Plus

	Genetic correlation with			
Linear type trait	Longevity	Milk yield (kg)	Fat yield (kg)	Protein yield (kg)
Stature	0.00	0.22	0.16	0.25
Chest width	–0.13	–0.14	–0.10	–0.13
Body depth	–0.03	0.24	0.25	0.24
Angularity	0.13	0.44	0.42	0.43
Rump angle	0.03	–0.11	–0.08	–0.11
Rump width	–0.11	–0.01	0.01	0.01
Rear legs, side view	–0.05	0.07	0.06	0.07
Foot angle	0.09	0.02	0.05	0.07
Fore udder attachment	0.06	–0.29	–0.23	–0.28
Udder depth	0.14	–0.48	–0.40	–0.44
Central ligament[1]	0.02	0.10	0.16	0.15
Front teat placement, rear view	0.03	–0.05	–0.03	–0.01
Teat placement, side view	–0.02	0.38	0.35	0.36
Teat length	–0.19	0.18	0.12	0.17

[1] Formerly called udder support

Table 6.12 Genetic correlations between linear type traits, longevity and milk production traits in UK Holstein Friesians. ([7]; Dr S Brotherstone.)

(80–84 points), Good (75–79 points), Fair (65–74 points) and Poor (64 points and below) [23].

A great deal of effort has been made over the last few years to harmonise the methods of linear type classification used internationally, to make it easier to interpret foreign bull proofs. There is now good evidence that trained operators produce linear type classifications which provide heritable and repeatable descriptions of the appearance of cows. Also, there is growing evidence that some linear type traits are genetically correlated with economically important traits such as milk production, longevity and disease resistance, live weight and feed intake (see Tables 6.12 and 6.17). Until recently, most breeders or commercial producers using type information have based their selection on the animals' scores for individual type traits, or on one of the overall scores of type merit. Recently, new indexes have been developed in several countries which attempt to make better use of type traits, by including them along with production traits in indexes of overall merit. Some of these developments are discussed later in this chapter.

Reproduction, health and workability

Most milk recording agencies collect details of inseminations made, either for management purposes or to provide parentage information for the resulting offspring when they produce milk records themselves. In some countries (e.g. Denmark, the Netherlands, Norway) these records are also used to produce routine genetic evaluations of cow (and bull) fertility. Several studies show an unfavourable genetic correlation between fertility and milk production. For example, an early review in this area showed genetic correlations averaging about 0.35 between fat corrected milk yield and the interval between calving and first insemination (though this may be influenced by management practices), and about 0.22 between fat corrected yield and number of inseminations [46]. So there is growing interest in predicting breeding values for cow fertility traits to assist selection decisions. Table 6.13 shows the heritabilities of some measures of fertility. In most cases these are very low, indicating the importance of management. Consequently, large numbers of daughters must be recorded to produce accurate bull proofs for fertility traits.

Most progeny testing agencies collect information on the calving difficulty of bulls undergoing test. The results of these surveys are often presented in semen sales catalogues as percentages of calvings classified as veterinary assisted, very difficult, slightly assisted, or unassisted. More sophisticated analyses are gradually being introduced which do a better job of adjusting the results for breed and parity of the mate etc. However, selection for ease of calving is more complicated than it seems at first sight. Bulls whose progeny are born with little or no calving difficulty generally produce smaller progeny. But, when these small daughters calve themselves, they are more likely to have difficulties than larger cows. In several countries there are two separate evaluations of calving ease for bulls: firstly a 'direct' evaluation of the bull as a father and, secondly, a 'maternal' evaluation of the bull – an evaluation of the bull based on the calving performance of his daughters. The typical range of heritabilities for calving difficulty and stillbirths are shown in Table 6.13. As with fertility traits, the heritabilities of these traits are usually very low.

Dairy cattle breeding programmes in several countries now include selection for resistance to some common diseases such as mastitis and ketosis. Selection for disease resistance has been practised for many years in the Scandinavian countries. Genetic evaluations there are based on comprehensive national databases of individual cow health events maintained by farmers and veterinary surgeons. Similar records have been collected for over a decade in the UK by the DAISY (Dairy Information System) recording service now operated by the University of

Trait	Heritability (se)	Source
Days to first service	0.03	[19]
Days open	0.05	"
	0.04 (0.02)	[21]
Services per conception	0.03	[19]
	0.01 – 0.02 (0.002 – 0.005)	[59]
Calving interval	0.03 (0.02)	[21]
Dystocia (calving difficulty)	0.03 – 0.20[1]	[33, 47 – reviews]
	0.05 – 0.34[2]	"
Stillbirth	0.00 – 0.05[1]	"
	0.04 – 0.14[2]	"

[1] Heritability on the categorical scale i.e. as measured. The range of heritabilities is similar for both direct and maternal effects on dystocia and stillbirth (i.e. whether they are considered as traits of the calf or as traits of the dam)

[2] Heritabilities on the underlying continuous scale

Table 6.13 Estimates of the heritability of some measures of fertility, calving difficulty and stillbirth. (Standard errors (se) are also given – these give an indication of the precision of the estimates; see the standard quantitative genetics textbooks referred to in Chapter 2 for more details.)

Trait	Heritability (se)	Source
Mastitis	0.0 – 0.37	[13]
Ketosis	0.05 – 0.10	"
Laminitis	0.22	[12 – review]
Sole contusion	0.21	"
Heel horn erosion	0.15	"
Interdigital dermatitis	0.09 – 0.13	"
Hyperplasia interdigitalis	0.31	"

Table 6.14 Estimates of the heritability of some diseases in dairy cattle.

	Temperament		Ease of milking	
Heritability	**0.11**		**0.21**	
Correlation with	**Phenotypic**	**Genetic**	**Phenotypic**	**Genetic**
Milk yield (kg)	0.09	0.40	0.08	– 0.07
Fat yield (kg)	0.09	0.59	0.09	0.16
Protein yield (kg)	0.03	0.46	0.07	0.01

Table 6.15 Heritabilities of temperament and ease of milking, and correlations with other production traits in first lactation UK Holstein Friesian cows. Both temperament and ease of milking were subjectively scored by farmers on a five-point scale, at the time of type classification by HFS. ([6]; Dr S Brotherstone, personal communication.)

Reading/National Milk Records, although there are relatively few animals involved. UK Livestock Services and several other UK milk recording organisations have recently launched, or are considering, simplified health recording schemes. As with reproduction traits, evidence of unfavourable genetic correlations between milk production and some diseases is stimulating interest in genetic evaluations for resistance to some common diseases in dairy cattle. Table 6.14 shows the heritabilities of some of these diseases. The heritabilities of most disease traits are fairly low.

Although direct evaluations of health traits are being practised or considered in some countries, in many others there is greater dependence on indicator traits. For example, somatic cells appear in the milk in response to udder infections such as mastitis. Somatic cell counts in milk have been shown to be heritable (e.g. heritability of 0.11-0.12 in different dairy breeds in the UK [34]) and genetically associated with the incidence of mastitis (e.g. genetic correlations of 0.4 to 0.8 with various measures of infection [10, 48]). Selection for reduced somatic cell counts has been shown to reduce the incidence of mastitis. Also, a growing number of milk payment schemes attach direct economic value to somatic cell counts as a measure of the hygienic quality of milk. Consequently, genetic evaluations for somatic cell counts have been launched, or are about to be launched, in several countries (e.g. Denmark, Finland, Sweden, the Netherlands, the UK, Canada, the US). Similarly, linear type traits can be useful indicators of problems of the mammary system or feet and legs [17]. The most useful of these indicator traits are already beginning to appear in indexes of overall merit, as explained later.

Good temperament and ease of milking have long been recognised as important to those at the sharp end of milk production. In the past, most selection pressure on these 'workability' or 'parlour' attributes has been achieved by culling the worst offenders. However, there is growing interest in national recording and evaluation schemes for these traits. For instance, in the UK and Eire HFS operate a five-point scoring system, based on the owner's opinion of the cow's temperament and ease of milking, alongside their routine type classification scheme. Table 6.15 shows some results from an initial analysis of these data. Temperament has a fairly low heritability, and ease of milking a moderate heritability. There are fairly strong genetic correlations between temperament and production, indicating that temperament is likely to improve with selection for yield. The genetic correlations between ease of milking and production are not significantly different from zero. These traits are incorporated into indexes of overall economic merit in several countries.

Feed intake, live weight and body condition

The importance of feed costs in the overall profitability of dairying was highlighted earlier. However, feed intake has been ignored in most improvement programmes until recently. This is partly because of the difficulty of recording feed intake in commercial herds. Also, there is a fairly strong genetic correlation between yield and gross efficiency, which means that selection for yield is expected to lead to improvements in efficiency. So, many testing agencies have relied on this correlated response, rather than undertaking direct selection. Central nucleus herds provide an ideal opportunity to record intake, and the renewed interest in these schemes in the early 1980s led to increased interest in feed intake in dairy cattle. Several studies have demonstrated that food intake is fairly highly heritable (0.16 to 0.44) [61]; and that, at least under *ad libitum* feeding of a complete diet, feed intake is not entirely a function of yield [43] (see Table 6.16). Also, total intake over the whole lactation can be predicted fairly well from several weeks of intake recording [42]. This is potentially useful for making selection decisions about embryo donors early in lactation in MOET schemes which use

	Milk yield (kg/day)	Fat yield (g/day)	Protein yield (g/day)	Dry matter intake (kg/day)	Average live weight (kg)	Average condition score (0–5)
Milk yield	**0.27**	0.74	0.90	0.43	–0.20	–0.36
Fat yield	0.55	**0.40**	0.78	0.50	–0.13	–0.27
Protein yield	0.80	0.75	**0.35**	0.50	–0.13	–0.28
Dry matter intake	0.59	0.76	0.64	**0.44**	0.19	–0.12
Average live weight	–0.09	0.03	–0.02	0.27	**0.44**	0.63
Average condition score	–0.46	–0.31	–0.35	0.00	0.67	**0.35**

Table 6.16 Estimates of the heritability of average daily milk production, average daily dry matter intake, average live weight and average condition score over the first 26 weeks of lactation, in Holstein Friesian heifers and cows at the SAC/University of Edinburgh Langhill Dairy Cattle Research Centre, and estimates of the phenotypic and genetic correlations among these measurements. Heritabilities are on the diagonal in bold, phenotypic correlations are above the diagonal and genetic correlations are below it [62].

	Heifers			Heifers + cows		
Linear type trait	Dry matter intake (kg/day)	Average live weight (kg)	Average condition score	Dry matter intake (kg/day)	Average live weight (kg)	Average condition score (0–5)
Stature	0.18	0.64	0.32	0.13	0.52	0.13
Chest width	0.25	0.86	0.57	0.28	0.79	0.73
Body depth	0.20	0.81	0.22	0.34	0.69	0.24
Angularity	–0.21	–0.56	–0.51	0.07	–0.43	–0.77
Rump Width	0.11	0.74	0.14	0.24	0.70	0.29

Table 6.17 Estimates of genetic correlations between some linear type traits, average daily dry matter intake, average live weight and average condition score over the first 26 weeks of lactation. Linear type data were from dairy cows at the SAC/University of Edinburgh Langhill Dairy Cattle Research Centre and the UK national data set; dry matter intakes, live weights and condition scores were from Langhill animals only [62].

feed intake information in selection. However, it remains difficult to present and use information on feed intake in isolation in national breeding programmes.

It is common for most mammals to lose weight (mainly fat) during lactation. However, over the last decade or so there has been concern that, at least when very high yielding North American Holsteins are kept in some lower input systems, the increasing milk yields of high genetic merit cows are being 'fuelled' by progressively greater losses of body condition, rather than by higher feed intake. This has led to wider interest in including feed intake as part of the breeding goal. The difficulties of direct measurement remain, but there are prospects for indirect measurement on two fronts. The first is via measurements of live weight, condition score and those linear type traits which are related to body size and shape. Tables 6.16 and 6.17 shows that there are quite strong genetic associations between some of these traits and feed intake. Potential uses of these measurements in indexes of overall economic merit are being investigated in the UK. The second approach is to predict intake from measurements of 'markers' in feed and faeces. Measurement of the concentrations of introduced or naturally-occurring (e.g. n-alkanes) markers in the forages fed, and subsequently in the faeces produced, allows prediction of the total amount of feed consumed. In expert hands these techniques are producing promising results. If field-scale application of these techniques becomes feasible, even in a selected group of 'test' herds, they could provide valuable additional information for future improvement programmes.

Methods and results of genetic evaluation

Milk production traits

Early methods used to evaluate bulls were based on the comparison of the performance of a bull's daughters with that of their dams. In the 1950s and 1960s the emphasis changed to comparing a bull's daughters with their contemporaries, milked in the same herd, year and season. This is still the basis of most modern methods of evaluation, although there have been major increases in sophistication, in particular the development of the BLUP procedures outlined in Chapter 5. These procedures have been adopted, and progressively upgraded, by most major dairying countries over the last couple of decades. For example, in the UK sire–maternal grandsire model BLUP evaluations were first done in 1979. This improved the accuracy of evaluations compared to earlier methods by recognising that not all cows were of equal merit, and hence the daughters of bulls being evaluated could be at an advantage or disadvantage as a result of having a genetically good or bad mother. However, this BLUP model only accounted for differences in the merit of cows via their sire. Indexes were produced for cows by combining their sire's and maternal grandsire's PTAs with their own production records. Individual animal model BLUP evaluations were introduced in the UK in 1992, providing PTAs directly for bulls and cows.

The steps involved in producing genetic evaluations include:

- Collation and checking of milk records by milk recording agencies and transfer of these records, together with the relevant pedigree information, to the agency responsible for genetic evaluation (if this is different). In most countries there are between one and four genetic evaluations per annum, so this step is repeated from one to four times per annum. (For example, in the UK, at the time of writing, two evaluations are done per annum by the Animal Data Centre (ADC).) However, there is a trend towards more frequent evaluations. These allow breeding organisations and producers to make their selection decisions earlier, or on the basis of better information, although obviously there are extra costs in doing evaluations more frequently. In New Zealand, for instance, evaluations are performed at three-weekly intervals during the milking season [18].

- In the past, evaluations were often based on heifer lactations only. However, increasing computer power has made it possible to include more lactations in evaluations. Records in progress from heifers are increasingly being used to allow earlier prediction of breeding values for bulls. (For example, up to five lactations for each cow are included in UK evaluations. Completed lactation records collected since 1977 are included in the evaluation. Pedigree information goes back much further, e.g. to the 1950s in the case of the Holstein Friesian breed. Records in progress from heifers have been included in evaluations since January 1995. Heifers must have a minimum of three official milk records in order to be included. Table 6.18 shows the total number of records involved in the February 1997 evaluations of dairy breeds in the UK.)

- Evaluations are usually done for single traits (i.e. the traits are evaluated one at a time). As explained earlier, there are benefits from multi-trait evaluations, but in many countries the large amount of data precludes this. Increases in computer power will enable wider use of multi-trait evaluations in future.

Breed	Number of lactations	Number of cows	Number of bulls
Holstein Friesian	11,528,862	4,998,851	59,307
Ayrshire	522,533	215,565	5,603
Jersey	298,158	122,193	3,498
Guernsey	200,395	81,818	1,579
Dairy Shorthorn	81,016	32,699	1,204

Table 6.18 Number of lactation records and number of bulls and cows evaluated in the February 1997 genetic evaluation of dairy breeds in the UK by the ADC [1].

- Evaluations are usually done for one breed at a time. However, across-breed evaluations have been introduced in New Zealand recently. This is feasible in New Zealand because of the relatively common occurrence of herds comprising several breeds and crosses.

- In many countries with large dairy cattle populations milk production records are pre-adjusted for some environmental effects. That is, these adjustments are made before BLUP evaluations are done. Ideally, all of these effects would be estimated simultaneously with the breeding values. Simultaneous estimation of environmental effects and breeding values is one of the major benefits of BLUP evaluations, resulting in higher accuracy. However, the large amount of data involved in many dairy cattle evaluations often makes it impractical to estimate all environmental effects simultaneously with breeding values. Improvements in computing power and computing strategies should make this more feasible in future. (In the UK, for example, milk records are pre-adjusted to account for the effects of lactation number (all records are converted to heifer equivalents), age within lactation, calving interval, month of calving, different levels of variation in production between herds and, in the Holstein Friesian breed, for the effects of heterosis and recombination in crosses between Holsteins and Friesians.)

- Breeding values or transmitting abilities are then predicted. Increasingly dairy cattle breeding values or transmitting abilities are predicted using individual animal model BLUP, which includes all relationships between relatives and predicts breeding values for all animals. The statistical model used in the BLUP evaluations accounts for differences in management between and within herds by assigning animals to contemporary groups. An animal's milk record is usually assigned to a different contemporary group depending on the herd, year and season of calving in which it occurred. (For example, in the UK animals are initially grouped in calving seasons of two months, but these are expanded if there are too few animals in a group. Heifers are grouped separately from animals in later lactations. Animals with imported dams are also assigned to different contemporary groups from those with UK-bred dams, to account for potential preferential treatment of the daughters of expensive imported cows. The model used in the UK accounts for sire x herd interactions – these occur if the daughter groups from one or more sires are managed differently from the rest of the herd. The model also accounts for permanent environmental effects, which were described in Chapter 4.)

● Unknown parents are usually assigned to genetic groups. (For example, in the UK unknown parents are assigned to groups according to the year of birth and sex of their offspring, and their own sex and country of origin.)

● In countries with wide use of imported animals or semen, foreign information is often used before local information becomes available, or it is combined with local information. This requires converted proofs or international evaluations for foreign animals, produced in the manner described in Chapter 5. For example, at the time of writing, if a foreign bull used in the UK has a local proof with a reliability below 50%, only a converted foreign proof is published. Converted foreign proofs of bulls are combined with local evaluations, once the local proof has a reliability of 50% or greater. The emphasis on foreign versus local information in combined proofs is proportional to the reliabilities of the two proofs: the one with the highest reliability contributes most information to the combined proof. Evaluations are based on a combination of foreign and UK information until the reliability of the UK proof reaches 90%, at which point foreign information is no longer used. This produces a smooth change from converted foreign proofs to UK proofs. Similar procedures are used for cows with foreign and local proofs, except that combined proofs are produced once UK proofs have a reliability of 30% or greater. When foreign information has been incorporated into the PTAs of parents in the manner described, the results for their progeny must then be adjusted too. Many countries are now using or planning to use INTERBULL MACE international evaluations, as described in Chapter 5. At first sight this would appear to remove the need to produce combined proofs. However, information from countries other than the one in which the bull was first tested is only used in INTERBULL evaluations once there are large numbers of daughters recorded in many herds. So it is still relevant to combine local and INTERBULL proofs until local information is included in the INTERBULL proof. As is currently the case with converted proofs, most countries will use INTERBULL evaluations only until bulls have a very reliable local proof. National evaluations in countries which are heavy importers of semen are usually timed to make the best use of new foreign or international evaluations.

● PBVs or PTAs (and indexes based on them) of the animals evaluated are expressed relative to some reference population of animals called the base. In most dairy cattle evaluations results are expressed relative to a fixed base, which is updated every few years. For example, at the time of writing, the base used in the UK is the average merit of cows born in 1990. This base was introduced in 1995, so PTAs expressed relative to it are denoted PTA95. Bulls or cows which have positive PTA95s for any of the five milk production traits are expected to have genetically higher yields or percentages for the traits concerned compared to the average merit of cows born in 1990. Bulls or cows with negative PTA95s are expected to be of lower genetic merit than those born in 1990. For example, a bull with PTAs of +800 kg milk, +35 kg fat, +30 kg protein, –0.05% fat and 0.00% protein is expected to leave daughters which yield 800 kg more milk, 35 kg more fat and 30 kg more protein at 0.05% lower fat content and the same protein content as the average of cows born in 1990. The alternative to a fixed base is a rolling base, such as that used in Canada. With a rolling base the predicted transmitting values are expressed relative to the merit of the current population, and so they change with each new evaluation. There are pros and cons to each type of base. For example, it is easier to compare the results of successive evaluations with a fixed base than with a rolling base. But a fixed base can give a false impression of progress if animals are selected on the basis of the sign of their PTA alone. If the population is improving rapidly some animals with positive PTAs compared to a base chosen a few years earlier will be worse than the current breed average. A frequently updated fixed base is a common compromise.

- The results of the evaluations for bulls and cows which have achieved some threshold level of reliability are sent to the owners of the animals concerned, milk recording agencies, breed societies, breeding companies and other interested parties. (Usually results with low reliability are restricted to the owners and recording agencies.) In most countries, predicted transmitting abilities are published for kg milk, kg fat, kg protein, % fat and % protein (see Table 6.19).

- In many countries, selection index scores are produced as well as PTAs. These are of three main types: (i) economic indexes based on milk production traits only, accounting for expected differences in market values of milk, fat and protein, and possibly for different costs

Name of bull	Reliability %	Predicted Transmitting Abilities (PTA95)					PIN95 £	ITEM £
		Milk kg	Fat kg	Protein kg	Fat %	Protein %		
Singing-Brook N-B Mascot ET	97	1290	33.3	39.5	–0.29	–0.04	141	141
Southwind Bell of Bar-Lee	99	834	28.4	27.8	–0.09	0.01	104	109
Kashome Bell Jurist	97	943	35.9	28.0	–0.05	–0.04	106	106
MOET Elsas Crewman ET	84	1055	31.2	27.0	–0.19	–0.11	96	103
MOET Flirt Dilemma ET	85	787	28.1	27.6	–0.07	0.03	105	102

Table 6.19 An example of the way PTAs are presented for bulls. The bulls shown were the top 5 Holstein Friesian bulls from the February 1997 evaluation in the UK, ranked on their ITEM index. (N.B. bull rankings and PTAs soon become outdated; the values shown in this table are for illustration only) [1].

of production (e.g. the PIN index used in the UK is based solely on PTAs for kg milk, fat and protein, weighted according to their expected future value and the feed and quota costs associated with their production); (ii) indexes which use arbitrary weightings to combine PTAs for milk production and type traits and (iii) broader economic selection indexes including direct measures or predictors of traits such as longevity, fertility, disease resistance or live weight as well as milk production (e.g. the ITEM index used in the UK which, at the time of writing, includes the same production PTAs as PIN, plus PTAs for linear type traits associated with longevity). In the third type of index, milk production and other traits are combined in an objective way, based on the strength of the genetic associations with traits in the breeding goal and the relative economic values of these. The principles of economic selection indexes were introduced in Chapter 5, but more detailed examples are given later in this chapter.

Other traits

As indicated earlier, a growing number of traits other than milk production traits are being recorded and evaluated in temperate dairying countries. Type assessment takes place in virtually all these countries, and there is increasing interest in recording and evaluating traits associated with reproduction, health, workability and longevity. In some countries there is also interest in traits associated with live weight, feed intake and body condition. Table 6.20 shows the extent to which these other traits are recorded and evaluated in various countries.

In some countries a single agency is responsible for genetic evaluation of all traits, in others a number of organisations are involved. For example, breed societies are responsible for genetic evaluation of type traits in many countries. Obviously, the more agencies involved in the recording and evaluation of traits, the greater the importance of good communication and coordination.

The principles and methods of genetic evaluation of these traits are usually similar to those for milk production traits. In some cases the methods used are less sophisticated than those used for milk traits, since the traits concerned are considered to be less important. In other cases the smaller size of the data sets involved permits the use of more sophisticated methods. For example, genetic evaluations for linear type traits scored by the HFS in the UK and Eire are done at the University of Edinburgh using multi-trait animal model BLUP, although the analyses are performed for groups of associated traits, rather than all traits simultaneously, to reduce computing requirements. For example, all udder traits are evaluated simultaneously, feet and leg traits are evaluated simultaneously, and so on.

The nature of many non-production traits means that they require some form of transformation before analysis or special consideration on the method of presentation of PTAs. For example, the units of measurement for linear type traits are arbitrary (usually scores from 1 to 9), so the results of genetic evaluations for type traits are often presented in standard deviation units rather than in the units of measurement. Hence breeding values range from about –3 to about +3 standard deviations about the base. Figure 6.13 gives an example of the usual method of presentation. Traits such as ease of calving, disease incidence and some measures

Country	Reproduction				Health			Workability			Longevity
	Calving ease [1]	Calf mortality [1]	Nonreturn rate [2]	Other cow fertility	Somatic cell count	Mastitis	Other disease	Milking speed	Temper-ament	Other	
Australia	• (d)							•	•	•	•
Canada	• (d, m)				•			•			•
Denmark	• (d, m)	• (d, m)	• (c)	•	•	•		•	•		
Finland		• (d, m)	• (b)	•	•	•	•	•	•	•	
France								•			
Germany	• (d, m)	• (d, m)	• (b, c)		•			•			•
Ireland	• (d)	• (d)						•	•		
Italy	• (d)							•			
Netherlands	• (d)		• (b, c)	•	•			•	•		
New Zealand								•	•	•	
Norway	• (d, m)	• (d, m)	• (c)			•	•	•	•	•	
Sweden	• (d, m)	• (d, m)		•	•	•	•	•	•	•	•
UK								•	•		
USA	• (d)				•						•

[1] d = direct, m= maternal
[2] b = bull, c = cow

Table 6.20 Summary of some traits, other than milk production and type, evaluated for dairy sires in various countries in 1995. Note that this provides only a 'snapshot' – evaluation of additional traits may have started since the survey was completed. The original report also contains full details of trait definitions and evaluation procedures and the breeds concerned [3].

of fertility are recorded on a categorical scale. That is, there are only a small number of discrete categories into which the records fall. As a result, the distributions of these traits may differ from the normal distribution on which the conventional methods of predicting breeding values depend. So, records of these traits are often transformed in some way prior to analysis, as mentioned in Chapter 2. Similarly, somatic cell counts usually show an extended distribution caused by a few animals or herds having very high counts. So, predictions of genetic merit are usually based on transformed data, often logarithms of the actual cell counts. PTAs can either be transformed back to the cell count scale or, perhaps more usefully, expressed in terms of the expected percentage increase or decrease in cell counts expected in a bull's daughters.

Apart from milk production and most type traits, many of the traits mentioned above have low heritabilities. This means that much larger daughter groups are needed to get reliable progeny test results. For example, to achieve a progeny test with a reliability of 0.8 requires records from around 40 daughters for a trait with a heritability of 0.35 (e.g. first lactation milk production), but over 150 daughters for a trait with a heritability of 0.1 (e.g. some measures of reproduction and resistance to disease). Since the resources available for testing are limited, this means that fewer bulls can be tested if the aim is to produce reliable progeny test results for traits which have a low heritability. However, even if testing is geared towards higher heritability traits, moderately reliable evaluations of these other traits on young bulls are still better than none. Also, higher reliabilities will eventually be achieved on widely used bulls, and this in turn will increase the reliability of evaluations on their relatives.

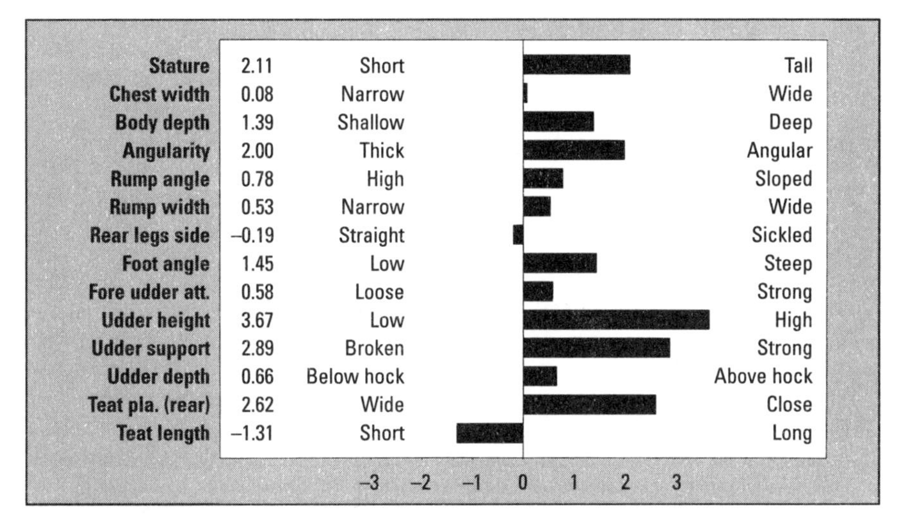

Figure 6.13 An example of how breeding values for linear type traits are usually presented (HFS).

To date, the lack of uniformity in recording non-production traits has hindered the objective use of information across countries. Although great strides have been made in harmonising the way linear type traits are scored in different countries, and which traits are scored, some differences remain. As a result, the accuracy of foreign conversions for some type traits, and some countries, is lower than that usually expected for milk production traits. Some of these difficulties will be reduced once international evaluations are available for type, and possibly other traits, as explained in Chapter 5.

Indexes of overall economic merit

In the past, most dairy cattle breeding schemes worldwide were geared towards increasing output. There is now growing interest in broader selection goals, stimulated by at least three factors. The first of these is the surplus of dairy products in many temperate countries. Quotas or price constraints have made higher production *per se* less attractive, and reducing costs of production more attractive. Secondly, there is growing public concern for the health and welfare of farm animals, as well as a wider appreciation of the direct economic consequences of disease. There is evidence from several countries that selection for milk production alone tends to lead to a deterioration in udder quality, a higher incidence of mastitis and some other diseases, and poorer reproductive performance. While some of these problems can be offset by improved management, it makes sense to try to prevent this decline in the first place. Thirdly, the formation of nucleus breeding schemes in some countries has facilitated direct recording of some new traits for selection, e.g. feed intake, health events.

Economic indexes which combine production PTAs have become quite widely used in many countries over the last decade or so in an attempt to address the first of the issues listed above (e.g. PIN in the UK, INET in the Netherlands, INEL in

Country	Index	Traits included in index						
		Milk, protein, or fat production (one or more yields or %)	Repro-duction	Health (including SCC)	Workability	Longevity	Type traits (in own right, or as predictors)	Beef production growth
Canada	Total Economic Value	•		•		•		
Denmark	S-index	•	•	•	•		•	•
Finland		•	•	•			•	
Norway		•	•	•	•		•	•
Sweden	Swedish bull index	•	•	•	•		•	•
UK	ITEM	•					•	
USA	USDA net merit $	•		•		•		

Table 6.21 Examples of several indexes of economic merit in use in 1995. Note that this provides only a 'snapshot' – indexes in some countries will have been altered since the survey was completed. The original report also contains full details of trait definitions and evaluation procedures and the breeds concerned [3].

France, ILQ in Italy, MF$ (milk, fat) and MFP (milk, fat, protein) values – in the US). The PIN index in the UK is fairly typical of this type of production index. PIN is based on PTAs for kg milk, kg fat and kg protein, each weighted by their expected future value. These weightings take into account the increasing value of protein compared to fat (a relative value of 1.5:1 is used in the current version of PIN), the extra transport and cooling costs of high volume–low solids milk, and the cost of leasing extra quota to match the higher productivity of daughters of high PTA bulls.

Farmers have long appreciated the need for balanced breeding objectives, and have sought to breed cows which will be functionally sound as well as productive. In many countries there are combined production and type indexes in use in dairy cattle breeding with this aim in mind. However, until recently, most of these have been based on arbitrary weightings on production and type. There is now growing interest in extending production indexes to include measures of reproduction, health, workability and longevity, with components of these overall indexes weighted according to sound economic and scientific principles. In some of these indexes type traits feature as predictors of traits contributing to profit, in others functional type traits (i.e. those relating to the mammary system or feet and legs) are given direct economic values. Table 6.21 gives some examples of this type of index in use in several countries in 1995. This table gives only a snapshot of what is a fast-changing field. In some countries the indexes mentioned have changed since the information was compiled, and indexes have been introduced in other countries not shown in the table. For instance, a new index (Breeding Worth) was introduced in New Zealand in 1996, based on breeding values for milk, fat and protein production, survival and live weight [18].

The development of the ITEM index in the UK by SAC and the University of Edinburgh provides a good example of some of the issues involved in producing these broader indexes. ITEM (index of total economic merit) is based on exactly the same principles as PIN, and attaches the same economic values to PTAs for kg milk, fat and protein as those used in PIN (see Table 6.22). It is relatively

Trait	Economic value per unit (£)	Economic value per sd (£)	Relative economic value
Milk (kg)	–0.03	–18	–0.25
Fat (kg)	+0.60	+15	+0.21
Protein (kg)	+4.04	+73	+1.00
Longevity (%)	+3.37	+40	+0.55

Table 6.22 Economic values for milk, fat and protein yield used in PIN95; the same values, plus the economic value shown for longevity are used in ITEM. Longevity is defined as the % of cows not involuntarily culled in the first four lactations, corrected for genetic differences in yield [64].

straightforward to calculate the economic values of higher production, but more difficult to assess the benefits of longevity. However, detailed studies of costs and returns in dairy herds, and computer based simulation of herds indicate that the net effect of improving longevity by 1 % (i.e. 1 % fewer cows culled involuntarily in the first four lactations) is around £3.37 per cow per annum [64]. This arises through lower costs for heifer rearing, replacement and culling, and having more cows producing at the mature yield level.

The version of ITEM used from 1995 to 1997 included breeding values for four linear type traits associated with cow longevity. These were udder depth, teat length, angularity and foot angle. The choice of type traits to go into ITEM was based on analysis of all available linear type assessment records ([7] and unpublished results). This analysis examined the longevity of cows in relation to linear type scores – the best possible 'test' of the value of these scores in measuring the functional soundness of cows under practical dairy farming conditions. The four type traits chosen had the closest genetic associations with longevity once the effects of differences in milk production were removed. Genetically shallower udders, shorter teats, greater angularity and steeper foot angles were all associated with greater longevity. Several other traits were linked more weakly to longevity but they added little to the overall picture once these four main traits had been accounted for. Because there have been such major changes in the UK black and white population over the last couple of decades as a result of the widespread use of imported Holstein semen, these relationships between type traits and longevity will be re-examined at regular intervals, and the ITEM index revised if necessary.

Direct measurement and evaluation of cows' longevity is planned in the UK at the time of writing. PTAs will be produced which are based on a combination of these direct measurements of longevity together with those linear type traits which are useful predictors of longevity. Because it takes such a long time to record longevity directly, type traits will still make a useful contribution. For example, early in a bull's life his PTA for longevity will be based mainly on direct information on the survival of ancestors, and indirect information on survival via type traits measured on a range of relatives. As the bull gets older, direct information on the survival of his daughters will accumulate and improve the accuracy of his evaluation. Once these combined proofs for longevity (i.e. survival plus type) are available, they will replace the individual type traits used to predict longevity in ITEM.

Tables 6.23 and 6.24 show the index weights for the first version of ITEM (used in 1995 to 1997), and the expected responses to selection on this compared to selection for production only (PIN) or longevity only (hypothetical index; LIN). Selection on this version of ITEM is expected to increase milk yields, while maintaining fat % and protein % at about current levels. Putting less emphasis on composition would lead to a decline in protein % because of a negative genetic correlation between yield and protein %. Selection on ITEM is expected to keep both udder depth and teat length at approximately the current breed average, whereas selection on PIN would result in increases in both of these traits. There has been some concern at the fact that ITEM attaches a negative weighting to teat length scores since, understandably, most producers would be unhappy about milking heifers with very short teats. However, this negative weighting on teat length is simply counterbalancing the effects of including milk production in the index, which leads to longer teats. The net effect expected following selection on ITEM is roughly no change in teat length.

The overall effect of including these type traits in this version of ITEM will be to increase genetic improvement of profitability by about two per cent per annum compared to selection on PIN. That is, ITEM is still identifying high producing animals, though there will be modest but important changes in the rankings of bulls and cows which reflect likely differences in longevity. Research in progress, as well as future changes in the economics of milk production, will lead to further refinements in ITEM in future. One of the earliest additions will be PTAs for lifespan based on a combination of type traits plus direct measurement of survival. Later additions are likely to include direct or indirect measures of live weight, condition score, feed intake, reproductive performance and disease resistance. Including additional measures in ITEM will lead to greater differences in the ranking of animals on PIN and ITEM in future and substantially greater economic benefits from selection on ITEM.

Trait	Index weight
milk PTA (kg)	–0.03
fat PTA (kg)	+0.60
protein PTA (kg)	+4.04
angularity PTA (sd)	+1.80
foot angle PTA (sd)	+1.10
udder depth PTA (sd)	+2.70
teat length PTA (sd)	–2.50

Table 6.23 ITEM index weights used in the Holstein Friesian breed in the UK in 1995 to 1997 [1, 64].

Breeding goal	Yield and longevity	Yield	Longevity
Index on which selection is based	**ITEM**	**PIN**	**(LIN)**
Estimated annual response			
Margin (£ per cow)	15.6	15.3	2.7
Longevity (%)	0.14	0.00	0.81
Milk (kg per cow)	117	119	0
Fat (kg per cow)	4.9	5.0	0.0
Protein (kg per cow)	3.9	3.9	0.0

Table 6.24 Estimated selection responses following selection on three indices; ITEM, PIN and a third hypothetical index with a breeding goal of increasing longevity only (LIN). Longevity is defined as the % of cows not involuntarily culled in the first four lactations [64].

As the array of information available on AI bulls worldwide increases, it is becoming more important to have these different pieces of information combined into indexes which gives them appropriate emphasis. The development and wider use of the sort of index described here should simplify selection, and help breeding organisations and individual producers to breed cows which are both healthy and more profitable.

Evidence of genetic improvement and its value

In several countries selection experiments in dairy cattle have shown that substantial rates of improvement can be achieved by selection on PTAs or their equivalent. In several cases rates of genetic change of about two per cent per annum have been achieved [58]. While these experiments served a useful purpose in the early days of progeny testing and national genetic evaluation schemes, there is now more direct evidence available. The widespread use of AI in most countries has created strong genetic links across herds and years. Hence, the mean PTAs or PBVs from national BLUP evaluations, for animals born in successive years, provide estimates of national genetic trends. These provide direct evidence of the rates of genetic improvement that are being achieved in practice.

Figure 6.14 for example, shows the genetic trends in milk, fat and protein production in milk recorded Holstein Friesian cows in the UK and the US over a 20-year period. The graph shows cumulative change in cows' predicted breeding values, which equates to the expected genetic change in performance of the cows themselves. Substantial rates of genetic change of over 110 kg milk per annum were sustained in the US over this 20-year period [69, 70]. The equivalent trend in the UK was about 35 kg per annum over the whole period, but rates of change in PBVs were very low until the mid-1980s.

Evidence from the US shows that environmental and management influences on yield there have remained fairly similar since the mid-1970s, so genetic improvement has accounted for most of the increase in production since then. This is probably due to a combination of improved methods of genetic evaluation, greater awareness of the benefits of selective breeding, and a plateau in the responses being achieved to improved feeding and management.

Rates of genetic change have been accelerated in many countries over the last few decades by the importation of North American genetic material. Figure 6.15 shows the trends in genetic merit of bulls in some major dairying countries. The convergence of genetic merit over the last decade is largely due to these importations, but also partly due to improvements in local testing programmes.

Clearly, achieving genetic improvements in yield is only useful if it leads to improved profitability. Over the last couple of decades more research has been undertaken to investigate this. In the UK much of this effort has been at the University of Edinburgh/SAC Langhill Dairy Cattle Research Centre. From the early 1970s cows in the Langhill herd have been bred to bulls with the highest possible PTAs for kg fat plus protein. Since the mid-1970s a control herd of around national average PTA for kg fat plus protein has also been maintained at Langhill, with management and feeding identical to that of the selection herd.

Until the late 1980s cows at Langhill were fed during the housed period (October to May) on a complete diet, based on silage, brewers' grains and concentrates. This feed was offered *ad libitum* and the cows consumed, on average, 2.5 tonnes of concentrate per annum. By the late 1980s, yields of milk and milk solids in the selection and control herds differed by about 20 per cent, or around 90 kg of fat plus protein – very close to the amount expected from differences in the pedigree indexes of the two herds [44]. More recently, the performance of the cows has been evaluated under two different feeding systems, to test for a genotype x feeding system interaction (see Plate 29). Cows in the low forage group receive a diet similar to that fed since the outset of the trial (i.e.

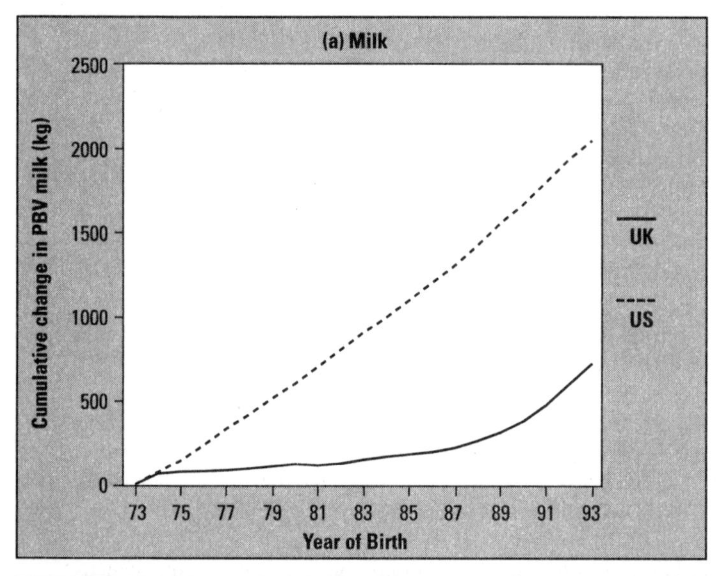

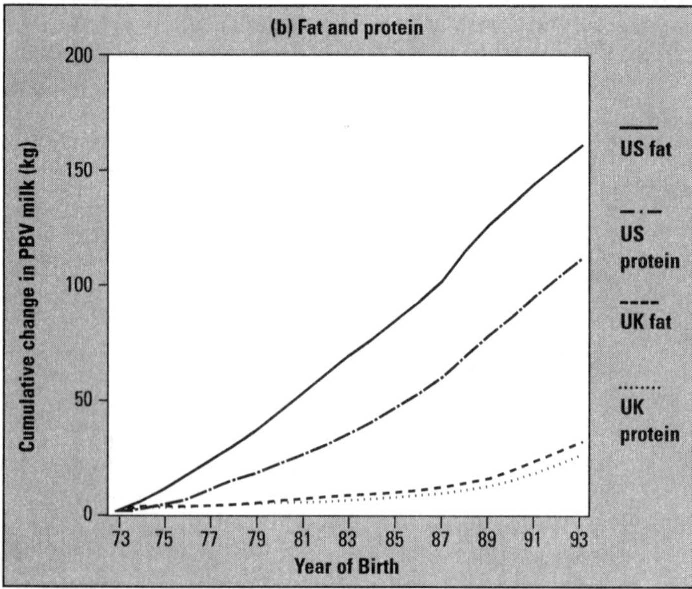

Figure 6.14 Cumulative change in cows' predicted breeding values for (a) milk, and (b) fat and protein production, in milk-recorded Holstein Friesian cows in the UK and the US over a 20-year period. US PBVs have been converted from pounds to kg (by multiplying by 0.4536). N.B. the graph shows *cumulative change* in PBV since 1973, and does not allow comparison of the average merit of cows in the two countries. (ADC; Dr G R Wiggans.)

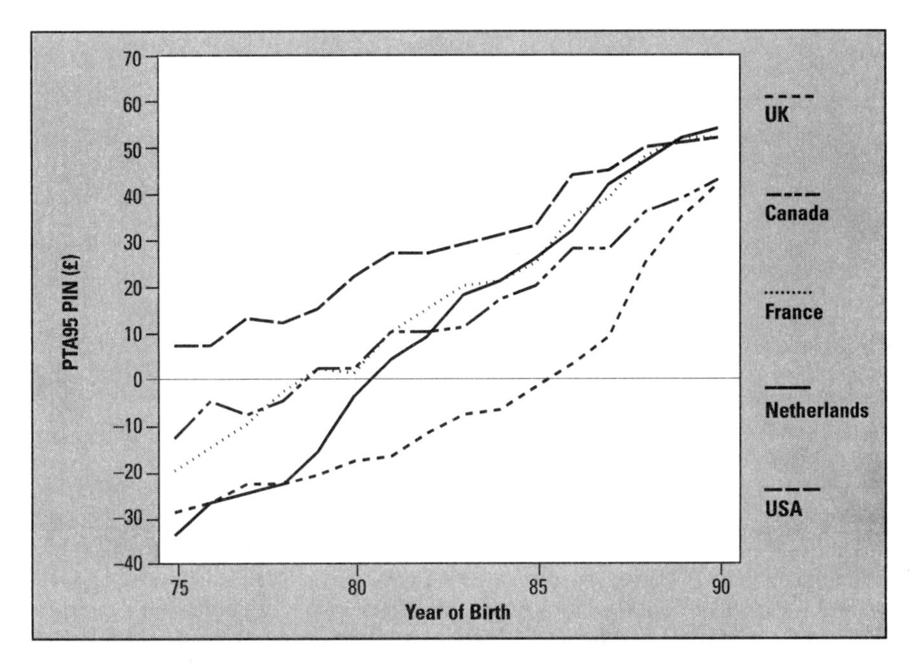

Figure 6.15 Trends in genetic merit of bulls in some major dairying countries. Trends are shown in terms of the UK production index PIN, which is based on PTAs for kg milk, kg fat and kg protein. This graph *does* allow direct comparison of estimated genetic merit of bulls among countries, since the data came from INTERBULL international evaluations. All the bulls used in a country and included in the evaluation, i.e. including foreign bulls, contribute to the mean PTA for that country. (ADC; INTERBULL.)

averaging about 2.5 tonnes of concentrates per cow per year). The rest of the animals are maintained on a higher forage system with around 1 tonne of concentrates per cow per annum, still fed as part of a complete diet. Results from the first four years of these feeding systems, for cows recorded over the first 38 weeks of lactation, are shown in Table 6.25. These results demonstrate that in both high and lower input systems cows of high genetic merit generally have higher physical and economic performance than their contemporaries of only average genetic merit.

Calculations based on the results above show that using bulls with high PTA values in a single year can lead to a £17,000 higher margin over food, health and reproduction costs in a 100-cow herd over a ten-year period. This benefit arises because of the improved performance of the bulls' daughters, and any of their offspring which enter the herd in the ten-year period. Continuous selection year on year would lead to even higher cumulative benefits. (These returns are discounted to account for the fact that semen is bought today, but the benefits from this investment occur in the future. The discounting assumes a 5% interest rate and 30% herd replacement rate).

Producers often question whether high-priced semen is good value, so these results have been used to derive break-even purchase prices for semen. The calculations used the same assumptions about discounting and replacement rates, and assumed that on average 4.4 straws of semen are needed to produce a replacement heifer. Applying these figures gave a discounted return per straw of

	Low Forage		High Forage	
	Selection	Control	Selection	Control
Milk yield (kg)	7569	6537	6372	5360
Fat %	4.19	4.20	4.54	4.37
Protein %	3.09	3.19	3.07	3.09
Fat plus protein yield (kg)	550	481	482	398
Dry matter intake (kg)	4803	4603	4149	3948
Gross energetic efficiency (MJ milk/MJ feed)	0.418	0.377	0.440	0.374
Live weight (kg)	610	610	601	590
Condition score (5 point scale)	2.55	2.70	2.45	2.59
Margin over all feed costs (£)	1008	825	914	712

Table 6.25 Performance and efficiency of selection and control line animals over the first 38 weeks of lactation, during four years of the Langhill feed intake trial from 1988/89 to 1991/92 [55].

semen of £2.60 per kg PTA for fat plus protein [65]. In other words, producers could afford to pay £2.60 more per straw for every extra kg PTA for fat plus protein and still break even in terms of margin over feed costs. This figure provides a useful guide when choosing semen from high merit bulls. For example, bulls could be ranked first on PTAs or index scores with semen prices alongside. Any bulls which have excessively high prices in relation to their fat and protein PTAs could then be excluded. All these figures are based on margins over feed costs, so the costs of acquiring additional quota need to be accounted for. Also, few producers aim only to break even, so they will have to pay less than £2.60 extra per straw per kg PTA for fat plus protein to achieve a return on their investment.

Genotype x environment interactions

The long-running debate on the importance of genotype x environment interactions (whether you need to 'breed horses for courses') has been refuelled by the massive importation of North American Holstein semen into Europe and elsewhere over the last couple of decades. Most experimental studies have indicated that, in temperate dairying systems, breed or strain x feeding system interactions are of little or no importance, at least for production per cow. In other words, the breeds or strains rank similarly in different feeding systems [22, 25], although the ranking of individual sires may still change.

The bigger the difference in performance between breeds or strains, and the bigger the difference between the environments in which they are compared, the greater the likelihood of finding important interactions. The differences between typical milk production systems in Canada and New Zealand are among the widest in temperate areas. Most dairy herds in Canada feed relatively high levels of concentrates, while most in New Zealand feed none at all. For this reason there was particular interest in the results of a trial to compare the performance of the progeny of Canadian and New Zealand Holstein Friesian sires in each of these two countries. Some of the results of this trial, which began in 1984/85, are shown in Table 6.26. There were major differences in milk production and live weight

	Nationality of dam and country in which daughters were reared			
	Canada		New Zealand	
	Canadian sires	NZ sires	Canadian sires	NZ sires
Milk yield (kg)	6699[1]	6392	3389[1]	3155
Fat yield (kg)	241	242	139	138
Protein yield (kg)	215	212	108[1]	105
Fat content (%)	3.64[1]	3.85	4.12[1]	4.38
Protein content (%)	3.22[1]	3.34	3.20[1]	3.33
Live weight at 18 mo. (kg)	447[1]	445	357[1]	347
Height at 18 mo. (cm)	131[1]	129	120[1]	118

[1] Significant difference between the two means within the country of birth, rearing and milking.

Table 6.26 A comparison of the performance of heifers sired by Canadian or New Zealand Holstein Friesian bulls, and born, reared and milked in Canada or New Zealand. ([22, 26, 45]; Dr R G Peterson.)

between heifers reared in the two countries, but these were largely due to differences in management and environment. Compared to the daughters of NZ sires, the daughters of Canadian sires produced slightly more milk of lower fat and protein content, and were slightly heavier and taller at eighteen months of age (though not at some other ages) in both Canada and New Zealand. So there is no evidence for a strain x environment interaction for production. However, despite the similar rankings of strains in the two environments, there were poor correlations between individual sire proofs estimated within country. In other words, individual sires re-ranked, depending on the country in which they were evaluated. Unfortunately, there is no published information available yet on the health, fertility and longevity of the two genotypes in the two countries. It is perhaps in these traits that important interactions are most likely to occur.

While the Langhill results in Table 6.25, and the results of similar studies in Northern Ireland, Eire and New Zealand [5], show no evidence of a major genotype x feeding system interaction, comparing the average production of the four groups of animals is a fairly crude test. Table 6.27 shows the results of a more recent analysis of the changes in production in relation to pedigree index for fat plus protein. These show that responses to selection on the low input (high forage) system tend to be lower than those in the high input system. (Similar results have been reported from a study in Australia [15].) This may indicate the beginning of a genotype x feeding system interaction. It appears that high merit cows may be unable to eat enough high forage diet to keep pace with their potential extra yield, especially their potential extra yield of protein. Despite mobilising more body tissue, they are still unable to match the increase in yield seen in their contemporaries on a higher concentrate diet. Research at Langhill, and elsewhere, will continue to monitor the potential development of an interaction, and to investigate new breeding strategies which increase emphasis on feed intake rather than tissue loss to fuel future increases in yield. Research is also in progress to examine the consequences of current breeding and feeding practices on the health, reproduction and welfare of dairy cattle. This may help to formulate guidelines on minimum standards of nutrition for cows of high genetic merit. Also, it will help in the formulation of more sustainable breeding goals for dairy cattle in future.

	Change in performance per kg increase in pedigree index for fat plus protein (regression coefficients with se in brackets)	
	Low Forage	High Forage
Milk yield (kg/day)	0.20 (0.03)	0.13 (0.03)
Fat %	0.001 (0.004)	0.000 (0.004)
Protein %	–0.002 (0.001)	–0.005 (0.001)
Fat yield (g/day)	7.8 (1.1)	5.4 (1.2)
Protein yield (g/day)	5.6 (0.8)	2.9 (0.9)
Dry matter intake (kg/day)	0.04 (0.01)	0.01 (0.01)
Live weight (kg)	0.1 (0.3)	0.0 (0.3)
Condition score (5 point scale)	–0.008 (0.002)	–0.009 (0.002)

Table 6.27 Changes in performance of Langhill cows in two feeding systems per kg increase in pedigree index for fat plus protein [66].

The wider use of international MACE evaluations, described in Chapter 5, will be helpful if genotype x environment interactions become more important in future. These evaluations allow for different genetic correlations between production in different countries. This should ensure that the ranking of imported sires, based on MACE proofs, is more closely tailored to the average production system in the country concerned. Also, in the absence of progeny tests in the relevant environment, the impact of interactions may be reduced through the use of broader indexes including affected traits (e.g. reproduction, feed intake).

Practical guidelines on selection

Pedigree breeders often have strong views on the sort of animal they wish to breed. They also have the interest and knowledge to wade through the vast quantities of information available in an increasingly international market to find animals suited to their breeding programme. It is generally healthy to have individualists and risk-takers in this sector of a livestock breeding industry, as long as there are universal yardsticks available for their customers to compare their products (e.g. the PTAs and indexes described above).

In contrast, many commercial dairy farmers are interested in a simpler set of rules to help them make sense of the advertising material, to become better informed in discussions with persuasive semen sales staff, and ultimately to breed more profitable cows. Although the tools available for selection are being updated continually, some general guidelines for selection to improve profitability are given below. These are aimed primarily at commercial herds, but most are equally relevant to pedigree breeders whose priority is to improve overall merit in commercially important traits:

- Check that the breed you are using is the most appropriate for your production circumstances by examining the results of objective breed comparisons and recording schemes such as those outlined in this chapter.

- Always use progeny tested AI bulls with reliable proofs, or teams of young or sib-tested AI bulls, to breed replacement heifers.

- Make a shortlist of bulls ranked in descending order on an appropriate index of overall economic merit (e.g. ITEM in the UK).

- If you firmly believe that your future markets will differ from the mainstream, for example because of a special milk payment scheme, or a strong trade for surplus heifer calves, then eliminate bulls from this list which have unfavourable production PTAs or type characteristics, or promote those with favourable characteristics. Similarly, if you have a particularly high incidence of mastitis you may wish to pay particular attention to bulls with favourable PTAs for SCC (somatic cell counts). However, this second-stage screening should be fairly light, or the benefits of ranking bulls on an economic index will be severely diluted.

- Eliminate bulls from the list if their semen is very highly priced relative to their genetic merit.

- From the remaining animals select a few (e.g. two to four) reliably proven progeny tested bulls (reliabilities over 75%), or more (e.g. eight or more) young or sib-tested bulls for use each year, or use bulls of both types. In theory, a single reliably tested bull is sufficient, but selecting more bulls gives scope to avoid matings between close relatives, and reduces the risks of all of the semen coming from a chance low quality batch.

- Avoid matings between close relatives (see Chapter 4 for more details). (Breed societies or recording agencies may provide a service for members producing mating lists to help avoid matings between cows currently in the herd and a nominated panel of bulls.)

- Avoid using bulls with poor PTAs for calving ease when these are available, or with a track record of calving difficulties.

- If there are extra heifer calves available as potential replacements, calculate a pedigree index for them for an appropriate index of economic merit and, subject to functional fitness, select those with the highest pedigree indexes as replacements.

- Manage the herd to achieve high fertility and, where appropriate, a compact calving pattern. This will maximise the opportunity to breed replacements from the top heifers and cows in the herd, as identified by PTAs or indexes from recording/evaluation agencies. It also allows a higher proportion of the herd to be mated to a beef bull, where appropriate.

Summary

● Returns from milk production are the most important returns in both high and low input dairying systems. Feed costs are the most important variable costs in both high and low input systems, although the relative importance of concentrate versus forage costs differs substantially between systems.

● Generally the highest yielding breeds, and the highest yielding individuals within breeds, have the highest gross efficiency of conversion of feed energy to milk energy and produce the highest margins over feed (and other) costs. While production per cow is a major component of profitability in higher input dairying systems, production per hectare is usually more important in lower input pastoral systems. The ranking of breeds on production per hectare is less clear than that on production per cow.

● In most temperate dairying countries the Holstein Friesian breed is predominant. However, the Jersey breed is also numerically important in some countries with mainly pastoral production, such as New Zealand and Australia. In most countries, there is little use of crossbreeding between dairy breeds, other than to enable breed substitution. Over the last few decades the local strains of black and white cattle in many temperate countries have been largely substituted by higher-yielding North American Holstein Friesian strains.

● In the past, most within-breed selection programmes have concentrated on increasing yield, with some attention being paid also to the type or conformation of cows. However, there is now increasing interest in broader breeding goals and indexes because of: (i) the increased emphasis in payment schemes in many countries on the yield (or content) of components, particularly protein; (ii) the introduction of milk quotas in some countries and (iii) the actual or perceived deleterious effect of selection for yield on the health, fertility and welfare of cows.

● Progeny testing schemes are central to genetic improvement of dairy cattle in most countries. The basis of progeny testing schemes is the comparison of the milk production of the daughters of young bulls being tested with the daughters of other bulls recorded in the same herd, year and season. Results are then combined across herds, years and seasons and used to predict the genetic merit of the bulls being tested, and of other animals in the recorded population.

● The main advantage of progeny testing is that it produces accurate predictions of a bull's genetic merit, based on the performance of daughters in commercial herds. The disadvantages are that progeny testing schemes are costly to run, and the results take many years to materialise. Progeny testing schemes also depend on access to a large population of milk recorded cows, and so they are difficult to sustain in numerically small breeds.

- In the late 1970s and early 1980s nucleus breeding schemes using multiple ovulation and embryo transfer (MOET) and sib testing were proposed as an alternative to progeny testing. Production records from a bull's sisters are available much sooner than those from his offspring. These schemes reduced generation intervals substantially compared to those possible in progeny testing, but they also reduced the accuracy of selection on males. However, on balance, the gains from shorter generation intervals outweighed the losses from lower accuracy, and the proposed schemes were predicted to give higher rates of genetic gain.

- MOET nucleus schemes have been established in several countries now. However, most schemes have evolved into 'fast track' testing schemes which have become integrated into progeny testing schemes, rather than replacing them as originally proposed. Keeping nucleus herds open to the use of elite progeny tested bulls helps to maximise genetic gain and control inbreeding in the nucleus. There are also benefits for progeny testing schemes from this integration: (i) MOET nucleus herds can provide unbiased and more accurate assessment of potential dams of young bulls for progeny testing and (ii) MOET nucleus herds provide a valuable direct source of young bulls for progeny testing.

- Progeny testing schemes for dairy bulls in most countries are based on national or regional milk recording schemes. Traditionally, milk records of cows have only been used for genetic evaluation once lactations are complete. However, there is a growing use of 'records in progress' to predict total lactation yields for genetic evaluations. This can reduce substantially the time lag between making test inseminations from a young bull and getting an initial proof.

- Yields of milk, fat and protein are all moderately highly heritable, while fat and protein content are both highly heritable. There are high genetic correlations among milk, fat and protein yields, indicating that selection for any one of these three traits is expected to lead to increases in the other two. However, there are negative genetic correlations between all three yield traits and protein %, and between yields of milk or protein and fat %, indicating that selection for yield is likely to lead to a reduction in protein or fat content, or both.

- Assessments of type or conformation are widely used in dairy cattle breeding. Most schemes now use linear assessment, in which the characteristics of the cow are scored in relation to biological extremes. The results of linear assessment are more repeatable and more amenable to scientific analysis than those from earlier schemes. Most linear type traits have moderately high heritabilities, and many are useful predictors of traits of economic importance.

- A growing number of countries are recording and evaluating various measures of reproduction, health, workability and longevity, or traits associated with them. Most of these traits have relatively low heritabilities. However, there are unfavourable associations between some of them and production, so genetic evaluations are valuable. PTAs for these traits are incorporated into indexes of overall economic merit in several countries. In a few countries measurements of feed intake, live weight and body condition, or predictors of these traits, are also being used in indexes or are being investigated with this in mind.

- BLUP methods of predicting breeding values or transmitting abilities for milk production have been adopted by most major dairying countries over the last couple of decades. Increasingly, individual animal model BLUP is used. This includes all relationships between relatives and predicts breeding values for all animals. Evaluations are usually based on multiple lactations. The large amount of data involved in many dairy cattle evaluations often makes it impractical to estimate all environmental effects simultaneously with breeding values, so data are often pre-adjusted for some effects. For similar reasons, only single-trait evaluations are currently feasible in many larger dairy populations.

- The principles and methods of genetic evaluation of type, reproduction, health, workability and longevity traits are usually similar to those for milk production traits. The nature of many non-production traits means that they require some form of transformation before analysis, or special consideration on the method of presentation of PTAs.

- In countries with wide use of imported animals or semen, foreign information plays an important role in dairy cattle breeding. Converted foreign production proofs or international evaluations for production are often used before local information becomes available, or they are combined with local information. Procedures for using foreign information are already well developed for milk production traits, are under development for type traits, and are the least developed for traits other than production and type.

- In many countries PTAs on individual traits are combined using selection indexes. As the array of information available on AI bulls worldwide increases, it is becoming more important to have these different pieces of information combined into indexes which gives them appropriate emphasis. The development and wider use of indexes of overall economic merit should simplify selection, and help to breed cows which are both healthier and more profitable.

- Selection experiments in dairy cattle have shown that substantial rates of genetic improvement in production (about two per cent per annum) can be achieved by selection on PTAs or their equivalent. Lower but still substantial rates of change have been made for several decades in some industry populations, e.g. in North America and New Zealand. Estimates of industry genetic trends have been much lower in many other countries (e.g. the UK) until recently. In many countries trends have increased now as a result of importation of genetic material (mostly directly or indirectly from North America), improvements in local breeding programmes, or both. Most studies show that these changes in production are associated with higher margins over feed and other costs.

● Most experimental studies indicated that, in temperate dairying systems, breed or strain x feeding system interactions are of little or no importance for production. However, several studies show that the ranking of sires, or the scale (but not the direction) of the response to selection, differs between feeding systems or environments. There are also some indications of interactions for traits other than milk production, including reproduction, health and feed intake. In the absence of progeny tests in the relevant environment, the impact of such interactions may be reduced through the use of broader indexes including affected traits (e.g. reproduction, feed intake) or predictors of these, and the use of MACE international evaluations.

References

1. Animal Data Centre. 1997. *UK Statistics for Genetic Evaluations (Dairy). February 1997.* ADC, Rickmansworth.
2. Boutflour, R. 1967. *The High Yielding Dairy Cow.* Crosby Lockwood, London.
3. Brandsma, J. and Banos, G. 1996. *Sire evaluation procedures for non-dairy production and growth and beef production traits practised in various countries 1996.* INTERBULL Bulletin No. 13, International Bull Evaluation Service, Department of Animal Breeding and Genetics, Uppsala, Sweden.
4. Brigstocke, T.D.A., Lindeman, M.A., Cuthbert, N.H., et al. 1982. 'A note on the dry matter intake of Jersey cows.' *Animal Production*, 35:285–7.
5. British Grassland Society. 1996. *Grass and forage for cattle of high genetic merit. Papers presented at the British Grassland Society Winter Meeting 1996.* British Grassland Society, Reading, Berks.
6. Brotherstone, S. 1995. 'Estimation of genetic parameters for the parlour traits in Holstein-Friesian dairy cattle.' *Book of Abstracts of the 46th Annual Meeting of the European Association for Animal Production.* Poster G1.24, p.12.
7. Brotherstone, S. and Hill, W.G. 1991. 'Dairy herd life in relation to linear type traits and production. 2. Genetic analyses for pedigree and non-pedigree cows.' *Animal Production*, 53: 289–97.

8. Bryant, A. M. 1993. 'Friesians and Jerseys.' *Dairying Research Corporation, Hamilton, Research Update*, June 1993, pp. 3–4.

9. Bryant, A. M., Cook, M.A.S. and MacDonald, K.A. 1985. 'Comparative dairy production of Jerseys and Friesians.' *Proceedings of the New Zealand Society of Animal Production*, 45:7–11.

10. Coffey, E.M., Vinson, W.E. and Pearson, R.E. 1986. 'Potential of somatic cell concentration in milk as a sire selection criterion to reduce mastitis in dairy cattle.' *Journal of Dairy Science*, 69:2163–72.

11. Cunningham, E.P. 1983. 'Structure of dairy cattle breeding in Western Europe and comparisons with North America.' *Journal of Dairy Science*, 66:1579–87.

12. Distl, O., Koorn, D.S., McDaniel, B.T., et al. 1990. 'Claw traits in cattle breeding programmes: report of the EAAP working group "Claw quality in cattle".' *Livestock Production Science*, 25:1–13.

13. Emanuelson, U. 1988. 'Recording of production diseases in cattle and possibilities for genetic improvements. A review.' *Livestock Production Science*, 20:89–106.

14. Food and Agriculture Organisation of the United Nations. 1996. *FAO Production Yearbook 1995. Vol 49*. Food and Agriculture Organisation of the United Nations, Rome.

15. Fulkerson, W., Davison, T. and Goddard, M. 1996. 'The genetic merit by level of feeding study at Wollongbar.' *Research to farm*. NSW Agriculture, Wollongbar, Australia.

16. Genus. 1996. *Genus Milkminder Annual Report* 1995–96. Genus Management, Wrexham, Clwyd.

17. Groen, A.F., Hellinga, I., and Oldenbroek, J.K. 1994. 'Genetic correlations of clinical mastitis and feet and legs problems with milk yield and type traits in Dutch Black and White dairy cattle.' *Netherlands Journal of Agricultural Science*, 42:371–8.

18. Harris, B.L., Clark, J.M. and Jackson, R.G. 1996. 'Across breed evaluation of dairy cattle.' *Proceedings of the New Zealand Society of Animal Production*, 56:12–15.

19. Hayes, J.F., Cue, R.I. and Monardes, H.G. 1992. 'Estimates of repeatability of reproductive measures in Canadian Holsteins.' *Journal of Dairy Science*, 75:1701–6.

20. Hinks, C.J.M. 1978. 'The use of centralised breeding schemes in dairy cattle improvement.' *Animal Breeding Abstracts*, 46:291–7.

21. Hoekstra, J., van der Lugt, A.W., van der Werf, J.H.J. and Ouweltjes, W. 1994. 'Genetic and phenotypic parameters for milk production and fertility traits in upgraded dairy cattle.' *Livestock Production Science*, 40:225–32.

22. Holmes, C.W. 1995. 'Genotype x environment interactions in dairy cattle: a New Zealand perspective.' In T.L.J. Lawrence, F.J. Gordon and A. Carson (eds), *Breeding and Feeding the High Genetic Merit Dairy Cow*. Occasional Publication No. 19, British Society of Animal Science, pp. 51–8.

23. Holstein Friesian Society of Great Britain and Ireland. 1995. 'Focus on type.' *Holstein Friesian Journal*, 77:50–9.

24. Huizinga, H. 1995. 'Delta – a review of the current situation.' *British Cattle Breeders Club Digest* No. 50, pp. 64–71.

25. Jasiorowski, H., Reklewski, Z. and Stolzman, M. 1983. 'Polish experiment on the comparison of ten Friesian cattle strains.' *International Dairy Federation, document 165*, pp. 7–19.

26. Kakwaya, D. and Peterson, R.G. 1991. 'Comparison of Canadian and New Zealand Holstein heifers for growth traits in Canadian herds.' *Journal of Dairy Science*, 74 (Supplement 1):231.

27. Korver, S. 1984. 'Efficiency of breeds for milk and beef production.' *British Cattle Breeders Club Digest No. 39*, pp. 21–6.

28. Livestock Improvement. 1996. *Dairy Statistics* 1995–1996. Livestock Improvement Corporation Limited, New Zealand Dairy Board, Hamilton, New Zealand.

29. Lohuis, M.M. 1993. *Strategies to improve efficiency and genetic response of progeny test programs in dairy cattle.* Doctoral thesis, University of Guelph, Canada.

30. Lohuis, M.M., Smith, C. and Dekkers, J.C.M. 1993. 'MOET results from a dispersed hybrid nucleus programme in dairy cattle.' *Animal Production*, 57:369–78.

31. McGuirk, B.J. 1990. 'Operational aspects of a MOET nucleus dairy breeding scheme.' *Proceedings of the 4th World Congress on Genetics Applied to Livestock Production*, Vol. XIV, pp. 259–62.

32. McGuirk, B.J. and Gimbert, A. 1996. 'The Genus sire improvement programme.' *British Cattle Breeders Club Digest* No. 51, pp. 1-4.

33. Meijering, A. 1984. 'Dystocia and stillbirth in cattle – a review of causes, relations and implications.' *Livestock Production Science*, 11:143–77.

34. Mrode, R.A., Swanson, G.J.T. and Winters, M.S. 1995. 'Genetic parameters for somatic cell counts (SCC) for three dairy breeds in the United Kingdom.' *Animal Science*, 60:545.

35. National Dairy Council. 1997. *Dairy Facts and Figures 1996.* National Dairy Council, London.

36. Nicholas, F.W. 1979. 'The genetic implications of multiple ovulation and embryo transfer in small dairy herds.' Paper presented at the *Annual Conference of the European Association for Animal Production*, Harrogate.

37. Nicholas, F.W. and Smith, C. 1983. 'Increased rates of genetic change in dairy cattle by embryo transfer and splitting.' *Animal Production*, 36:341–53.

38. Oldenbroek, J.K. 1984. 'Differences in efficiency for milk production between four dairy breeds.' Paper presented at the *Annual Conference of the European Association for Animal Production*, The Hague, Netherlands.

39. Oldenbroek, J.K. 1984. 'Holstein Friesians, Dutch Friesians and Dutch Red and Whites on two complete diets with a different amount of

roughage: performance in first lactation.' *Livestock Production Science*, 11:401–15.

40. Oldenbroek, J.K. 1986. 'The performance of Jersey heifers and heifers of larger dairy breeds on two complete diets with different roughage contents.' *Livestock Production Science*, 14:1–14.

41. Pander, B.L., Hill, W.G. and Thompson, R. 1992. 'Genetic parameters of test day records of British Holstein-Friesian heifers.' *Animal Production*, 55:11–21.

42. Persaud, P. and Simm, G. 1991. 'Genetic and phenotypic parameters for yield, food intake and efficiency of dairy cows fed *ad libitum*. 2. Estimates for part lactation measures and their relationship with 'total' lactation measures.' *Animal Production*, 52:445–50.

43. Persaud, P., Simm, G. and Hill, W.G. 1991. 'Genetic and phenotypic parameters for yield, food intake and efficiency of dairy cows fed *ad libitum*. 1. Estimates for 'total' lactation measures and their relationship with live-weight traits.' *Animal Production*, 52:435–44.

44. Persaud, P., Simm, G., Parkinson, H. and Hill, W.G. 1990. 'Relationships between sires' transmitting ability for production and daughters' production, food intake and efficiency in a high-yielding dairy herd.' *Animal Production*, 51:245–53.

45. Peterson, R.G. 1991. 'Evidence of a GxE with Canadian and New Zealand Holsteins.' Paper presented at the *Annual Conference of the European Association for Animal Production*, Berlin, Germany.

46. Philipsson, J. 1981. 'Genetic aspects of female fertility in dairy cattle.' *Livestock Production Science*, 8:307–19.

47. Philipsson, J., Foulley, J.L., Lederer, J., et al. 1979. 'Sire evaluation standards and breeding strategies for limiting dystocia and stillbirth.' Report of an EEC/EAAP working group. *Livestock Production Science*, 6:111–27.

48. Philipsson, J., Ral, G., and Berglund, B. 1995. 'Somatic cell count as a selection criterion for mastitis resistance in dairy cattle.' *Livestock Production Science*, 41:195–200.

49. Rendel, J.M. and Robertson, A. 1950. 'Estimation of genetic gain in milk yield by selection in a closed herd of dairy cattle.' *Journal of Genetics*, 50:1–8.

50. Residuary Milk Marketing Board of England and Wales. 1995. *EC Dairy Facts and Figures. 1994 edition*. The Residuary Milk Marketing Board of England and Wales, Thames Ditton, Surrey.

51. Residuary Milk Marketing Board of England and Wales. 1995. *UK Dairy Facts and Figures. 1994 edition*. The Residuary Milk Marketing Board of England and Wales, Thames Ditton, Surrey.

52. Robertson, A. and Rendel, J.M. 1950. 'The use of progeny testing with artificial insemination in dairy cattle.' *Journal of Genetics*, 50:21–31.

53. Russell, K. 1974. *The Principles of Dairy Farming*. 7th ed. Farming Press, Ipswich.

54. SAC. 1995. *SAC Farm Management Handbook 1995/96*. SAC, Edinburgh.

55. SAC/University of Edinburgh. 1994. *Langhill 94. Report and herd brochure from the Langhill Dairy Cattle Research Centre*. SAC, Edinburgh.

56. SAC/University of Edinburgh. 1995. *Langhill 95. Report and herd brochure from the Langhill Dairy Cattle Research Centre*. SAC, Edinburgh.

57. Simm, G., Veerkamp, R.F. and Persaud, P. 1994. 'The economic performance of dairy cows of different predicted genetic merit for milk solids production.' *Animal Production*, 58:313–20.

58. Smith, C. 1984. 'Rates of genetic change in farm livestock.' *Research and Development in Agriculture*, 1:79–85.

59. Swalve, H.H., Topf, C. and Langholz, H.-J. 1992. 'Estimation of genetic parameters for differently recorded fertility data in dairy cattle.' *Journal of Animal Breeding and Genetics*, 109:241–51.

60. Swanson, G. 1995. 'Reliability: what does it mean?' *British Dairying*, 2:8–10.

61. van Arendonk, J.A.M., Groen, A.F., van der Werf, J.H.J. and Veerkamp, R.F. 1995. 'Genetic aspects of feed intake and efficiency in lactating dairy cows.' *Proceedings of Symposium on Utilization of local feed resources in dairy cattle. Perspectives of environmentally balanced production systems*. European Association for Animal Production publication, Wageningen Institute of Animal Science, Wageningen, Netherlands.

62. Veerkamp, R.F. and Brotherstone, S. 1997. 'Genetic correlations between linear type traits, food intake, live weight and condition score in Holstein Friesian dairy cattle.' *Animal Science*, 64: 385-92.

63. Veerkamp, R.F. and Emmans, G.C. 1995. 'Sources of genetic variation in energetic efficiency of dairy cows.' *Livestock Production Science*, 44:87–97.

64. Veerkamp, R.F., Hill, W.G., Stott, A.W. et al. 1995. 'Selection for longevity and yield in dairy cows using transmitting abilities for type and yield.' *Animal Science*, 61:189–97.

65. Veerkamp, R.F. and Simm, G. 1992. *How much can you afford to pay for high genetic merit semen and high genetic merit heifers?* SAC Technical Note T318. SAC, Edinburgh.

66. Veerkamp, R.F., Simm, G. and Oldham, J.D. 1995. 'Genotype by environment interactions: experience from Langhill.' In T.L.J. Lawrence, F.J. Gordon and A. Carson (eds), *Breeding and Feeding the High Genetic Merit Dairy Cow*. Occasional Publication No. 19, British Society of Animal Science, pp. 59–66.

67. Villanueva, B. and Simm, G. 1994. 'The use and value of embryo manipulation techniques in animal breeding.' *Proceedings of the 5th World Congress on Genetics Applied to Livestock Production*, Vol. 20, pp. 200–207.

68. Visscher, P.M. and Thompson, R. 1992. 'Univariate and multivariate parameter estimates for milk production traits using an animal model. I. Description and results of REML analyses.' *Genetics, Selection, Evolution*, 24:415–30.

69. Wiggans, G.R. 1991. 'National genetic improvement programs for dairy cattle in the United States.' *Journal of Animal Science*, 69:3853–60.

70. Wiggans, G.R. and VanRaden, P.M. 1991. 'Method and effect of adjustment for heterogeneous variance.' *Journal of Dairy Science*, 74:4350–57.

71. Woolliams, J.A. and Smith, C. 1988. 'The value of indicator traits in the genetic improvement of dairy cattle.' *Animal Production*, 46:333–45.

Further reading

Animal Science (formerly *Animal Production*; Journal of the British Society of Animal Science)

Groen, A.F., Sölkner, J., Strandberg, E. and Gengler, N. (compilers). 1996. *Proceedings International Workshop on Genetic Improvement of Functional Traits in Cattle*, Gembloux, Belgium January 1996. INTERBULL Bulletin No. 12, International Bull Evaluation Service, Department of Animal Breeding and Genetics, Uppsala, Sweden.

Groen, A.F., Steine, T., Colleau, J.-J., et al. 1997. 'Economic values in dairy cattle breeding, with special reference to functional traits. Report of an EAAP working group.' *Livestock Production Science* (in press).

INTERBULL Bulletins. International Bull Evaluation Service, Department of Animal Breeding and Genetics, Uppsala, Sweden.

Journal of Animal Science (Journal of the American Society of Animal Science)

Journal of Dairy Science (Journal of the American Dairy Science Association)

Livestock Production Science (Journal of the European Association for Animal Production).

Owen, J.B. and Axford, R.F.E. (eds). 1991. *Breeding for disease resistance in farm animals*. CAB. International, Wallingford.

Philipsson, J., Banos, G. and Arnason, T. 1994. 'Present and future uses of selection index methodology in dairy cattle.' *Journal of Dairy Science*, 77:3252–61.

*Proceedings of the Association for the Advancement of Animal Breeding and Genetics (*formerly the *Australian Association of Animal Breeding and Genetics)*

Proceedings of the New Zealand Society of Animal Production

Proceedings of the World Congresses on Genetics Applied to Livestock Production

Visscher, P.M., Bowman, P.J. and Goddard, M.E. 1994. 'Breeding objectives for pasture based dairy production systems.' *Livestock Production Science*, 40:123–37.

Wiggans, G.R., Misztal, I. and Van Vleck57, L.D. 1988. 'Implementation of an animal model for genetic evaluation of dairy cattle in the United States.' *Journal of Dairy Science*, 71, (Supplement 2):54–69.

CHAPTER 7
Beef cattle breeding

Introduction

The aim of this chapter is to discuss the breeding goals, the breeds and crosses used, the selection criteria and the methods of testing and genetic evaluation used in beef cattle breeding. As before, the emphasis is on genetic improvement in temperate production systems. Figure 7.1 shows the world production of beef and veal by continent. North and Central America have the highest production at about 30% of the world total, and Europe follows with about 21% of global production. Figure 7.2 shows details of the production in the world's fifteen highest producing countries. The USA has by far the highest production, followed by Brazil, China, the Russian Federation, and Argentina. Several European countries also appear in the list.

Beef cattle breeding in temperate countries is less homogeneous than dairy cattle breeding, so it is worth setting the scene a bit further at this stage. In most European countries over 50% of beef production is from pure dairy or dual-purpose breeds, either from cull cows, from male calves or from surplus female calves not required as dairy herd replacements (see Figure 7.3). So this chapter includes discussion of beef breeding goals and criteria in dairy and dual-purpose breeds, as well as considering these issues in specialised beef breeds. In addition to this direct contribution from dairy herds, there is an indirect contribution to beef production through crossing of dairy cows to beef bulls. This produces beef x dairy calves for slaughter and, in some countries, beef x dairy suckler cows (i.e. cows kept for rearing beef calves).

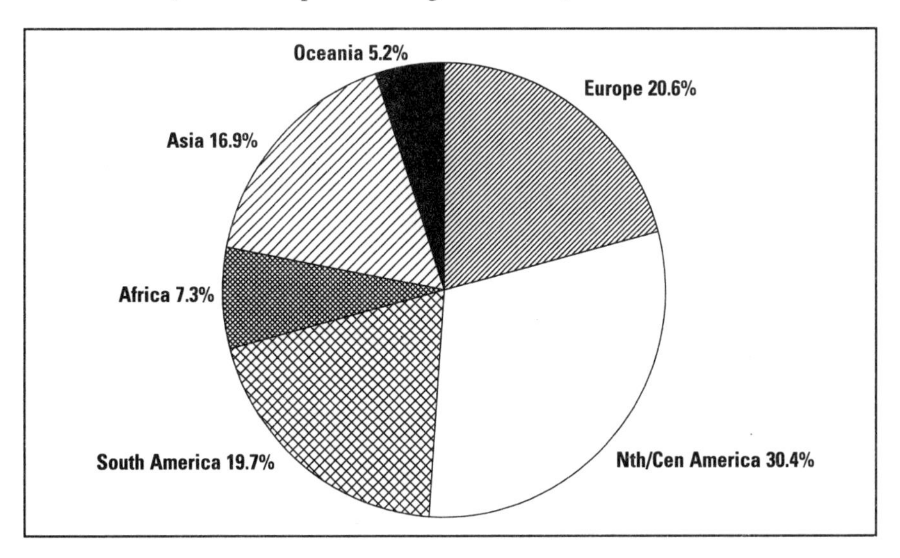

Figure 7.1 Proportion of world beef and veal production by continent in 1995 [18].

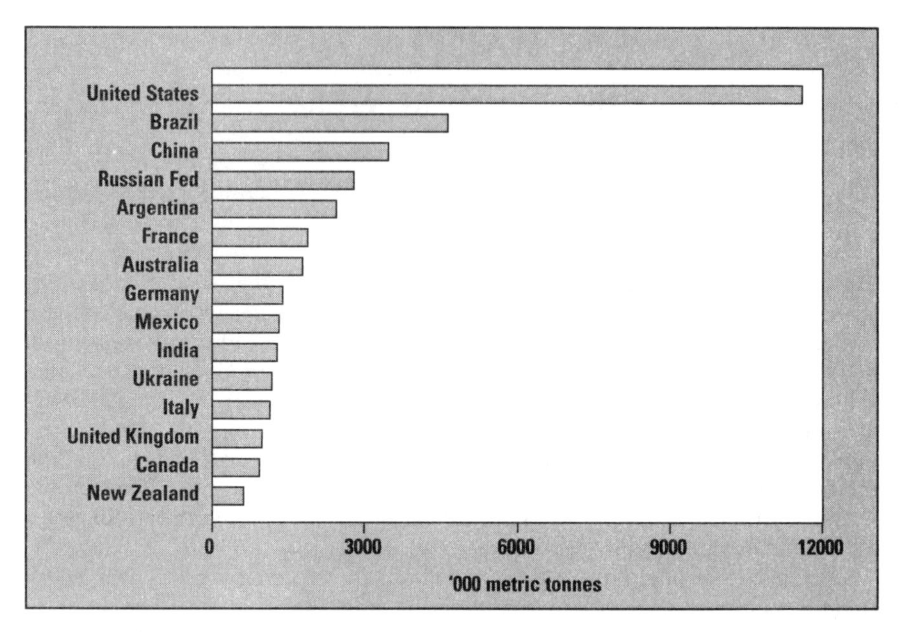

Figure 7.2 Beef and veal production in 1995 in the world's fifteen highest producing countries [18].

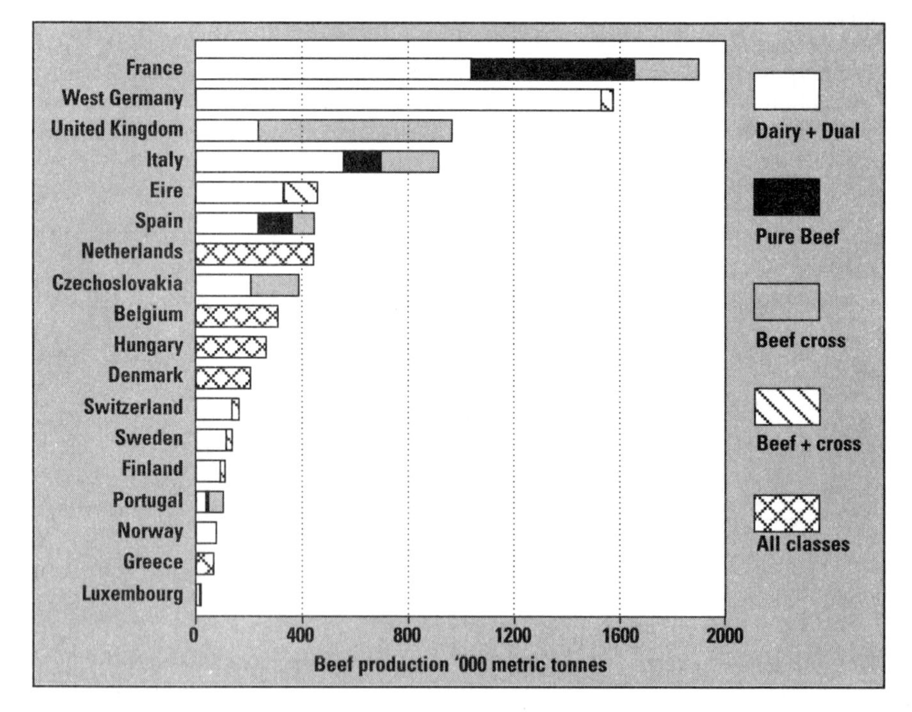

Figure 7.3 Total annual indigenous beef production in eighteen European countries in the late 1980s. Where the information was available, the graph shows the contribution from dairy and dual-purpose breeds, specialised beef breeds, and crosses between them [45].

Many European countries have only a small specialised beef cattle breeding industry; in many cases this comprises purebred terminal sire breeds to supply beef bulls for crossing in dairy or dual-purpose herds. In contrast, pure beef breeds account for a high proportion of total production in France and, to a lesser extent, in Italy and Spain. In the UK and Eire suckler herds of beef x dairy cows, derived as a by-product of the use of beef bulls in dairy herds, make an important contribution to total output. In other major temperate beef producing countries such as the US, Canada, parts of South America, New Zealand and parts of Australia, beef production is based on extensively grazed or ranched cows, mainly of pure British beef breeds like the Hereford, Aberdeen Angus and Shorthorn, or crosses among them (see Plates 31 and 32). In some of these countries, like the US, Canada and parts of Australia, this extensive pre-weaning regime is usually followed by a more intensive finishing period in feedlots (see Plate 33). Hence, there is a much wider range of production systems, breeding goals and testing programmes than those in temperate dairy cattle breeding. The more extensive nature of many production systems, and the widespread use of crossbred animals in the commercial sector of most beef industries, also means that performance recording and genetic improvement are concentrated in a much smaller sector of the population than in dairy cattle breeding.

Breeding goals

From the brief introduction above it is apparent that there are two broad categories of beef production in many countries: (i) beef production from dairy and dual-purpose herds and (ii) beef production from specialised beef herds. Within the specialised beef sector there is further differentiation into terminal sire and maternal breeds, crosses or lines. Terminal sire breeds also get used in dairy and dual-purpose herds. Each of these categories of use requires a distinct set of beef breeding goals, or at least different priorities, and these are discussed below.

Beef breeding goals in dairy and dual-purpose breeds

At first sight it seems efficient to breed for both milk and meat production from the same type of animal. This is probably true in small-scale production systems, but it appears to be less true in large, specialised farming systems. Most of the evidence suggests that there is an unfavourable genetic correlation between milk production and growth or carcass characteristics [38]. In other words, selection for either milk or beef merit tends to cause a deterioration in the other characteristics. It is possible to select for improvement in both characteristics at once, but the rate of progress which can be achieved is much lower than that possible with single-trait selection, or selection for two traits which are favourably correlated (as explained in Chapters 4 and 5).

Some breeds or strains, like the Simmental strains in several continental European countries, have achieved fairly high productivity in both milk and beef traits as a result of many generations of selection. Even for these strains, it is difficult to compete nationally and internationally with both specialised milk and specialised beef breeds. As a result there is a general trend towards milk production from more specialised dairy cattle breeds and strains. In some countries there is still an attempt to limit the expected deterioration in beef merit by performance testing dairy bulls for growth and conformation, and pre-selecting bulls on these traits prior to progeny testing for milk production. In other countries, the deterioration in beef merit of the specialised dairy strains is compensated for, at least partially, by crossing those females not required to breed replacement dairy heifers to specialised beef breeds. So, in temperate dairying countries with large-scale specialised industries, breeding goals in dairy breeds have little or no emphasis on beef traits. Even in dual-purpose breeds, the emphasis on beef traits is likely to be secondary to that on milk traits. In some countries with a strong history of using dual-purpose breeds (e.g. Switzerland), additional support via government has had to be introduced to maintain their role in traditional farming systems.

Terminal sires for use in dairy herds and specialised beef herds

Terminal sire beef breeds (i.e. those specially selected to sire the slaughter generation of animals) are used in dairy herds for two main purposes. The first is to mate to dairy heifers to reduce the incidence of calving difficulties compared to that following matings to a dairy sire. The second is to mate to mature dairy cows which are not required to breed replacement dairy heifers.

The main priority of dairy farmers selecting a beef bull for use on heifers is calving ease. Difficult calvings are costly, both directly and because they delay rebreeding, depress milk production and compromise both cow and calf survival and welfare. Hence, dairy heifers have often been mated to bulls from one of the easier-calving beef breeds, such as the Hereford, Aberdeen Angus and Limousin. However, mating dairy heifers to a beef bull is becoming less common as more dairy producers realise that their heifers are often the highest genetic merit animals in the herd, and hence valuable as dams of replacements. Also, the wider availability of calving ease evaluations in dairy breeds means that it is easier to select a dairy sire suitable for mating to heifers.

Although calving ease is still important when beef sires are mated to mature dairy cows, the incidence of calving difficulties is lower in mature cows than heifers. Hence, there is more scope to select beef bulls for other attributes to maximise returns from calf sales. Many beef cross calves born on dairy farms are sold at a young age. So increasing calf weight and conformation (muscularity or shape) are important breeding goals for dairy farmers choosing a beef breed, or individual beef sire – though increasing weight and conformation tends to conflict with the aim of reducing calving difficulties.

The performance of beef cross calves in later life is of little direct concern to most dairy farmers, although, in theory, sire breeds or individual sires with high genetic merit for later performance ought to result in higher rewards in the market place. These market signals work reasonably well at the level of sire breed, especially if these produce easily recognised, colour-marked calves. There is less widespread discrimination among sires within a breed, although in some countries AI companies, beef breed societies or recording agencies have schemes to identify and promote beef sires for use in dairy herds which combine acceptable calving ease with good growth and carcass characteristics (e.g. in Britain the Lim-Elite and Blue-Elite progeny testing schemes aim to identify and promote high merit Limousin and Belgian Blue bulls; also, several AI companies actively evaluate beef bulls for use in dairy herds). The more comprehensive identification of cattle, together with the adoption of improved technologies for storing and transferring breeding values and performance records, may help to improve communication and market signals between different sectors of the industry in future.

In many of the specialised beef production systems in temperate countries there is widespread use of crossbreeding. Often this is to achieve complementary use of breeds, as explained in Chapter 3. Usually small or medium-sized breeds or crosses are used as dam lines, and larger breeds are used as terminal sires. Larger breeds are valuable as terminal sires as they usually have faster growth rate, and produce leaner carcasses at a given weight than smaller breeds. (Generally as animals grow towards maturity they get fatter. When different breeds are compared at similar degrees of maturity in live weight – say when each breed reaches 60% of its expected mature weight – they tend to have similar proportions of fat. However, breeds differ in fatness quite markedly when they are compared at the same weight. At a given slaughter weight, progeny of larger breeds, or larger sires within a breed, tend to be leaner. Alternatively, at a given level of fatness, progeny of larger breeds, or larger sires within a breed, generally produce heavier carcasses.) Although ease of calving is still important when terminal sire breeds are used in specialised beef breeding herds, their main role is to improve the growth and carcass characteristics of their crossbred offspring. Hence, there is interest in growth, carcass merit and calving ease whether terminal sire breeds are being used in dairy or beef herds, but the emphasis on these traits varies between these uses.

The definition of carcass merit depends to some extent on whether commercial animals are sold at live auctions or directly to abattoirs, but it usually encompasses some measure of weight, fatness and conformation. In theory, good communication between sectors of the industry should mean that breeding goals are similar whether animals are marketed dead or alive. However, in practice they often differ. For example, in the UK a premium per kg live weight is paid in live markets for animals of continental breeds and crosses, and especially those with good conformation. Except in extreme cases, fatness is a secondary concern, although buyers use breed, weight or age and sex as indirect indicators of fatness. For animals sold directly to abattoirs, overall returns will depend largely on carcass weight and visual scores for fatness and conformation, although breed and sex may modify the carcass price.

Table 7.1 shows the fat and conformation grid used to classify beef carcasses by the Meat and Livestock Commission (MLC) in Britain. (This method of classification is based on a standard one employed across the EU, but some fat and conformation classes are subdivided in the British version. MLC is an agency funded mainly by a levy on carcasses, and is involved in marketing, promotion, research and technology transfer in the meat and livestock industries.) The table also shows the average price differentials paid for carcasses in different fat and conformation classes in 1993/94. This sort of information is used to derive economic values for selection indices for beef cattle. In general, the relative economic value for conformation derived from market information is lower than the value perceived by breeders, and to some extent lower than the value attached to conformation in live markets. This may be explained partly by the value of

Conformation class ↓ Poorer conformation		Fat class → Fatter 1 or 2	3	4L	4H	5L	5H
	E	103.2	103.2	103.4	103.5	98.5	97.8
	U	102.2	102.2	102.4	102.5	97.5	96.9
	R	99.7	100.0	100.1	99.8	95.9	91.9
	O+	97.1	97.3	97.2	96.4	92.7	90.6
	O–	93.1	93.9	93.5	92.7	90.0	87.6
	P	89.3	90.0	89.6	88.8	86.2	84.0

Table 7.1 The fat and conformation grid used to classify beef carcasses by the MLC in Britain. The figure also shows the price differentials for carcasses of different fat and conformation classes in 1993/94. Prices are expressed in percentage units, relative to the price of an R3 carcass. ([32]; Dr B J McGuirk and Dr P R Amer, based on MLC statistics.)

conformation scores in the live animal in predicting killing out percentage, but it probably also reflects the widely held belief in many meat industries that high conformation is of direct value, apart from its value as a predictor of additional meat yield. In many North American and Australasian markets, a premium is paid for high marbling – that is, high levels of visible intramuscular fat in the eye muscle. Particularly in North America, this premium for marbling is based on its value as an indicator of good eating quality. Recently, interest in marbling in several exporting countries has been fuelled by its importance in the lucrative Japanese beef market.

Meat eating quality is becoming an increasingly important issue with consumers and the meat industry in richer countries. The post-slaughter treatment of carcasses, especially chilling rate, ageing and method of hanging, are known to have important effects on eating quality [15, 17]. However, there is less information on pre-slaughter effects on beef eating quality, such as breed, breeding value within breed, or production system. The information that is available suggests that there are breed differences in indirect measures of meat quality, especially marbling, colour and fibre type. There are differences in tenderness between breed types: double-muscled breeds generally have the most tender meat, followed by other *Bos taurus* breeds, with *Bos indicus* breeds ranking lowest. There are less consistent differences in tenderness between the non-double muscled *Bos taurus* breeds, or between any of the breed types, in juiciness and flavour. Despite this, there are consistent reports of substantial within-breed genetic variation in both indirect and direct measures of eating quality [24]. This indicates that there is scope for improvement through within-breed selection, though in the absence of good live animal predictors of eating quality this is difficult to achieve without progeny testing. Strong economic incentives would be required to justify the introduction of progeny testing, if this is not already used. However, with better communication between sectors of the industry and better continuity of identification of animals and carcasses useful information for second-stage selection could be generated gradually. For example, laboratory or taste-panel assessments of eating quality could be obtained on samples of commercial cattle to produce EBVs for purebred relatives. In future, molecular markers of eating quality may allow more efficient selection programmes (see Chapter 9).

Breeding replacement females for specialised beef herds

The main breeding goals for cows in specialised beef herds, in addition to adequate growth and carcass merit, are good fertility, ease of calving, good maternal ability (which includes adequate milk production and good mothering ability) and low or intermediate mature size, in order to reduce cow maintenance requirements. These individual goals are sometimes aggregated into measures such as **weight of calf weaned per cow per annum,** or **weight of calf weaned per kg cow mature weight per annum.**

The ability of animals to withstand extreme climates and to tolerate low quality feed and periods of feed shortage is also important in some areas, and there is often concern about possible genotype x environment interactions for these 'adaptation' traits. These traits are often difficult to define, and the most practical route towards achieving within-breed improvement is often simply to record and select on performance in the harsh environment concerned [43] (see Plate 34). The emphasis on each of these traits will vary depending on the production system and breed or crossbred type of cow used. In some cases the traits of importance will be best improved by selection, in others they will be best improved by crossbreeding. For instance, the fertility of crossbred cows is usually high as a result of heterosis, and so is of less concern in selection within the component breeds. Also, in beef x dairy suckler cows milk production is usually adequate and so other traits assume greater importance in the overall breeding goal. The mature size of specialised beef cows is usually kept in check by choosing a small or medium-sized breed for use either in the purebred form, or as a component breed of the final cross.

Beef x dairy heifers have made an important contribution to suckler herds in the UK and Eire in the past. However, these heifers have been produced largely as a by-product of crossing dairy cows to beef bulls to reduce calving difficulties or to produce calves with improved growth and carcass merit for finishing, rather than to produce breeding beef females *per se*. The wider use of Holstein sires in dairy herds in the UK and Eire is leading to an increase in the size and a reduction in the beef conformation of dairy cows. Also, the number of dairy cows in most European countries is declining rapidly as a result of quotas on total production and increasing production per cow. At the same time, there is wider use of large continental beef breeds as crossing breeds in dairy herds, rather than traditional British beef breeds. Together these factors threaten the traditional supply of medium-sized, well-conformed replacement beef x dairy females. This may create opportunities for some breeds or individual breeders to concentrate on breeding goals relevant for beef sires to cross to the more specialised dairy cow in use today. However, this alone is unlikely to meet the demand for replacement females, and more beef herds in these countries are already breeding their own replacements. This often involves the use of one of the rotational crossing systems described in Chapter 3.

Despite these apparently different roles for beef cattle, in practice many beef breeds and individual breeders have attempted to fulfil all of them. For example, in response to the dramatic increase in the use of the large continental beef breeds such as the Charolais, Simmental and Limousin in many temperate beef-producing countries over the last few decades, breeders of the traditional British beef breeds have concentrated on increasing growth rate. They have achieved some success in this, but in most cases they have still lost a large share of their former market, at least as terminal sires. At the same time, this emphasis on increasing growth rate has probably led to a deterioration in other attributes, like intermediate mature size, ease of calving and milk production, which made them obvious contenders for crossing to dairy heifers and breeding replacement beef cows. It would be easy to blame this on a lack of long-term vision. However, in many countries the increase in the popularity of continental breeds threatened the survival of many traditional British breeds unless they competed more effectively. Also, market signals to breeders of bulls for crossing to heifers or breeding replacement suckler cows were clearly not strong enough to sustain breeding programmes for these markets alone. In future, greater clarity about breeding goals, better objective information on the merits of different breeds and crosses, and better tools to select individual animals within breeds will help producers both to match breeds and sires to particular roles more effectively, and to improve the efficiency of beef production. Some developments in these areas are discussed later in this chapter.

Tables 7.2 and 7.3 show some results from a survey of agencies responsible for beef breeding schemes in eighteen European countries. This survey was carried out at the end of 1989, which makes it a bit dated, but it is probably the most recent survey of its kind to cover both specialised beef breeds and dairy and dual-purpose breeds. (See reference [11] for more up-to-date information on

Specialised beef and local breeds	No. of schemes	Dairy and dual-purpose breeds	No. of schemes
Charolais	9	Holstein Friesian	9
Limousin	9	Simmental[1]	5
Aberdeen Angus	7	Red and white	4
Hereford	6	Local breeds	3
Simmental[1]	4	Other	6
Local breeds	14		
Other	8		
Total	57	Total	27

[1] The Simmental breed is considered a specialised beef breed in some countries, and a dual-purpose breed in others.

Table 7.2 Details of the breeds involved in a survey of beef breeding schemes in specialised beef, dairy and dual-purpose breeds in eighteen European countries [45].

Specialised beef and local breeds		Dairy and dual-purpose breeds	
Trait	Number of schemes in which this was a main breeding goal (from total of 57)	Trait	Number of schemes in which this was a main breeding goal (from total of 27)
Growth rate	49 (86%)	Growth rate	22 (81%)
Carcass quality	37 (65%)	Carcass quality	16 (59%)
Calving ease	31 (54%)	Calving ease[2]	0 (0%)
Maternal ability[1]	22 (39%)	Maternal ability[3]	1 (4%)
Milk production	0 (0%)	Milk production	26 (96%)

[1] Probably includes milk production in most cases; [2] probably under-reported as it may be considered as a dairy trait; [3] maternal ability other than milk production.

Table 7.3 The main breeding goals in the cattle breeding schemes surveyed in 18 European countries. Column totals exceed the total number of schemes surveyed, as most schemes had more than one main beef breeding goal [45].

recording and evaluation of beef traits in dairy and dual-purpose breeds in some INTERBULL member countries.) Table 7.2 gives details of the schemes surveyed – 57 breeding schemes in specialised beef breeds and 27 in dairy or dual-purpose breeds were involved. Table 7.3 shows the main beef selection objectives in these schemes. Most of the schemes in all breed types aimed to improve growth and carcass attributes. Improving calving ease and maternal ability was part of the breeding goal in 54% and 39% of the specialised beef breeding schemes. These traits appeared to be less important in dairy and dual-purpose breeds. (Although the importance of calving ease may be underestimated, as the survey asked about beef traits in the breeding goal, and calving ease may have been considered as a dairy trait in dairy and dual-purpose breeds.)

The breeds and crosses used in beef production, and how they match the breeding goals mentioned above, are discussed in the following section. The strategies used for within-breed selection for these goals, the heritabilities of traits in beef breeding goals and the genetic correlations among them are examined in a later section.

Breeds and crosses used in beef production

It is more difficult to assess the relative contribution of different breeds and crosses to beef production than it is for dairying in Europe. This is partly because of the contribution which dairy and dual-purpose breeds make to beef output, the widespread use of crossbreeding, and the generally poorer systems for recording and collating national data on breed use in beef production. However, the following series of tables attempts to piece together the information that is available.

Clearly the predominance of black and white strains in the dairy industry (as discussed in Chapter 6) means that they are major contributors to beef output both directly through surplus calves and cull cows and, in some countries, indirectly through their contribution to the genetic make-up of suckler cows. However, the increasing specialisation for milk production in black and white strains means that their predominance is often seen as a disadvantage in beef production. Because of the economic incentive towards specialisation for milk production in most temperate countries, the biggest opportunity to improve beef output from dairy breeds is through crossing surplus females to specialised beef breeds.

Table 7.4 shows the numbers of purebred females of the major specialised beef breeds in 1989 in some of the larger beef-producing countries in Europe. This table illustrates the predominance of the specialised French beef breeds, particularly the Charolais and Limousin, and to a lesser extent the British breeds, particularly the Hereford and Angus. The popularity of the French breeds is probably due to their high growth rates or high lean meat yield, while the popularity of the British breeds is probably due to their relatively low incidence of calving difficulties. Also, the traditional British breeds, especially the Aberdeen Angus, have had something of a renaissance recently because of perceived benefits in eating quality. The results of one comparison of the growth and carcass attributes of these breeds were shown in Table 3.4. References to some other comparisons, and a summary of results from several of these, are given at the end of this chapter [2, 14, 21, 29, 46].

The increased use of the specialised French beef breeds as terminal sires in Europe, often at the expense of the traditional British breeds, is mirrored in many other temperate beef-producing countries. However, the British breeds remain

Country	Breed	No. breeding females (000s)		Herd size		% progeny from AI sires	% progeny from ET donors
		Total	Recorded	Average	% herds > 60		
France	Charolais	1490.0	101.0	27	–	10.2	< 0.2
	Limousin	560.0	49.0	32	–	16.4	< 0.1
	Blonde d'A	250.0	22.0	17	–	48.5	< 0.2
	Salers	164.0	17.5	42	–	4.2	< 0.1
	Aubrac	55.0	12.2	42	–	4.4	< 0.1
	Gasconne	18.0	4.7	17, 90[1]	–	7.1	–
W. Germany	Charolais	4.1	4.1	8	–	< 10	–
	Galloway	1.7	1.7	5	–	< 5	–
	Angus	1.6	1.6	19	–	< 5	–
	Limousin	0.9	0.9	6	–	< 5	–
	Highland	0.7	0.7	4		–	–
UK	Simmental	10.0	3.7	7	0.3	50.0	3
	Charolais	10.0	3.2	5	5.0	50.0	< 4
	Limousin	9.0	3.2	6	–	30.0	8
	Angus	8.0	1.1	12	2.0	20.0	5
	Hereford	7.0	2.3	9	5.0	45.0	1
Italy	Piemontese	303.0	33.0	20.0	3.1	28.5	< 1
	Marchigiana	99.7	30.4	4.4	0.1	36.0	–
	Chianina	84.0	15.3	9.7	2.0	64.0	< 1

[1] Two distinct groups according to herd size.

Table 7.4 The numbers of purebred females of the major specialised beef breeds in some of the larger beef-producing countries in Europe in 1989 [45].

important in breeding herds, either as purebreds or as components of crossbred maternal lines, in many of these countries (e.g. the US, Canada, Australia, New Zealand). Although in some of these countries the traditional supremacy of the British breeds in this maternal role is being challenged by other breeds.

Although Table 7.4 shows some of the most numerous beef breeds in Europe, several less numerous breeds have a disproportionate influence through the use of AI, especially in dairy herds. For example, in the UK there are relatively small numbers of purebred Belgian Blue cattle, but this breed was responsible for the second largest number of beef inseminations made by the main AI organisations in 1993/94. The growth in importance of this breed is due to its ability to leave high conformation crossbred calves, with acceptable levels of calving ease, when mated to dairy cows. The numbers of inseminations to other beef breeds throughout the UK are shown in Table 7.5.

As mentioned earlier, a relatively high proportion of beef produced in France and Italy is from purebred specialised beef herds. In France, the Charolais, Limousin, Blonde d'Aquitaine and Salers breeds are most numerous, while the Piemontese, Marchigiana and Chianina breeds are most numerous in Italy. The breed composition of cows in specialised beef herds in the UK is rather poorly documented. Although comprising a small sample, MLC recorded suckler herds probably give the best indication of breed composition in lowland, upland and hill areas. Tables 7.6 and 7.7 show the breeds of cow and bull used in these MLC recorded herds in 1991 [31, 42]. Hereford crosses, mainly out of dairy cows, dominated in the lowland sector. Upland suckler herds were predominantly

Breed of bull	Number of inseminations (000s)		
	England and Wales	Scotland	Northern Ireland
Aberdeen Angus	41	5.5	8.9
Belgian Blue	147	5.3	14.1
Blonde d'Aquitaine	21	2.7	19.0
Charolais	106	15.4	30.2
Hereford	43	0.6	2.0
Limousin	142	21.2	12.2
Simmental	71	10.4	15.5
Welsh Black	10	–	–
All beef breeds	591	63.1	102.5

Table 7.5 Number of inseminations to beef breeds by the former milk board AI centres (the largest AI organisations) in England and Wales and Scotland, and by AI Services (Northern Ireland) Ltd. in 1993/94. Only inseminations to the most numerous breeds are shown. The figures exclude DIY inseminations. The figures include inseminations in pedigree beef and suckler herds, but the majority were in dairy herds [39].

Type of herd	Lowland	Upland	Hill
No. of herds	131	102	14
Breed of cow	Percentage of Herds		
Aberdeen Angus and crosses	2	20	0
Blue Grey (Shorthorn x Galloway)	1	3	43
Hereford and crosses	35	42	0
Limousin and crosses	9	8	0
Simmental and crosses	10	18	14
South Devon and crosses	5	0	0
Welsh Black	0	5	0
Others[1] and Mixed	60	42	100

[1]Others include Sussex, Devon, Lincoln Red and crosses.

Table 7.6 Breed or crossbred type of cow in MLC recorded suckler herds in 1991. The table shows the percentage of herds with cows of the breed shown. Percentages do not add up to 100 because some herds have cows of more than one breed or cross [31].

Hereford or Angus cross cows, usually out of dairy cows. Blue Grey cows (F1 crossbred females from matings between a white (Whitebred) Shorthorn bull and a Galloway cow) were the most common type in hill suckler herds. However, there has been a marked increase in the use of continental cross cows in all sectors over the past few years [30, 31]. These include both bought-in replacements and homebred replacements. The increasing use of these continental cross suckler cows is partly a deliberate management policy to offset the influence of more specialised dairy cattle. It is also partly a consequence of their ready availability, following the increasingly widespread use of continental crossing sires in both dairy and suckler herds. Earlier results from MLC recorded suckler herds showed a small number of pure traditional hill breeds, such as the Galloway, which do not feature in Table 7.6, although the most recent results for hill herds are based on only fourteen herds.

Type of herd	Lowland	Upland	Hill
No. of herds	**131**	**102**	**14**
Breed of bull	**Percentage of Herds**		
Aberdeen Angus	3	3	7
Blonde d'Aquitaine	13	3	7
Charolais	26	52	14
Limousin	28	27	14
Simmental	29	15	36
South Devon	8	0	0
Welsh Black	1	1	0
Others[1]	8	2	7

[1]Others include Gelbvieh, Sussex, Belgian Blue and Hereford.

Table 7.7 Breed of bull in MLC recorded suckler herds in 1991. The table shows the percentage of herds with bulls of the breed shown. Percentages do not add up to 100 because some herds have bulls of more than one breed [31].

Selection within breeds

Systems of testing

Most beef cattle genetic improvement programmes are based on performance testing or progeny testing. Both of these depend on performance recording. Essentially this involves recording the identity, pedigree, birth date, sex, and performance (e.g. live weights) of individual animals, plus any major management groupings or treatments likely to influence performance. In performance testing schemes these records are then used to predict the genetic merit of the recorded animals themselves, as explained in Chapter 5. In progeny testing schemes the records are usually used to predict the genetic merit of sires. The main features of these two main types of testing are described below.

Performance testing

Since many of the traits of interest in beef cattle can be recorded in both sexes prior to sexual maturity there is a fairly long history of performance recording and performance testing in beef breeding. This dates from the 1940s and 1950s in the US, and from slightly later in many other countries. Often, recording associations were established specifically to operate early performance recording schemes. Today it is usually the responsibility of breed associations (e.g. in the US), government departments or agencies receiving some government support (e.g. in many European countries) or private agencies, either alone or in partnership with each other. In some countries (especially where herds are large and geographically dispersed, such as in the US) performance is usually measured by the breeders themselves, and then sent to the recording agency. In others, such as the UK, France and some other European countries, some or all of the measurements are made by field staff from the recording agency. The approach used also depends on the value breeders and their customers attach to independent authentication of records.

The evolution of performance recording and testing of pedigree beef cattle in Britain is fairly typical of that in many temperate beef producing countries.

	Specialised beef + local breeds	Dairy and dual purpose breeds
No. breeding schemes	**57**	**27**
Performance test:		
on farm	40 (70%)	7 (26%)
on station	46 (81%)	25 (93%)
Progeny test:		
on farm	31 (54%)	10 (37%)
on station	27 (47%)	10 (37%)

Table 7.8 Details of the testing schemes used for beef traits in specialised beef, dairy and dual-purpose breeds in several European countries in 1989 [45].

Performance recording began in earnest about 30 years ago. At that time the main organisation responsible for recording was the Beef Recording Association. This organisation was subsequently incorporated into the Meat and Livestock Commission which continued to be the major performance recording agency in Britain until 1995. This role then passed to a new company, Signet, which is under joint MLC and SAC ownership.

Compared to the situation in dairy cattle breeding, a relatively low proportion of beef cattle are performance recorded. This is partly because of the greater distinction between commercial and breeding herds in the beef than the dairy industry, especially in countries where crossbreeding is widespread. For example, performance recorded animals comprise less than two per cent of the total beef cattle population in the US [34], Australia and the UK. However, even within the purebred sector, there is usually a much lower proportion of recording than in the dairy sector (see Table 7.4, for example). This is partly due to the greater additional effort involved in recording beef cattle. It is also due to the less direct route for the rewards to accrue: milk recording helps to improve the profitability of milk production directly on the recording farm; in contrast, most of the economic benefits of performance recording beef cattle come through improved sales of breeding stock. In the past the less widespread conviction of the benefits of recording among beef cattle breeders and their

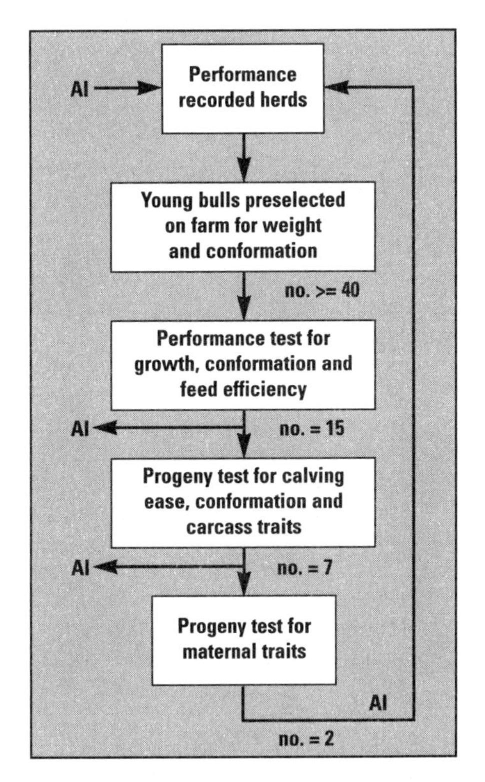

Figure 7.4 The sequential testing of beef bulls as practised in specialised beef breeds in France. The numbers shown are typical annual numbers tested in the Limousin or Blonde d'Aquitaine breeds; in the Charolais breed the numbers tested are about three times higher. ([33]; Dr F Ménissier.)

Trait 1 measured on bulls in central performance test station	Trait 2 measured on 30 male progeny	Genetic correlation between traits 1 and 2
Weight at approx. 480 days	Weight at 500 days	+0.40
Muscular development (live)	Muscular development (live)	+1.00
Muscular development (live)	Muscular development (carcass)	+0.97
Skeletal development	Skeletal development	+0.75

Table 7.9 Genetic correlations between the performance of Charolais bulls in French central testing stations, and the subsequent performance of their progeny in test stations [10, 23].

customers affected the size and consistency of these rewards. However, there is now a growing awareness of the value of performance recording in many countries.

Most performance testing schemes involve recording the pre-weaning performance of all animals on-farm. In some countries, post-weaning performance continues to be measured on farm. In others, central performance testing is used. Central testing of beef cattle has been quite widely used worldwide since the 1950s, especially in the US, Canada and Europe (see Table 7.8 for details of some schemes in Europe). It involves submitting some animals, especially higher-performing bulls, from the breeders' own farms to a central station, where they are compared with bulls from other herds in a uniform environment. Disentangling true genetic merit from the effects of good feeding and management was particularly difficult before the availability of BLUP methods, and central testing was designed to reduce this problem. Also, it can create larger groups of contemporary animals for comparison – small pedigree herd sizes in many countries limit selection intensities (see, for example Table 7.4). Central testing also allows more comprehensive measurements of performance.

Despite the potential benefits of central testing, the correlations between the performance of bulls in central stations and the subsequent performance of their progeny is often lower than expected. In other words, central test results can be poor indicators of a bull's breeding value. This is often attributed to large pre-test environmental effects. These may be reduced by starting tests at younger ages and putting greater emphasis on traits measured later in the testing period [3, 44]. Partly because of these complications, the central testing of beef bulls has ceased in some countries (for example central testing in Britain by the MLC ceased in the early 1980s). Recent results from France show reasonable correlations between the weight of bulls tested in central stations and the live weight of their progeny. There were higher correlations still between sire and progeny performance for skeletal and muscular development (see Table 7.9). This may reflect more uniform rearing conditions on French farms in the regions concerned than in other studies. It is also likely that muscular development is less affected by the rearing environment than live weight is.

Progeny testing

Even when the emphasis is on selection on individual animal's performance records there is an automatic, but often slow, accumulation of progeny records from the sires used in performance-recorded herds. However, in many countries there is a more deliberate strategy of first performance testing, then progeny testing bulls, with selection at each stage. As with performance testing, progeny testing schemes either operate on-farm or at central testing stations (see, for example Table 7.8).

Sequential testing is particularly well organised in the specialised beef breeds in France (see Figure 7.4). Large numbers of purebred animals are performance recorded on-farm for weights at birth, 120 and 210 days, and for muscular and skeletal development at weaning [10, 33]. In 1992 over 130,000 bull calves were recorded in this way. The best males from on-farm recording are brought to central testing stations after weaning and tested further from eight to fourteen months of age. Over 1,800 young bulls were tested in this way in 1992. About 125 beef bulls are progeny tested for growth and carcass traits annually in France. Most of these are selected on the basis of their performance in central test stations. About 35 of these bulls per annum go on to be progeny tested to assess their daughters' maternal ability in central progeny test stations. By law, bulls must be progeny tested to qualify for widespread AI use in France – less than 30% of proven bulls are selected for this purpose.

Less structured schemes operate in other countries to accelerate completion of on-farm progeny tests for bulls in the specialised beef breeds. For example, in Britain MLC and Signet support Young Bull Proving Schemes in association with several breed societies. These schemes are designed to identify promising young performance-recorded bulls and to get rapid progeny test results for them in pedigree herds.

In countries where beef bulls are widely used in dairy herds, breed societies or AI organisations often support progeny tests of beef bulls in dairy herds. For example a scheme for progeny testing Limousin bulls in dairy herds (Lim-Elite) was established by the British Limousin Cattle Society several years ago, and is now operated by a commercial company. Selection of bulls is based on Signet on-farm performance records. Several hundred straws of semen from selected bulls are used at random in participating dairy herds. Calving ease is recorded in these herds, and about 30 progeny from each bull are transferred to beef finishing units to obtain growth and carcass records. The results are analysed and semen from the best bulls is marketed to pedigree and commercial producers. Ten teams of bulls have completed tests at the time of writing. A similar scheme in the Belgian Blue breed, called Blue-Elite, was established recently. As the results of these schemes are not yet analysed using BLUP, they cannot be compared across teams. However, this development is likely in future. In Britain, several of the larger AI companies operate similar progeny testing schemes for beef bulls in dairy herds. Many of the bulls selected for progeny testing come from Signet-recorded herds. Most of the schemes involve on-farm recording of calving ease, calf size and conformation. The largest company, Genus, also records the growth and carcass characteristics of the progeny of some beef bulls at a central testing station.

Cooperative breeding schemes

Although most breeding schemes revolve around performance testing or progeny testing, as outlined above, there are some variations which deserve special mention here. The first of these are cooperative breeding schemes, such as group breeding schemes and sire referencing schemes. The benefits of these schemes were outlined in Chapter 3. Several sheep group breeding schemes and a few cattle schemes were established in New Zealand and Australia in the late 1960s and during the 1970s. Following the early success of these schemes others were established elsewhere. For example, a scheme was established to improve weaning weight in the Welsh Black cattle breed in Britain. This scheme has now ceased, but another cattle group breeding scheme (ASDIG) was initiated a few years ago by several South Devon breeders, which uses both embryo transfer techniques and BLUP evaluation procedures [40].

Perhaps because of the relatively high legal and financial commitment required, and the growth in uptake of national across-herd genetic evaluation procedures, there seems to have been a decline in interest in cattle group breeding schemes over the last decade or so. However, formal or informal sire referencing schemes have been established in several breeds in France, Denmark, Britain and the US, either before or during this period. These schemes involve the use of an agreed panel of sires on a proportion of the cows in each member's herd, usually by AI. In some cases these schemes have been formed specifically to create or strengthen genetic links between herds to allow more accurate across-herd or across-test genetic evaluations. In other cases – such as the cooperative schemes established recently in Britain by small groups of Charolais (GLB) and Limousin (White Rose) breeders – there are additional aims, such as to record more comprehensively than many other herds (see Plate 35). Similar schemes would be eminently suitable in other beef cattle breeds and other countries, particularly those in which the low use of AI leads to weak genetic links across herds.

MOET nucleus schemes

The theoretical benefits of central nucleus herds using multiple ovulation and embryo transfer (MOET), were described for dairy cattle in the last chapter. Similar benefits are expected in beef cattle; in fact the potential value of MOET in accelerating response to selection was first reported for beef cattle by Drs Roger Land and Bill Hill in 1975 [27]. They estimated that responses to selection for growth rate could be doubled by the use of MOET, albeit with higher rates of inbreeding. As in dairy cattle, these original estimates of the benefits of MOET are now believed to be on the high side. Recent estimates suggest that 30% extra progress is possible, compared to a conventional scheme of similar size and with the same rate of inbreeding [47].

While MOET has been used widely in beef cattle both as a means of importing and exporting genetic material, and to multiply newly imported breeds or valuable individuals more rapidly than possible with natural reproduction, it has not been used widely in structured breed improvement programmes to date. An experimental MOET scheme in Simmental cattle operated at SAC in Aberdeen until recently [12]. The aim was to test and develop, on a field scale, some of the reproductive technologies involved in these schemes. Although selection was practised for a relatively short time, good responses were achieved in an index of growth and carcass traits. MOET is being used on a smaller scale in several of the industry cooperative breeding schemes in Britain which were outlined above.

Traits recorded

Generally, on-farm performance recording schemes around the world have concentrated on measuring live weights at regular intervals (or growth rates between these), together with visual scores of muscularity and measurements or scores of height or skeletal development. The development of mobile, reasonably accurate ultrasonic scanners in the 1970s and 1980s allowed measurements of fat and muscle depths or areas to be included in some on-farm recording schemes. Typically these measurements are taken on or over the eye muscle at one of the last ribs, or in the loin region of animals at about a year or 400 days of age. At least in theory, one of the benefits of central testing is that it permits more frequent and more comprehensive measurements to be made. For example, it is rarely practical to measure feed intake of individual animals on farms, but it is fairly common in central performance test stations. Similarly, progeny testing allows actual carcass measurements to be obtained.

As an example, the traits recorded in beef cattle breeding schemes in the European survey mentioned earlier are shown in Table 7.10. These tend to be far less comprehensive than expected from the breeding goals shown in Table 7.3, especially for dairy and dual-purpose breeds. At the time of the survey, the most commonly measured traits were live weights or growth traits, visual scores on the live animal or carcass (in progeny testing schemes) and calving ease. Feed intake or efficiency and ultrasonic measurements of fat and muscle were also measured in a small number of schemes.

The evolution of on-farm performance recording in Britain is fairly typical of that in many other countries. Beef performance recording was based originally on records of birthweight, and live weights recorded at about 100-day intervals thereafter until 400 or 500 days of age. Since the 1970s other measurements have been added periodically, including assessments of calving ease on a five-point scale, ultrasonically measured fat depth over the eye muscle, a visual assessment of muscularity on a fifteen-point scale, and more recently ultrasonically measured eye muscle depth. Feed intake was recorded at MLC central performance testing stations, and is still measured on a small number of privately tested bulls. Initially all weights other than birthweight were authenticated by MLC staff. Now the minimum service offered by Signet involves a single authenticated weighing per annum, with a further three annual weighings recommended, which may be done by the breeders. Currently, recording live weights is the minimum requirement for membership of the Signet Beefbreeder service. However, more comprehensive recording (more traits and more animals) is being encouraged by Signet, MLC, some individual breeders and breed societies. Breeders who record more

	Specialised beef + local breeds	**Dairy and dual-purpose breeds**
No. breeding schemes	57	27
Live or carcass weights	49 (86%)	25 (93%)
Birthweight	19 (33%)	0 (0%)
Visual scores	39 (68%)	16 (59%)
Calving difficulty	25 (44%)	1 (4%)[1]
Feed intake	19 (33%)	5 (19%)
Ultrasonic measurements	14 (25%)	4 (15%)
Cow fertility	5 (9%)	0 (0%)[1]

[1]Probably under-reported as it may be considered a dairy trait.

Table 7.10 The main traits recorded in beef cattle breeding schemes in several European countries [45].

comprehensively, and have a level of performance above breed average, may now join 'Excel' schemes designed to promote their animals as being particularly accurately recorded and evaluated.

Generally, terminal sire characteristics have dominated beef breeding schemes in Europe. With the exception of some breeding schemes in France, few of the maternal characteristics, such as fertility, mentioned earlier, have been recorded. As a result, what little objective selection there has been for maternal characteristics has been on traits like calving ease, birthweight, and 200-day weight, which are of importance in both terminal sire and maternal lines. However, until recently, methods of separating direct and maternal genetic influences on these traits have not been in widespread use. Maternal traits have received more attention in North America, Australia and New Zealand, where specialised beef herds account for a far higher proportion of beef output. Genetic evaluations for scrotal size (which is an indicator of both male and female fertility, and age at puberty), female fertility (measured as days from the start of the mating period to calving) have been introduced recently for some breeds in Australia and New Zealand. Evaluations for scrotal size and mature cow weight have been introduced for some breeds in the US.

The following series of tables shows means, standard deviations and estimates of the heritabilities of many of the traits mentioned as breeding goals or criteria, together with estimates of phenotypic and genetic correlations among some of them. These are taken from a recent, very comprehensive review of the international scientific literature [25, 26]. An important finding of this review was that there are

Trait	Unit of measurement	Mean	Phenotypic sd	Mean heritability (se)
Age at first calving – direct	days	682	39	0.06 (0.020)
Age at first calving – maternal	"	–		0.19 (0.035)
Conception rate – cows, direct	% calving	} 76	} 47	0.17 (0.015)
Conception rate – heifers, direct	"			0.05 (0.010)
Conception rate – cows, maternal	"	–	–	0.02 (0.019)
Conception rate – heifers, maternal	"	–	–	0.02 (0.009)
Calving ease – cows, direct	% unassisted	90.6	15.7	0.13 (0.002)
Calving ease – heifers, direct	"	91.2	26.7	0.10 (0.002)
Calving ease – cows, maternal	"	98.2	15.3	0.12 (0.002)
Calving ease – heifers, maternal	"	90.6	31.7	0.09 (0.002)
Pelvic area, constant age	cm^2	221	58	0.58 (0.058)
Perinatal mortality – cows, direct	% mortality	3.1	20.9	0.10 (0.001)
Perinatal mortality – heifers, direct	"	9.8	31.7	0.15 (0.003)
Perinatal mortality – cows, maternal	"	4.2	20.3	0.11 (0.001)
Perinatal mortality – heifers, maternal	"	6.9	41.3	0.11 (0.001)
Calving rate	% calves weaned	–	–	0.17 (0.010)
Scrotal circumference, constant age	cm	33.9	2.7	0.48 (0.019)

Table 7.11 Means, phenotypic standard deviations and weighted mean estimates of heritabilities for a range of reproductive traits in beef cattle. These come from an extensive review of published values, but estimates of means, sds and heritabilities are not necessarily from the same source. The standard errors (se) of heritability estimates provide a measure of their reliability, and a means to test whether or not estimates are significantly different from zero, or from each other. See the original review for sources and full details of the methods used to weight estimates and obtain standard errors [25, 26].

significant effects of breed and country on estimates of heritabilities and correlations. If very reliable local estimates are available, these should be used to design breeding schemes and criteria, and perform genetic evaluations. In other cases it would be better to use published values, or to combine local estimates with the overall means of published estimates.

Table 7.11 shows heritability estimates for reproductive traits. These are often shown separately for heifers and older cows. This is because the traits concerned are believed to be affected by different genes in heifers and cows, or have different

Trait	Unit of measurement	Mean	Phenotypic sd	Mean heritability (se)
Birthweight – direct	kg	35.1	4.3	0.31 (0.003)
Birthweight – maternal	kg	34.9	4.4	0.14 (0.002)
Weaning weight – direct	kg	203	25	0.24 (0.002)
Weaning weight – maternal	kg	217	29	0.13 (0.002)
Yearling weight – direct	kg	345	38	0.33 (0.004)
Postweaning gain	kg/day	0.978	0.134	0.31 (0.004)
Yearling gain – direct	kg/day	0.852	0.082	0.34 (0.017)
Feed intake	kg DM/day	6.48	0.62	0.34 (0.025)
Feed conversion ratio (feed (or energy)/gain)	kg (or unit energy)/kg	6.00	0.66	0.32 (0.024)
Mature cow weight	kg	446	54	0.50 (0.021)

Table 7.12 Means, phenotypic standard deviations and weighted mean estimates of heritabilities for a range of growth traits in beef cattle. These come from an extensive review of published values, but estimates of means, sds and heritabilities are not necessarily from the same source. See the original review for sources and full details of the methods used to weight estimates and obtain standard errors [25, 26].

Trait	Unit of measurement	Mean	Phenotypic sd	Mean heritability (se)
Backfat depth, constant age[1]	mm	10.6	2.6	0.44 (0.019)
Carcass weight, constant age	kg	313	19	0.23 (0.011)
Carcass weight, constant finish	kg	–	–	0.36 (0.111)
Dressing %, constant age	%	60.1	1.9	0.39 (0.021)
Cutability (meat yield), constant age	%	54.8	20.1	0.47 (0.024)
Lean:bone ratio	–	4.07:1	0.29	0.63 (0.041)
Lean %	%	62.3	2.8	0.55 (0.028)
Marbling, constant age	score	7.36	2.51	0.38 (0.034)
Eye muscle area, constant age	cm^2	68.9	7.0	0.42 (0.023)
Tenderness	Warner-Bratzler shear force	5.48	1.00	0.29 (0.038)

[1] For most traits heritability estimates at a constant weight or finish are similar

Table 7.13 Means, phenotypic standard deviations and weighted mean estimates of heritabilities for a range of carcass traits in beef cattle. These come from an extensive review of published values, but estimates of means, sds and heritabilities are not necessarily from the same source. See the original review for sources and full details of the methods used to weight estimates and obtain standard errors [25, 26].

incidences, or both, and so heritabilities may differ. Also, in this and some of the following tables, both direct and maternal estimates of heritabilities are shown. As mentioned in Chapter 5, direct heritabilities refer to the trait as measured on the animals recorded. For instance, the direct heritability of calving ease is a measure of the genetic variation in calving ease due to the size and shape of calves themselves, the direct influence calves have on gestation length, plus any other calf factors affecting calving ease. In contrast, the maternal heritability of calving ease is a measure of genetic variation among cows in traits affecting calving ease, such as pelvic size and shape, body condition, and maternal influence on gestation length. It is often important to make a distinction between direct and maternal genetic influences in evaluating and selecting animals because they can be

Trait 1	Trait 2	Mean correlation between traits	
		Phenotypic	Genetic
Age at first calving, direct	Birthweight, direct	0.05	–0.14
	Postweaning gain	0.10	0.48
	Mature cow weight	–0.36	–0.10
Calving ease, direct	Calving ease, maternal	–	–0.30
	Birthweight, direct	–0.28	–0.74
	Weaning weight, direct	0.04	–0.21
	Postweaning gain	–0.09	–0.54
	Yearling weight, direct	0.01	–0.29
	Mature cow weight	–0.03	–0.23
Birthweight, direct	Birthweight, maternal	–	–0.35
	Weaning wt, maternal	0.34	–0.14
	Weaning wt, direct	0.46	0.50
	Postweaning gain	0.19	0.32
	Yearling wt, direct	0.38	0.55
	Mature cow weight	0.34	0.67
Weaning weight, direct	Birthweight, maternal	0.34	–0.05
	Weaning wt, maternal	–	–0.16
	Postweaning gain	0.09	0.44
	Yearling wt, direct	0.71	0.81
	Feed intake	0.47	–
	Feed conversion ratio	0.00	–0.50
	Mature cow weight	0.45	0.57
Yearling weight, direct	Postweaning gain	0.74	0.81
	Feed intake	0.64	0.16
	Feed conversion ratio	–0.46	–0.60
	Mature cow weight	0.54	0.72
Market weight, constant age	Dressing %, const. age	0.06	–0.23
	Backfat, constant age	0.19	0.00
	Eye muscle, const. age	0.33	0.43
	Marbling, const. age	0.15	0.34
	Cutability, const. age	–0.22	–0.19

Table 7.14 Mean estimates of phenotypic and genetic correlations between pairs of traits in beef cattle from an extensive review of published values. Correlations with age at calving are unweighted, all others are weighted. See the original review for sources and full details of the methods used to weight estimates [25, 26].

antagonistic. For example, as mentioned in Chapter 6, selecting a smaller bull may improve calving ease when this bull's offspring are born. However, when the bull's daughters calve themselves, calving ease may deteriorate, because they are smaller adults as well as smaller calves. Table 7.11 shows that many of the traits concerned with reproduction have fairly low heritabilities. However, many are economically important, and there is substantial variation in them, so there is both the incentive and scope for genetic improvement.

Table 7.12 shows mean estimates of the heritabilities of growth traits. Direct heritabilities tend to be moderately high, while maternal heritabilities tend to be slightly lower. Also, the direct heritability of live weights tends to be fairly high at birth, lower at weaning and higher again at yearling or later ages. Table 7.13 shows the heritabilities of a range of carcass measurements. These tend to be even higher than those for growth traits. However, they have to be assessed either indirectly on live candidates for selection (e.g. by ultrasonic measurements), or directly on progeny or other relatives of the candidates for selection, so they are not as easy to improve as it seems at first sight.

Table 7.14 shows mean estimates of phenotypic and genetic correlations between some reproduction, growth and carcass traits. There are unfavourable associations between several reproduction and growth traits, which complicates selection for both characteristics. Generally, there are strong phenotypic and genetic associations between weights at different ages – the closer the age, the stronger the association. This means that selection for growth characteristics alone usually leads to unfavourable responses in birthweight (i.e. increased birthweight which generally leads to more difficult calving), and heavier mature cow weight. Unfavourable responses in birthweight can be reduced by selecting individual animals with EBVs which depart from this general trend (e.g. low birthweight EBVs, but high EBVs for later weights), or by including birthweight and later weights in an index with negative and positive economic values respectively. Similar approaches can be used to reduce unfavourable responses in cow mature weight, although using separate terminal sire and maternal breeds or crosses is probably more efficient.

Methods and results of genetic evaluation

Overview

Until the early 1970s the genetic evaluation methods employed in most countries were fairly simple. These usually produced adjusted records of performance, contemporary comparisons or predicted breeding values which could only be compared within herd (or within a central test team). However, from the 1970s onwards BLUP methods of evaluation began to be adopted. The benefits of BLUP evaluations were discussed fully in Chapter 5. BLUP provides more accurate prediction of breeding values than other methods. Also, BLUP PBVs can be compared across herds and years, providing that there are genetic links between herds and years. This is particularly important in those countries where pedigree herd sizes are small, and so selection intensities are usually low.

Sire model BLUP evaluations for beef cattle across herds and years were first used in the early 1970s in the US [6]. The uptake of BLUP has been slower in beef

	Number of schemes using each method of genetic evaluation	
	Specialised beef and local breeds (total no. schemes = 57)	Dairy and dual-purpose breeds (total no. schemes = 27)
Contemporary comparison	49 (86%)	9 (33%)
Selection index	28 (49%)	16 (59%)
BLUP		
sire model	11 (19%)	14 (52%)
sire maternal grandsire model	4 (7%)	1 (4%)
animal model	1 (2%)	0 (0%)

Table 7.15 Methods of genetic evaluation of beef merit used in specialised beef, dairy and dual-purpose breeds in several European countries in 1989. Totals exceed 100%, as some schemes use several methods of evaluation, e.g. within-herd contemporary comparisons of individual traits, plus an index based on these [45].

cattle than in dairy cattle in most other countries. However, advances in computer power and computing strategies mean that those countries which have changed to BLUP evaluations recently have often adopted animal model evaluations immediately rather than progressing through simpler models first. Often these initial across-herd and generation evaluations for sires were run independently from the existing within-herd evaluations. Later, in some countries, methods were developed to integrate these two approaches, to make use of national sire evaluations in within-herd evaluations. The wider adoption of national animal model BLUP evaluations, producing PBVs for all animals not just sires, has solved this problem in some countries. In others, national evaluations and within-herd performance information are still combined to provide up-to-date predictions of genetic merit between national evaluations.

Initially the methods used to evaluate beef traits in dairy and dual-purpose breeds were more sophisticated than those used in specialised beef breeds – probably because the more advanced methods were being used already to evaluate dairy traits. Table 7.15 shows some more results of the survey of European beef breeding schemes which illustrate this point. In 1989 most specialised beef breeding schemes relied on contemporary comparisons or selection indexes calculated within-herd or within-performance or progeny test. Only 28% of the specialised beef schemes surveyed were using BLUP methods then. However, BLUP methods are being used more widely in specialised beef breeds, and under more sophisticated models, than at the time of the survey (e.g. multi-trait animal model BLUP is now in widespread use in specialised beef breeds in Britain and France). A survey of evaluation methods in dairy and dual-purpose breeds in some INTERBULL member countries in 1995 showed that the majority of these countries used BLUP evaluations for beef traits in dairy and dual-purpose breeds. About half the countries used animal model BLUP, and several did multi-trait evaluations for beef traits [11].

The evolution of genetic evaluation methods for specialised beef breeds in Britain is fairly typical of that in many countries. Until the mid-1980s beef cattle performance records in Britain were adjusted for age and environmental (or fixed) effects, such as age of dam, and results were presented as within-herd or within-test contemporary comparisons. From the mid-1980s, records were adjusted

Plate 33 Hardy beef breeds such as the Galloway have undergone many years of natural and artificial selection for adaptation to harsh environments. (Courtesy of Dr Peter Amer.)

Plate 34 Charolais cow and calf from the GLB Charolais Improvement Group in Yorkshire – the group aims to accelerate improvement of beef cattle through comprehensive use of recording and the creation of strong across-herd links through the use of common sires.

Plate 35 Beef sire summaries from several countries.

Plate 36 Angus cow and calf in Messrs W and D McLaren's Netherton herd, Perthshire, Scotland.

Plate 37 Sheep being gathered in Central Otago, New Zealand.

Plates 38 & 39 Hill sheep breeds such as the Scottish Blackface (left) and Swaledale (right) have undergone many years of natural and artificial selection for adaptation to harsh environments. (39 courtesy of Dr Tony Waterhouse.)

Plate 40 Fine wool Merino flock rams at pasture on the New England Tablelands, Armidale, northern New South Wales. (Courtesy of Dr Sandra Eady.)

Plate 41 Sheep under extensive range conditions in Alberta, Canada.

Plate 42 Yearling Rambouillet rams in Madison, Wisconsin, US. The breed is a descendent of the Merino.

Plate 43 South Australian strain Merino stud ram, housed and fed in preparation for showing and sale; Mt. Bryan, mid-north South Australia. (Courtesy of Dr Sandra Eady.)

Plate 44 Buyers inspecting Medium wool Merino flock rams run under commercial paddock conditions; Mungindi, north-western New South Wales. (Courtesy of Dr Sandra Eady.)

Plate 45 Dorset Downs from the Lincoln University flock, Canterbury, New Zealand. Rams of this breed are used as terminal sires on dual-purpose ewes in New Zealand.

Plate 46 Lacaune dairy ewes in south-central France. (Courtesy of Professor David Thomas.)

Plate 47 Lacaune dairy ewes in the milking parlour on a farm in south-central France. (Courtesy of Professor David Thomas.)

Plate 48 Sarda sheep being hand milked in Sardinia, Italy. The Sarda is the most numerous of the Italian milk sheep breeds. Sheep milk production in many Mediterranean countries is based on traditional husbandry practices, including hand milking. (Courtesy of Antonello Cannas.)

Plate 49 A flock of Comisana sheep. The Comisana is the second most numerous dairy breed in Italy. (Courtesy of Antonello Cannas.)

Plate 50 Welsh Mountain ewes from the CAMDA group breeding scheme nucleus flock in North Wales.

Plate 51 Members of the Elite Texel Sire Referencing Scheme in Britain selecting rams for use as reference sires. (Courtesy of Peter Johnson.)

Plate 52 Merino ewes being drenched to control internal parasites at 'Nerstane' stud, Woolbrook, northern New South Wales. The resistance of internal parasites to anthelmintics is of major concern in parts of Australia and New Zealand, and increasingly in other sheep-producing countries too. However, selection of sheep for resistance to the parasites appears to be effective. (Courtesy of Dr Sandra Eady.)

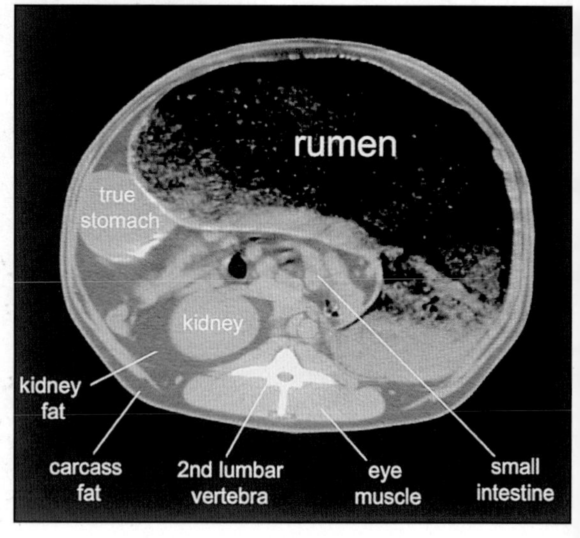

Plate 53 An image obtained by computer tomography, showing a 'cross section' through a sheep lying on its back in a scanning cradle. Air appears black in this image, fat dark grey, muscle light grey and bone white. Key organs and tissues are labelled. Advanced imaging techniques such as CT are very useful tools in research, and have the potential to accelerate selection for altered carcass composition. (Courtesy of Dr Mark Young.)

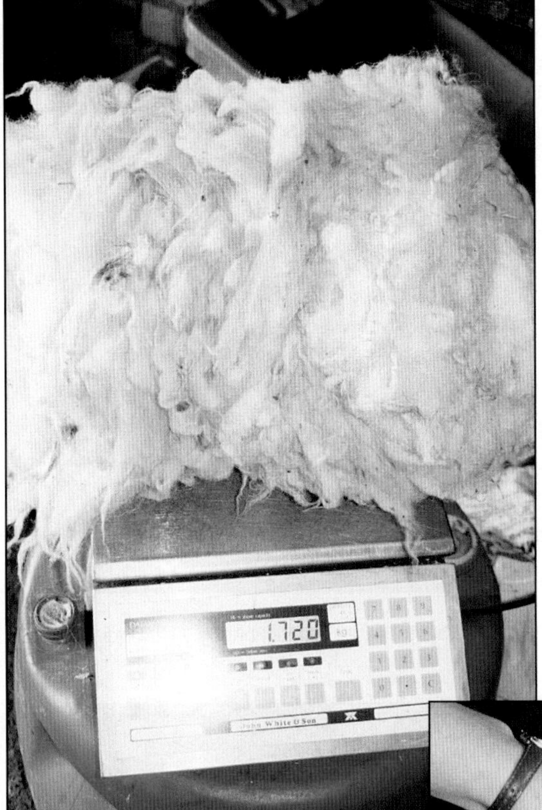

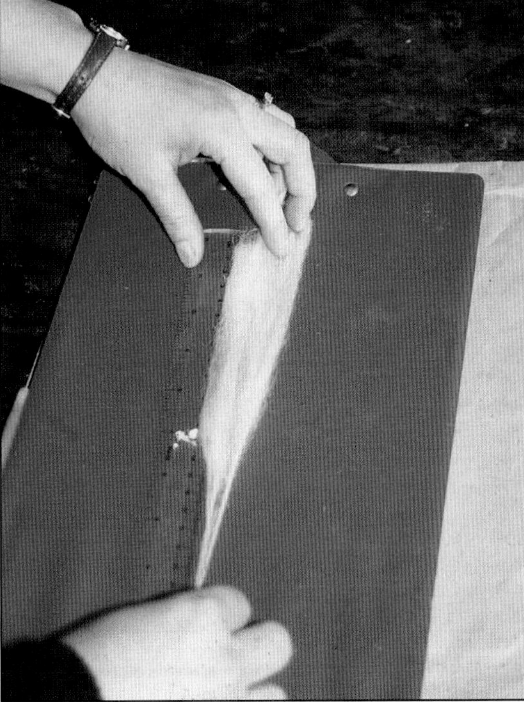

Plates 54 & 55 Objective recording of fleece weight (above) and staple length (right) from sheep in the SAC/Roslin Institute Hill Sheep Breeding Project.

Plate 56 Ram lambs from the SAC Suffolk Selection Experiment. The larger lamb is from the selection line, the smaller one from the control line.

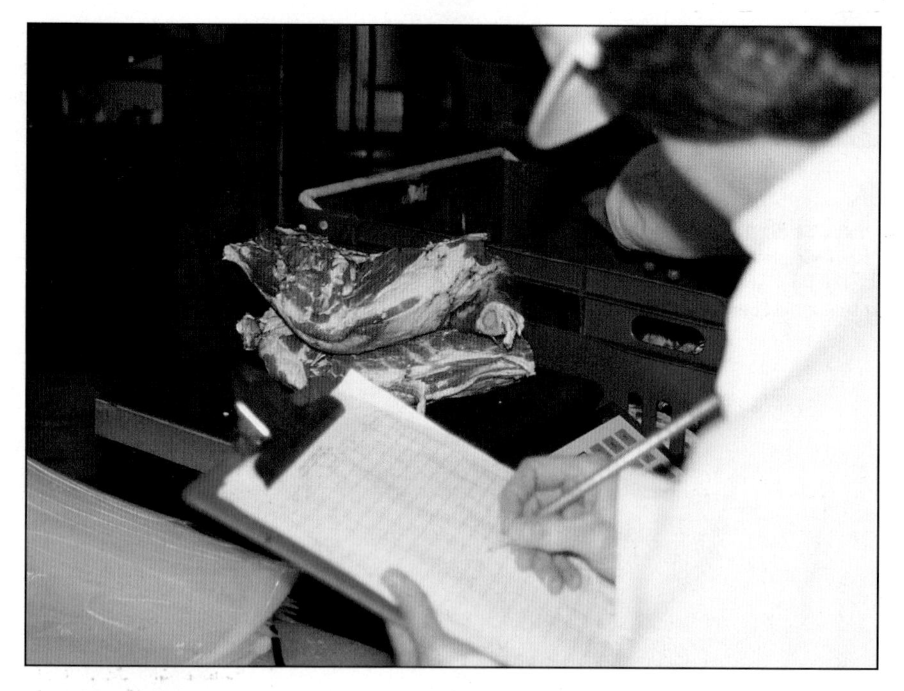

Plate 57 Recording saleable meat yields as part of a progeny test of high and low index rams from the SAC Suffolk Selection Experiment.

Plate 58 'Coring' of a fleece sample from a Merino, prior to automatic measurement of fibre diameter.

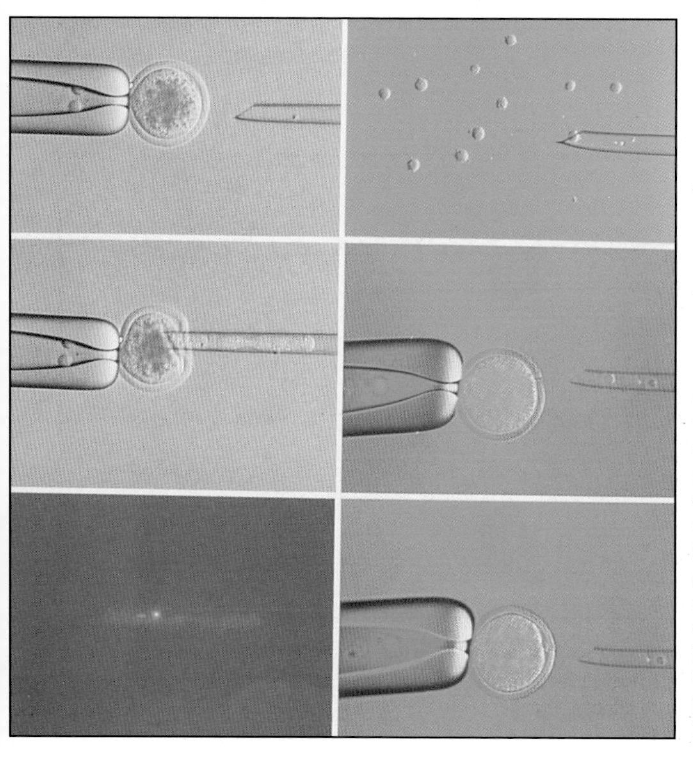

Plate 59 Some of the stages involved in nuclear transfer: (i) (top left) an oocyte (unfertilised egg) held by suction from a pipette; (ii) (middle left) a smaller pipette is inserted to remove the chromosomes; (iii) (bottom left) the presence of the chromosomes in the pipette is confirmed under special lighting; a cell to be transplanted is (iv, v) (top right and middle right) picked up by pipette, inserted into the oocyte, and (vi) (bottom right) the pipette is removed. (Courtesy of Dr Ian Wilmut.)

Plate 60 Cloned sheep. The two sheep on the left (Megan and Morag) were produced by nuclear transfer of cultured cells originally derived from an embryo [7]. The sheep on the right (Dolly) was produced by nuclear transfer of a cultured cell originating from an adult ewe – the first time that a mammal had been produced from a cell derived from an adult [60]. (Courtesy of Dr Ian Wilmut.)

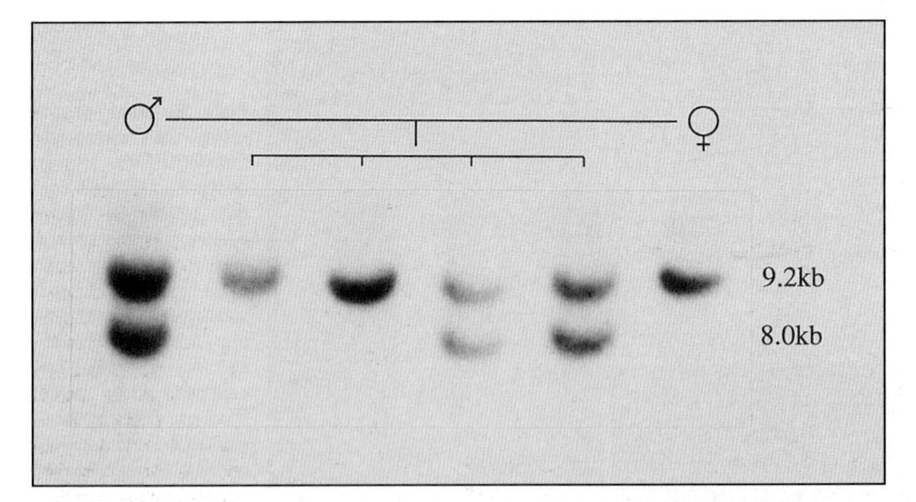

Plate 61 Detection of variation in DNA fragment size in a sire (left lane), a dam (right lane) and their four offspring (middle lanes), using a technique known as Southern blotting. The fragments of different sizes are created by the use of a restriction enzyme; these are then separated by gel electrophoresis, 'blotted' onto a membrane, and detected using a DNA probe. (Courtesy of Dr Alan Archibald.)

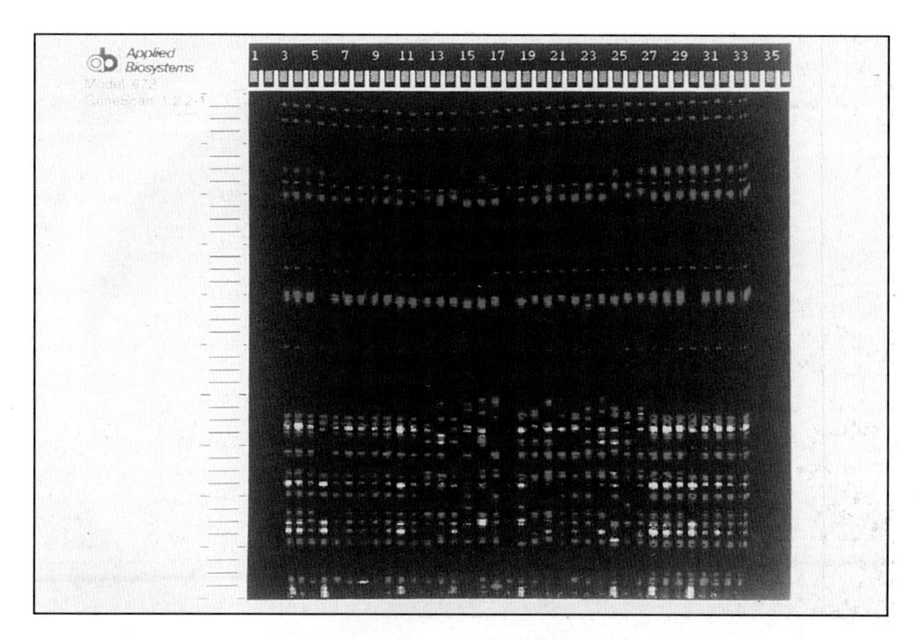

Plate 62 A multiple-lane gel, showing variation at eleven different microsatellite loci in cattle. The red bands, which appear in the same location in all lanes, are from a constant marker, and provide a benchmark. The presence or absence of the yellow, green and blue bands in particular locations on the gel indicates the genotype of the animal with respect to the eleven markers. Several lanes at the far left-hand side and several at the far right-hand side of the gel are repeat samples from the same bull, which explains why the patterns are identical in these lanes. The centre rows are from samples from several different bulls. Determination of animals' marker genotypes is becoming increasingly automated. (Courtesy of Dr John Williams.)

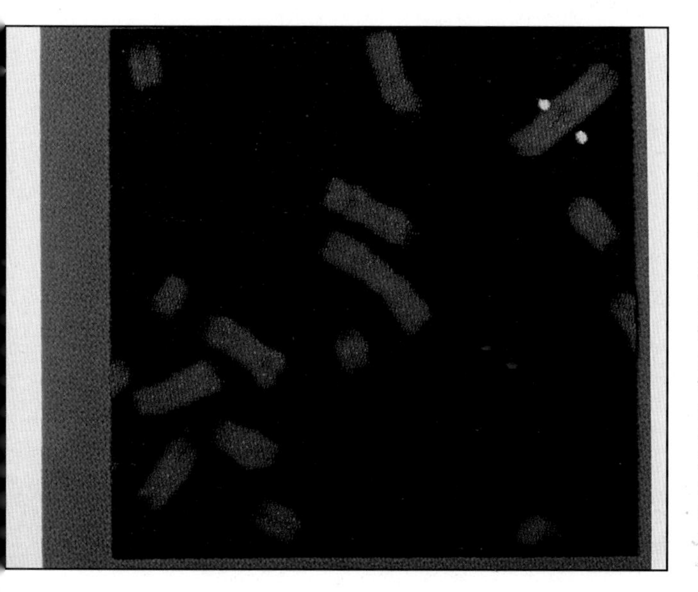

Plate 63 An example of fluorescence *in situ* hybridisation. A 'spread' of bovine chromosomes appears red in this plate. The bright yellow specks indicate the position of a fluorescently labelled probe which has bound to a complementary sequence of DNA on the chromosome, thus showing the physical location of a particular sequence of DNA on that chromosome. (Courtesy of Dr L Ferretti.)

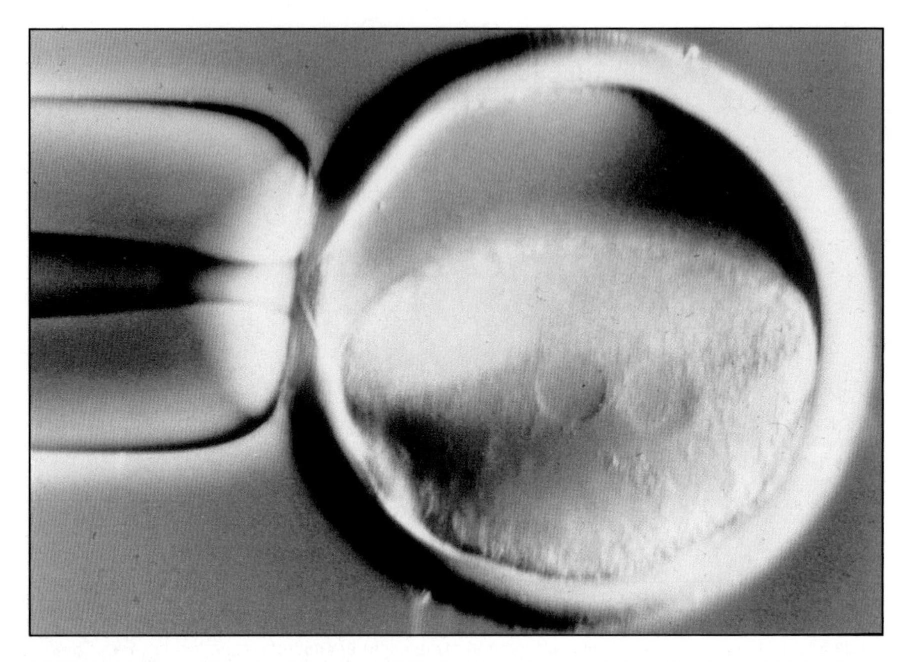

Plate 64 Gene transfer by microinjection of DNA. The plate shows a single-cell embryo which has been treated by centrifugation to allow the pronuclei to be seen clearly (these are the two separate structures containing the chromosomes from the sperm and the egg, which will later form the single nucleus in the cell; they appear as two circles in the centre of the picture). DNA is about to be inserted into one of these pronuclei by micropipette. (Courtesy of Dr Ian Wilmut.)

by standardising as described in Chapter 5. Breeders were provided with adjusted contemporary comparisons for the recorded traits and, providing sufficient traits were recorded, a selection index combining these adjusted records. Between 1991 and 1994 multi-trait animal model BLUP evaluations for most traits were introduced in all performance-recorded beef breeds in Britain.

Steps involved in genetic evaluation

The steps involved in genetic evaluation of beef cattle typically include:

- Collation and checking of performance records by the recording agency, and transfer of these to the agency responsible for genetic evaluation, if this is different. (In some countries, such as the US, national evaluations of beef breeds are performed at universities; in most others, evaluations are performed by the recording agency itself, or by a government agency.) Evaluations are usually done separately for each breed. National BLUP evaluations often include a large amount of historical performance data, and require pedigree as well as performance records. This step is repeated each time evaluations are performed. In most countries with national evaluations these are done from one to three times per annum, often following the busiest recording seasons, or in time for major shows or sales. Within-herd evaluations are often done more frequently. (For example, national BLUP evaluations in Britain include performance records from about 1970 onwards for most breeds. Most of the historical records are for birthweights, 200- and 400-day weights, but increasing numbers of records are being collected on ultrasonic measures of backfat and muscle depth, visual muscle scores and calving ease. Immediate ancestry is recorded by Signet along with performance data. Table 7.16 shows the numbers of records for weight traits included in BLUP evaluations of the most numerous breeds in the early 1990s, together with the means and standard deviations of these traits. Additional pedigree information, and for some breeds additional records of birthweights, are obtained from the breed societies. For much of the recording period, breed 'standard' birthweights were assigned, when breeders did not record actual birthweights, prior to within-herd evaluation. These standard weights, together with outlying records for any traits, were removed from the data before BLUP evaluations began, and outlying records continue to be removed from new data. It is likely that many of the remaining birthweight records are estimated rather than actually measured, but it is more difficult to account for these.)

- Unless BLUP methods are being used, performance records are adjusted for environmental effects, such as age at measurement or dam age, using one of the methods outlined at the beginning of Chapter 5. However, even when BLUP methods are being used, records of performance are sometimes pre-adjusted for some environmental effects to reduce computing demands, as explained in the last chapter.

Trait	Breed	No. observations	Mean (kg)	sd (kg)
Birthweight	Charolais	84,073	42.9	4.6
	Simmental	40,748	40.3	3.9
	Limousin	49,846	37.0	3.4
	Aberdeen Angus	5,898	31.7	3.5
	South Devon	4,110	43.8	5.8
200-day weight	Charolais	48,238	282	37
	Simmental	34,923	286	33
	Limousin	31,046	251	29
	Aberdeen Angus	22,019	208	29
	South Devon	20,629	253	36
400-day weight	Charolais	23,699	522	49
	Simmental	18,613	518	44
	Limousin	17,375	459	39
	Aberdeen Angus	11,711	390	38
	South Devon	8,314	441	50

Table 7.16 The numbers of records for weight traits included in the initial BLUP evaluations of the most numerous breeds in Britain in the early 1990s together with the means and standard deviations (after standardisation of within herd variance) of these traits [13].

(For example, in Britain the Signet recording scheme involves weighing animals in batches at regular intervals. Few animals are weighed at exactly 200 and 400 days of age. So their actual live weight records together with their ages at weighing are used to estimate their weight at 200 and 400 days, prior to national BLUP evaluations. To get estimated 200- and 400-day weights, animals must have at least two weight records between 170 and 300 days, and two between 300 and 500 days. Estimating weights in this way is equivalent to using all records to plot a growth curve, and reading off the weights at the required ages. It makes full use of all available weight information, and is more accurate than using single weights, which can be strongly influenced by recent episodes of feeding, drinking or excreting!)

- If traditional evaluation methods are being used, then breeding values may be predicted from the animals' own adjusted records of performance alone. Or, the selection index methods described in Chapter 5 may be used to combine the animals' own adjusted records with those from relatives.

- If BLUP evaluation methods are being used then (at least some) environmental effects are estimated and breeding values are predicted simultaneously. As explained in Chapter 5, accounting for most of the environmental effects at the same time as predicting breeding values allows environment and genetic merit to be separated more effectively. The statistical model used determines which environmental effects are accounted for. Some of these are identified explicitly (these are discussed below), others are accounted for by assigning animals to contemporary groups. The importance of assigning animals to a group in which all of them have been managed similarly, and experienced similar climatic conditions etc. was discussed in Chapter 5. There is often a dilemma, especially when herds or flocks are small, about whether to put animals in larger groups where they can be compared more accurately, or in smaller groups where they are much more likely to be with true contemporaries. This dilemma is dealt with in British beef evaluations by: (i) creating contemporary groups within each herd, as different management and feeding practices between herds are likely to account for much of the non-genetic variation between animals; (ii) assigning animals to separate contemporary groups for each trait – few animals have records on all traits, so it is better to put them in the most appropriate group for each trait evaluated rather than form groups which are only appropriate for one recorded trait. (For example, most animals have live weight records, but fewer have fat depths. If allocations of animals to contemporary groups for both traits were based on the numbers of contemporaries available for weight traits, there would probably be too few animals in each group with fat measurements to make accurate comparisons among them); (iii) forming contemporary groups flexibly, so that clusters of calves which are born together in a herd end up in the same group, rather than risking separation because of a rigid definition, such as calendar month of birth (e.g. this would put calves born on 31 March and 1 April in different groups) and (iv) subsequently merging adjacent contemporary groups if there are too few animals [13]. Additionally, if animals within a herd have been managed in different groups by breeders, and this is recorded, then management group codes are also used to assign animals to appropriate contemporary groups.

● Sometimes there are wide differences between herds in the range of weights or other measurements. (This is called **heterogeneity of variance**, and it was mentioned in Chapter 5.) This can occur simply because of differences in the management and feeding policies of different herds, but it can also be a consequence of preferential treatment of some groups of animals within a herd. On the one hand, unless something is done about this, animals from the most variable herds will tend to get the highest EBVs. On the other hand, if the records are adjusted so that all herds have the same range in performance (but still have different means) there is a risk of downgrading some animals that really are genetically superior. In some countries records are adjusted to attempt to deal with the problem. For example, in Britain a middle-of-the-road approach is used, which scales the variation in each herd towards an overall average range. But, the amount of scaling that goes on depends on the size of the contemporary group which animals belong to. Small groups with either extremely low or extremely high variation get scaled up or down more than big groups with low or high variation, as there is a much higher risk that extreme variation in a small group is a fluke, or the result of accidental or deliberate differences in management.

- The environmental or fixed effects to be accounted for explicitly in BLUP evaluations need to be identified. These often include month or season of birth, whether or not the calf was born as a result of embryo transfer, whether or not it was fostered, birth type (single or twin), age and breed of dam (if crossbred dams are present) and age of measurement. For example, Table 7.17 shows the environmental or fixed effects which are accounted for in BLUP evaluations of different traits in Britain. Pre-weaning management of animals is usually similar regardless of the sex of the calf. For this reason contemporary groups are formed across sexes for pre-weaning traits, and sex is fitted as a separate effect. Since post-weaning management of calves usually differs for the different sexes, forming contemporary groups across sexes would muddle up the true effects of sex with differences in performance which are really due to different management of bulls and heifers. So, contemporary groups are formed within sex for traits recorded after weaning.

- BLUP evaluations are then done. The particular model used determines which relationships among animals are accounted for, and which animals get PBVs (see Chapter 5 for more details on BLUP models). For example, with individual animal model BLUP the relationships between all animals are recognised and accounted for, and all animals get PBVs. For traits with important maternal genetic components, such as birth and weaning weights, maternal breeding values (and permanent environmental effects, which account for the non-genetic effects of particular dams on calf weights) are often estimated in addition to direct breeding values.

Effect	Birth-weight	200-day weight	400-day weight	Muscle score	Fat depth
CG[1] excluding sex from definition	•	•			
CG[1] including sex in definition			•	•	•
Month of birth (1-12)	•	•	•		
Sex (male/female)	•	•			
Embryo transfer calf (yes/no)	•				
Foster calf (yes/no)		•			
Birth type (single/twin)	•	•	•	•	•
Breed of mother at birth (1-5)[2]	•				
Breed of mother at weaning (1-5)[2]		•	•	•	•
Percentage purebred (1-4)[2]	•	•	•	•	•
Age of dam at birth (linear and quadratic terms)	•	•	•		
Age at measurement (linear and quadratic terms)				•	•

[1] CG = Contemporary group [2] Effects fitted in four or five categories denoting different breed groups or increasing proportions of breed being evaluated

Table 7.17 The environmental or fixed effects which are accounted for in the BLUP evaluations of different traits in the UK. (MLC/Signet; Dr R E Crump.)

● In some cases, especially when data sets are very large, traits are evaluated one at a time. In others, especially when data sets are smaller, or when computing resources permit, some or all traits are evaluated together. Single-trait evaluations require heritability estimates or variances relevant to the population of animals being evaluated. Multi-trait evaluations require estimates of heritabilities and of correlations among traits, or variances and covariances. (Table 7.18 shows the phenotypic and genetic parameters used in multi-trait animal model BLUP evaluations in Britain since 1997. Ideally, parameters would be estimated separately for each breed, but in most cases there is too little data for traits other than live weights to provide reliable breed-specific estimates.)

As long as reliable estimates of genetic parameters are available, multi-trait evaluations of associated traits are more accurate than single-trait evaluations. This is because measurements on correlated traits help to predict breeding values, even when a measurement is available on the trait concerned. An added benefit of multi-trait animal model BLUP evaluations is that they allow breeding values to be predicted even though some traits have not been measured on all animals, or cannot be recorded until later in life. Using clues from associated traits is not as good as measuring the trait directly – but on the whole it helps. In dairy cattle, the EBVs of bulls for milk production are evaluated on the basis of their daughters' milk records. In beef cattle we have to rely on the weaning weights of calves as an indirect measure of milk yield, as explained in Chapter 5. Getting information on weaning weights of descendants is time consuming. However, EBVs for live weights can also provide indirect information on EBVs for milk yield. In general, there is a negative genetic association between milk yield and postweaning growth rate in beef cattle. That is, families which have *genetically* high later growth rate often have below average *genetic merit* for milk yield. (This generalisation is also

	Gest. length	Calving ease	Birth-weight	200-day weight (direct)	200-day weight (maternal)	400-day weight	Muscling score	Muscle depth	Fat Depth
Gestation length	**0.30**	–0.10	0.24	0.07	–	0.05	0.20	0.20	0
Calving ease	–0.21	**0.12**	–0.30	–0.05	–	–0.10	–0.20	–0.20	0
Birthweight	0.55	–0.58	**0.24**	0.28	–	0.19	0.10	0.10	0.05
200-day weight (direct)	0.10	–0.29	0.50	**0.34**	–	0.79	0.40	0.31	0.16
200-day weight (maternal)	0	0	–0.10	–0.15	**0.07**	–	–	–	–
400-day weight	0.05	–0.10	0.41	0.85	0	**0.40**	0.44	0.43	0.22
Muscling score	0.20	–0.20	0.39	0.43	0	0.55	**0.26**	0.49	0.12
Muscle depth	0.20	–0.20	0.47	0.75	0	0.55	0.65	**0.26**	0.16
Fat Depth	0	0	0.09	0.22	0	0.12	0	0.18	**0.29**

Table 7.18 The phenotypic and genetic parameters used in BLUP evaluations in Britain at the time of writing. Heritabilities are on the diagonal, phenotypic correlations above it and genetic correlations below the diagonal. (MLC/Signet; Dr R E Crump.)

true across breeds – generally good dairy breeds are poorer than average for growth, and vice versa). Although this generalisation does not hold for every family, it is a reasonable guide in the absence of the sort of information on weaning weights of large numbers of grandoffspring (and other relatives). As an animal gets older and records from grandoffspring and other descendants become available, then EBVs will increasingly be based on these.

- The results of evaluations are expressed as predicted (or estimated) breeding values (PBVs or EBVs), predicted (or estimated) transmitting abilities (PTAs or ETAs), or expected progeny differences (EPDs). As explained in Chapter 5, PBVs or EBVs express genetic merit in terms of the recorded animal itself, while PTAs, ETAs or EPDs all express merit in terms of expected performance of progeny (i.e. a PTA, ETA or EPD for any trait is half the PBV or EBV for that same trait). In North America the results of across-herd evaluations of sires are usually presented as EPDs, though in the US within-herd results are usually expressed as EBVs. In most other countries the results of beef cattle evaluations are expressed as EBVs. (Table 7.19 shows the traits for which EBVs are currently produced by Signet/MLC in Britain. Evaluations for gestation length, calving ease, birthweight and 200-day weight are split into direct and maternal components, though maternal evaluations for gestation length, calving ease and birthweight are not presented at the moment. The direct EBVs in the Signet scheme are called 200-day growth EBVs, and the maternal EBVs are called 200-day milk EBVs.)

Traits currently evaluated
Calving ease (direct)
Gestation length (direct)
Birthweight (direct)
200-day growth (direct)
200-day milk (maternal)
400-day growth
Muscling score
Ultrasonic muscle depth
Ultrasonic backfat depth

Table 7.19 Individual traits currently evaluated by BLUP as part of the Signet Beefbreeder scheme in Britain. Indexes are also produced from these EBVs as explained later in this chapter.

- EBVs or EPDs are expressed relative to some reference population of animals, called the base. In most countries a fixed base is used, i.e. EBVs or EPDs are expressed relative to those of a group of animals born in a particular year, though this base is updated from time to time. For example, at the time of writing, EBVs for all beef breeds in Britain are expressed relative to the average performance of calves born in the breed concerned in 1980. In other words, animals with positive EBVs are expected to have genetically higher weights, fat depths, muscle scores or index scores than those born in 1980. Conversely, animals with negative EBVs are expected to have genetically lower weights, fat depths, muscle scores or index scores than those born in 1980. Animals with EBVs of zero are expected to be of similar genetic merit in the trait concerned to those born in 1980. Fixed bases ought to be updated at regular intervals to reduce the risk of breeders and their customers selecting animals which have positive EBVs but are of lower genetic merit than the *current* breed average. Current breed averages are often published in sire summaries, in order to help protect against inadvertent selection of animals of lower merit. In some countries EBVs or EPDs are expressed relative to a rolling base. For example, in Canada EPDs are expressed relative to the average merit of calves born in the last three years. In France EBVs are expressed relative to the merit of calves born in the last five years.

- In some countries EBVs or EPDs for individual traits are combined into indexes of overall economic merit. These are discussed in more detail in a later section.

- Accuracies (or reliabilities, which are squared accuracies, as explained in Chapter 5) are usually produced for each of the EBVs (or EPDs). The level of accuracy depends on the heritability of the trait concerned and the amount of performance information used in the evaluation. In animal model BLUP evaluations, performance records from the animal itself as well as all its relatives contribute, and so influence the accuracy. Also, in multi-trait evaluations, the accuracy of an EBV for a particular trait depends not only on the amount of information available on this trait, but also on the amount available on correlated traits. There is a natural tendency for accuracies of EBVs to increase as an animal gets older. Until an animal has performance records itself, its EBVs will be based mainly on information from ancestors, and

Trait	Young bull with own records only, in a group with 5 contemporaries	Young bull with no records itself, but with 5 recorded progeny	Widely used bull with no records itself, but 50 recorded progeny
Birthweight	51%	60%	92%
200-day milk	21%	29%	69%
200-day growth	42%	52%	89%
400-day growth	51%	60%	92%
Backfat depth	43%	53%	89%
Muscling score	40%	50%	88%

Table 7.20 Typical accuracies of EBVs for different traits from multi-trait animal model BLUP evaluations for bulls with different sources of records available. See Table 7.18 for details of the heritabilities of these traits and correlations among them. (Dr R E Crump; MLC/Signet.)

Accuracy of current EBV	95% chance that final EBV is in range:
20%	–8 to +28
40%	–7 to +27
60%	–4 to +24
80%	–1 to +21
90%	+2 to +18
100%	10

Table 7.21 Possible range of final very accurate EBVs for an animal with a current EBV for 200-day milk of +10 kg (see Figure 5.6 also). (Dr R E Crump.)

hence be of relatively low accuracy. Accuracies will increase markedly when the animal has performance records of its own (and from its collateral relatives) included in an evaluation. Accuracies will increase even more dramatically if records become available on large numbers of progeny. Table 7.20 shows typical levels of accuracy for different traits for animals with different amounts of information recorded.

BLUP EBVs are already scaled to take account of their accuracy – the less information used in producing an EBV, the closer it will be to the mean for the population concerned. However, accuracies are still a useful guide to the risk that an individual animal's EBV will change in future. They are particularly useful in distinguishing between animals whose EBVs are based mainly on ancestors' records, or indirect measurements, and those with direct information recorded. Table 7.21 shows the importance of checking accuracies of EBVs, especially for low accuracy traits. For example, because EBVs for milk (or more properly the maternal genetic component of weaning weight) are based on indirect information, and the trait has a relatively low heritability, relatively few animals have EBVs with high accuracies. Low accuracy figures imply that the eventual very accurate EBV for milk, obtained after collecting many records, *could* be quite different from the initial EBV. This leads many breeders to ask: why bother publishing low accuracy EBVs? The answer is that they are still the best guide available, and that across a group of animals, they will be accurate on average. For as long as we rely on natural methods for breeding cattle, there will be a very large element of chance in which genes calves get from their parents – all EBVs can do is load the dice in the breeders' favour, to

Bull name Sire Dam	Breeder/ owner/ herd prefix	Identity no./ date of birth		Birth-weight (kg) (+ = heavier)	200-day weight – maternal (kg) (+ = heavier)	200-day weight – direct (kg) (+ = heavier)	400-day weight (kg) (+ = heavier)	Ultrasonic fat depth (mm) (+ = fatter)	Ultrasonic muscle depth (mm) (+ = deeper)
Glendale Bill Glendale Fred Glendale Mary	A. Smith Glendale	ABC1567 20/03/97	**EBV** **Accuracy**	+2.0 45%	−1.5 20%	+20.5 40%	+35.5 50%	−0.1 39%	+0.3 40%
Glendale Bob Glendale Fred Glendale Daisy	A. Smith Glendale	ABC1569 27/03/97	**EBV** **Accuracy**	−1.0 40%	+1.0 18%	−0.5 38%	−5.5 45%	−0.2 38%	−0.3 39%
High Farm Felix Glendale Fred High Farm Sue	B. Brown High Farm	DEF1234 15/02/97	**EBV** **Accuracy**	+4.0 47%	−3.5 21%	+25.0 41%	+40.0 52%	+0.3 40%	+0.5 42%
Newfield Superman Newfield Ted Newfield Blossom	G. White Newfield	GHJ3456 11/03/95	**EBV** **Accuracy**	+2.5 92%	−1.0 69%	+22.5 89%	+37.5 92%	0.0 89%	+0.4 85%

Table 7.22 An example of the typical layout of information in beef sire summaries.

help them make more informed selection decisions. If the accuracy of EBVs is low, then the dice is only slightly loaded, if the accuracy is high, then the dice is more heavily loaded in the breeders' favour.

- Results of evaluations are sent to individual breeders after each new evaluation. Usually these include summaries of the predicted merit of the current calf crop, and the sires and dams they represent. Also, annual (or more frequent) national sire summaries are often produced for each breed from one of these evaluations (see Plate 35). These national sire summaries list EBVs (or EPDs) and accuracies for sires which are currently in use, and whose EBVs (or EPDs) have accuracies above an agreed threshold level. Lists of trait leaders (sires ranked on EBVs) are usually presented for most traits. Additionally, these national summaries sometimes give details of promising young bulls and top cows. Table 7.22 shows the typical layout of information in sire summaries.

Evaluations across herds, breeds and countries

In order to be able to compare BLUP PBVs fairly across contemporary groups and years, genetic links are needed between groups and years, as explained in Chapter 5. In dairy herds, strong links occur automatically because of the very widespread use of AI. In some countries there is little use of AI in specialised beef breeds, and this has limited the introduction of national across-herd genetic evaluations. Sire referencing schemes have been introduced to create links between herds in some of these countries.

However, AI use is higher in other countries, especially those with smaller herds where bull ownership is less cost-effective than AI, and where a high turnover of bulls is needed to control inbreeding. For example, between 20% and 50% of births in pedigree herds of the major beef breeds in Britain are the result of AI (see Table 7.4). Also, the recent introduction of foreign breeds to a country, or the popularity of imported strains within a breed, tend to increase the use of AI. In such cases there will often be strong enough genetic links between herds and years to make reliable comparisons of EBVs across herds and years. For example, any two contemporary groups are considered to be linked in beef evaluations in Britain if there is at least one animal in each group with a common parent, e.g. calves in two groups with the same sire. On this fairly liberal definition, only about three numerically small breeds have a high proportion of unconnected contemporary groups. Breeders and advisers can use this information to help target the use of particular sires in these breeds to create better links.

Similarly, a major technical limitation to performing evaluations across breeds is that animals of different breeds are rarely kept as contemporaries under similar management and feeding systems. However, across-breed evaluations are becoming feasible using information from crossbred animals, or from designed breed comparisons, together with estimates of genetic trends in each of the purebred populations since the breed comparison was made [2, 7]. Dealing properly with heterosis when records from crossbred animals are used is another complication of across-breed genetic evaluations.

Compared to the situation in dairy cattle, there has been less effort to date in developing international conversions of EBVs or EPDs for beef cattle, or performing international genetic evaluations. However, there is growing interest in this area. For example, international conversions have been produced for some beef breeds in use in Canada and the US. Also, across-country evaluations are being investigated or performed routinely for several breeds in the US and Canada, France and Luxembourg, and Australia and New Zealand [7, 19, 22].

Indexes of overall economic merit

The theoretical benefits of using selection indexes were described in Chapter 5. Briefly, these apportion selection emphasis in the most appropriate way, based on the relative economic importance of traits in the breeding goal, and on the strength of genetic associations between measured traits and breeding goal traits (if these differ). Until recently, the emphasis in beef cattle breeding in North America has been on using sophisticated methods to produce individual trait EBVs. In contrast, in Europe, while less sophisticated methods of evaluation were used until recently, selection indexes have been quite widely used in both specialised beef breeds and in dairy and dual-purpose breeds (see Table 7.15).

Much of the emphasis in Europe has been on producing indexes for terminal sire characteristics. For example, a terminal sire index was introduced in Britain in the mid-1980s and used in most breeds until 1997. The selection objective of this index was to maximise the margin between saleable meat yield and feed costs, taking into account the costs of difficult calvings [1]. Index scores were calculated from the animal's own records of calving difficulty score, 200- and 400-day weight and a visual muscling score. If they were recorded, additional measurements of birthweight, feed intake and ultrasonically measured fat thickness were included to increase the accuracy of the index.

Until the introduction of across-herd BLUP evaluations in the early 1990s, the terminal sire index was based on measurements on individual recorded animals alone. Following the introduction of BLUP evaluations, a slightly modified version of this index, called the Beef Value, was introduced, based on EBVs for most of the original set of measured traits. As these are multi-trait animal model BLUP EBVs, all available performance records of relatives were used, as well as records from the animal itself, and so the resulting index was more accurate. The Beef Value was designed to provide a simple tool to rank bulls (or cows) on the expected profitability of their crossbred offspring, taking into account likely saleable meat yield, feed costs and calving difficulty costs. Neither saleable meat yield nor feed intake are measured directly in pedigree herds, and calving difficulty is accurately assessed in relatively few. However, many investigations worldwide have shown that bigger animals generally have a higher saleable meat yield, but eat more. Fatter animals generally have a lower yield of saleable meat, and they eat more than leaner animals. Also, higher calf birthweights are generally associated with greater calving difficulties. Hence, the traits that are recorded are used to provide clues about those that are not recorded *but influence profit*, as explained in Chapter 5.

New indexes were developed in 1997 for Signet performance-recorded beef herds in Britain. These are more closely linked to market returns (i.e. using associations with carcass weight, fat class and conformation class rather than with

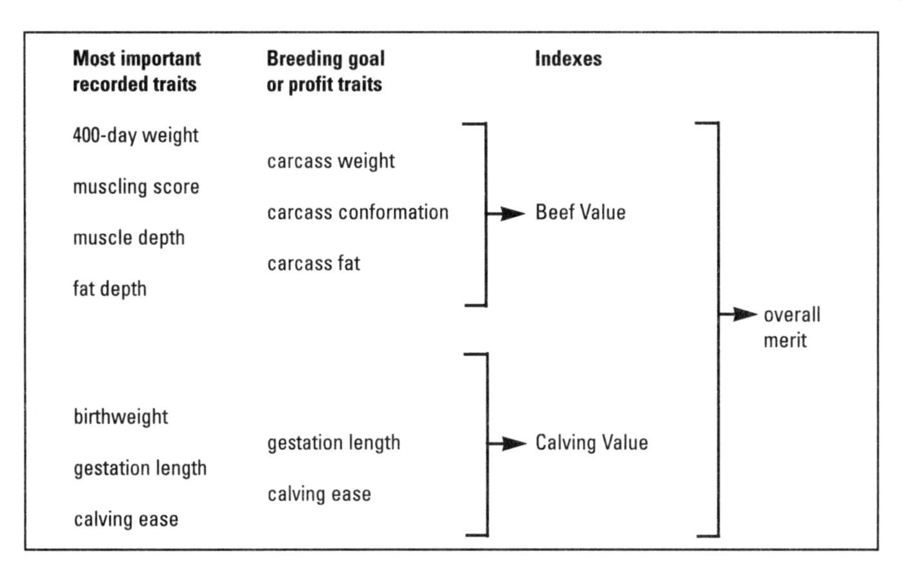

Figure 7.5 The structure of the indexes produced for Signet recorded pedigree beef herds in Britain. (Dr P R Amer.)

saleable meat yield). Separate indexes are available for calving performance and for growth and carcass performance of terminal sires. The Calving Value ranks animals on genetic merit for calving ease, based mainly on records of birthweight, calving ease and gestation length, while the Beef Value ranks them on genetic merit for growth and carcass traits, based mainly on records of weights, fat depth, muscle depth and muscle score. These two indexes can be added together to rank animals on overall merit for calving ease and production (see Figure 7.5). The contributions which the Calving Value and Beef Value make to overall merit vary depending on the importance of calving ease and on variation in the component traits in the breed concerned. However, typically, Calving Value accounts for about 16% of the variation in overall merit.

The indexes described above were produced to aid selection of terminal sire bulls. Fewer indexes have been developed for maternal breeds or lines (e.g. including EBVs for fertility, maternal component of calving ease, direct and maternal components of weaning weight, mature size). The simplest index of this type, which is used in several countries, is to combine direct and maternal EBVs for weaning weight. The weight of calf weaned by daughters of bulls in a maternal breed or line will be maximised by selecting bulls on their maternal EBV for weaning weight (or milk EBV) plus *half* their direct EBV for weaning weight (this is because calves benefit from *all* their mother's genes for milk, but they only benefit from *half* her genes for growth). However, other important maternal traits need to be considered separately if this approach is used.

Indexes combining BLUP EBVs for reproduction, growth and carcass traits have been developed in Australia. An important feature of these indexes is that the economic values applied can be tailored or customised to individual breeder's requirements. This is achieved via a computer software package which uses data on returns and costs of beef production for individual producers or production systems [5]. In a study of seventeen cases, relative economic values for reproduction varied from 0.9 to 6.5, and those for carcass traits varied from 1.1 to 2.5, at a constant value of 2 for growth traits. In theory, unless systems differ dramatically, the use of customised indexes has little effect on expected response. However, in practice, the use of customised indexes appears to improve both the understanding and the uptake of indexes.

Evidence of genetic improvement and its value

Estimates of genetic change

In theory, changes of at least one per cent of the mean per annum are possible following selection for weight or growth traits in beef cattle. However, in practice, rates of change are often lower than this. For example, a review of several beef cattle selection experiments showed that average changes of 0.6% and 0.8% per annum were achieved with selection for weaning and yearling weight respectively [37].

The increased uptake of across-herd BLUP genetic evaluations over the last decade has permitted more widespread estimation of genetic trends in industry breeding schemes. When evaluations are made within breeds, EBVs and trends cannot be used to compare the *absolute* performance of different breeds. However, they do indicate the genetic *changes* which have occurred within breeds over a given time period. For example, Figure 7.6 shows estimated genetic trends in birthweight, 200- and 400-day growth since 1980 for the most numerous performance-recorded beef breeds in Britain. (The trends in other traits are not shown as they are less reliable because of the much smaller number of records available to date.) Genetic trends are simply average EBVs plotted by year of birth. They give a useful retrospective view of the changes which have occurred within a breed, providing that the genetic parameters used are appropriate. Hence, these graphs show the breed-wide changes which have occurred as a result of selection on within-herd performance records, as well as other subjective methods of selection. The changes shown in these graphs have occurred without the benefit of BLUP EBVs, which can be compared across herds, so it will be interesting to see whether rates of change increase now that these tools are available. Genetic changes in weight in the Aberdeen Angus breed in Britain were about double those in the other breeds over the same time period (see Plate 36). This is probably due to regular importations of larger Angus bulls to the UK population, especially from Canada.

Trends similar to, or lower than, those shown have been reported in several breeds in Canada and Australia [16, 20]. Slightly higher trends in weaning weight have been reported in the US Angus and Hereford breeds [7]. This may be explained partly by the earlier availability of BLUP methods in the US beef industry. (An increase in genetic trends in milk production was evident in the US shortly after BLUP evaluations became available in the dairy industry there). It is probably also partly due to the higher herd and population sizes for these breeds in the US. Similar trends in weaning weight (from about 0.2 to 1.1 kg per annum), and positive trends in muscularity, were reported for the major French breeds between 1991 and 1995 [22].

In most of these studies of industry trends, the rates of change achieved are

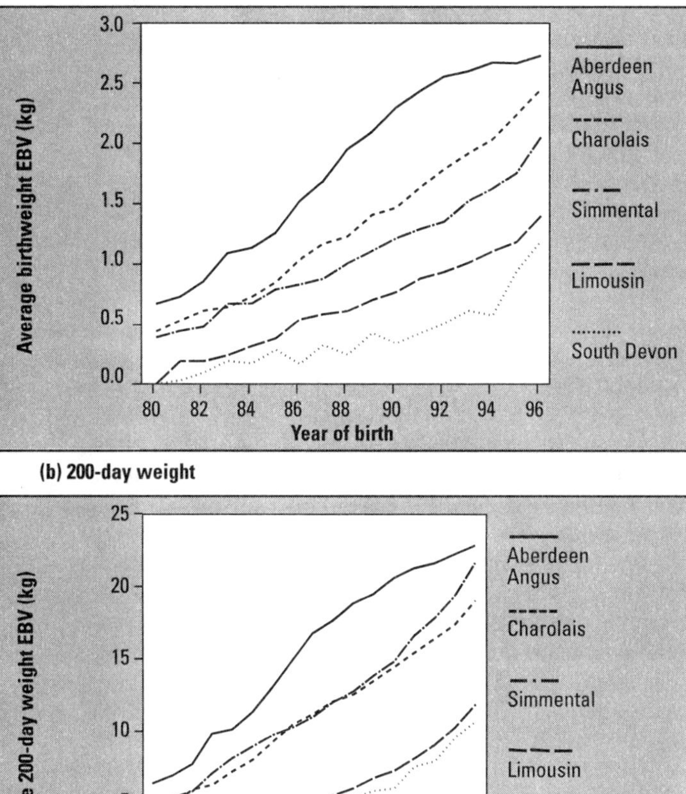

(b) 200-day weight

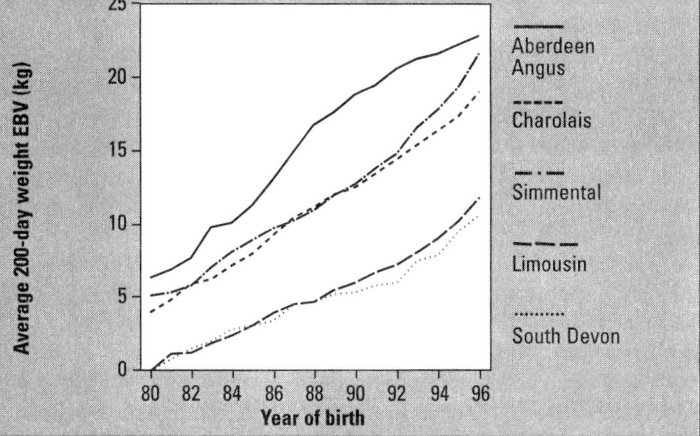

(b) 400-day weight

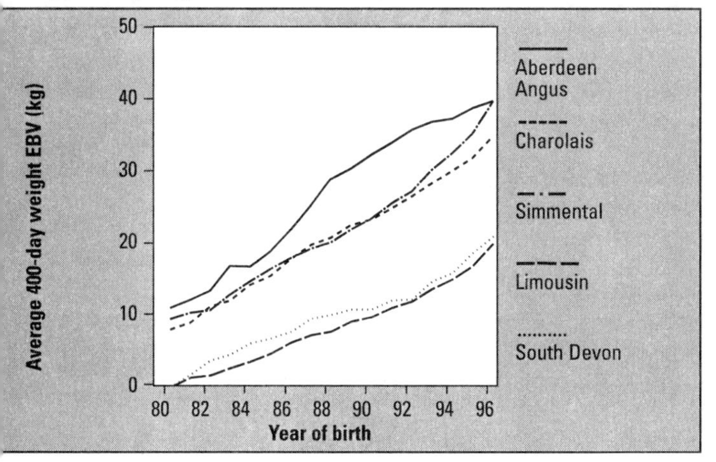

Figure 7.6 Trends in direct EBVs for (a) birthweight, (b) 200-day weight, and (c) 400-day weight, in the five most numerous performance-recorded beef breeds in Britain. (MLC/Signet; [13].)

well below those theoretically possible, and below those actually achieved in selection experiments. For example, the changes in 200- and 400-day weights in Britain ranged from 0.15% to 0.5% of the breed mean per annum for the different breeds. The higher values in this range are from breeds which have imported foreign bloodlines of quite different performance, and so they exaggerate the changes made by local within-breed selection. The apparently low rate of change is partly explained by the fact that selection has not been solely for weight traits. However, it is also partly due to the relatively low use of objective methods of selection, and the fact that, in at least some of the countries mentioned, only within-herd comparisons could be made for most of the period concerned.

Genetic increases in birthweight have occurred in most of the studies of industry trends mentioned above. This is largely a consequence of selection for higher weights at later ages, rather than a desire to increase birthweights *per se*. That is, the change in birthweight is a result of the positive genetic correlations between weights at different ages. Most breeders would like to limit increases in birthweight, because of its association with calving difficulties. Multi-trait animal model BLUP EBVs can help to identify individual animals which depart from the general trend, e.g. by having low birthweight EBVs, but high EBVs for later weights.

Value of genetic improvement

Compared to the situation in dairy cattle, there have been relatively few studies of the value of genetic improvement in beef cattle, though these do show favourable estimates of cost:benefit [4]. A recent study of the costs and benefits of implementation of across-herd BLUP and index selection in the terminal sire sector of the British beef industry showed that estimated discounted returns exceeded the costs of implementation, including research, within a few years of introduction. Estimated annual discounted returns are expected to reach about £20 million per annum, and exceed annual costs of implementation by a factor of 30:1, about 20 years after introduction of these technologies [41].

The costs and benefits of genetic improvement in meat animals have been most comprehensively investigated in pigs. For example, there is good evidence from the pig industry in Britain that genetic improvement is highly cost beneficial at a *national* level. A study in the early 1980s estimated benefits of about £100 million per annum from annual investments of £2 million [35]. While these absolute values are now out of date, the cost:benefit ratio is probably less so. Although this cost: benefit ratio for genetic improvement of pigs appears to be even more favourable than that for beef cattle, both represent very substantial returns on investment.

Despite the slower rates of improvement in beef cattle, compared to that possible in pigs, and differences between countries in the scale of the industries, the costs of improvement

Profit trait	Economic response (£) in progeny
Carcass weight (kg)	2.38
Carcass conformation (pts.)	0.51
Carcass fatness (pts.)	–0.05
Gestation length (days)	0.15
Calving ease (units)	0.41
Total economic response per calf	£3.39

Table 7.23 Expected responses following selection of bulls ranked in the top 40% on the combined new Beef Value and Calving Value indexes, described in the text. The relative contributions of Calving Value and Beef Value to overall merit in this example were 16:84. (Dr P R Amer.)

schemes, and the value of the products, it is likely that there are usually major *national* benefits from beef improvement. However, it is often difficult for *individual* commercial producers to measure the increased margins from genetically superior stock and, in turn, to pay an appropriate premium to breeders of replacement stock. In Britain an MLC study relating the economic performance of crossbred cattle to the index scores of sires has helped to convince producers of the value of within-breed selection. The study involved pairs of high and low index bulls of several breeds, selected on within-herd indexes prior to the introduction of BLUP evaluations. A high index natural service bull, producing about 120 offspring over his working life, was expected to leave progeny worth a total of about £2500 more than a low index sire [28]. The availability of across-herd BLUP evaluations means that high merit animals should now be identified more accurately, and much higher selection intensities can be achieved. Hence the financial benefits of within-breed selection should be even higher than this study suggested. The expected benefits in a commercial herd from selecting bulls ranked in the top 40% on the combined new indexes mentioned earlier, are shown in Table 7.23.

Genotype x environment interactions

There have been a number of experimental investigations, and analyses of field data to investigate genotype x environment interactions in beef cattle [8, 9, 36]. Several studies indicate that small or medium-sized breeds have higher overall productivity or profitability in extensive grazing systems than larger breeds. But, larger breeds tend to do best in more intensive systems with high levels of concentrate feeding. These differences may be most pronounced in composite traits such as weight of calf weaned per cow mated, or per unit of cow weight.

In general, if there is a big enough difference between genotypes, or environments, or both, then it is likely that interactions will be found. The overall message as far as interactions are concerned is: (i) to be aware that they can exist; (ii) to take action if there is sound evidence that they exist in the traits of interest in the breed or system concerned (e.g. by restricting the choice of sires to those measured in similar conditions, or those evaluated in a wide range of conditions) and (iii) not to let the possibility of interactions in some cases detract from the value of genetic improvement in most cases.

Practical guidelines on selection

Some practical guidelines on selection to improve performance in objectively measured traits are given below. Most of these are equally relevant in purebred and commercial herds, and whether the selection is for carcass or maternal traits. However, pedigree breeders using these methods often wish to pay more attention to individual trait EBVs than commercial producers do. Also, they may be willing to take greater risks by selecting animals with low accuracy EBVs, or by selecting occasional unrecorded animals if levels of recording are low in the breed concerned. The guidelines on selection between breeds or crosses are aimed primarily at commercial producers, since most pedigree breeders will already be firmly committed to their current breed.

Selection between breeds or crosses

- Clearly define the role of the animals you are selecting, and identify the animal characteristics which are economically important. Choose an appropriate breed or cross, based on the sort of objective comparisons of performance discussed earlier.

Selection of bulls (or semen) for use in a purebred herd

- Set your breeding objective – the priority should be on traits expected to be of most economic importance in a few (cattle) generation's time.

- Identify the selection index or set of EBVs which is most relevant for your breeding objective.

- Produce a shortlist of bulls which can be fairly compared, ranked on this index or on EBVs for the most important individual traits.

- If national across-herd BLUP evaluations are available, this shortlist might be compiled from a national sire summary, or from EBVs presented in bull sale or semen catalogues. Include homebred bulls on the shortlist if their index scores or EBVs warrant it.

- If there are no national across-herd evaluations in the breed of your choice, select bulls from cooperative schemes or individual herds which have a history of using objective selection for the index or set of EBVs most relevant to your breeding goal. (Try to compare the performance of purchased and homebred bulls, or new and existing purchased bulls, by mating them to groups of cows balanced for genetic merit. If the purchased bulls are inferior to homebred bulls, change sources, or stick to homebred bulls.)

- If there are economically important traits in your breeding goal which are not included in the index available, or for which no EBVs are available, then eliminate any animals which do not meet your minimum standards. Culling for these secondary traits should be in proportion to the economic importance of the trait, and the scope for genetic change in them. Selection for traits of minor economic importance, or traits which have a small genetic component will dilute overall progress.

- Eliminate any animals which are not functionally sound, for example those which have unacceptable locomotion, testicle size or jaw structure. Culling should be confined to important characteristics which have a genetic component.

- If accuracies are available for the index or EBVs of your choice, and you are concerned about the risk of using low accuracy bulls, eliminate those with very low values, or spread the risk by using more bulls. (As explained in Chapter 5, BLUP EBVs already account for accuracy, and so *on average* progress in a herd or breed is maximised by selecting on EBVs regardless of accuracy. However, breeders are often concerned because large sums of money are at stake on *individual* animals. In these situations the accuracy provides a useful measure of the risk that an EBV will change in future. Choosing high EBV bulls with high accuracies is a less risky strategy than choosing high EBV bulls with low accuracy, but if there are few high EBV, high accuracy bulls available, choosing several bulls is less risky than gambling on only one.)

● Within reason, choose the highest index bull left on the list that you can afford.

● Avoid matings between close relatives. (Computer programs may be available via the recording agency or breed society to assist with this task).

● Genetic gain will be maximised by replacing bulls whenever a bull of higher merit is available within this herd or elsewhere (though this may conflict with shorter term profitability). If there are no national evaluations to allow you to compare the merit of your own bulls with others, or to compare the merit of your own bulls of different ages, calculate the male selection intensities and generation intervals which will maximise progress in your breeding goal (as explained in Chapter 4) and aim to replace bulls accordingly.

Selection of replacement females in a purebred herd

● When selecting replacement females to maximise genetic gain in a pedigree herd, rank them in a similar way to that outlined for bulls for the relevant breeding goal (e.g. terminal sire traits or maternal traits, depending on the role of the breed concerned).

● Unless the herd is of low genetic merit, it will usually be most cost-effective to select among homebred females only, rather than buying in replacement females.

● If BLUP EBVs are available then the genetic merit of females of all ages within the herd can be compared directly. Rank the potential female replacements on the chosen index, or on the individual BLUP EBVs of most relevance. Genetic gain will be maximised by replacing cows of any age which have low EBVs by replacements which have higher EBVs. (In a closed herd which has been selecting for a few years, selection of replacement females and culling on BLUP EBVs is unlikely to alter the age structure of the herd dramatically. However, when selection on BLUP EBVs first starts, some attention may need to be paid to the short-term economic effects of dramatic changes in age structure. For example, if it turns out that most older cows in the herd have very low EBVs, although it would be genetically preferable, it may not be economically justifiable to cull all of these in one year because of the effect this may have on herd output.)

- If BLUP EBVs are not available, then it is difficult to compare the genetic merit of cows of different ages. The simplest policy is to calculate the optimum female generation interval in order to maximise genetic progress in your breeding goal (as explained in Chapter 4), and then cull cows on age to achieve this.

- Similar rules on culling for functional fitness or traits of secondary economic importance to those described above for bulls should be applied.

Selection of replacement bulls and cows for a commercial herd

- Genetic progress in commercial herds will be maximised by following guidelines on bull selection similar to those above.

- Usually, crossbred suckler cows do not have EBVs available. In choosing cows to breed replacement females, rank animals on half their sire's EBVs for important traits, if these are available. Alternatively rank them on their actual performance, adjusting where possible for age etc. Whenever there is scope to do so, try to breed replacements from the highest ranking cows on this list, subject to functional soundness.

- In selecting cows to breed replacements there may be a conflict between maximising genetic progress and maximising short-term economic performance in commercial (and pedigree) herds. For example, it may be more profitable in the short term to select replacements which are likely to calve at the optimum time, rather than to select strictly on estimated genetic merit. Hence, most scope for genetic improvement will be through selection of bulls and initial selection of the breed or cross of cows to be used.

Summary

- Beef cattle breeding in temperate countries is less homogeneous than dairy cattle breeding. In most European countries over 50% of beef production is from pure dairy or dual-purpose breeds. In other major temperate beef producing countries, beef production is based on extensively grazed or ranched cows, mainly of pure British beef breeds like the Hereford, Aberdeen Angus and Shorthorn, or crosses among them. In some of these countries, this extensive pre-weaning regime is usually followed by a more intensive finishing period in feedlots.
- Hence, there are two broad categories of beef production in many

countries: (i) beef production from dairy and dual-purpose herds and (ii) beef production from specialised beef herds. Within the specialised beef sector, there is further differentiation into terminal sire and maternal breeds, crosses or lines. Terminal sire breeds also get used in dairy and dual-purpose herds. Each of these categories has distinct beef breeding goals, or at least different priorities.

- While it may be efficient to breed for both milk and meat production from the same type of animal in small-scale production systems, it appears to be less efficient in large, specialised farming systems. As a result there is a general trend towards milk production from more specialised dairy cattle breeds and strains.

- Terminal sire beef breeds are used in dairy herds for two main purposes. The first is to mate to dairy heifers to reduce the incidence of calving difficulties, compared with that following matings to a dairy sire. The second is to mate to mature dairy cows which are not required to breed replacement dairy heifers.

- In many of the specialised beef production systems in temperate countries there is widespread use of crossbreeding – often to achieve complementary use of breeds. Although ease of calving is still important when terminal sire breeds are used in specialised beef breeding herds, their main role is to improve the growth and carcass characteristics of their crossbred offspring.

- The main breeding goals for cows in specialised beef herds, in addition to adequate growth and carcass merit, are good fertility, ease of calving, good maternal ability and low or intermediate mature size, to reduce cow maintenance requirements. The ability of animals to withstand extreme climates and to tolerate low quality feed and periods of feed shortage is also important in some areas. In some cases the traits of importance will be best improved by selection, in others they will be best improved by crossbreeding.

- The predominance of black and white strains in the dairy industry means that they are major contributors to beef output, both directly through surplus calves and cull cows and, in some countries, indirectly through their contribution to the genetic make-up of suckler cows. In many temperate countries the predominant specialised beef breeds are the French breeds, particularly the Charolais and Limousin, and to a lesser extent the British breeds, particularly the Hereford and Angus. However, several less numerous breeds have a disproportionate influence through the use of AI, especially in dairy herds.

- Most beef cattle genetic improvement programmes are based on performance testing or progeny testing. Most performance testing schemes involve recording the pre-weaning performance of all animals on-farm. In some countries, post-weaning performance continues to be measured on-farm. In others, central performance testing is used. Even when the emphasis is on selection on individual animal's performance records, there is an automatic accumulation of progeny records from the sires used in performance-recorded herds. However, in many countries there is a more deliberate strategy of first performance testing, then progeny testing bulls, with selection at each stage. As with performance testing, progeny testing schemes either operate on-farm or at central testing stations.

- In some countries there are cooperative breeding schemes, such as group breeding schemes and sire referencing schemes, though often these are based on national recording and testing schemes. Despite the potential value of MOET nucleus schemes in accelerating response to selection, to date MOET has been used mainly in beef cattle as a means of importing and exporting genetic material, and to multiply newly imported breeds or valuable individuals more rapidly than possible with natural reproduction.

- Generally, on-farm performance recording schemes around the world have concentrated on measuring live weights at regular intervals (or growth rates between these), together with visual scores of muscularity and measurements or scores of height or skeletal development. Ultrasonic measurements of fat and muscle depths or areas have also become more widely used over the last decade. More frequent and more comprehensive measurements are often made in central tests.

- Reproductive traits have fairly low heritabilities. However, many are economically important, and there is substantial variation in them, so there is both the incentive and scope for genetic improvement. Direct heritabilities of growth traits tend to be moderately high, while maternal heritabilities tend to be slightly lower. The heritabilities of carcass measurements tend to be even higher than those for growth traits. However, they have to be assessed either indirectly on live candidates for selection (e.g. by ultrasonic measurements), or directly on progeny or other relatives of the candidates for selection, so they are not as easy to improve as it seems at first sight.

- There are unfavourable associations between several reproduction and growth traits, which complicates selection for dual-purpose characteristics. Generally there are strong phenotypic and genetic associations between weights at different ages – the closer the age, the stronger the association. This means that selection for growth characteristics alone usually leads to unfavourable responses in birthweight (i.e. increased birthweight which generally leads to more difficult calving), and heavier mature cow weight.

- Until the early 1970s the genetic evaluation methods employed in most countries were fairly simple. They usually produced adjusted records of performance, contemporary comparisons or predicted breeding values which could only be compared within herd (or within a central test team). However from the 1970s onwards, BLUP methods of evaluation began to be adopted. Sire model BLUP evaluations for beef cattle across herds and years were first used in the early 1970s in the US. The uptake of BLUP has been slower in beef cattle than in dairy cattle in most other countries. However, advances in computer power and computing strategies mean that those countries which have changed to BLUP evaluations recently have often adopted multi-trait animal model evaluations immediately.

● The results of evaluations are usually expressed as estimated breeding values (EBVs), or expected progeny differences (EPDs). These are expressed relative to a fixed base in many countries, and a rolling base in a few countries.

● To be able to compare BLUP PBVs fairly across contemporary groups and years, genetic links are needed between groups and years. In some countries there is little use of AI in specialised beef breeds, and this has limited the introduction of national across-herd genetic evaluations. Sire referencing schemes have been introduced to create links between herds in some of these countries.

● The main technical limitation to performing evaluations across breeds is that animals of different breeds are rarely kept as contemporaries under similar management and feeding systems. However, across-breed evaluations are becoming feasible using information from crossbred animals, or from designed breed comparisons, together with estimates of genetic trends in each of the purebred populations.

● Compared to the situation in dairy cattle, there has been less effort to date in developing international conversions of EBVs or EPDs for beef cattle, or performing international genetic evaluations. However, there is growing interest in this area.

● Until recently, the emphasis in beef cattle breeding in North America has been on using sophisticated methods to produce EBVs for individual traits. In contrast, in Europe, while less sophisticated methods of evaluation were used until recently, selection indexes have been quite widely used in both specialised beef breeds and in dairy and dual-purpose breeds. Much of the emphasis in Europe has been on producing indexes for terminal sire characteristics. Fewer indexes have been developed for maternal breeds or lines.

● Indexes combining BLUP EBVs for reproduction, growth and carcass traits have been developed in Australia. An important feature of these indexes is that the economic values applied can be tailored or customised to individual breeder's requirements.

- In theory, changes of at least one per cent of the mean per annum are possible following selection for weight or growth traits in beef cattle. However, in practice, rates of change in selection experiments and in industry are often lower than this. The wider use of more effective genetic evaluation techniques should allow more rapid improvement in a range of traits in future.

- It is likely that there are major *national* benefits from genetic improvement of beef cattle in many countries. However, it is difficult for *individual* commercial producers to measure the increased margins from genetically superior stock and, in turn, to pay an appropriate premium to breeders of replacement stock. Experimental evidence of these benefits is helping to increase awareness of the value of genetic improvement among producers.

References

1. Allen, D.M. and Steane, D.E. 1985. 'Beef selection indices.' *British Cattle Breeders Club Digest No. 40*, pp. 63–70.

2. Amer, P.R., Kemp, R.A. and Smith, C. 1992. 'Genetic differences among the predominant beef cattle breeds in Canada: An analysis of published results.' *Canadian Journal of Animal Science*, 72:759–71.

3. Andersen, B.B., de Baerdemaeker, A. , Bittante, G. et al. 1981. 'Performance testing of bulls in AI: report of a working group of the Commission on Cattle Production.' *Livestock Production Science*, 8:101–19.

4. Barlow, R. and Cunningham, E.P. 1984. 'Benefit-cost analyses of breed improvement programmes for beef and sheep in Ireland.' *Proceedings of the Second World Congress on Sheep and Beef Cattle Breeding*, Vol. II, P 31.

5. Barwick, S.A., Henzell, A.L. and Graser, H.-U. 1994. 'Developments in the construction and use of selection indexes for genetic evaluation of beef cattle in Australia.' *Proceedings of the 5th World Congress on Genetics Applied to Livestock Production*, Vol. 18, pp. 227–30.

6. Benyshek, L.L. and Bertrand, J.K. 1990. 'National genetic improvement programmes in the United States beef industry.' *South African Journal of Animal Science*, 20:103–9.

7. Benyshek, L.L., Herring, W.O. and Bertrand, J.K. 1994. 'Genetic evaluation across breeds and countries: prospects and implications.' *Proceedings of the 5th World Congress on Genetics Applied to Livestock Production*, Vol. 17, pp. 153–60.

8. Bertrand, J.K., Berger, P.J. and Willham, R.L. 1985. 'Sire x environment interactions in beef cattle weaning weight field data.' *Journal of Animal Science*, 60: 1396–402.

9. Bishop, S.C. 1993. 'Grassland performance of Hereford cattle selected for rate and efficiency of lean gain on a concentrate diet.' *Animal Production*, 56:311–19.

10. Bonnett, J.N., Journaux, L., Mocquot, J.C. and Rehben, E. 1994. 'Breeding cattle for the next millennium.' *British Cattle Breeders Club Digest No. 49*, pp. 10–18.

11. Brandsma, J. and Banos, G. 1996. *Sire evaluation procedures for non-dairy production and growth and beef production traits practised in various countries 1996.* INTERBULL Bulletin No. 13, International Bull Evaluation Service, Department of Animal Breeding and Genetics, Uppsala, Sweden.

12. Broadbent, P.J. 1990. 'The NOSCA Simmental MOET project.' *British Cattle Breeders Club Digest No. 45*, pp. 44–8.

13. Crump, R.E., Simm, G., Nicholson, D. et al. 1997. 'Results of multivariate individual animal model genetic evaluations of British pedigree beef cattle.' *Animal Science,* 65: 199–207.

14. Cundiff, L.V., Gregory, K.E., Koch, R.M. and Dickerson, G.E. 1986. 'Genetic diversity among cattle breeds and its use to increase beef production efficiency in a temperate environment.' *Proceedings of the 3rd World Congress on Genetics Applied to Livestock Production*, Vol. IX, pp. 271–82.

15. Cuthbertson, A. 1994. 'Enhancing beef eating quality.' *British Cattle Breeders Club Digest No. 49*, pp. 33–7.

16. de Rose, F.P. and Wilton, J.W. 1988. 'Estimation of genetic trends for Canadian station-tested beef bulls.' *Canadian Journal of Animal Science*, 68:49–56.

17. Dikeman, M.E. 1990. 'Genetic effects on the quality of meat from cattle.' *Proceedings of the 4th World Congress on Genetics Applied to Livestock Production*, Vol. XV, pp. 521–30.

18. Food and Agriculture Organisation of the United Nations. 1996. *FAO Production Yearbook 1995. Vol 49.* Food and Agriculture Organisation of the United Nations, Rome.

19. Graser, H.-U., Goddard, M.E. and Allen, J. 1995. 'Better genetic technology for the beef industry.' *Proceedings of the Australian Association of Animal Breeding and Genetics*, Vol. 11, pp. 56–64.

20. Graser, H.-U., Hammond, K. and McClintock, A.E. 1984. 'Genetic trends in Australian Simmental.' *Proceedings of the Australian Association of Animal Breeding and Genetics*, Vol. 4, pp. 86–7.

21. Gregory, K.E., Cundiff, L.V. and Koch, R.M. 1991. 'Breed effects and heterosis in advanced generations of composite populations from preweaning traits of beef cattle.' *Journal of Animal Science*, 69:947–60. (Also, see other papers by the same authors in Volumes 69 and 70.)

22. Journaux, L., Rehben, E., Laloë, D. and Ménissier, F. 1996. *Main results of the genetic evaluation IBOVAL96 for the beef cattle sires.* Edition 96/1. Institut de l'Elevage, Département Génétique Identification et Contrôle de Performances, and Institut National de Recherche Agronomique, Station de Génétique Quantitative et Appliquée, Jouy-en-Josas.

23. Journaux, L. and Renand, G. 1992. 'Genetic relationship between station performance testing and station progeny testing of Charolais bulls.' *Book of Abstracts from the 43rd Annual Meeting of the European*

Association for Animal Production, Madrid, Spain, Paper G.Va.37, p. 210.

24. Kemp, R.A. 1994. 'Genetics of meat quality in cattle.' *Proceedings of the 5th World Congress on Genetics Applied to Livestock Production*, Vol. 19, pp. 439–45.
25. Koots, K.R., Gibson, J.P., Smith, C. and Wilton, J.W. 1994. 'Analyses of published genetic parameter estimates for beef production traits. 1. Heritability.' *Animal Breeding Abstracts*, 62:309–38.
26. Koots, K.R., Gibson, J.P. and Wilton, J.W. 1994. 'Analyses of published genetic parameter estimates for beef production traits. 2. Phenotypic and genetic correlations.' *Animal Breeding Abstracts*, 62:825–53.
27. Land, R.B. and Hill, W.G. 1975. 'The possible use of superovulation and embryo transfer in cattle to increase response to selection.' *Animal Production*, 21:1–12.
28. Lewis, W.H.E. 1992. 'Beef index validation – national results from suckler herds.' *British Cattle Breeders Club Digest No. 47*, pp. 24–7.
29. Liboriussen, T. 1982. 'Comparison of paternal strains used in crossing and their interest for increasing production in dairy herds.' *Proceedings of the 2nd World Congress on Genetics Applied to Livestock Production*, Vol. V, pp. 469–81.
30. Meat and Livestock Commission. 1983. *Beef Yearbook 1983*. Meat and Livestock Commission, Milton Keynes.
31. Meat and Livestock Commission. 1992. *Beef Yearbook 1992*. Meat and Livestock Commission, Milton Keynes.
32. Meat and Livestock Commission. 1995. *Beef Yearbook 1995*. Meat and Livestock Commission, Milton Keynes.
33. Ménissier, F. 1988. 'La sélection des races bovines à viande spécialisées en France.' *Proceedings of the 3rd World Congress on Sheep and Beef Cattle Breeding*, Vol. 2, pp. 215–36.
34. Middleton, B.K. and Gibb, J.B. 1991. 'An overview of beef cattle improvement programs in the United States.' *Journal of Animal Science*, 69: 3861–71.
35. Mitchell, G., Smith, C., Makower, M. and Bird, P.J.W.N. 1982. 'An economic appraisal of pig improvement in Great Britain. 1. Genetic and production aspects.' *Animal Production*, 35:215–24.
36. Morris, C.A., Baker, R.L., Hickey, S.M. et al. 1993. 'Evidence of genotype by environment interaction for reproductive and maternal traits in beef cattle.' *Animal Production*, 56:69–83.
37. Mrode, R.A. 1988. 'Selection experiments in beef cattle. Part 2. A review of responses and correlated responses.' *Animal Breeding Abstracts*, 56:155–67.
38. Pirchner, F. 1986. 'Evaluation of industry breeding programs for dairy cattle milk and meat production.' *Proceedings of the 3rd World Congress on Genetics Applied to Livestock Production*, Vol. IX, pp. 153–64.
39. Residuary Milk Marketing Board of England and Wales. 1995. *UK Dairy Facts and Figures. 1994 edition*. The Residuary Milk Marketing Board of England and Wales, Thames Ditton, Surrey.

40. Scott, D. 1992. 'Accelerated development of a British beef breed.' *British Cattle Breeders Club Digest No. 47*, pp. 62–6.

41. Simm, G., Amer, P.R. and Pryce, J.E. 1998. 'Benefits from genetic improvement of sheep and beef cattle in Britain.' In *SAC Animal Sciences Research Report 1997*. SAC, Edinburgh, (in press).

42. Simm, G., Conington, J. and Bishop, S.C. 1994. 'Opportunities for genetic improvement of sheep and cattle in the hills and uplands.' *In Livestock Production and Land Use in Hills and Uplands*. T.L.J. Lawrence, D.S. Parker and P. Rowlinson (eds), Occasional Publication No. 18, British Society of Animal Production, pp 51–66.

43. Simm, G., Conington, J., Bishop, S.C., Dwyer, C.M. and Pattinson, S. 1996. 'Genetic selection for extensive conditions.' *Applied Animal Behaviour Science*, 49:47–59.

44. Simm, G., Smith, C. and Prescott, J.H.D. 1985. 'Environmental effects on bull performance test results.' *Animal Production*, 41:177–85.

45. Simm, G., Steane, D.E. and Wray, N.R. 1990. 'Developments in beef cattle breeding programmes in Europe.' *Proceedings of the 4th World Congress on Genetics Applied to Livestock Production*, Vol. XV, pp. 231–43.

46. Thiessen, R.B., Hnizdo, E., Maxwell, D.A.G., et al. 1984. 'Multibreed comparisons of British cattle. Variation in body weight, growth rate and food intake.' *Animal Production*, 38:323–40. (Also, see other papers by the same authors in later volumes of Animal Production.)

47. Villanueva, B., Simm, G. and Woolliams, J.A. 1995. 'Genetic progress and inbreeding for alternative nucleus breeding schemes for beef cattle.' *Animal Science*, 61:231–9.

Further reading

Animal Science (formerly *Animal Production*; Journal of the British Society of Animal Science)

Canadian Journal of Animal Science (Journal of the Agricultural Institute of Canada/Canadian Society of Animal Science)

Journal of Animal Science (Journal of the American Society of Animal Science)

Livestock Production Science (Journal of the European Association for Animal Production)

Proceedings of The Association for the Advancement of Animal Breeding and Genetics (formerly *The Australian Association of Animal Breeding and Genetics)*

Proceedings of the New Zealand Society of Animal Production

Proceedings of the World Congresses on Genetics Applied to Livestock Production

CHAPTER 8

Sheep breeding

Introduction

Sheep are probably the most versatile of the domestic animal species. In temperate countries today they are kept mainly for the production of meat, wool and milk. However, particularly in some of the world's poorest countries, they have additional, important roles, including the provision of pelts, fertiliser and fuel, providing a four-legged form of financial investment, and using resources unsuitable for other forms of agriculture.

Figures 8.1 to 8.3 show the world's major producers of sheep meat, milk and wool by continent. Asia and Europe have the highest production of both meat and milk, while Oceania has by far the highest production of wool. Figures 8.4. to 8.6 show the fifteen highest producing countries for each commodity. China, Australia and New Zealand head the list of individual countries producing most meat and wool. Turkey, Italy, China and Greece head the list of countries producing most sheep milk.

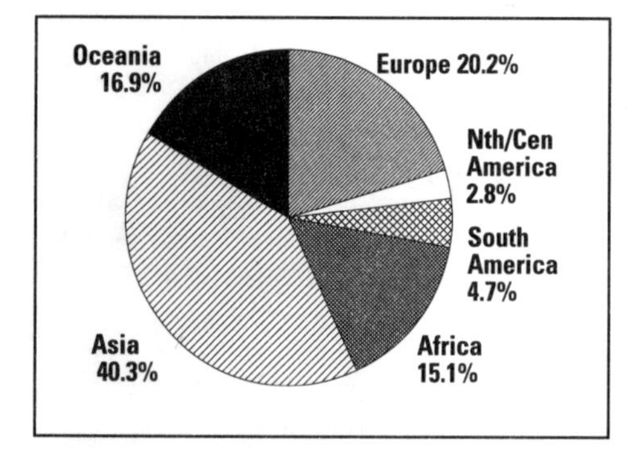

Figure 8.1 Proportion of world production of mutton and lamb by continent in 1995 [16].

In northern Europe, despite their historical importance as wool producers – a role in which they had a major impact on the social and industrial shape of Britain in particular – sheep are kept primarily for meat production today. For example, wool accounts for only about 2.5% to 4% of average gross output from sheep enterprises in Britain [31, 36]. In several southern European countries, and in parts of Asia and Africa, sheep

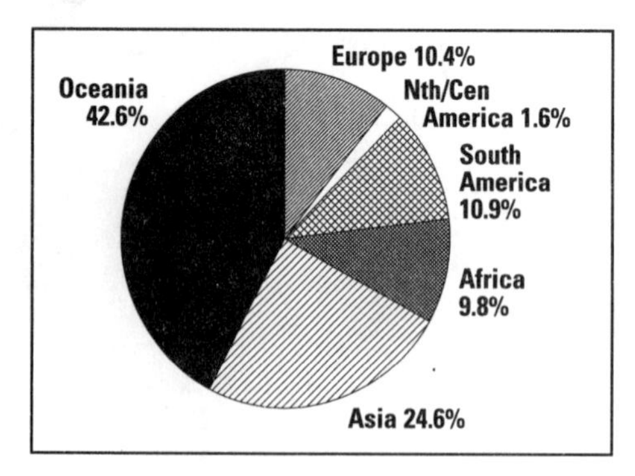

Figure 8.2 Proportion of world production of greasy wool by continent in 1995 [16].

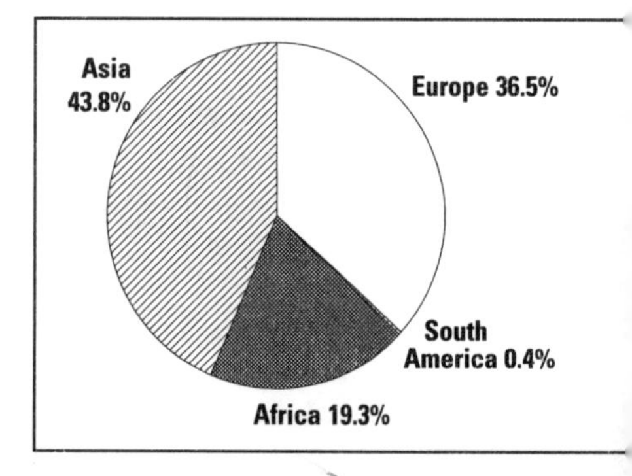

Figure 8.3 Proportion of world production of sheep milk by continent in 1995 [16].

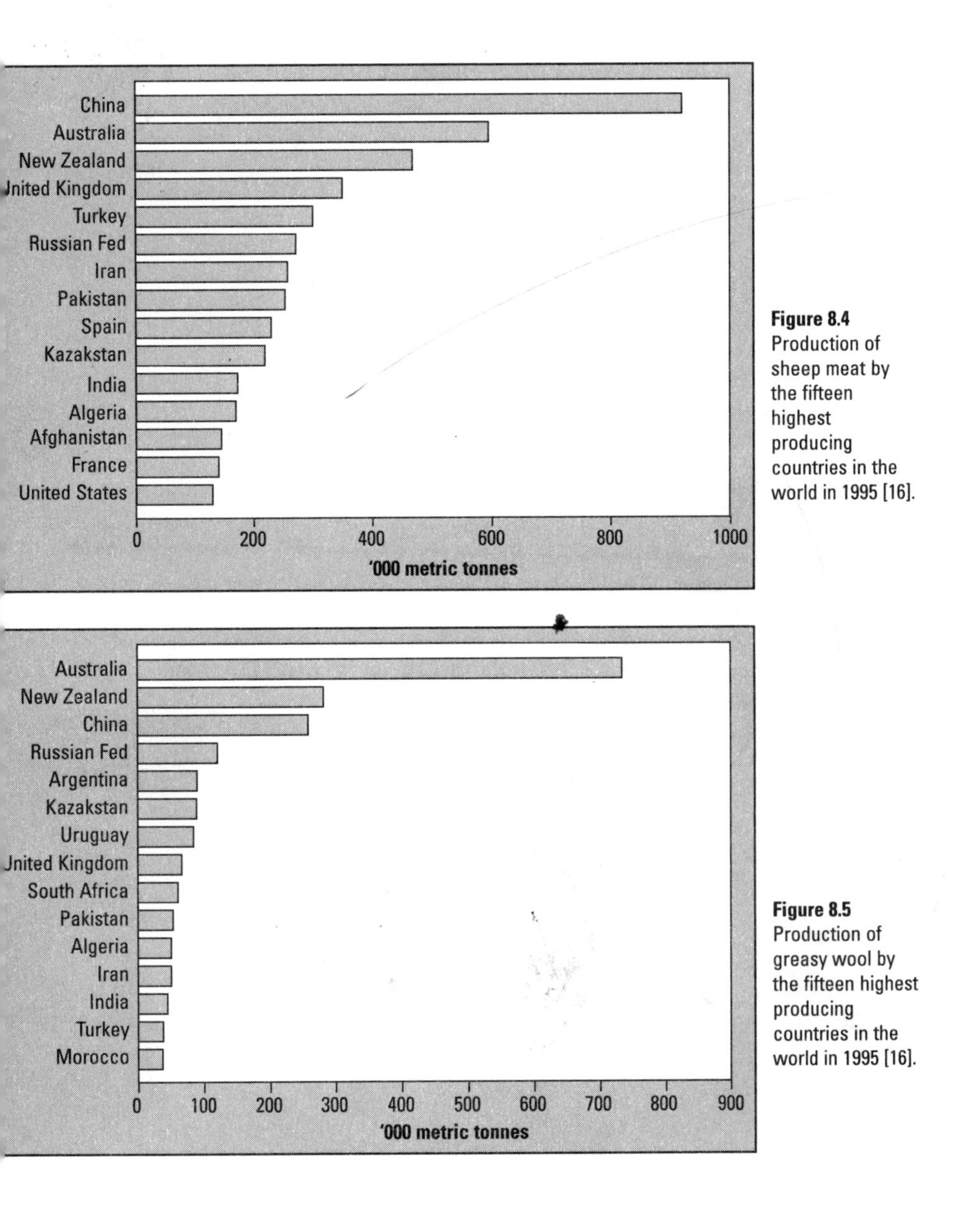

Figure 8.4 Production of sheep meat by the fifteen highest producing countries in the world in 1995 [16].

Figure 8.5 Production of greasy wool by the fifteen highest producing countries in the world in 1995 [16].

are kept primarily for milk production, and meat is produced as a by-product, albeit a valuable one. Although most wool is used in the northern hemisphere, most production is in the southern hemisphere. Australia has the world's largest specialised wool production industry, but meat is an important by-product – especially indirectly from dual-purpose F1 ewes resulting from crosses between Border Leicester sires and Merino ewes. In New Zealand, the substantial sheep industry is based largely on dual-purpose breeds for the production of both meat and wool (see Plate 37).

The aim of this chapter is to discuss the breeding goals, the breeds and crosses,

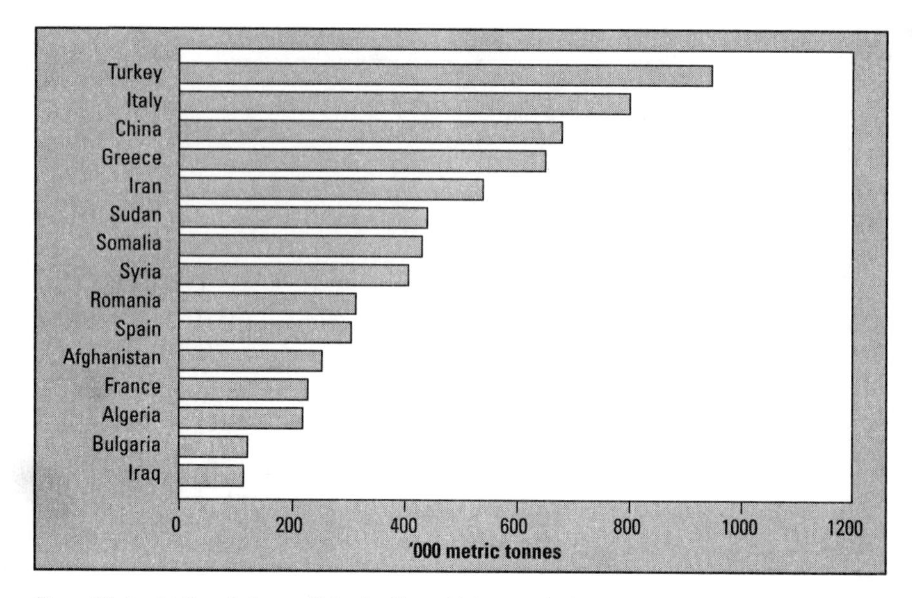

Figure 8.6 Production of sheep milk by the fifteen highest producing countries in the world in 1995 [16].

the selection criteria, and the methods of testing and evaluation used in sheep breeding in temperate areas. The emphasis is on sheep meat production and wool production, but milk production and multi-purpose production are discussed briefly too.

Breeding goals

Meat production

Figure 8.7 summarises the contribution of different measures of performance to the superiority in gross margin of the top third of commercial flocks recorded by Signet/MLC in Britain for several production systems. These charts help to identify the animal characteristics which affect profitability in a range of meat sheep production systems. Although the relative importance of these varies between systems, and between countries, the key animal characteristics are given below.

The number of lambs reared to slaughter per ewe per annum. Lamb output per ewe is obviously strongly influenced by litter size and lamb survival. There is variation both between and within breeds in litter size and to a lesser extent survival. These two characteristics are related at the phenotypic level, as higher litter sizes generally result in smaller individual lamb birthweights, and higher rates of lamb mortality. Hence, there is usually an intermediate optimum number of lambs born for a given production system. For example, in lowland meat production systems, the optimum number of lambs per litter may be two or higher, but in very harsh hill or range conditions it may be closer to one lamb. In most sheep production systems ewes lamb once per annum. However, because of a relatively short gestation length, there is an opportunity to increase output by lambing more than once per annum, and frequent lambing systems have been adopted in a few areas (the targets are usually three lambings in two years or five lambings in three years). Sheep are seasonal breeders in countries with varying day length, but there is variation both between and within breeds in the onset and the duration of the breeding season. These animal characteristics influence the ability to lamb more frequently, and also affect output in systems with annual lambing.

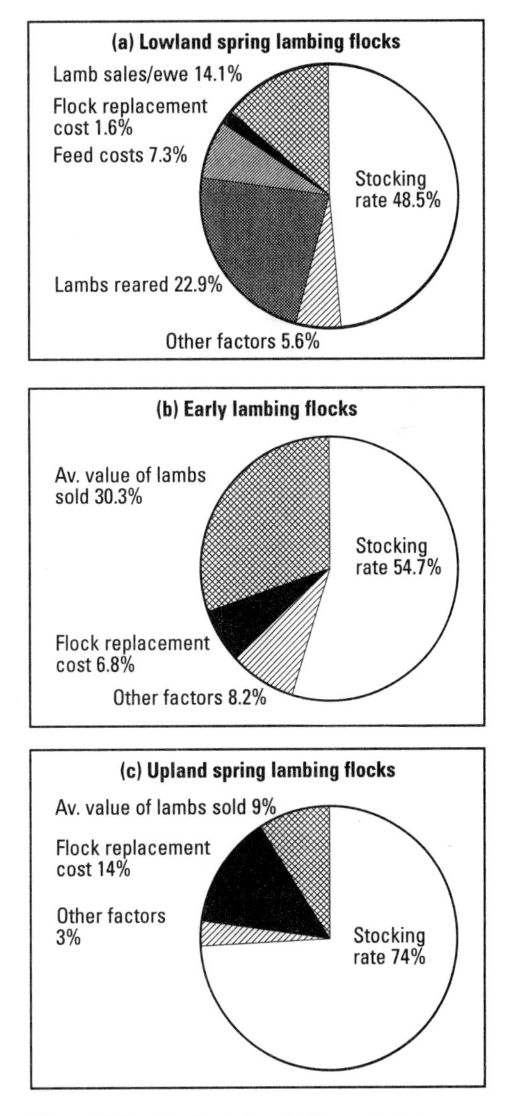

Figure 8.7 Contribution to top-third superiority in gross margin per hectare for Signet Flockplan recorded flocks in different sectors (a) lowland spring lambing flocks, selling most of their lambs off grass in summer and autumn 1995; (b) early lambing flocks, 1995; (c) upland spring lambing flocks selling most of their lambs off grass in summer and autumn 1995 [31].

The weight and carcass attributes of the lambs produced. In some countries and production systems, some or all of the lambs are sold after weaning for finishing on other farms (or in feedlots). The proportion of lambs which are sold directly for slaughter, rather than sold for finishing elsewhere, is an important factor affecting profitability. For example, in Britain, the Signet/MLC results show that flocks achieving a higher proportion of lambs sold directly for slaughter produce higher returns per lamb. Obviously there are many management and environmental influences on the proportion of lambs sold for slaughter, but lamb growth and carcass composition are important animal factors which also influence it.

In systems producing lambs directly for slaughter, carcass value has an important impact on gross margins. Carcass value is usually a function of carcass weight, fatness and conformation, although breed, age and sex can also influence carcass prices in their own right. The relative importance of these measures of carcass value varies between regions and countries. Often there are strong local preferences for carcasses of a certain weight, fat or conformation range, and those falling outside this range will be penalised. For example, in Britain most lamb carcasses fall in the weight range of 16 to 22 kg [31]. In Mediterranean countries the preference is for much lighter, leaner carcasses. This market for lighter carcasses has led to a large increase in exports of carcasses from the smaller hill breeds in Britain over the last few years.

In many Western countries consumers have shown a strong preference for leaner meat over the last few decades. Hence, for these countries, efficient production of *lean* meat is a primary breeding goal. It may not be biologically or economically efficient to attempt to match consumer preferences precisely in the live animal, but they are often too far apart. For example, in 1986 the average lamb carcass in Britain contained 20 percentage units of fat in excess of that desired by consumers [20]. There has been some improvement since then in the composition of carcasses produced, but even now payment schemes probably do not discriminate strongly enough between carcasses of different fatness, and so there is still a mismatch between consumer demand and market supply.

In many countries carcasses are scored visually for fat cover and conformation. For example, Table 8.1 shows the fat and conformation grid used to classify lamb carcasses by MLC in Britain, together with the proportions of carcasses falling into each class in 1995. This method of classification is based on a standard one employed across the EU, but fat classes 3 and 4 are subdivided into low and high bands in the British version. The target specification for most major buyers in Britain is for carcasses in the leaner fat classes 1, 2 and 3L and in conformation classes E, U or R. While there has been a steady improvement in the proportion of carcasses falling within the target sector, this was still only 49.3% in the sample classified by MLC in 1995 [31]. To some extent price differentials reflect consumer preference for leaner meat. But the true market signals for leaner meat are blunted by overcapacity in the British abattoir sector, which means that achieving high throughput of animals often overrides payment on quality.

Choosing a breeding goal to select between or within terminal sire breeds is complicated by the association between mature size and carcass composition. Generally, as all animals grow towards maturity they get fatter. Breeds differ in fatness quite markedly when they are compared at the same weight. However, when they are compared at similar degrees of maturity in live weight, most of the these differences in carcass composition disappear [23]. Differences in mature size may also explain much of the within-breed variation in carcass composition at a given weight. So, the simplest way to breed leaner lambs is probably to select larger terminal sire breeds and larger sires within these breeds. (This assumes that

		Fat class			→		Fatter	
		1	2	3L	3H	4L	4H	5
Conformation class	E	–	0.3	1.3	0.9	0.3	0.1	0.1
↓	U	–	2.2	9.8	5.8	1.7	0.8	0.3
	R	0.1	8.6	27.8	10.9	2.5	1.0	0.3
Poorer	O	0.7	8.1	9.1	2.0	0.2	0.1	–
conformation	P				4.9			

Table 8.1 Sheep carcass classification grid used by MLC in Britain, showing the distribution of lamb carcasses classified by them in 1995 (about 27% of total slaughterings). Fat classes range from 1, the leanest, to 5 the fattest. Conformation is scored on a five-point scale, from E, the highest conformation class to P, the poorest. Similar grids are used throughout the EU [31].

all commercial tier progeny are slaughtered – if some female progeny are kept for breeding then there will be a penalty because of their larger mature size and hence higher maintenance costs.)

Although mature size is a useful concept when considering breeding goals, it is difficult to apply in practice in within-breed selection, because of the time and expense involved in keeping animals until they are mature before making selection decisions. In practice, early growth rates are usually used as indicators of eventual mature size. Also, although mature size is important it does not explain *all* the differences in fatness, so it is worth paying attention to levels of fatness in immature animals too [38]. Hence, the breeding goal is often to increase the rate of gain of lean tissue to an immature weight, or to increase the proportion of lean in the carcass at an immature weight. While these broad definitions of breeding goals have been applied to breeds used in a range of production systems so far, they could be refined to match particular systems more closely in future. For example, some breeds or sectors of a breed may choose to go for rapid early lean growth to produce maximum returns from early grass-based finishing systems. In this case, growth rate or mature size may be more important than fatness in the overall breeding goal. Others may opt for maximising returns from longer finishing systems, in which case controlling fatness may be more important than growth rate or mature size.

Carcass conformation may be a useful indicator of differences in saleable meat yield *among* breeds. However, most studies *within* breeds, except double-muscled breeds or strains, show that conformation is closely associated with fatness, and is of little or no value in predicting saleable meat yield, or proportion of higher-priced cuts [39]. Despite this, payment schemes in many countries include premiums and penalties for high and low conformation carcasses respectively, so it can be argued that conformation has a direct economic value.

Feed consumption. As with other livestock, feed costs are the most important variable costs in sheep meat production. In some countries or regions, or in some seasons, sheep meat production is based on concentrate feeding of lambs, and so lamb growth and feed conversion efficiency can have a big influence on profitability. However, in most temperate countries sheep meat production is based on extensive grazing of grass or forage crops. Feed costs and feed efficiency are still important here, but they are most commonly measured indirectly via output of lamb per hectare and the stocking rate of ewes achieved. Stocking rate has an important effect on gross margin per hectare. Although the stocking rate achieved is probably mainly a function of land type and grassland management, animal characteristics such as ewe mature weight and litter size also have an important effect on it (i.e. because feed requirements are proportional to weight, it is easier to achieve high stocking rates with ewes of low mature weight, than those of higher mature weight).

Health and longevity. Figure 8.7 shows that ewe replacement costs have an important bearing on gross margins achieved. Ewe longevity is affected by a wide range of both management and animal factors, including voluntary or involuntary culling due to disease (e.g. tooth loss, mastitis). Sheep are notable for the large number of diseases to which they can succumb(!), and so veterinary and medicine costs also have a direct effect on profitability. There is genetic variation between and within breeds in susceptibility to many common diseases.

The profitability of sheep meat production is affected by a mixture of maternal characteristics such as ewe fertility, litter size, rearing ability, mature size and milk production, and terminal sire characteristics such as growth and carcass quality. In some countries, such as France, these breeding goals are often pursued together in the same pure breeds. In other countries, especially Britain, there is widespread use of crossbreeding to achieve the complementary use of breeds with different growth and reproductive characteristics, and to allow the pursuit of different breeding goals in specialised maternal and terminal sire breeds.

Wool production

The main uses of wool are in the production of garments, furnishing or other fabrics, fillings and carpets. The fibre diameter and staple length (the length of the shorn fibres) are the main wool characteristics determining the end-use, and hence the price of wool. Table 8.2 illustrates the typical fibre diameter of wool of different categories, and gives examples of the major breeds which produce each type, as well as the end-uses of the wool. The table illustrates the clear demarcation between breed types in terms of fibre diameter, but there are also big differences in fleece weight. For example, Merinos typically produce fleeces weighing about 5 kg (greasy fleece weight, i.e. as shorn), compared to fleece weights of about 3 to 4 kg for dual-purpose breeds like the Coopworth, and 2 or 2.5 kg for the larger of the British hill breeds.

Only the finest wool is suitable for use in garment manufacture, and it is this type of wool which is most valuable. The vast majority of wool produced for garment manufacture comes from Merino sheep, or derivatives of this breed. Within this finewool sector there are further strong links between fibre diameter and price. Coarser wool from dual-purpose breeds or meat breeds is less valuable, and gets used for the other purposes mentioned. The coarsest wool of all gets used in the manufacture of carpets, because of its hard-wearing properties. An

Wool category	Fibre diameter (microns)	Major breeds	End-use of wool
Fine	Up to 25 μ	Merino, Polwarth	Garment manufacture – light weight, high quality clothing fabric, high quality knitwear.
Medium	25 to 30 μ	Corriedale, other dual-purpose breeds (usually 50% Merino)	Garment manufacture – medium weight clothing fabric, machine and hand knitting yarns.
Coarse	30 μ and over	Romney, Coopworth, Perendale, Border Leicester	Garment manufacture – heavy weight clothing fabric. Tweeds. Blankets. Thick machine and hand knitting yarns. Furnishing fabric. Carpets.
Speciality carpet	Medullated, mean 40 μ	Drysdale, Tukidale, Carpetmaster, Elliotdale, Scottish Blackface	Carpets. Upholstery fillers.

Table 8.2 Typical fibre diameter of wool of different categories, examples of the major breeds which produce each wool type, and end-uses of the wool [33].

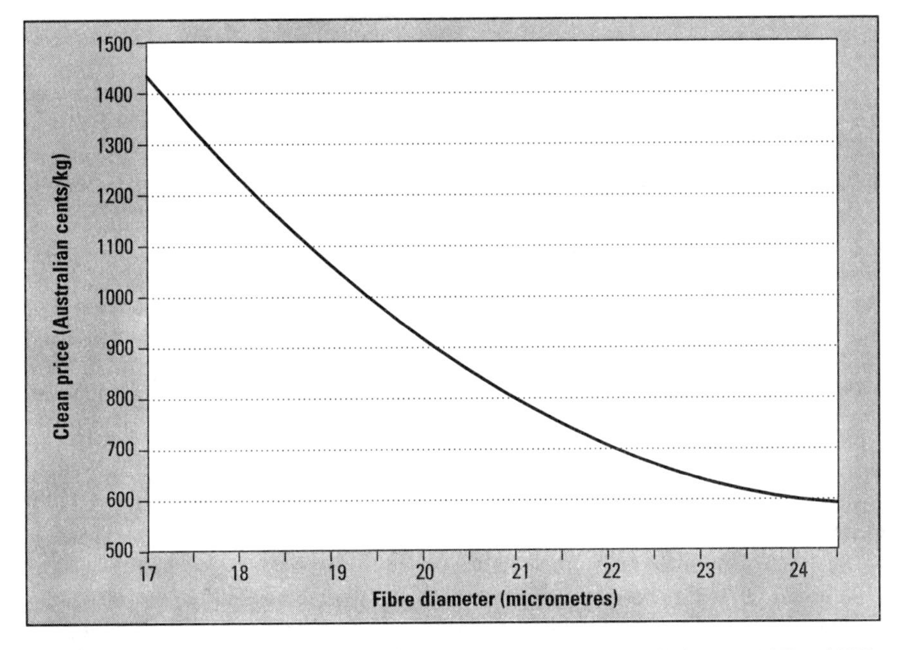

Figure 8.8 The relationship between wool fibre diameter and price in Australia between 1975 and 1996. The line shown is a line of 'best fit' through the actual points for each year. ([2]; Dr K D Atkins.)

additional important property of speciality carpet wool from some breeds, such as the Drysdale or Scottish Blackface, is the high proportion of medullated fibres – fibres with a hollow core of large fragile cells, within a sheath of the normal dense cortical cells [33]. Medullation affects the appearance of dyed wool and the wool handling characteristics – medullated fibres feel 'crisp' rather than soft. Although fibre diameter is important in determining the category of use, there is little or no premium for finer fibre *within* these non-garment sectors.

The main factors affecting income in specialised garment wool production systems are mean fleece weight and mean fibre diameter. Fleece weight is expressed either as greasy or clean. Greasy fleece weight is the weight as clipped. Clean fleece weight is the weight after scouring to remove natural oils and vegetable matter, dust etc. The two measures are highly correlated genetically. Clean fleece weight expressed as a percentage of greasy fleece weight is termed yield – typically this is about 68%, but it can vary markedly depending on the environment in which the sheep are kept. There is also substantial genetic variation in yield within a given environment. The mean fibre diameter is important because it determines both the weight of the fabric (finer fibres produce lighter weight cloth) and its comfort (finer fibres result in a 'softer' finish, which is especially important for fabrics worn next to the skin). Figure 8.8 shows the relationship between fibre diameter and price in Australia between 1975 and 1996. Other wool characteristics such as the range in fibre diameter, staple length, staple strength, 'style' (visual characteristics of the fleece associated with processing quality), wool colour and level of contamination can also be important, although several of these are related to fibre diameter. Other animal factors, such as the incidence of fleece rot, fly

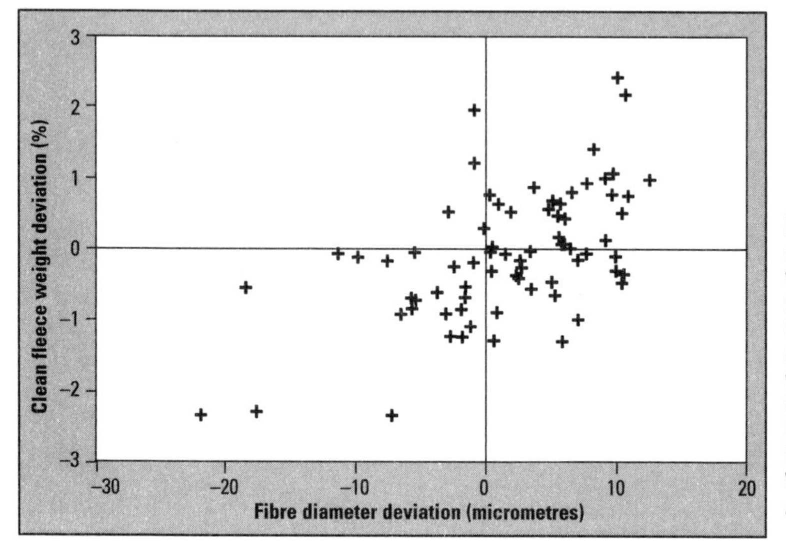

Figure 8.9 The average fleece weight and the average fibre diameter of groups of wethers from different Merino studs in New South Wales. (Dr K D Atkins; [4].)

strike, and internal parasitism are important in certain areas – either because they reduce productivity and welfare or, in extreme cases, cause higher mortality.

Historically, most Merino breeding programmes have focused on increasing fleece weights while maintaining fibre diameter. However, the trend towards lighter weight fabrics in garment manufacture means that fibre diameter is likely to become more important in future [3]. One of the fundamental problems in setting and achieving breeding goals for fine wool production is that there is a moderately antagonistic relationship between fleece weight and fibre diameter, both across strains and within strains. In other words, higher producing strains or individuals tend to produce wool of higher fibre diameter. This is illustrated in Figure 8.9, which shows both the average fleece weight and the average fibre diameter of groups of wethers (castrated males) from different Merino studs in New South Wales, Australia.

Milk production

Sheep milk production in Mediterranean countries usually involves out-of-season lambing from October to January, rather than the more typical lambing season in northern Europe of March to April. Lambs are usually reared on their mothers for a month or so and then weaned. Most lambs are slaughtered at weaning, but in some countries they are reared for slaughter at heavier weights. In most systems milking of ewes commences after weaning, but in others, ewes are also milked during the suckling period. Ewes are milked for up to seven months after weaning. Most sheep milk produced in the EU is used for further processing, especially the manufacture of cheese, and so compositional quality of the milk is important. In the main milk producing countries in the EU the contribution of milk sales to total income ranges from about one third in those countries, systems or breeds with lower milk yields, up to two thirds of total income when milk yields are high [8].

Hence the animal factors influencing profitability in milk production systems include the ability to lamb out of season, the yield and quality of milk produced, the number of lambs weaned and, to a lesser extent, the growth rate of these lambs.

Multi-purpose breeding goals

The profitability of animal production usually depends on an array of animal characteristics rather than on any one alone. The options for selection for more than one trait were discussed in Chapters 3 to 5. For within-breed selection, the most efficient option is usually to select on a multi-trait index, where the breeding goal comprises the sum of breeding values for all important traits, each weighted by its relative economic value. So, for sheep enterprises where any combination of meat, milk and wool production is important, it is sensible to consider index selection for this combination. In multi-purpose or maternal breeds, crosses or lines, it is important to consider lifetime profitability, to account for differences in the timescale over which traits are expressed, and the number of expressions of each trait. For instance, in dual-purpose breeds producing meat and wool, meat output is a function of the number and value of lambs reared at each lambing over the ewe's lifetime, together with her cull value. In a breed which averages 1.5 lambs reared, and typically has 5 lamb crops, ewes would rear an average of 7.5 lambs in their lifetime. So the economic value of a trait which genetically improves value per lamb needs to be multiplied by 7.5 to account for multiple expressions, but then divided by 2 to account for the fact that the ewe contributes only half the genes of her lambs.

Wool traits have multiple expressions too – probably five or six shearings per ewe lifetime in this example – so the economic value of improvement per fleece needs to be multiplied by 5 or 6 to account for this. Since genetic improvements in wool traits are expressed by the ewe herself, the relative economic values remain on this scale, rather than dividing by 2 as for offspring traits. Similarly, for selection between breeds or crosses, it is sensible to compare these for lifetime physical performance in each of the important characteristics, weighted by the economic values – in other words ranking breeds or crosses on their expected total lifetime profit.

In addition to these traits directly influencing returns from meat, milk or wool, there are other important animal characteristics worth considering, both in specialised and multi-purpose breeding goals. Reproduction traits obviously have an important bearing on both meat and milk production. The number of lambs born and, especially, the number of lambs reared has an important influence on profit in meat and milk production. In both meat and milk production, regularity of breeding is also an important goal.

In some countries there has been concern recently over the increasing resistance of internal parasites of sheep to anthelmintic treatment. This, together with concerns over possible chemical residues in meat and the environmental impact and cost of using anthelmintics, have led to interest in genetic selection for resistance to parasites as an alternative or supplement to the use of anthelmintics. So, in areas where internal parasites are a major problem, and especially where parasite resistance to drugs is emerging, selection for sheep which are resistant to parasites is likely to be an important part of the breeding goal.

Sheep are often kept in harsh environments, whether these are the cold and wet conditions typical of the hills and mountains of Britain, or the hot and dry conditions typical of many sheep producing areas of Australia (see Plates 38 to 41). It is likely that through a combination of natural and artificial selection, some breeds have become better adapted to these harsh conditions than others. A better understanding of the traits involved in adaptation to harsh environments is needed, both to prevent possible deleterious effects of selection for production alone, and possibly to allow more objective genetic improvement in future. For example, it is likely that a range of physical (e.g. lamb birthcoat and adult fleece types), behavioural (e.g. lamb vigour and sucking behaviour, maternal behaviour and grazing behaviour) and disease resistance traits is important in conferring good adaptation to harsh environments. A better knowledge of the heritabilities of the traits involved, and the genetic associations among them, and with other production traits, would help in developing more sustainable breeding goals in future [41].

Some of the approaches to matching breeds and crosses to particular breeding, and within-breed selection for these goals, are discussed in the next two sections.

Breeds and crosses used in sheep production

The vast number of breeds and crosses of sheep involved worldwide in the production of meat, milk and wool makes a comprehensive discussion here impossible. Instead, the aim is to outline briefly some of the breeds, crosses and industry structures involved in a few major sheep producing countries.

Meat production

The UK has one of the largest specialised sheep meat production industries in the world, with a breeding flock of around 20.5 million ewes. MLC did comprehensive surveys of the British sheep industry in 1971 and 1987. The later survey showed that, while there are many regional variations, the much publicised stratification of the industry remains important [29]. This stratified system involves largely purebred flocks of the hill breeds in harsher areas, which are usually bred pure for the first four lamb crops. Older ewes of these breeds are then drafted to better ground for crossing, particularly with longwool sires, e.g. from the Bluefaced Leicester and Border Leicester breeds. The resulting crossbred ewes form the basis of the sheep industry in much of the uplands and lowlands, and are usually mated to rams of the terminal sire breeds.

This system has evolved for a variety of reasons over the decades. But, as discussed in Chapter 3, it does make very efficient use of the complementarity of breeds, and of both maternal and individual heterosis. The hardiness and relatively small mature size of hill breeds is complemented by the larger size and greater prolificacy of the longwool crossing breeds to create an F1 female of intermediate size with good maternal characteristics which shows heterosis for survival and reproduction. These ewes in turn are mated to rams from larger terminal sire breeds with improved growth and carcass attributes to produce a three-way crossbred lamb which itself shows heterosis for survival.

Table 8.3 shows the number of ewes of the six most numerous hill breeds,

Breed	Total number of ewes mated ('000) 1971	1987	No. ewes mated as a % of national flock (1987)
Scottish Blackface	2338	2567	14
Welsh Mountain	1974	1627	9
All Cheviots	807	685	3
Swaledale	504	1209	7
Hardy Speckled Face	327	815	4
Beulah	206	508	3
Total	6156	7411	40

Table 8.3 Estimated total number of ewes mated in the most numerous hill sheep breeds in Britain in 1971 and 1987 [29].

Ewe type	Ram type	% of hill ewes mated to this ram type
Hill ewes	Same hill breed	57.4
	Other hill breed	2.9
	Longwool/crossing breed	23.0
	Terminal sire breed	15.2
	Other	1.5

Table 8.4 Mating structure of purebred hill ewes in Britain in 1987 [29].

Crossbred type	Total number of ewes mated ('000) 1971	1987	No. ewes mated as a % of national flock (1987)
Longwool x hill[1]			
North Country Mule (BfL x Sw)	311	3233	17
Welsh Mule (BfL x WM)	0	370	2
Scottish Mule (BfL x SB)	0	502	3
Greyface (BL x SB)	214	332	2
Welsh Halfbred (BL x WM)	304	410	2
Scottish Halfbred (BL x Ch)	590	346	2
Masham (T (or W) x Sw (or D)	406	249	1
F1 Hill breeds			
Hill x hill	323	631	4
Suffolk x hill	154	182	1
Total	2302	6255	34

[1] BfL = Bluefaced Leicester, Sw = Swaledale, WM = Welsh Mountain, SB = Scottish Blackface, BL = Border Leicester, Ch = Cheviot, T = Teeswater, W = Wensleydale, D = Dalesbred.

Table 8.5 Estimated number of crossbred ewes of the most numerous types mated in Britain in 1971 and 1987. [29]

which alone account for around 39% of the British national flock. The most dramatic change in absolute terms between 1971 and 1987 was in the Swaledale breed, which increased by 0.7 million ewes. Table 8.4 shows that the majority of hill ewes are bred pure. A small proportion are crossed to other hill breeds, but more commonly crosses are to longwool sires, or to terminal sires, to produce lambs for further breeding or slaughter respectively. Statistics for the most numerous crossbred ewe types are shown in Table 8.5. North Country Mule ewes, derived from Swaledale ewes crossed to Bluefaced Leicester rams, increased in number by almost 3 million between 1971 and 1987 – probably because of the higher number of lambs weaned by this cross than its competitors, as outlined in Chapter 3. Smaller increases also occurred in other crossbred ewes resulting from matings with Bluefaced Leicester rams. The survey also showed the importance of terminal sire breeds in the British sheep industry. Despite accounting for only about two per cent of the breeding ewe flock, rams from the terminal sire breeds sired 69% of all lambs slaughtered in Britain. The most numerous terminal sire breeds were the Suffolk, Texel and Charollais.

In contrast to the UK situation, most specialised meat production in other European countries is from purebred sheep. In many of these countries there are strong regional allegiances to local meat breeds, such as the Ile de France, Berrichon du Cher, Charollais, and the hardy Préalpes du Sud in France.

Wool production

In most specialist wool producing countries, the vast majority of wool production is from Merino ewes and wethers. Specialised wool breeds are believed to have originated in the Middle East and to have been spread around the Mediterranean by Phoenicean, Greek and Roman traders [26]. After the fall of the Roman Empire, they survived in the Iberian Peninsula. There, under the influence of the Moors, and later the Spanish aristocracy, these Merinos became the major source of high quality woollen cloths for the expanding economies and populations of Europe. Exports of Merinos were strictly controlled, but elite flocks were established elsewhere in the late eighteenth century by way of royal favour. At about this time famous flocks were established at Rambouillet, France and in Saxony. The Peninsular War in the early nineteenth century broke the Spanish monopoly on fine wool and led to a decline in the local woollen industry. As a result of this, and the growing demand for fine wool by textile mills in Britain and elsewhere, large numbers of Spanish Merinos were driven across the Pyrenees to France. Even greater numbers were shipped to North and South America, South Africa and Australia, where their influence remains today, either directly or indirectly (see Plate 42).

There were, and still are, many strains of Merinos with different production capabilities. There are meat-producing and dual-purpose strains (e.g. the German Mutton Merino and the South African Dohne Merino respectively), but it is the specialised wool-producing strains which are the most numerous worldwide today. The largest population of specialised wool-producing Merinos occurs in Australia (about 41 million ewes and 32 million wethers in 1996). There, and elsewhere, the main groups of strains are classified as fine-, medium- or broad-woolled, with typical fibre diameters of about 20, 22 and 24 microns (μ) respectively (see Plates 43 and 44). However, there are many more subdivisions of these groups, down to the level of **bloodlines**, which are effectively animals from the same stud. Traditionally, commercial wool producers in Australia have shown great loyalty to particular studs, and so they have been tied to producing a particular type of wool. There is currently a great deal of extension effort being made in Australia to encourage commercial producers to become more mobile in their choice of stud in order to exploit the large differences in fibre diameter and fleece weight between bloodlines. Similarly, there are efforts to encourage stud breeders to use both between- and within-bloodline selection to accelerate improvements in wool quality while maintaining fleece weight [3].

Dual-purpose meat and wool production

Both New Zealand and Australia have large populations of sheep used in dual-purpose production of meat and wool. Merinos were introduced to New Zealand from New South Wales in 1842 [35]. However, these were not very well-suited to the wetter climate and improved pastures of much of New Zealand. So only a small proportion of the current sheep population comprises purebred Merinos. However, they did contribute to the formation of several synthetic breeds, such as the Corriedale. This was developed in the mid- to late 1800s by crossing the Merino primarily with the Lincoln breed. The majority of the approximately 50 million breeding ewes in New Zealand are purebred dual-purpose sheep which produce relatively high fleece weights of intermediate quality (e.g. for the production of heavier garments and furnishings), as well as producing lambs for meat production. The majority of these dual-purpose sheep are Romney, Coopworth, Perendale and Corriedale ewes. Typically, these flocks are bred pure, or part of the flock is crossed to terminal sires. Historically, the Southdown was the most important of these crossing sire breeds, but the demand for larger, leaner carcasses has led to substitution by other breeds such as the Dorset Down, the Suffolk and the recently introduced Texel (see Plate 45). (In Australia, New Zealand and other major wool-producing nations, white breeds are generally preferred for crossing in situations where some females may be retained for breeding, because of the penalties for coloured fibres in fleeces.)

In Australia a proportion of draft Merino ewes are mated to longwool sires, especially Border Leicester sires, in a system which is analogous to the crossing of draft hill ewes in Britain. The resulting F1 females form the backbone of the commercial lamb production industry in Australia, which is concentrated in the wetter parts of the country. Typically these F1 ewes are mated to terminal sire breeds such as the Poll Dorset or Suffolk, and the newly introduced Texel. As in Britain, this system allows complementary use of breeds – for wool traits in the case of the Merino, and reproduction traits in the case of the Leicester – as well as maximising the benefits of maternal and individual heterosis. A smaller proportion of the Australian national flock comprises dual-purpose breeds like the Corriedale and Polwarth. (The Polwarth is also a synthetic breed, developed in Australia by backcrossing the Corriedale to the Merino.) These dual-purpose breeds are either bred pure or, as with Merino crossbred ewes, a proportion are crossed to terminal sires.

Milk production

In Southern Europe, most sheep milk production, and associated meat production, is from pure breeds with strong regional ties. For example, the Lacaune breed from the Massif Central area of France is famous for the production of Roquefort cheese. In Greece, Italy, Spain and Portugal the most numerous breeds include the Sarda (Italy), Churra (Spain), Comisana (Italy), Latxa (Spain), Serra da Estralla (Portugal), Karagouniki and Lesvos (both Greece; [8]). (See Plates 46 to 49.) There are smaller numbers of dairy sheep enterprises in northern Europe. These are usually based on Friesian ewes, or derivatives of it such as the British Milksheep, which is a synthetic breed formed from crosses with the Bluefaced Leicester, Lleyn, Polled Dorset and other breeds.

Selection within breeds

Meat production

Systems of testing

In many respects the development of breeding programmes and testing systems for specialised meat breeds, or for meat traits in dual-purpose breeds, has shadowed the development of the equivalent systems in beef cattle which were discussed in the last chapter. The majority of improvement programmes are based on on-farm recording of performance. However, central performance tests also operate in France, Canada, the US, Finland, Denmark, and in a few other European countries. Central testing of lambs from small groups of breeders is also practised in a few breeds in the UK. As explained in the last chapter, central tests involve several breeders bringing their animals together under uniform feeding and management to increase contemporary group size and increase the accuracy of comparisons between animals. However, as discussed for beef cattle, there can be problems with this type of scheme if there are large pre-test environmental effects on animal performance. These can obscure true genetic differences between animals, even at the end of test. In France, Norway, Iceland, to a small extent in some other European countries, and in Australia, progeny testing for carcass composition operates in conjunction with recording performance on-farm or in a central station [38].

Individual flock recording schemes. The development of on-farm recording in Britain is probably fairly typical of that in most meat or dual-purpose sheep industries. National recording of sheep began in the early 1970s, shortly after the formation of MLC, and most individual animal performance recording of sheep breeds in Britain remained under the auspices of MLC until 1995. Since then Signet, a new company jointly owned by MLC and SAC, has taken over operation of these services. The Signet Sheepbreeder scheme is based on on-farm recording of pedigree information, litter size, live weights at a range of ages from weaning to breeding ages – these are used as measures of both direct and maternal performance – and, most recently, ultrasonic measurements of carcass merit. Breeders are responsible for recording pedigree and birth details, and records of weights, while ultrasonic measurements are obtained by Signet staff.

About 600 flocks participate in the Signet scheme in Britain. About 75% of performance-recording flocks are from the terminal sire breeds. The average size of recorded flocks in these breeds is about 70 ewes – around two to three times larger than the average non-recorded flock. There are smaller numbers of performance-recorded hill flocks, but these have larger average flock sizes, usually of about 200 ewes. Few flocks from the traditional longwool crossing breeds are performance recorded. In total, less than 1% of purebred ewes are performance recorded in Britain, although around 8% to 15% of pedigree registered sheep in the main terminal sire breeds are recorded. A relatively low proportion of sheep are performance recorded in most other major lamb-producing countries too.

There are several reasons for this low level of performance recording. There is probably less conviction of the benefits of genetic improvement of objective measures of performance among pedigree sheep (and beef) breeders and their customers than among dairy farmers. This often results in stock of high predicted genetic merit being undervalued. Also, the additional effort and indirect costs involved in recording individual ewe and lamb performance are probably greater than those in dairy cattle recording. Extra input is needed at lambing time and other key measurement times to ensure that accurate records of pedigree and performance are obtained. Often recording has to be done outside, which requires additional commitment, especially in extreme climates. In both sheep and beef cattle recording schemes, the fact that terminal sire breeds are the most commonly recorded breed type is partly a function of the more or less single-purpose role, and hence relatively simple selection goal, in these breeds. Also, recording is often easier due to more favourable locations and smaller flock or herd sizes.

Group breeding schemes. Cooperative breeding schemes, such as group breeding schemes and sire referencing schemes, are important in several countries. They are usually based on individual flock recording. Group breeding schemes were initiated in the 1960s by commercial sheep and beef producers in New Zealand and Australia, who were dissatisfied with the quality of breeding stock produced by elite studs at the time. In their heyday, several million sheep and cattle were involved in these schemes in New Zealand and Australia – including one Australian Merino group alone with over a million ewes. These schemes usually involve the creation of a nucleus flock of elite animals, drawn from cooperating members' flocks (or elsewhere). The nucleus may then be closed to the importation of males or females from other flocks (e.g. to help maintain high health status), or it may remain open to further importations of animals of sufficiently high merit [19, 34]. Animals in the nucleus flock are comprehensively performance recorded and intensely selected to maximise rates of genetic gain. Usually the purpose of the nucleus flock is to breed high genetic merit replacement breeding stock, especially rams, for use in members' own flocks. The screening of elite animals into the nucleus, the more comprehensive recording, and the larger size of the nucleus, usually mean that rates of gain are higher than those which could be achieved by individual members, especially when their flocks are small.

The success of these original schemes in the 1960s and 1970s led to the establishment of similar schemes elsewhere. For example, group breeding schemes were established in several sheep breeds in Britain. The longest established group is the CAMDA Welsh Mountain group breeding scheme in North Wales. The scheme started in 1976 when ten Welsh Mountain breeders formed a nucleus flock of 400 ewes after recording their individual flocks with MLC for two years (see Plate 50). An open nucleus was maintained until 1983 when the flock was closed for health reasons. Two 50-ewe control flocks were also established, one 'commercial' control using industry rams, and one conventional genetic control to allow genetic improvement to be measured over time. Several smaller group breeding schemes have been established in Britain since the CAMDA scheme – mainly in hill and crossing breeds [31].

While group breeding schemes can undoubtedly help to accelerate genetic improvement, there has been a relatively low involvement in this type of scheme

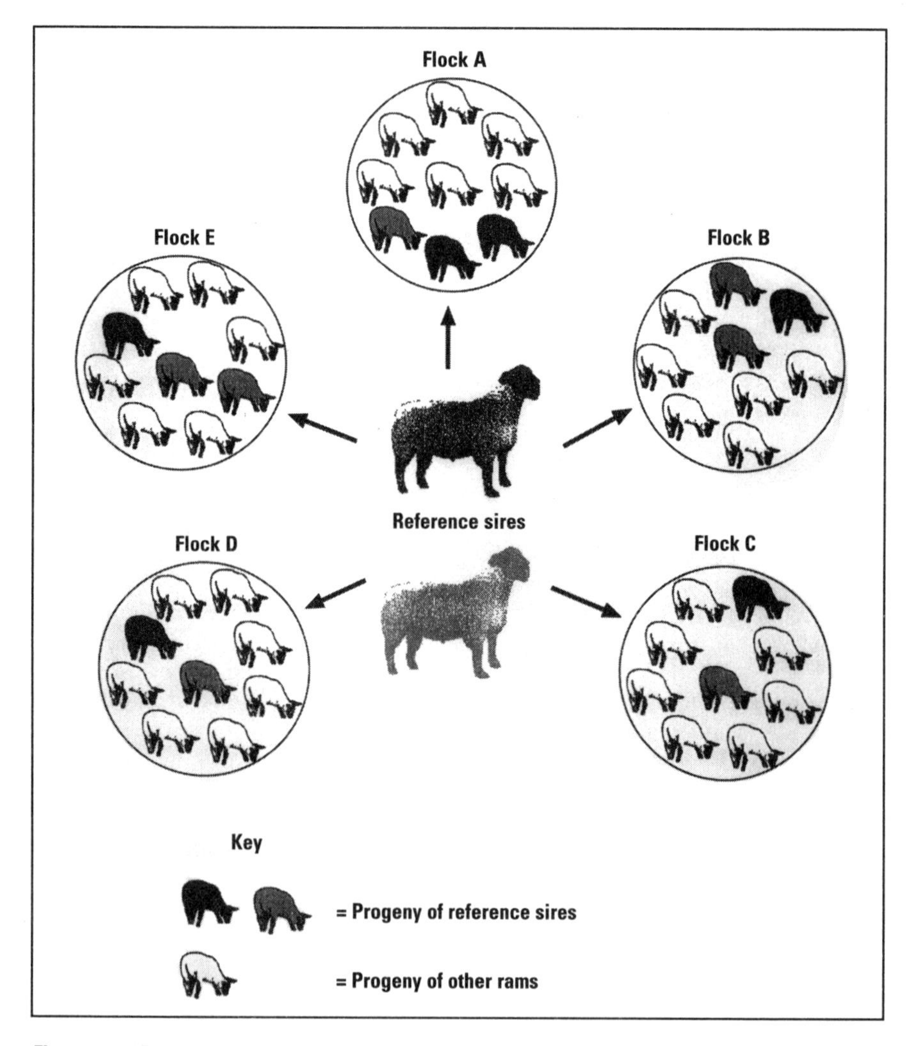

Figure 8.10 The principle of sire referencing schemes. These rely on the creation of genetic links across flocks, usually through the use of a panel of AI rams in common across flocks [37, 44].

in many countries, compared to that seen originally in New Zealand and Australia. Even in these countries, the popularity of these schemes has declined recently [32]. This may be due in part to the high level of cooperation and the financial and legal commitment required to make the schemes work. However, some of these obstacles may be overcome by sire referencing schemes. In several cases sire referencing schemes have evolved from what were previously group breeding schemes.

Sire referencing schemes. Sire referencing schemes have similar aims to group breeding schemes but they do not require the formation of central nucleus flocks or herds, which can be expensive to maintain. Instead, genetic links are created across members' flocks by the use of a panel of AI rams on a portion of the ewes in each flock (see Figure 8.10), or by sharing rams across flocks. Because of these links between flocks, across-flock BLUP methods can then be used to produce EBVs which can be compared fairly across cooperating flocks (see Chapter 5). Sire referencing schemes have been in operation for sheep or beef cattle in Australia, New Zealand, the US, South Africa and France for over a decade now. The development of these schemes has been assisted in several countries recently by improvements in artificial insemination techniques for sheep, the wider accessibility of BLUP evaluation procedures, and the availability of cheaper, more powerful computers. For example, sire referencing schemes have been established in about fifteen sheep breeds in the UK since 1990 [17]. About half the Signet-recorded flocks are now members of these schemes.

The details of some of the UK schemes are shown in Table 8.6. The operation of these schemes usually involves:

- **Selection of a panel of reference rams for use across members' flocks.** To qualify as potential reference sires in most schemes, animals must have high EBVs or index scores and be functionally sound. Qualifying animals are often brought to a central location where all members gather to view them and vote for their choice, based on their own preferred combination of index score, pedigree, conformation and

Breed	Year scheme started (year of first lambing)	No. member flocks evaluated in 1996	Total no. ewes in 1996	Reference sires used primarily by AI or natural service
Suffolk	1990	48 (64)[1]	3964	AI
Charollais	1990	23 (45)	1610	AI
Meatlinc	1991	3	475	NS
Texel	1992	43 (65)	3519	AI
Vendeen	1993	11	665	AI/NS
Scottish Blackface	1994	4 (7)	1275	AI
North Country Cheviot	1994	5	592	NS
Polled Dorset	1995	10	2173	AI/NS

[1] Numbers in brackets are number of member flocks in 1997, if different from 1996

Table 8.6 Details of some sheep sire referencing schemes in Britain. (Signet; MLC.)

breed type (see Plate 51). Most schemes have a panel of about six reference sires in use at any one time, with two or three of these replaced annually, depending on their latest EBVs and popularity with members. In most of these schemes the reference sires originate in members' flocks. Depending on the scheme, semen may be purchased from the owner of the selected reference sire without transfer of ownership, or the scheme may purchase the ram, or a share in him.

- **Use of two or three reference sires, by AI or natural mating, on a proportion of the ewes in each member's flock.** In schemes with a wide geographic spread of members, insemination with fresh semen is impractical, and so laparoscopic AI with frozen semen is often used. Laparoscopic AI involves deposition of fresh or previously frozen semen directly into the uterus via a small incision in the abdomen. Conception rates following cervical insemination of sheep with frozen semen are usually low. In contrast, conception rates following laparoscopic insemination with frozen semen are usually much higher. A total of 30 ewes per flock is usually recommended for mating to reference sires. This number provides a reasonable compromise between maximising rates of genetic gain and minimising AI use (to minimise possible welfare problems, as discussed in Chapter 9, and to reduce costs), or the use of nominated sires [21].

- **Recording performance in appropriate traits.** This is done via the Signet Sheepbreeder recording scheme outlined earlier. Most of the UK schemes are in terminal sire breeds, so selection is based on a lean growth index described below.

- **Evaluation of performance records using across-flock, multi-trait animal model BLUP**. This produces EBVs which can be compared fairly across flocks and across years. This allows animals of different age groups to be compared, and genetic trends to be estimated (examples of these are given in the next section).

- **Use of these results to select the next generation of potential reference sires and to select sires and replacement females for the members' flocks.**

Traits recorded

The animal characteristics important in meat production were identified earlier – these include measures of growth, carcass composition and reproduction. Some of these can be measured directly on candidates for selection. In these cases, the traits in the breeding goal and those actually recorded (the selection criteria) are often the same. For example, records of litter size, number of lambs reared and weights of lambs at various ages can all be obtained relatively easily. However, carcass characteristics have to be measured indirectly, or on relatives of the candidates for selection. Over the last few years ultrasonic measurements have become quite widely used in several countries to predict the carcass composition

Trait	Coefficient of variation (%)	Heritability (no., sd)	Comments
Birthweight	17	0.12 (7, 0.12)	Meat breeds
		0.19 (19, 0.09)	Dual-purpose
Weaning weight	15	0.21 (13, 0.18)	Meat breeds
		0.20 (42, 0.09)	Dual-purpose
Post weaning weight	13	0.28 (15, 0.09)	Meat breeds
		0.26 (26, 0.12)	Dual-purpose
Yearling weight	11	0.22 (6, 0.14)	Meat breeds
		0.33 (17, 0.17)	Dual-purpose
Fat depth, live animal	30	0.28 (30, 0.13)	Unadjusted for
Fat depth, carcass	36	0.31 (24, 0.09)	weight or age at
Eye muscle dimensions[1], live animal	11	0.24 (20, 0.19)	slaughter
Eye muscle dimensions, carcass	11	0.29 (18, 0.07)	"

[1] Eye muscle depth; width or area

Table 8.7 Estimates of coefficients of variation and weighted mean heritabilities for growth and carcass traits in sheep, from an extensive survey of published results. The number of estimates contributing to the weighted mean heritability, and the standard deviation of estimates, are shown in brackets. The coefficient of variation is a measure of the variation in the trait concerned (see Chapter 2). It is calculated as the standard deviation of the trait, divided by its mean. In this case it has been multiplied by 100 so that it is expressed as a % [15].

	Single record of performance		Average of several records of performance		Repeatability
Trait	CV (%)	Heritability (no., sd)	CV (%)	Heritability (no., sd)	
Ewe fertility	47	0.06 (18, 0.07)	40	0.07 (5, 0.03)	0.09 (27, 0.05)
Ovulation rate	30	0.21 (9, 0.20)	–	–	0.37 (10, 0.15)[1]
					0.31 (9, 0.11)[2]
No. lambs born/ewe joined	58	0.08 (22, 0.08)	36	0.15 (18, 0.08)	0.11 (35, 0.04)
No. lambs born/ewe lambing (or litter size)	36	0.10 (53, 0.07)	24	0.21 (9, 0.16)	0.14 (50, 0.07)
No. lambs weaned per ewe joined	73	0.05 (18, 0.04)	43	0.15 (14, 0.09)	0.08 (20, 0.04)
No. lambs weaned per ewe lambing	51	0.05 (25, 0.05)	–	–	0.08 (17, 0.03)
Lamb survival (maternal)	46	0.07 (12, 0.08)	36	0.11 (4, 0.10)	0.09 (18, 0.05)
Weight of lamb weaned /ewe joined	82	0.13 (4, 0.12)	43	0.18 (7, 0.11)	0.10 (4, 0.05)
Weight of lamb weaned /ewe lambing	51	0.14 (8, 0.06)	–	–	0.15 (6, 0.04)
Testis diameter (or scrotal circumference)	12	0.24 (14, 0.16)	–	–	–

[1] Repeatability within a year; [2] repeatability across years

Table 8.8 Estimates of coefficients of variation, weighted mean heritabilities and repeatabilities for reproduction traits in sheep, from an extensive survey of published results. Repeatabilities measure the extent to which the performance of an animal is similar at successive recordings, as a result of both genetic effects and permanent environmental effects – see Chapter 4 for more details. The number of estimates contributing to the weighted means, and the standard deviation of estimates, are shown in brackets [15].

of candidates for selection. (The assumption in this chapter is that recording and evaluation schemes are geared towards improving characteristics under the control of many genes, as well as non-genetic effects. There has been much interest recently in single genes with a major effect on carcass composition, such as the callipyge gene mentioned in Chapter 2. With the advances in molecular genetics outlined in Chapter 9, it is likely that more genes with large effects on carcass composition, and other characteristics, will be identified. When these genes are of interest in improvement programmes, current recording and evaluation methods may need modification, e.g. to account for the genotype of individual animals at the locus of interest, as well as recording overall performance.)

The heritabilities of some of the important traits in meat or dual-purpose breeds, and the correlations among them, are shown in Tables 8.7 to 8.10. Generally, growth and carcass traits are moderately variable and moderately highly heritable. In contrast, most reproductive traits are more variable but have low heritabilities. The repeatabilities of most reproductive traits, except ovulation

Trait 1	Trait 2	Mean correlation between traits	
		Phenotypic	Genetic
Birthweight	Weaning weight	0.30	0.39
	Post weaning weight	0.32	0.07
	Yearling weight	0.31	0.32
	Hogget weight	0.32	0.29
Weaning weight	Post weaning weight	0.68	0.87
	Yearling weight	0.57	0.86
	Hogget weight	0.54	0.72
	Fat depth, live	0.25	0.31
	Fat depth, carcass	0.46	0.50
	Eye muscle[1], live	0.34	0.46
	Eye muscle, carcass	0.42	0.63
Post-weaning weight	Yearling weight	0.78	0.89
	Hogget weight	0.69	0.89
	Fat depth, live	0.50	0.46
	Fat depth, carcass	0.57	0.26
	Eye muscle, live	0.53	0.51
	Eye muscle, carcass	0.64	0.78
Yearling weight	Hogget weight	0.74	0.97
	Fat depth, live	0.51	0.42
	Eye muscle, live	0.60	0.46
Fat depth, live	Fat depth, carcass	0.50	0.70
	Eye muscle, live	0.37	0.33
	Eye muscle, carcass	0.45	0.65

[1] Eye muscle depth, width or area

Table 8.9 Mean estimates of phenotypic and genetic correlations between live weights measured at different ages, and carcass traits in sheep from an extensive review of published values. N.B. some estimates are based on a small number of studies – see original review for details. Correlations with fat and muscle measurements are unadjusted for live or carcass weight [15].

rate, are fairly low. So there is scope for increasing the accuracy of selection by using repeated measures of performance when these are available (as explained in Chapter 4). This is reflected by the fact that heritabilities based on the average of several repeated measures of performance are slightly higher than those based on single measures of performance. The correlations among live weights measured at different ages are generally high, especially later in life when the strong maternal influence on birthweights and other early weights has declined. Generally, the shorter the interval between measurements, the higher the correlations are. Correlations between weight and carcass dimensions are usually positive. There are usually small positive phenotypic correlations between weight and various measures of reproductive performance. The genetic associations between weight and fertility (conception rate and related traits) tend to be negative, but those between yearling or hogget live weight and numbers of lambs born or reared are usually positive.

In addition to the traits mentioned above, there is growing interest in selection for disease resistance, especially resistance to intestinal worms, in meat and dual-purpose breeds in New Zealand and Australia (see Plate 52). This interest is spreading elsewhere as the resistance of worms to drenches appears to be spreading. Faecal egg counts (FEC) are being used as an indicator of resistance to intestinal parasites. FEC has been shown to be a good indicator of worm burdens and to be heritable (heritability of about 0.25 – 0.3 [24, 48]). Similarly, selection for tolerance to facial eczema, a fungal infection of the skin, is part of the breeding goal in some meat and dual-purpose breeds in affected areas of New Zealand. In animals which have been exposed to the disease agent, either naturally or by an artificial challenge, tolerance can be measured from the concentration of a liver enzyme in the blood. In parts of Australia there is also interest in selection for resistance to fly strike.

		Mean correlation between traits	
Trait 1	**Trait 2**	**Phenotypic**	**Genetic**
Weaning weight	Fertility	0.03	–0.16
	No. lambs born/ewe joined	0.08 (0.12)	0.20 (0.02)
	No. lambs born/ewe lambing	0.01	–0.10
	No. lambs weaned/ewe joined	0.09 (0.16)	0.34 (0.03)
	No. lambs weaned/ewe lambing	–0.07	–0.27
Yearling weight	Fertility	0.06	–0.34
	No. lambs born/ewe joined	0.09 (0.17)	–0.17 (0.24)
	No. lambs born/ewe lambing	0.09	0.13
	No. lambs weaned/ewe joined	(0.15)	(0.35)
Hogget weight	Fertility	0.02 (0.07)	–0.06 (0.29)
	No. lambs born/ewe joined	0.11 (0.14)	0.27 (0.09)
	No. lambs born/ewe lambing	0.20 (0.09)	0.59 (0.33)
	No. lambs weaned/ewe joined	0.06 (0.12)	0.28 (0.23)
	No. lambs weaned/ewe lambing	0.15	0.34

Table 8.10 Mean estimates of phenotypic and genetic correlations between live weights measured at different ages, and reproduction traits in sheep from an extensive review of published values. N.B. some estimates are based on a small number of studies – see original review for details. Correlations in brackets are with average performance over a number of records [15].

Similarly, in Britain there is a great deal of interest, in several meat and dual-purpose breeds, in selection for resistance to scrapie. Resistance to this disease is effectively controlled at a single locus, but the number of alleles segregating at this locus varies between breeds. For example, in the Suffolk breed there are three common genotypes, whereas in the Shetland, Cheviot and Swaledale breeds there are many more genotypes present. The degree of resistance conferred by the different genotypes varies. The developments in molecular genetics described in Chapter 9 now allow the scrapie-resistance genotypes of sheep to be determined directly from a blood sample. As a result, some breeders are testing their animals, especially rams, to ensure that only those with the more resistant genotypes are used for breeding.

Methods and results of genetic evaluation

Until recently, the results of most meat sheep recording schemes have been evaluated by contemporary comparisons, or indexes based on these. For example, in Britain, performance records from the MLC/Signet Sheepbreeder recording scheme were allocated to contemporary groups based on lamb sex, litter size and dam age. Records were then standardised within contemporary groups, and the individual animal's performance in each trait was presented as a deviation from its contemporary group mean in sd units (see Chapter 5). Multi-trait indexes were derived from these standardised deviations.

Over the last few years there has been growing interest in many countries in applying more sophisticated methods of genetic evaluation. The advantages of BLUP methods were described fully in Chapter 5. In particular they offer more accurate estimates of breeding value than other methods, and a more flexible method of combining information on different traits from different classes of relatives. Also, where sufficiently strong genetic links occur, they allow EBVs to be compared fairly across flocks and years. This has several important benefits. Firstly, the number of candidate animals for selection which can be directly compared is greatly increased. This has direct benefits on the rate of response which can be achieved. Secondly, if relevant estimates of heritabilities and correlations are used, the genetic trend in performance can be charted year by year by comparing the average EBVs of animals born in the breed or flock in successive years. This provides a valuable check on genetic progress, both for breeders themselves and for their customers. However, compared with the situation in dairy and beef cattle there is relatively little use of AI in sheep in most countries, and so genetic links between flocks are usually weaker. This has limited the uptake of across-flock evaluations, except where these links have been created deliberately, such as in sire referencing schemes.

The steps involved in genetic evaluation of meat breeds, or evaluation of meat traits in dual-purpose sheep breeds, are generally very similar to those outlined for beef cattle in the last chapter. Briefly, these include:

- Collation and checking of performance records by the recording agency, and transfer of these to the agency responsible for genetic evaluation, if this is different. (In most countries evaluations are performed by the recording agency itself, a government agency or university.) This step is repeated each time evaluations are performed.

Usually this is once or twice per annum for across-flock or national evaluations, and more frequently for within-flock evaluations, but this varies depending on the seasonality of the breed concerned and the timespan over which traits of interest are recorded. The use of computerised records is less widespread and more recent in sheep breed societies than it is in cattle breed societies. Hence pedigree records are often maintained by the performance recording agency.

- If traditional evaluation methods are being used, performance records are first adjusted for environmental effects. For example, lamb weight records might be adjusted for month or season of birth, litter size at birth and during rearing, age or parity of dam and age of the animal itself at measurement. Then breeding values are predicted from the animal's own adjusted records of performance, either with or without adjusted records from relatives.

- If BLUP evaluation methods are being used then (at least some) environmental effects are estimated and breeding values are predicted simultaneously. The statistical model used determines which environmental effects are accounted for. Some of these, such as the ones listed above, are identified explicitly, others are accounted for by assigning animals to contemporary groups.

- BLUP evaluations are then done. The particular model used determines which relationships among animals are accounted for and which animals get PBVs. For example, with individual animal model BLUP the relationships between all animals are recognised and accounted for and all animals get PBVs.

- The results of evaluations are expressed as predicted (or estimated) breeding values (PBVs or EBVs), predicted (or estimated) transmitting abilities (PTAs or ETAs), or expected progeny differences (EPDs).

- EBVs or EPDs are expressed relative to some reference population of animals called the base. In most countries a fixed base is used for across-flock BLUP evaluations, i.e. EBVs or EPDs are expressed relative to those of a group of animals born in a particular year, though this base is updated from time to time.

- In some countries EBVs or EPDs for individual traits are combined into indexes of overall economic merit. These are discussed in more detail in the following section.

- Accuracies or reliabilities may be produced for each of the EBVs (or EPDs).

- Results of evaluations are sent to individual breeders after each new evaluation. Usually these include summaries of the predicted merit of the current lamb crop and the sires and dams they represent.

> National sire summaries are not yet used as widely as in beef breeds, but this is likely to occur when national evaluations become more widespread.

BLUP methods are used in either national or particular meat sheep breeding programmes in Canada, the US, Australia, New Zealand, Norway, Finland, Denmark, Sweden, France and the UK. In those countries which adopted BLUP evaluations relatively early, single-trait sire model evaluations were often used, mainly for within-flock evaluations. However, multi-trait animal model evaluations are becoming more widely used now, and these are gradually being extended to across-flock evaluations of groups of flocks or national populations. For example, in the UK multi-trait animal model BLUP evaluations are performed annually (or more often) for the sire referencing schemes mentioned earlier. Within-flock multi-trait animal model BLUP evaluations are done for other recorded flocks.

Use of indexes of overall economic merit

Selection indexes have become quite widely used in meat and dual-purpose breeds over the last decade or so. For example, in Britain since the mid-1980s participants in the Sheepbreeder recording scheme have had a choice of four breeding goals: lamb growth, ewe mature size, litter size and maternal ability (maternal influence on offspring eight-week weight). These are used individually or weighted in various combinations using multi-trait selection indexes. Table 8.11 shows some examples of the emphasis put on these traits in different breeds. Formerly, the choice of breeding goal traits dictated which records were used to calculate index scores. If selection was for lamb growth only, for example in a terminal sire breed, then index scores were based on early live weights both of the lamb itself and its relatives. If selection was for overall productivity of a maternal breed, index scores were derived from records of litter sizes, lamb and ewe weights on the candidate animals for selection, plus equivalent information from several classes of relatives [27, 30]. However, since the introduction of multi-trait animal model BLUP evaluations, index scores are derived solely from the animal's own EBVs for the relevant traits (i.e. all available performance information from relatives has already contributed to these EBVs).

Multi-trait indexes incorporating measures of reproduction and growth, or reproduction, growth and wool production, are used in specialised meat and

	Selection objectives			
Breed type	**Lamb growth**	**Mature size**	**Litter size**	**Maternal ability**
Terminal sire breeds	100	–	–	–
Lowland ewe breeds	17	–	66	17
Hill breeds	40	9	11	40

Table 8.11 Percentage contribution of different objectives to the main MLC/Signet multi-trait indexes for sheep in the UK. Several other indexes with different emphasis have been derived for particular breeds or groups of breeders [28].

dual-purpose breeds respectively in New Zealand (Animalplan and Flock-Linc recording schemes), Australia (LAMBPLAN recording scheme), South Africa and elsewhere [6, 11].

Since the mid-1980s ultrasonic measurements of fat and muscle have been obtained in testing programmes in a number of countries. As a result, new or expanded indexes have been produced to incorporate these measurements. For example, in Britain MLC launched an ultrasonic scanning service for performance-recorded sheep in 1988. This was based on research at SAC to test scanning techniques and derive and test a selection index for terminal sire breeds to combine measurements of weight and fat and muscle depths. The development and testing of this index are described below as they illustrate some general principles.

The breeding goal of this index comprises carcass lean weight and carcass fat weight at a constant age. The selection criteria are live weight, ultrasonic fat depth and ultrasonic muscle depth adjusted to a constant age. This index is very similar to the one described in Chapter 5 except that, in this case, the relative economic values were chosen to achieve 'desired gains' in the traits in the breeding goal, rather than being based on actual market returns. This approach was chosen because of the weak relationship between carcass price and fatness in Britain at the time the index was derived.

Figure 8.11 shows the expected responses in goal traits from selection on indexes combining measurements of live weight, ultrasonic fat and muscle depths

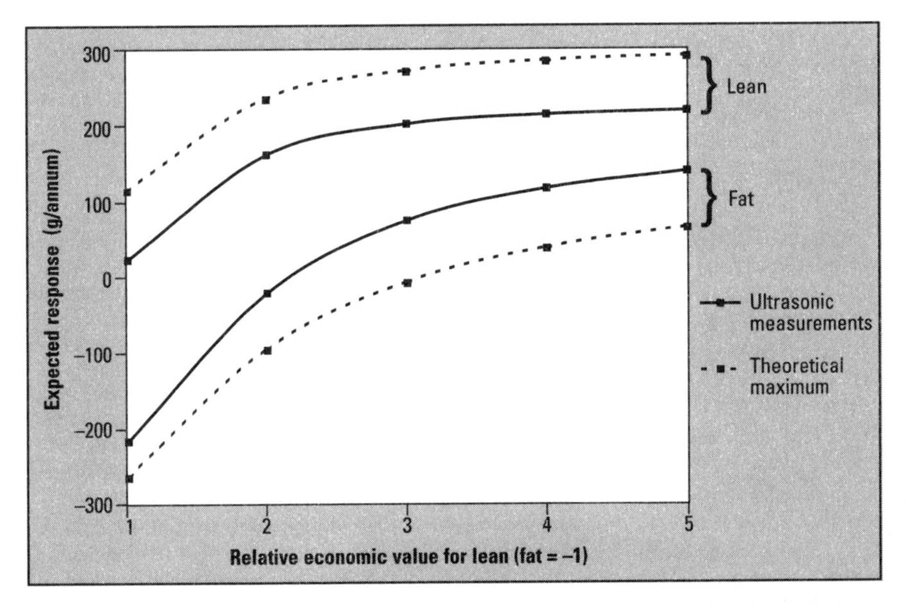

Figure 8.11 Expected responses in carcass lean weight and carcass fat weight following index selection, when the relative economic values of carcass lean and fat weights range from +1:–1 to +5:–1. Selection is either on indexes combining measurements of live weight, ultrasonic fat depth and ultrasonic muscle depth, or on indexes based on live weight and 100% accurate measurements of fat and lean content of the live animal. This provides an estimate of the theoretical maximum rate of change which could be achieved, e.g. by the use of more advanced scanning methods. None of the indexes in this example use information from relatives [42].

on the animal itself (i.e. excluding records from relatives), but with a range of different relative economic values for weight of lean and fat in the carcass. All the responses shown are at a constant age. The graph illustrates that with a high penalty on fat compared to the value of lean (the left-hand side of the graph), selection on the index derived will lead to large reductions in weight of fat, but little improvement in weight of lean at a constant age. Conversely, if the penalty for producing fat is low relative to the value of carcass lean weight (the right-hand side of the graph), selection on the index derived will lead to large increases in both carcass lean weight and carcass fat weight. (Although fat *weight* is increasing, fat *proportion* will decrease if there is a large enough increase in lean weight.)

These results might be confusing at first sight, but they are a direct consequence of the positive correlations between live weight, carcass fat weight and carcass lean weight. Within a breed, at a constant age, bigger animals generally have more lean *and* more fat. If selection is solely for weight at a constant age, we expect to get more lean and more fat. A selection index can help to restrict the increase in fat weight. But, if fat is heavily penalised, a lot of the potential improvement in lean weight has to be sacrificed, in order to avoid increases in fat weight. If fat is not heavily penalised, increases in fat weight can be tolerated in order to get greater increases in lean weight.

The graph also shows the responses expected from selection on an index of live weight and hypothetical measurements on the live animal which give perfect prediction of carcass composition. Some advanced imaging techniques used in human medicine, such as computed tomography (CT) may give close to this level of accuracy (see Plate 53). Following early investigations with CT in Norway, CT facilities to predict body composition in farm animals have been established in several locations in Australia, New Zealand, Hungary and the UK. In some cases,

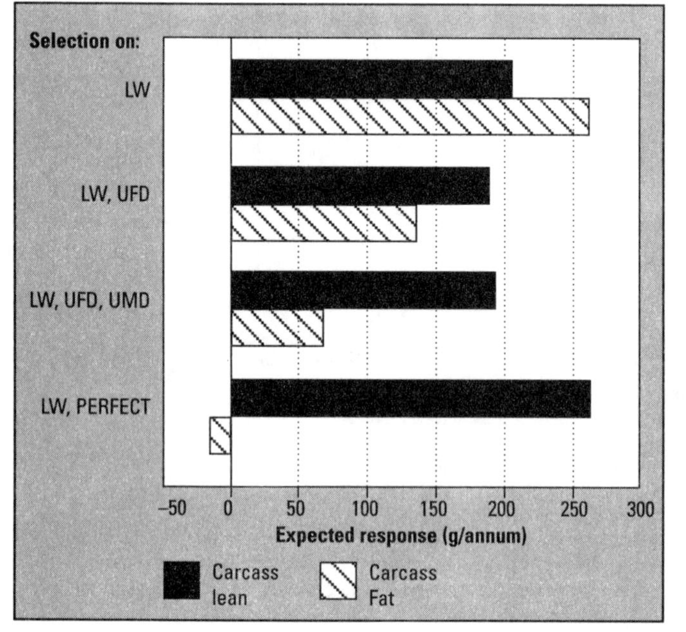

Figure 8.12 Expected responses in carcass lean weight and carcass fat weight following selection on various combinations of index measurements. Index measurements include live weight (LW), ultrasonic fat depth (UFD), ultrasonic muscle depth (UMD), or hypothetical measurements which produce 100% accurate prediction of fat and lean content of the live animal (Perfect). This provides an estimate of the theoretical maximum rate of change which could be achieved, e.g. by the use of more advanced scanning methods [38, 42].

Selection Goal[1]	Relative Economic Value
Carcass lean weight	+3
Carcass fat weight	–1
Index Measurements[1]	**Index Weightings**
Live weight	+0.10/kg = +0.55/sd
Ultrasonic fat depth	–0.41/mm = –0.51/sd
Ultrasonic muscle depth	+0.26/mm = +0.53/sd

[1] all at 150 days of age

Table 8.12 Details of the lean growth selection index used in the SAC Suffolk selection experiment and in the MLC/Signet Sheepbreeder recording scheme in Britain. (N.B. the index weights shown are only relevant when selection is based on measurements from candidates only, and not on records from relative) [42].

two-stage selection programmes are being developed for sheep, with CT measurements being made on rams with the best EBVs or index scores for ultrasonic measurements and live weight from on-farm recording programmes.

Intermediate relative economic values of +3 for lean and –1 for fat were chosen for use in an SAC selection experiment in Suffolk sheep, and for use in the Sheepbreeder scheme, as they are expected to give almost maximum responses in lean weight, while heavily restricting changes in fat weight. This is illustrated further in Figure 8.12, which shows expected responses in lean and fat weights following selection on weight alone, an index combining weight and fat only, the full index described above, or selection on an index with 100% accurate measurements on the live animal. Table 8.12 shows the index weights derived for use on ultrasonic and live weight measurements from candidates for selection alone (i.e. excluding records from relatives), assuming relative economic values of +3 and –1. (The index used in the Sheepbreeder scheme also uses information from relatives.)

In order to produce more comprehensive indexes for multi-purpose breeds, e.g. to account for new measurements of carcass merit, estimates of genetic parameters are required for the new traits. Estimates of economic values of new traits will also be required if conventional (rather than 'desired gains') selection indexes are being used. For example, new sets of genetic parameters have been produced recently to expand the range of indexes available in the Australian LAMBPLAN recording service [15]. Similarly, research is in progress at SAC and the Roslin Institute in Edinburgh and in ADAS, to estimate genetic parameters and economic values for survival, maternal behaviour, reproduction, wool characteristics, growth and carcass traits in hill sheep (see Plates 54 and 55). This will allow more comprehensive selection indexes for hill breeds to be constructed in future, which take into account the economic importance of growth and carcass composition but balance improvements in these traits with the need to maintain or improve ewe productivity and traits associated with welfare and adaptation to harsh environments. Similar work started recently in these institutions and the Welsh Institute of Rural Studies, to develop more comprehensive indexes for longwool or crossing breeds.

Evidence of genetic improvement and its value

In theory, if there is genetic variation in any animal characteristic, then there is scope for changing it through selection. So, estimates of variances and heritabilities provide evidence of the *potential* for genetic improvement. In practice, however, having evidence that genetic change has actually been made is far more convincing for breeders and their customers (and reassuring for scientists and advisers!). In the past, most evidence of this sort has come from self-contained selection experiments, or from related trials sampling animals from industry which have different EBVs or different selection histories. However, with the wider use of BLUP methods in sheep breeding, more evidence of this sort should come from industry breeding schemes themselves in future, via estimates of genetic trends. Impressive genetic trends from industry schemes are particularly valuable in encouraging wider uptake of objective selection methods, because genetic improvement in these cases has been achieved by breeders, on commercial farms rather than in experimental flocks. Some examples of each of these types of evidence are given below.

Selection in the CAMDA Welsh Mountain group breeding scheme, mentioned earlier, has been on a selection index designed to increase lamb weight, increase ewe mature size and maintain prolificacy at the initial level (since this was considered to be optimal for the hill environment). Since 1980, a clear divergence in performance of the nucleus and control flocks has been achieved as shown in Figure 8.13. An average increase of about 0.4 kg per year was achieved in 12-week lamb weights as a result of selection over the first eight years of operation [18] and improvement has continued at a similar rate since then.

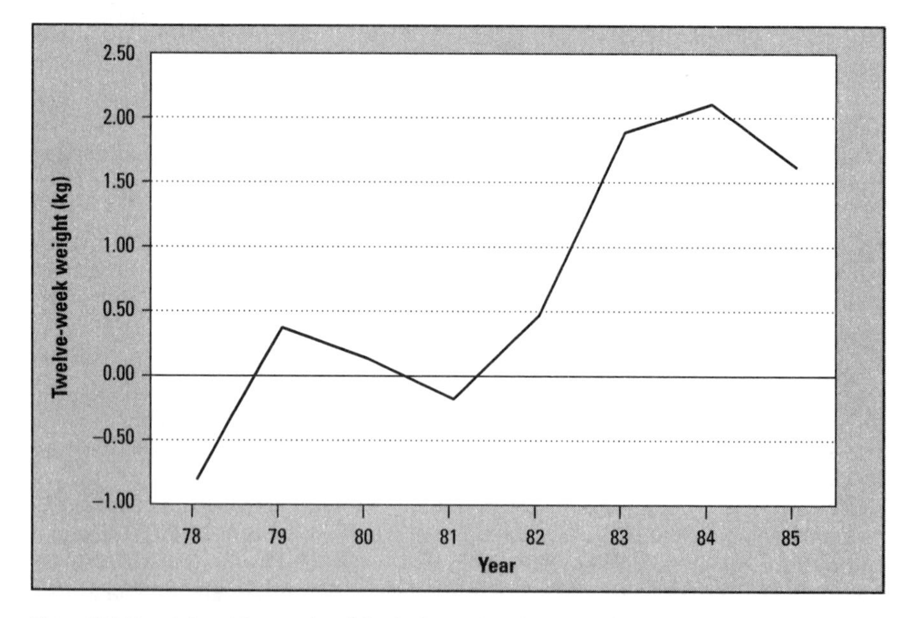

Figure 8.13 Trends in twelve-week weights in the nucleus flock of the CAMDA Welsh Mountain group breeding scheme over the first eight years of operation. Values shown are the differences between average weights in the nucleus and control flocks [27].

	Selection Line	Control Line	Difference (Sel-Con)%
Number of rams tested	113	48	–
Live weight (kg)	67.0	61.0	+10%
Fat depth (mm)	5.9	6.9	–13%
Muscle depth (mm)	28.7	25.4	+13%
Index score	180	100	+18%[1]

[1]Difference in index score measured before scaling – indexes were scaled so that the average score in the control line each year equalled 100 points, and so that the index had an sd of 40 points

Table 8.13 Performance test results for ram lambs tested in 1994, after nine years of selection in the SAC Suffolk selection experiment.

Over the past decade or so, selection experiments have been established for various carcass traits at a number of locations around the globe, but particularly in New Zealand and the UK. Most of the New Zealand selection lines have been selected for ultrasonic backfat depth, adjusted for live weight. However, index selection has been practised in a number of the more recent experiments in the UK and New Zealand. In most of these experiments rates of genetic change in excess of two per cent per annum have been achieved in fat depth or index score. These responses are close to the maximum expected values for the traits and flock sizes concerned [38, 39].

One example of this type of experiment is the SAC selection experiment in Suffolk sheep mentioned previously. Selection in the SAC flock was based on the index described before, which combines measurements of live weight, ultrasonic fat depth and ultrasonic muscle depth, all measured at 150 days of age. Compared to unselected control line ram lambs, selection line ram lambs performance tested in 1994, after nine years of selection, had about 6 kg higher live weight (+10%), 1.0 mm lower ultrasonic fat depth (–13%), 3 mm higher ultrasonic muscle depth (+13%) and 18% higher index score (see Table 8.13 and Plate 56). Similar proportional responses were obtained in ewe lambs. Work is now in progress to measure responses in the carcass composition of purebred lambs at a wider range of degrees of maturity, and on feeds differing in quantity and quality.

Estimated genetic trends in several of the UK sire referencing schemes described earlier are shown in Figures 8.14 and 8.15. Annual responses in lean growth index appear to be similar to those achieved in the SAC experiment. Because of the much larger size of the industry schemes, they can achieve these similar responses in the index of objectively measured traits while taking account of visually assessed breed characteristics and conformation. (Also, some breeders use the individual trait EBVs to put additional selection emphasis on a component of the index).

Often, pedigree animals are reared under more favourable conditions than their commercial counterparts. The SAC experiment involved *ad libitum* feeding of a high energy, high protein diet, partly to replicate levels of performance achieved in many industry pedigree flocks in the UK, and partly to increase variation in carcass composition in order to make it easier to detect differences between animals using ultrasonic scanning. However, the majority of lambs slaughtered for meat production in the UK are reared at grass. Two experiments have been conducted at SAC which provide evidence of the commercial value of objective genetic improvement programmes in pedigree flocks.

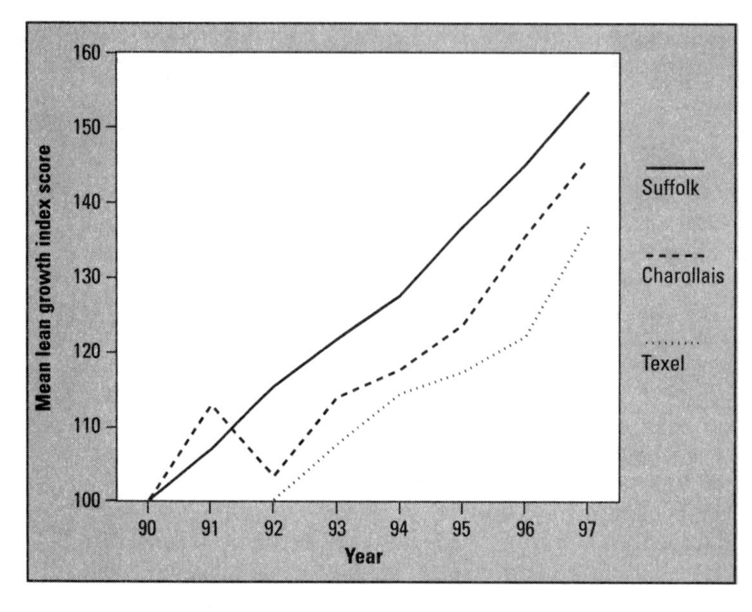

Figure 8.14 Genetic trends in lean growth index in the three largest sire referencing schemes in the UK. The index in all three breeds is similar to that used in the SAC flock, except that the index used in sire referencing schemes is based on multi-trait animal model BLUP EBVs, and hence includes records from relatives. The index has an sd of 40 points in each breed. (MLC; Signet; Suffolk Sire Reference Scheme Ltd.; Charollais Sire Referencing Ltd.; Elite Texel Sires Ltd.)

The first of these trials involved high or low index Suffolk rams mated to Scottish Mule ewes. Their lambs were reared at grass to produce carcasses of about 16.5, 20.0 or 23.5 kg. Sample joint dissections on these carcasses showed that the progeny of high index sires had about 1.0% more lean and about 3.5% less fat than the progeny of low index sires (sires differed by 100 index points, or 2.5 standard deviations in index score [22]). A more recent experiment showed significantly higher saleable meat yield from the carcasses of selection line progeny than from control line progeny, both at a constant carcass weight (+0.10 kg) and a constant level of fatness (+0.25 kg) (see Plate 57). On average, carcasses from selection line sires achieved prices about £1.50 higher than those for controls, and were slaughtered eleven days sooner; this earlier slaughter date would save about £0.60 per lamb in grazing costs [43]. This financial advantage is worth up to £600 over the working life of a ram. In both of these experiments the progeny of high index sires had apparently poorer conformation, but this difference appears to be almost entirely due to differences in fatness.

These results have been used to estimate national discounted returns in the British sheep industry as a result of the use of ultrasonic scanning and the lean growth index described. Based on information about the levels of uptake of scanning, and assumptions about the extent to which selection is based on this information, estimated discounted returns of about £16.6 million per annum are expected 20 years after selection commenced. Annual returns then are expected to exceed net annual costs of implementation, including research and development, by a factor of over 25:1 [40]. The wider use of selected rams in the crossbred sector could increase these benefits up to six-fold.

A study of the value of genetic improvement in the Australian lamb industry predicted net present values ranging from about Aus. $38 million to Aus. $160 million (about £19 million to £80 million) over a 30-year period, depending on the particular set of assumptions on the value of reducing fat, the levels of uptake and industry structures. With assumptions producing intermediate benefits of about

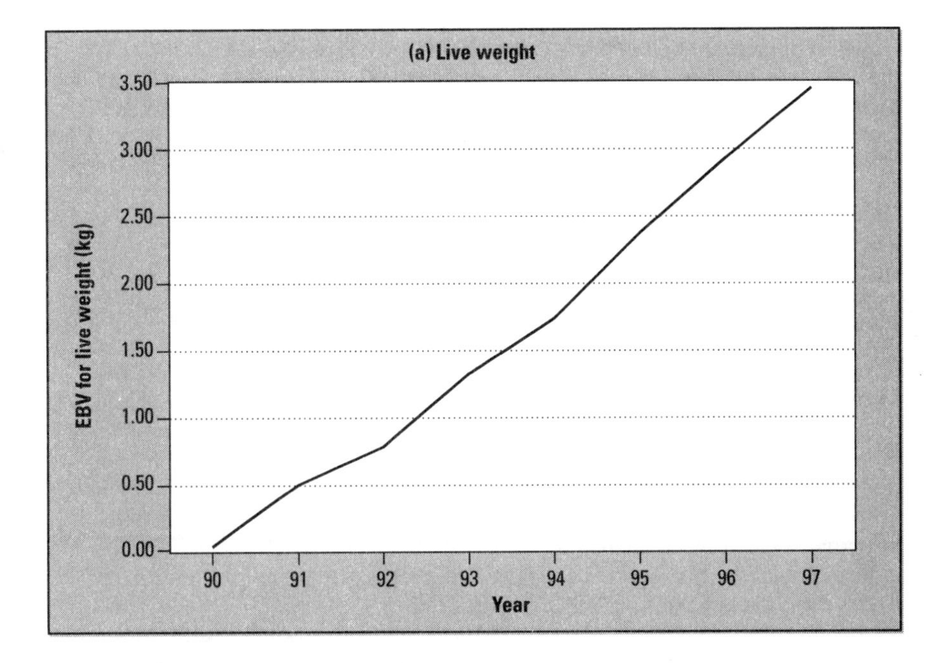

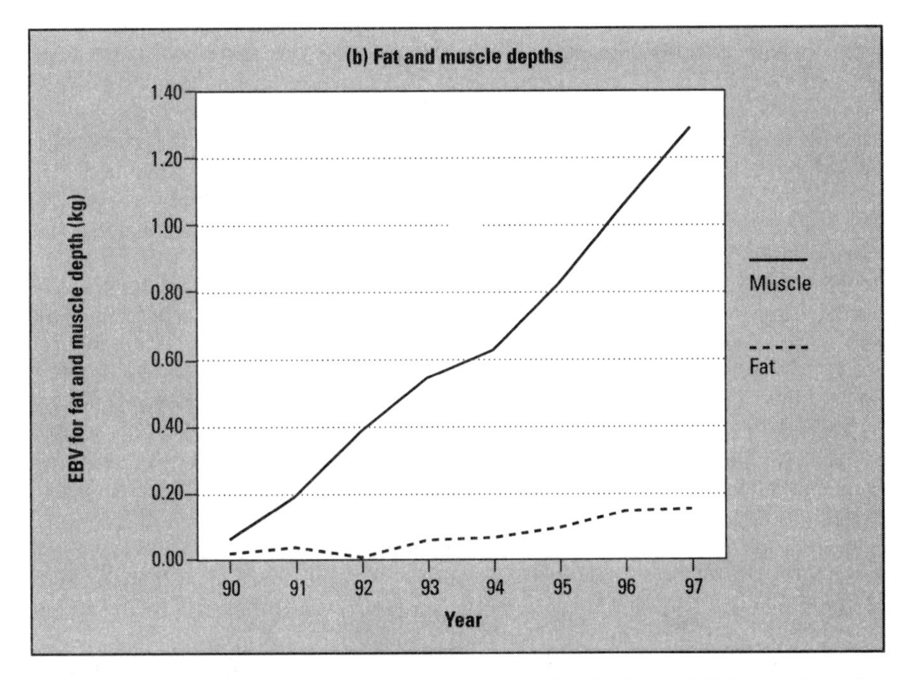

Figure 8.15 Genetic trends in components of the lean growth index in the Suffolk sire referencing scheme in the UK (a) live weight at scanning (about 20 weeks of age), and (b) ultrasonically-measured fat and muscle depths. Trends in components of the index are similar in the other large terminal sire schemes. (MLC; Signet; Suffolk Sire Reference Scheme Ltd.)

Aus. $90 million, a ratio of benefit to cost of 17:1 was predicted at year 30 [7]. Given the difference in time horizons, and the fact that the size of the Australian meat lamb sector is about half that of the British sector, these results are fairly similar.

Wool production

Systems of testing

The improvement of wool traits in specialised wool breeds and dual-purpose breeds is also based on on-farm recording in most countries. The most important objectively measured traits are fleece weight and fibre diameter. Fibre diameter is increasingly being measured at approved laboratories with sophisticated laser equipment to measure both the mean and coefficient of variation of fibre diameter (see Plate 58). Other fleece characteristics, such as staple strength, colour, resistance to compression and style (a visual appraisal of manufacturing properties), have usually been assessed subjectively in the past but objective techniques are now available for most of them. These characteristics are all moderately or highly heritable, but are of secondary economic importance to fleece weight and fibre diameter [33].

Despite the suitability of most of the important wool traits for objective measurement and selection, these techniques have been used only recently in most Merino industries. For example, performance recording schemes for wool traits were developed for Merino sheep in Australia in the 1950s. However, these were not widely used [12]. The introduction of fleece measurement services in New South Wales in the mid-1970s was more widely supported. Other fleece measurement services followed, but the link between these measurements and other aspects of performance and the methods of estimating breeding values varied between centres. A national performance recording scheme was launched in 1987 to provide a framework for the use of objective fleece measurements and a means for calculating EBVs for important wool traits in a standard way. Similar developments are occurring in most of the major finewool producing countries. It has been estimated that objective measurements are used by just over a third of Merino breeders in Australia and South Africa [26, 33].

Three other types of testing scheme in the Australian Merino industry deserve mentioning further. The first of these are **central wether tests**, which began in the 1970s. These are **commercial product evaluations** – a comparison of the merits of breeding stock available from different studs. These involve sampling, from commercial flocks, small teams of wethers sired by rams from identified bloodlines. The fleece characteristics of these wethers are then compared at a central station. (Similar tests have been used in the past to compare the commercial breeding stock sold by different pig breeding companies.) Individually these tests are probably not a very reliable way of comparing studs, because of the small numbers of animals involved. However, results of these tests are now being pooled to allow more reliable and wider comparisons, thanks to the links between tests created by the use of common bloodlines [2, 3].

The second type of testing scheme, which developed in the mid-1980s in Australia, is central progeny testing. This involves studs submitting a small number of sires for progeny testing in a central location [14]. Each sire is usually inseminated to 55 ewes, with a view to recording objective and visual wool characteristics on a minimum of 20 hogget or 15 adult progeny. Compared to wether tests, these central progeny tests have the advantage of producing results for individual sires, rather than studs, albeit over a longer period. Central tests also have a wider role in providing genetic links between studs, with the potential for across-stud BLUP evaluations [3]. Similar progeny testing schemes operate in South Africa and South America [26].

Cooperative breeding schemes are the third type of testing scheme employed in the genetic improvement of Merino sheep in Australia and elsewhere. Although group breeding schemes have been employed in Australia in the past, interest appears to have waned recently. In part this is probably due to a lack of agreement on breeding objectives. However, there is renewed interest in sire referencing in Australia, in some cases as an adjunct to central testing. A sire referencing scheme operates in the dual-purpose Dohne Merino breed in South Africa, along very similar lines to those described above for meat breeds [26].

Traits recorded

Tables 8.14 and 8.15 show the heritabilities of, and correlations among, some of the important characteristics in specialised wool breeds, or dual-purpose breeds where wool is an important component. The heritabilities of both greasy and clean fleece weights are moderately high; that of fibre diameter is higher still. There is a positive (i.e. unfavourable) correlation between fleece weight and fibre diameter. Also, there are moderate positive genetic correlations between live weight and fleece weight and lower but mainly positive (i.e. unfavourable) genetic correlations between live weights and fibre diameter. There are generally fairly weak associations between reproductive performance and fleece weight at both the phenotypic and genetic levels.

In addition to the traits listed in these tables, faecal egg count (FEC) is being used increasingly by Merino breeders as an indicator of resistance to intestinal parasites. FEC has been shown to be a good indicator of worm burdens and to be heritable (heritability of about 0.25 – 0.3). Recently much effort has gone into estimating associations with production traits and exploring ways of including FEC in the overall breeding goal [47, 48]. One attractive approach is to examine expected responses in resistance, and in production traits, following selection on indexes with different sets of economic values. (This is a similar approach to the one outlined earlier in this chapter for carcass traits; these are both informal

Trait	Coefficient of variation (%)	Heritability (no., sd)	Repeatability (no., sd)
Greasy fleece weight	15	0.34 (25, 0.14)	0.58 (19, 0.15)
Clean fleece weight	15	0.37 (28, 0.11)	0.52 (10, 0.20)
Fibre diameter	8	0.51 (27, 0.13)	0.70 (12, 0.07)

Table 8.14 Estimates of coefficients of variation and weighted mean heritabilities and repeatabilities of wool traits in specialised wool breeds of sheep, from an extensive survey of published results. The number of estimates contributing to the weighted mean heritability, and the standard deviation of estimates, are shown in brackets [15].

versions of what are called 'desired gains' selection indexes.) In the absence of precise estimates of the economic value of genetic improvement in resistance to worms, or where this varies between properties, this approach allows breeders to choose how much potential genetic gain in production they are willing to sacrifice in order to achieve a given amount of progress in resistance to worms [47]. A Merino Breeders Worm Control Network (Nemesis) has been established in Australia to provide information and coordinate services to assist Merino breeders in selecting for worm resistance.

		Mean correlation between traits	
Trait 1	**Trait 2**	**Phenotypic**	**Genetic**
Birthweight	Greasy fleece weight	0.24	0.25
	Clean fleece weight	0.25	0.12
	Fibre diameter	0.01	0.17
Weaning weight	Greasy fleece weight	0.29	0.33
	Clean fleece weight	0.29	0.24
	Fibre diameter	0.08	0.04
Yearling weight	Greasy fleece weight	0.38	0.28
	Clean fleece weight	0.29	0.09
	Fibre diameter	0.07	–0.11
Hogget weight	Greasy fleece weight	0.37	0.21
	Clean fleece weight	0.29	0.18
	Fibre diameter	0.13	0.10
Fertility	Greasy fleece weight	–0.23 (–0.01)	–0.29 (0.32)
	Clean fleece weight	–0.22	–0.57
	Fibre diameter	–0.07	0.26
No. lambs born/ewe joined	Greasy fleece weight	–0.10 (0.07)	–0.06 (0.12)
	Clean fleece weight	–0.04 (0.09)	–0.49 (0.12)
	Fibre diameter	–0.09 (0.12)	0.31 (–0.07)
No. lambs born/ewe lambing	Greasy fleece weight	0.03 (–0.02)	0.01 (0.06)
	Clean fleece weight	(–0.01)	0.14 (0.41)
	Fibre diameter	–0.04	0.08 (0.61)
No. lambs weaned/ewe joined	Greasy fleece weight	–0.10 (0.06)	–0.16 (0.11)
	Clean fleece weight	–0.13 (0.08)	–0.10 (–0.04)
	Fibre diameter	–0.04 (0.11)	0.13 (–0.12)
No. lambs weaned/ewe lambing	Greasy fleece weight	–0.01	–0.22
	Clean fleece weight	–0.03	0.28
	Fibre diameter	–0.04	0.35
Greasy fleece weight	Clean fleece weight	0.88	0.84
	Fibre diameter	0.28	0.17
Clean fleece weight	Fibre diameter	0.27	0.21

Table 8.15 Mean estimates of phenotypic and genetic correlations between live weights measured at different ages, wool and reproduction traits in sheep from an extensive review of published values. N.B. some estimates are based on a small number of studies – see original review for details. Correlations in brackets are with average performance over a number of records [15].

Methods of genetic evaluation

Most objective selection for wool traits in the past has depended on contemporary comparisons or EBVs for individual traits. BLUP evaluations are becoming more widespread in Merino breeding programmes, just as they are in other sheep breeding programmes worldwide. However, there are at least two limitations to their uptake in Merino breeding programmes in Australia, and probably elsewhere. The first is the lack of pedigree information in many stud flocks. It has been suggested that new molecular genetic techniques may allow animals' pedigrees to be established retrospectively from a blood sample [3]. If this approach proves to be cost-effective, it would increase the scope for using more sophisticated BLUP models. The second is the lack of links across flocks, which prevents across-flock BLUP evaluations. Despite the large size of most Merino studs compared to typical flock sizes in meat breeds, across-flock evaluations would still permit higher selection intensities, and hence higher rates of response. Although wider links between flocks would be beneficial, the central progeny testing of rams mentioned earlier is providing some links between studs. The steps involved in genetic evaluation are similar to those outlined earlier for meat breeding programmes.

Use of indexes of overall economic merit

Over the last few years the use of multi-trait indexes in Merino breeding programmes has increased. These indexes are particularly useful in selecting for unfavourably correlated traits, like fleece weight and fibre diameter. An additional aid in this area has been the development of computer software, such as the OBJECT software developed for Merino breeders by NSW Agriculture in Australia [5], which allows the formulation of personalised breeding objectives. This software allows breeders and their advisers to decide which traits to include in the overall objective. It also calculates relative economic values based on the size and composition of the flock concerned, the quantity and specifications of wool sold, the reproductive performance of the flock, and the age, weight and value of surplus sheep sold. Although there is flexibility in the way breeding objectives are formulated, the modifications have to fall within realistic ranges. Expected responses from selection on these flock-specific economic values can be compared with those expected from selection on a standard index.

Tailoring breeding objectives to individual breeder's (or group's) requirements is rarely expected to give very different responses from those following selection on a single national index. However, having the software and advisory support to explore the options is usually very informative itself. If it results in greater understanding of the scope and limits of objective selection by breeders, and a greater uptake of these techniques, then usually this will compensate for the cost and complication of having different indexes in use.

Evidence of genetic improvement and its value

Although the estimates of genetic variation and heritabilities indicate the considerable scope for genetic improvement of wool traits, there are only a few examples so far of responses being measured in experimental or industry breeding schemes [25, 45]. So the wider use of BLUP would also be beneficial in allowing genetic trends to be estimated in industry breeding programmes.

The value of genetic improvement in the Merino industry in Australia, has been estimated to be Aus. $3,500 million over 30 years, in present value (roughly £1,750 million [1]). Improvements in the structure of the industry and realistic levels of uptake of objective selection methods and new breeding technologies could add a further Aus. $1,900 million, at present value, over the next 30 years.

Milk production

The systems of testing for dairy sheep are usually very similar indeed to those for dairy cattle, as described in Chapter 6. These involve progeny testing of rams in milk-recorded flocks, through the use of AI. Rams are selected for further use, or culled, on the basis of their daughters' milk production, milk quality, and associated traits. Milk recording of dairy cattle usually takes place at monthly intervals, and both morning and afternoon milkings are usually recorded on test days. However, the costs of this level of recording are prohibitive for most sheep dairy enterprises and only 2% to 22% of sheep in the specialised dairy breeds in the EU are recorded. Simplified recording systems for sheep have been approved by ICAR (the International Committee on Animal Recording). Both of these involve recording only a single milking on each test day. The first method involves recording morning and afternoon milkings in alternate months. The second method involves recording at the same time each month, but adjusting records for the difference between morning and afternoon milkings based on total bulk tank readings.

The problem of low levels of recording has been tackled, in France in particular, by attempting to segment the population clearly into nucleus and commercial tiers. Milk recording is then concentrated in the nucleus tier, which comprises 10% to 20% of the total population [8]. Genetic improvement is disseminated to the commercial tier via the sale of rams from nucleus flocks, or via AI from nucleus rams. Most recording systems have concentrated on milk yield until recently. However, fat and protein concentrations are now being recorded more widely, and other traits such as functional type traits and resistance to mastitis are being examined in France and Spain. Estimates of the heritabilities of the milk production traits and the genetic correlations among them are shown in Table 8.16. These agree very closely with the equivalent estimates for dairy cattle shown in Chapter 6.

As for milk recording, methods of evaluation of dairy sheep have generally

Traits	Milk yield	Fat yield	Protein yield	Fat content	Protein content
Milk yield	**0.30**				
Fat yield	0.83	**0.28**			
Protein yield	0.91	0.89	**0.29**		
Fat content	–0.31	0.26	–0.06	**0.35**	
Protein content	–0.40	–0.04	–0.03	0.63	**0.46**

Table 8.16 Estimates of the heritabilities of milk production traits in sheep (on the diagonal, in bold) and the genetic correlations among them. Milk yields were recorded using the simplified method described in the text, based on morning milkings only. Fat and protein contents were recorded on only a subset of these milk recording occasions [8].

followed those used in dairy cattle. Animal model BLUP evaluations have been used for dairy sheep in France, Italy and Spain since 1991. Genetic trends in milk yield of 2.0% and 2.4% of the mean per annum have been reported for the Manech and Lacaune breeds in France from 1980 to 1992 [9].

Genotype x environment interactions

There has been as much, if not more, heated debate about genotype x environment interactions in sheep breeding as there has in other species. The golden rule mentioned earlier still applies – if there is a big enough difference between genotypes, or environments, or both, then it is likely that interactions will be found. A good example of this is the fact that most specialised meat breeds selected under intensive management simply would not survive in British mountains, or the Australian rangelands. However, within breeds the answers are less clear. The issue of testing meat breeds in the UK under intensive feeding when their progeny are expected to perform in an extensive environment was mentioned earlier. Although a more precise investigation of genotype x environment interactions is still in progress with the SAC Suffolk flock, it appears that selection under intensive feeding does improve performance in the extensive environment, albeit by less than expected in some traits.

Interactions may be more important within hill breeds, especially where live weight and fat levels are being measured. A recent study in Scottish Blackface sheep has shown a low genetic correlation between live weights measured in intensive and extensive feeding systems, and only a moderate genetic correlation between fat depths measured in these two environments [10]. Also, the heritability of ultrasonic fat depth measured on improved pastures was about double that measured on unimproved hill land [13]. These results mean that the optimum environment for selection of hill sheep may vary, depending on the importance of altering weight and reducing or controlling fat in the breeding goal.

In specialised wool breeds, there are few reports of interactions affecting individual production traits, although they are sometimes reported for overall economic merit and secondary traits such as fleece rot and some measures of wool quality [3, 46].

The overall message as far as interactions are concerned is: (i) to be aware that they can exist; (ii) to take action if there is sound evidence that they exist in the traits of interest in the breed or system concerned (e.g. by restricting the choice of sires to those measured in similar conditions, or those evaluated in a wide range of conditions) and (iii) not to let the possibility of interactions in some cases detract from the value of genetic improvement in most cases.

Practical guidelines on selection

Some practical guidelines on selection to improve performance in objectively measured traits are given below. Most of these are equally relevant in purebred and commercial flocks, and whether the selection is for meat, milk or wool traits. However, pedigree breeders using these methods often wish to pay more attention to individual trait EBVs than commercial producers. Also, they may be willing to take greater risks by selecting animals with low accuracy EBVs, or by selecting occasional unrecorded animals if levels of recording are low in the breed concerned. The guidelines on selection between breeds or crosses are aimed primarily at commercial producers, since most pedigree breeders will already be firmly committed to their current breed.

Selection between breeds or crosses

- Clearly define the role of the animals you are selecting, and identify the animal characteristics which are economically important. Choose an appropriate breed or cross, based on the sort of objective comparisons of performance discussed earlier.

Selection of rams (or semen) for use in a purebred flock

- Set your breeding objective – the priority should be on traits expected to be of most economic importance in a few (sheep) generation's time.

- Identify the selection index or set of EBVs which is most relevant for your breeding objective.

- Produce a shortlist of animals which can be fairly compared, ranked on this index or on EBVs for the most important individual traits.

- If national across-flock BLUP evaluations are available, this shortlist might be compiled from a national sire summary, or from EBVs presented at ram sales or in semen catalogues. Include homebred rams on the shortlist if their index scores or EBVs warrant it.

- If across-flock evaluations are only available for cooperative breeding schemes, then rank the animals available from the scheme making the best genetic progress in traits relevant to your breeding goal. (Often, access to high merit animals, and hence rate of genetic gain in purebred flocks, will be improved by joining an effective cooperative scheme.)

- If there are no across-flock evaluations in the breed of your choice, select rams from one or more flocks which have a history of using objective selection for the index or set of EBVs most relevant to your breeding goal. (Try to compare the performance of purchased and homebred rams, or new and existing purchased rams, by mating them to groups of ewes balanced for genetic merit. If the purchased rams are inferior to homebred rams, change sources, or stick to homebred rams.)

- If there are economically important traits in your breeding goal which are not included in the index available, or for which no EBVs are available, then eliminate any animals which do not meet your minimum standards. Culling for these secondary traits should be in proportion to the economic importance of the trait, and the scope for genetic change in them. Selection for traits of minor economic importance, or traits which have a small genetic component, will dilute overall progress.

- Eliminate any animals which are not functionally sound, for example those which have unacceptable locomotion, testicle size or jaw structure. Culling should be confined to important characteristics which have a genetic component.

- If accuracies are available for the index or EBVs of your choice, and you are concerned about the risk of using low accuracy rams, eliminate those with very low values, or spread the risk by using more rams. (As explained in Chapter 5, BLUP EBVs already account for accuracy, and so *on average* progress in a flock or breed is maximised by selecting on EBVs regardless of accuracy. However, breeders are often concerned because large sums of money are at stake on *individual* animals. In these situations the accuracy provides a useful measure of the risk that an EBV will change in future. Choosing high EBV rams with high accuracies is a less risky strategy than choosing high EBV rams with low accuracy, but if there are few high EBV, high accuracy rams available, choosing several rams is less risky than gambling on only one.)

- Within reason, choose the highest index ram left on the list that you can afford.

- Avoid making matings between close relatives in purebred flocks. (Computer programs may be available via the recording agency or breed society to assist with this task).

- Genetic gain will be maximised by replacing rams whenever a ram of higher merit is available within this flock, or elsewhere. If there are no national evaluations to allow you to compare the merit of your own rams with others, or to compare the merit of your own rams of different ages, calculate the male selection intensities and generation intervals which will maximise progress in your breeding goal (as explained in Chapter 4) and aim to replace rams accordingly.

Selection of replacement females in a purebred flock

- Progress will be maximised by selecting replacement females on the most relevant index or set of EBVs. Choose this index or set of EBVs as described above for rams.

- Unless the flock is of low genetic merit, it will usually be most cost-effective to select among homebred females only, rather than buying in replacement females.

- If BLUP EBVs are available then the genetic merit of females of all ages within the flock can be compared directly. Rank the potential female replacements on the chosen index, or on the individual BLUP EBVs of most relevance. Genetic gain will be maximised by replacing ewes of any age which have low EBVs by replacements which have higher EBVs. (In a closed flock which has been selecting for a few years, selection of replacement females and culling on BLUP EBVs is unlikely to alter the age structure of the flock dramatically. However, when selection on BLUP EBVs first starts, some attention may need to be paid to the short-term economic effects of dramatic changes in age structure. For example, if it turns out that most older ewes in the flock have very low EBVs, although it would be genetically preferable, it may not be economically justifiable to cull all of these in one year because of the effect this may have on flock output.)

- If BLUP EBVs are not available, then it is difficult to compare the genetic merit of ewes of different ages. The simplest policy is to calculate the optimum female generation interval in order to maximise genetic progress in your breeding goal (as explained in Chapter 4), and then cull ewes on age to achieve this.

- Apply similar rules on culling for functional fitness, or traits of secondary economic importance, to those described above for rams.

Selection of replacement rams and ewes for a commercial flock

- Selection of rams for a commercial flock should follow the same guidelines as those for purebred flocks, except that the choice of index or EBVs should focus on expected market returns over the working life of the ram (pedigree flocks should have longer timescales than this).

- Rams purchased to breed replacement females in self-replacing flocks should be selected on an index or set of EBVs for traits of importance in their daughters, e.g. litter size, lamb weaning weight, mature size.

- Crossbred replacement females are unlikely to have EBVs. However, whenever possible, buy replacement females from breeders who use objective information relevant to your breeding goal in their own breeding programme.

Summary

- Sheep are kept primarily for the production of meat, wool and milk in temperate countries. In some countries or systems only one of these products is important, but in others there are multi-purpose production systems, and hence multi-purpose breeding goals.

- Reproductive performance, growth and carcass composition are the most important components of the breeding goal for meat production. Fleece weight and fibre diameter are the most important components of the breeding goal for specialised wool production. Reproductive performance, milk yield and milk composition are the most important breeding goal traits for milk production. However, other traits influencing inputs, like feed consumption, disease resistance and longevity, are also important in most cases.

- Specialised meat breeds, such as the Suffolk, Texel, Charollais and Dorset Down are widely used as terminal sires in crossing systems (e.g. in the UK, New Zealand and Australia, where they are usually mated to longwool x hill ewes, purebred dual-purpose ewes and longwool x Merino ewes respectively). In other countries (e.g. France and several other European countries) meat production is based on pure breeds.

- Specialised wool production worldwide is dominated by the Merino breed, or breeds or crosses derived from it.

- Milk production is usually based on specialised regional breeds, such as the Lacaune (France), Sarda (Italy), Churra (Spain), and Karagouniki (Greece). In northern Europe the Friesian breed, or derivatives of it, are often used.

- The majority of improvement programmes in meat producing breeds are based on on-farm recording of performance. However, central performance testing and central progeny testing are important in some countries. Cooperative breeding schemes based on one or more of these methods of testing are important in several countries.

- The traits recorded in meat breeding programmes usually include litter size, number of lambs reared and weights of lambs at various ages. Also, ultrasonic measurements of fat and muscle depth or area are becoming quite widely recorded.

- Generally, growth and carcass traits are moderately variable and moderately highly heritable. In contrast, most reproductive traits are more variable but have low heritabilities.

- Until recently, the results of most meat sheep recording schemes were evaluated by contemporary comparisons, or indices based on these. However, BLUP methods are being adopted more widely now. The relatively low use of AI in sheep in most countries means that genetic links between flocks are weaker than those in dairy and beef herds. This has prevented the use of across-flock evaluations in some countries, and confined them to sire referencing schemes in others.

- Selection indexes have become quite widely used in meat and dual-purpose breeds over the last decade or so. These usually combine growth and ultrasonic measurements in terminal sire breeds, and often include reproduction or wool traits, or both, in dual-purpose breeds.

- There is evidence of substantial rates of genetic change in growth and carcass traits in selection experiments in several countries, and in industry schemes in a few countries. The national value of genetic improvement schemes has been estimated in several countries, and this is usually high in relation to their cost.

- The improvement of wool traits in specialised wool breeds and dual-purpose breeds is also based on on-farm recording in most countries. Central wether tests, central progeny testing and cooperative breeding schemes are also important in some countries.

- The most important objectively measured traits are fleece weight and fibre diameter, but others include staple strength, colour, resistance to compression and style.

- The main fleece characteristics are all moderately or highly heritable, but there is a positive (i.e. unfavourable) correlation between fleece weight and fibre diameter.

- Most objective selection for wool traits in the past has depended on contemporary comparisons or EBVs for individual traits. BLUP evaluations are becoming more widespread in Merino breeding programmes, but their uptake in some cases is limited by the lack of pedigree information in stud flocks. The lack of links across flocks also limits the use of across-flock BLUP evaluations.

- Multi-trait indexes are increasingly being used in Merino breeding programmes. Uptake is being stimulated in some cases by the availability of computer software and advice to help formulate personalised breeding objectives.

- There are only a few examples so far of responses being measured in experimental or industry breeding schemes in specialised wool breeds. However, the value of genetic improvement in the finewool industry in Australia has been estimated to be very high indeed.

- Improvement programmes for dairy sheep usually involve progeny testing of rams in milk-recorded flocks, through the use of AI.

- Recording systems are usually simpler than those used in dairy cattle. These concentrated on yield in the past, but in several countries these now include milk quality, somatic cell count and functional type traits.

- The heritabilities of milk production traits are moderately high.

Genetic correlations among yield traits are also high, but there are negative correlations between some yield traits and concentration of fat and protein, as in dairy cattle.

● Animal model BLUP evaluations are used for dairy sheep in several countries, and. substantial genetic trends in milk yield have been reported.

● Genotype x environment interactions appear to be important in some sectors, but there are well-established guidelines to minimise their impact.

References

1. Atkins, K.D. 1993. 'Benefits of genetic improvement to the Merino wool industry.' *Wool Technology and Sheep Breeding*, 41:257–68.
2. Atkins, K.D. 1995. 'Strategies to reduce the average fibre diameter of the Australian wool clip.' *Proceedings of the Australian Association of Animal Breeding and Genetics*, Vol. 11, pp. 580–6.
3. Atkins, K.D. and Casey, A.E. 1994. 'Recent developments in wool sheep breeding.' *Proceedings of the 5th World Congress on Genetics Applied to Livestock Production*, Vol. 18, pp. 23–30.
4. Atkins, K.D., Coelli, K.A., Casey, A.E. and Semple, S.J. 1995. 'Genetic differences among Merino bloodlines from NSW wether comparisons (1983-1993).' *Wool Technology and Sheep Breeding*, 43:1–14.
5. Atkins, K.D., Semple, S.J. and Casey, A.E. 1994. 'Object – Personalised breeding objectives for Merinos.' *Proceedings of the 5th World Congress on Genetics Applied to Livestock Production*, Vol. 22, pp. 79–80.
6. Banks, R.G. 1994. 'LAMBPLAN: genetic evaluation for the Australian lamb industry.' *Proceedings of the 5th World Congress on Genetics Applied to Livestock Production*, Vol. 18, pp. 15–18.
7. Banks, R.G. 1994. 'Structural effects on returns from genetic improvement programs: a case study.' *Proceedings of the 5th World Congress on Genetics Applied to Livestock Production*, Vol. 18, pp. 123–6.
8. Barillet, F., Astruc, J.M. and Marie, C. 1995. 'Breeding for the milk market: the French situation compared to the EU Mediterranean countries.' *Proceedings of the World Sheep and Wool Congress*. Royal Agricultural Society of England, Kenilworth.
9. Barillet, F., Sanna, S., Boichard, D. et al. 1993. 'Genetic evaluation of the Lacaune, Manech and Sarda dairy sheep with an animal model.' *Proceedings of the 5th International Symposium on Machine Milking of Small Ruminants*, (*Hungarian Journal of Animal Production, Supplement No. 1, 1993*), pp. 289–304.
10. Bishop, S.C., Conington, J., Waterhouse, A. and Simm, G. 1996. 'Genotype x environment interactions for early growth and ultrasonic measurements in hill sheep.' *Animal Science*, 62:271–7.

11. Blair, H.T. and McCutcheon, S. N. (eds). 1993. *Proceedings of the A.L. Rae Symposium on Animal Breeding and Genetics*. Department of Animal Science, Massey University, Palmerston North, New Zealand.

12. Brien, F.D. 1990. 'WOOLPLAN and its relationship to improvement of wool production.' *Proceedings of the 4th World Congress on Genetics Applied to Livestock Production*, Vol. XV, pp 181–4.

13. Conington, J., Bishop, S.C., Waterhouse, A.W. and Simm, G. 1995. 'A genetic analysis of early growth and ultrasonic measurements in hill sheep.' *Animal Science*, 61:85–93.

14. Cottle, D.J. and James, J.W. 1994. 'Australian Merino central test sire evaluation schemes – operational issues.' *Proceedings of the 5th World Congress on Genetics Applied to Livestock Production*, Vol. 18, pp. 35–8.

15. Fogarty, N.M. 1995. 'Genetic parameters for live weight, fat and muscle measurements, wool production and reproduction in sheep: a review.' *Animal Breeding Abstracts*, 63:101–43. (Also, see erratum preceding p. 935, Vol. 63)

16. Food and Agriculture Organisation of the United Nations. 1996. *FAO Production Yearbook 1995. Vol. 49*. Food and Agriculture Organisation of the United Nations, Rome.

17. Guy, D.R. and Croston, D. 1994. 'UK experience and progress with sheep sire referencing schemes.' *Proceedings of the 5th World Congress on Genetics Applied to Livestock Production*, Vol. 18, pp. 55–8.

18. Guy, D.R., Croston, D. and Jones, D.W. 1986. 'Response to selection in Welsh Mountain sheep.' *Animal Production*, 42:442.

19. James, J.W. 1977. 'Open nucleus breeding schemes.' *Animal Production*, 24:287–305.

20. Kempster, A.J., Cook, G.L. and Grantley-Smith, M. 1986. 'National estimates of the body composition of British cattle, sheep and pigs with special reference to trends in fatness. A review.' *Meat Science*, 17:107–38.

21. Lewis, R.M. and Simm, G. 1995. 'Optimum designs for sire referencing schemes in sheep.' *Book of Abstracts of the 46th Annual Meeting of the European Association for Animal Production*. Poster G3.17, p. 42.

22. Lewis, R.M., Simm, G., Dingwall, W.S. and Murphy, S.V. 1996. 'Selection for lean growth in terminal sire sheep to produce leaner crossbred progeny.' *Animal Science*, 63:133–42.

23. McClelland, T.H., Bonaiti, B. and Taylor, St. C.S. 1976. 'Breed differences in body composition of equally mature sheep.' *Animal Production*, 23:281–93.

24. McEwan, J.C., Dodds, K.G., Watson, T.G. et al. 1995. 'Selection for host resistance to roundworms by the New Zealand sheep breeding industry: the WormFEC service.' *Proceedings of the Australian Association of Animal Breeding and Genetics*, Vol. 11, pp. 70–3.

25. McGuirk, B.J. Rose, M. and Scott, R. 1982. 'Productivity of classers' grades and sire selection differentials for fleece weight in two Merino studs.' *Australian Journal of Experimental Agriculture and Animal Husbandry*, 22:274–80.

26. McMaster, J.C. 1995. 'Breeding for fibres.' *Proceedings of the World Sheep and Wool Congress*. Royal Agricultural Society of England, Kenilworth.

27. Meat and Livestock Commission. 1986. *Sheep Yearbook 1986*. Meat and Livestock Commission, Bletchley.

28. Meat and Livestock Commission. 1987. *Sheep Yearbook 1987*. Meat and Livestock Commission, Bletchley.

29. Meat and Livestock Commission. 1988. *Sheep in Britain*. Meat and Livestock Commission, Bletchley.

30. Meat and Livestock Commission. 1992. *Sheep Yearbook 1992*. Meat and Livestock Commission, Milton Keynes.

31. Meat and Livestock Commission. 1996. *Sheep Yearbook 1996*. Meat and Livestock Commission, Milton Keynes.

32. Parker, A.G. 1993. 'Application of performance recording in sheep – industry structure.' *Proceedings of the A.L. Rae Symposium on Animal Breeding and Genetics*. Blair, H.T. and McCutcheon, S. N. (eds). Department of Animal Science, Massey University, Palmerston North, New Zealand, pp. 120–6.

33. Ponzoni, R.W., Rogan, I.M. and James, P.W. 1990. 'Genetic improvement of apparel and carpet wool production.' *Proceedings of the 4th World Congress on Genetics Applied to Livestock Production*, Vol. XV, pp. 149–66.

34. Roden, J.A. 1994. 'Review of the theory of open nucleus breeding systems.' *Animal Breeding Abstracts*, 62:151–7.

35. Ryder, M.L. 1983. *Sheep and Man*. Duckworth, London.

36. Scottish Agricultural College. 1995. *Farm Management Handbook 1995/96*. SAC, Edinburgh.

37. Simm, G. 1988. *Artificial insemination and embryo transfer: how they can help sheep breeders*. SAC Technical Note T136. SAC, Perth.

38. Simm, G. 1992. 'Selection for lean meat production in sheep.' *In Recent Advances in Sheep and Goat Research*, A.W. Speedy (ed.), CAB International, pp. 193–215.

39. Simm, G. 1994. 'Developments in improvement of meat sheep.' *Proceedings of the 5th World Congress on Genetics Applied to Livestock Production*, Vol. 18, pp. 3–10.

40. Simm, G., Amer, P.R. and Pryce, J.E. 1998. 'Benefits from genetic improvement of sheep and beef cattle in Britain.' In *SAC Animal Sciences Research Report 1997*. SAC, Edinburgh, (in press).

41. Simm, G., Conington, J., Bishop, S.C. et al. 1996. 'Genetic selection for extensive conditions.' *Applied Animal Behaviour Science*, 49:47–59.

42. Simm, G. and Dingwall, W.S. 1989 'Selection indices for lean meat production in sheep.' *Livestock Production Science*, 21:223–33.

43. Simm, G. and Murphy, S.V. 1996. 'The effects of selection for lean growth in Suffolk sires on the saleable meat yield of their crossbred progeny.' *Animal Science*, 62:255–63.

44. Simm, G. and Wray, N.R. 1991. *Sheep sire referencing schemes – new*

opportunities for pedigree breeders and lamb producers. SAC Technical Note T264. SAC, Edinburgh.

45. van Wyk, J.B., Erasmus, G.J. and Olivier, J.J. 1994. 'Variance component estimates and responses to selection on BLUP of breeding values in Merino sheep.' *Proceedings of the 5th World Congress on Genetics Applied to Livestock Production*, Vol. 18, pp.31–4.

46. Woolaston, R.R. 1987. 'Genotype x environment interactions and their possible impact on breeding programs.' In B. J. McGuirk (ed.) *Merino Improvement Programs in Australia*, Australian Wool Corporation, Melbourne, pp.421-5.

47. Woolaston, R.R. 1994. 'Preliminary evaluation of strategies to breed Merinos for resistance to roundworms.' *Proceedings of the 5th World Congress on Genetics Applied to Livestock Production*, Vol. 20, pp. 281–4.

48. Woolaston, R.R. and Eady, S.J. 1995. 'Australian research on genetic resistance to nematode parasites.' In G.D. Gray, R.R. Woolaston and B.J. Eaton, (eds) *Breeding for Resistance to Infectious Diseases in Small Ruminants*, Australian Centre for International Agricultural Research, pp. 53–75.

Further reading

Animal Science (formerly *Animal Production*; Journal of the British Society of Animal Science).

Livestock Production Science (Journal of the European Association for Animal Production).

McGuirk, B.J. (ed.). 1987. *Merino Improvement Programs in Australia*. Australian Wool Corporation, Melbourne.

Morley, F.H.W. (ed.). 1995. *Merinos, Money and Management*. Post Graduate committee in Veterinary Science, University of Sydney, Australia.

Owen, J.B. and Axford, R.F.E. (eds). 1991. *Breeding for Disease Resistance in Farm Animals*. CAB International, Wallingford.

Piper, L. and Ruvinsky, A. (eds). 1997. *The Genetics of Sheep*. CAB International, Wallingford, (in press).

Proceedings of the Association for the Advancement of Animal Breeding and Genetics (formerly the *Australian Association of Animal Breeding and Genetics*).

Proceedings of the New Zealand Society of Animal Production

Proceedings of the World Congresses on Genetics Applied to Livestock Production

Wood, J.D. and Fisher, A.V. (eds). 1990. *Reducing Fat in Meat Animals*. Elsevier, London.

CHAPTER 9
New technologies

Introduction

Genetic improvement is one of the most effective strategies available for altering the performance of farm animals. It is relatively slow compared to some other methods, such as improved feeding, but it is permanent and cumulative, and in most cases it is highly cost-effective and sustainable. Populations of animals of high genetic merit are needed to achieve high efficiency and competitiveness in any livestock industry. Genetic improvement has generally been used very effectively in the pig and poultry industries of many countries. However, at least until recently, it has been used less effectively in some dairy industries, and usually much less effectively in beef and sheep industries. The aim of this chapter is to discuss some of the new technologies which could lead to more effective genetic improvement programmes in livestock. While most of these technologies are relevant in all species, many are likely to be of particular value in ruminants.

Rates of genetic improvement depend on four main factors, as outlined in Chapter 4: (i) the selection intensity achieved; (ii) the accuracy with which genetic merit in the trait of interest is predicted; (iii) the amount of genetic variation in the trait of interest; (iv) the generation interval. Generally speaking, the greater the selection intensity, accuracy and genetic variation, and the shorter the generation interval, the greater the annual rate of genetic improvement. The main opportunities for breeders to accelerate rates of improvement are through choice of the most accurate methods of predicting breeding values and by maintaining high selection intensities and low generation intervals. However, there are biological limits on the extent to which selection intensity and generation interval can be altered. As a result of their earlier sexual maturity, and their higher reproductive rates, it is possible to achieve greater selection intensities and shorter generation intervals in pigs and poultry than in ruminants. Largely as a result of this, annual rates of genetic improvement in pigs and poultry are often up to double those predicted or achieved in ruminants [48].

The scope for increasing rates of genetic gain through the wider use of BLUP methods of predicting breeding values, especially in beef cattle and sheep, has been discussed fully already. Rates of genetic gain in overall economic merit will be maximised by ensuring that breeding goals include all traits which are of major economic importance and are heritable, and include none that are of very minor importance, or are not heritable. Similarly, selection indexes should include all the available measurements which make a significant contribution to predicting merit in breeding goal traits. There is considerable scope for the use of more comprehensive breeding goals and criteria in most cattle and sheep breeding programmes. Some examples of developments in this area have been covered already. For instance, including traits affecting longevity and health in dairy cattle breeding indexes is likely to make an increasingly important contribution to overall economic progress, cow welfare, and the sustainability of breeding programmes. Similarly, a better understanding of the relationships between production traits and traits conferring adaptation to harsh environments could help in the production of more sustainable breeding programmes for hill sheep and suckler cows in these environments.

In most cases the tools to achieve these improvements are already available, but they need to be fine tuned and applied more widely. For instance, the methodology for producing more comprehensive breeding goals and indexes is well developed. However, indexes need to be tailored to the particular goal, by obtaining estimates of the relevant genetic parameters and economic values. Although the methodology is available, this is by no means a trivial task. In other cases, new technologies can contribute to accelerating genetic improvement. For example, new scanning techniques can improve rates of progress in carcass characteristics by providing more accurate predictions of the carcass composition of candidates for selection. New techniques for automatic identification and data capture could improve the accuracy of selection for milk characteristics. Similar techniques could provide vast amounts of additional information on the carcass weights and grades of commercial meat animals. This could be of considerable value in increasing the accuracy of evaluation of related purebred animals.

There are two other types of new technology which can have a major impact on rates of genetic improvement. These are **reproductive technologies** and **molecular genetic technologies**. The current and potential future reproductive and genetic technologies, together with their possible impact on genetic improvement of ruminants, are discussed in detail in the next two sections. Any new technique needs to have a favourable cost:benefit ratio for it to be widely adopted. These issues are not discussed in detail here, partly because many of the techniques are at an early stage of development, and so both success rates and costs are difficult to estimate. However, it is worth making two general comments about the cost:benefits of new technologies. Firstly, it will usually be easier to justify the use of more expensive technologies in order to accelerate genetic improvement programmes in elite herds or flocks than to justify their use for dissemination of improvement to commercial tiers. This is because of the higher average value of animals produced from elite herds or flocks. Secondly, it will be easier to justify the use of more expensive technologies in cattle than in sheep breeding or dissemination programmes. Again, this is because of the higher average *per capita* value of elite and commercial cattle than their sheep equivalents.

In many industrialised countries over the last few years there has been growing public interest in methods of food production, and particularly in the welfare of farmed animals. Particular concerns are expressed about the potential animal welfare or ethical implications of some of the reproductive and genetic technologies outlined in this chapter, so these issues are discussed in the final section.

Reproductive technologies

It is possible to achieve much higher selection intensities in species or breeds with a high reproductive rate than in those with lower reproductive rates. Similarly, shorter generation intervals can be achieved in those species or breeds which reach sexual maturity at a younger age. It is largely because of biological advantages in these reproductive characteristics that higher rates of genetic change are possible in pigs and poultry than in ruminants. In dairy cattle the main traits of interest are sex-limited and measured fairly late in life, which compounds the disadvantage that cattle suffer in reproductive rate. There are several reproductive technologies already available, or under development, which can accelerate progress in genetic improvement programmes in ruminants. These include artificial insemination (AI), multiple ovulation, embryo recovery and embryo transfer (MOET), *in vitro* production of embryos (i.e. production by laboratory culture), sexing of semen or embryos, and cloning (i.e. mass production of identical embryos). These techniques are outlined briefly below, and their potential impacts are discussed. More details on the techniques themselves are given in the reviews or books cited [19, 20, 21, 36, 46, 59, 63].

In addition to their potential value in accelerating response to selection in breeding programmes, many of these techniques also have the potential to accelerate dissemination of genetic improvement from the elite to the commercial tiers of livestock industries. Techniques of value in selection will generally be useful in dissemination, but their cost and ease of use may limit their role in dissemination. However, not all techniques of value in dissemination will be useful in selection programmes. So, the value of each of the techniques in dissemination is also discussed.

Artificial insemination

Artificial insemination (AI) has been available to cattle breeders for over 50 years to enhance the reproductive rate of males. The development of reliable techniques for extending (i.e. diluting to allow wider use) and freezing cattle semen has augmented this benefit. AI allows much higher selection intensities among males than those possible with natural mating. Also, the desired number of progeny can often be produced sooner by AI than by natural mating, so male generation intervals can be reduced. AI can contribute to more accurate evaluation of genetic merit as well by permitting large scale progeny testing in many herds. As a result substantial rates of genetic improvement have been achieved in several countries through the use of AI in well-designed dairy cattle breeding schemes. Progress has been particularly high when the use of AI has been coupled with accurate techniques for predicting breeding value, as outlined in Chapter 6.

Although the use of AI is less widespread in beef cattle breeding than in dairy cattle breeding, the technique can have a similar impact. In many countries one of the major contributions of AI to beef breeding programmes is to create genetic links between herds, which allow across-herd genetic evaluations.

The fact that AI in cattle is relatively cheap and simple, and often allows access to very reliably proven, high genetic merit animals means that it is currently the most effective method of dissemination of genetic improvement to commercial herds. This is particularly true in dairying, where commercial herds rely heavily on AI, and have as good access as elite breeders to high genetic merit bulls. In fact, in some countries, the average genetic merit of commercial herds for production traits exceeds the merit of elite breeders' herds, because breeders in the latter sector have put less emphasis on production and more emphasis on other characteristics such as breed type. Most commercial beef cows are kept in more extensive production systems than dairy cows. This makes oestrous detection more difficult, and hence limits the use of AI for dissemination. However, there is growing use of oestrus synchronisation in commercial beef herds to make AI more practical, and so allow access to bulls of higher merit.

To date, AI has had a much smaller impact in most sheep breeding programmes than in cattle breeding programmes. This is largely because of the poor conception rates which usually accompany the cervical insemination of previously frozen semen. However, the technique has had an important impact in schemes where the use of fresh semen is practical, including dairy sheep breeding programmes, or in schemes involving the laparoscopic intra-uterine insemination of frozen semen. (With laparoscopic AI the uterus is viewed, and semen is inserted into the uterus, via a small incision in the abdomen.) For instance, the strategic use of laparoscopic AI, and in some cases cervical AI, to create genetic links across flocks, is central to the success of sheep sire referencing schemes, as discussed in Chapter 8. There is no doubt that new techniques which produce high conception rates, and use less invasive insemination methods than laparoscopy, could have a major impact on both the rates of genetic gain, and the dissemination of genetic improvement in the sheep industries of many countries. The growing interest in the use of objective selection in sheep breeding programmes, and the evidence of high rates of improvement being achieved in several large cooperative breeding schemes, would make the development and application of improved techniques for dissemination particularly timely.

Multiple ovulation and embryo transfer

Over the last few decades increasingly reliable procedures have been developed for superovulation, embryo recovery, embryo freezing and embryo transfer in cattle [63]. More recently similar procedures have been developed for sheep but, unlike the situation in cattle, embryo recovery and transfer techniques are only practicable at present with laparoscopy or surgery [16, 21]. A number of applications of embryo procedures have been proposed or practised in cattle and sheep breeding. These include uses in: (i) within-breed genetic improvement programmes; (ii) the international trading of genetic material (offering potential advantages in economy, animal welfare and disease control); (iii) accelerating breed substitution by multiplication of newly introduced breeds and (iv) conservation of genetic material by freezing embryos from valuable individual animals, or from rare or endangered breeds or species. Additionally, several of the new reproductive or genetic procedures discussed in this chapter hinge on the use of embryo transfer.

Multiple ovulation and embryo transfer (MOET) potentially offers similar benefits in selection of females to those offered by AI in males (though in practice the benefits are often smaller). That is, female selection intensities can be increased, female generation intervals can be reduced, and the accuracy of evaluating embryo donors, or full sibs created by MOET, can be increased. Since the mid-1970s there have been several studies on the potential impact of MOET on the rates of genetic improvement in ruminants. The earliest of these studies predicted that rates of improvement in beef cattle, dairy cattle and sheep could be increased up to twofold compared to conventional breeding schemes [32, 40, 43, 49].

As a result of this work, commercial dairy cattle breeding schemes based on MOET were initiated in several countries in the mid- to late 1980s, as described in Chapter 6. Experimental breeding schemes involving MOET were also established in beef cattle and sheep at about the same time. However, further theoretical work from the mid-1980s onwards showed that the initial results were very sensitive to alterations in some key assumptions. For example, most of the early studies used optimistic success rates for MOET, and these have been difficult to achieve on a field scale in practice. Also, the early studies usually ignored the fact that genetic variation, and hence response to selection, is reduced by the process of selection, and by inbreeding – the latter being a usual consequence of effective selection in domestic animals. These factors appear to be particularly important in small, closed nucleus populations, such as those in which the use of MOET was first envisaged. The large variability in numbers of transferable embryos recovered per collection was also ignored in many early studies. This variability is expected to have an unfavourable impact on rates of genetic gain and, especially, on rates of inbreeding, since fewer animals make a bigger contribution to the genetic make-up of the next generation. Advances in the theory of predicting rates of inbreeding also showed that these had been severely underestimated in most of the early studies [41, 53].

Much effort has been directed at the design of breeding schemes to overcome these high rates of inbreeding, and several successful techniques have been identified [6, 23, 51, 53, 62, 64]. These include:

- the use of factorial mating designs (mating each donor cow to different selected bulls in successive matings, rather than to the same bull)
- using selected parents only once
- equalising family sizes (this will be easier to achieve with new techniques producing higher yields of embryos)
- using more than one male from each selected full sib family
- deliberately choosing less closely related animals
- reducing the emphasis on ancestors' performance when calculating BLUP EBVs (because BLUP methods use records of relatives they tend to lead to selection of more closely related animals, and so to higher rates of inbreeding; this effect can be reduced by altering the emphasis on records from relatives)

Use of some of these techniques alone or in combination is expected to reduce inbreeding substantially, with little or no change in genetic gain. So, with appropriate modifications to design MOET schemes probably can offer increased rates of response compared with conventional schemes. However, these rates of response are still unlikely to be as high as first predicted. For example, in beef MOET schemes rates of gain may be 30% higher than in conventional schemes, at acceptable levels of inbreeding, rather than 100% higher as first predicted [54]. In future, the advantage to MOET schemes may be augmented as a result of the use of techniques which result in improved yields of transferable embryos, or through the use of some of the new reproductive or genetic technologies outlined below.

There have been few recent studies of the benefits of MOET in sheep. However, for meat breeds it is probably reasonable to expect similar benefits to those in beef MOET schemes. In wool breeds the benefits appear to be similar to those outlined for beef cattle too [65].

In theory, MOET could be a highly effective technique for dissemination of genetic improvement to commercial sectors of the livestock industries. Since embryos have already passed the hurdle of fertilisation, they have the potential to produce higher calving or lambing rates than AI, though this potential is still far from being realised. Similarly, embryos already contain the full complement of chromosomes necessary for their development. So, MOET can deliver populations of commercial animals with 100% of their genes from elite sector parents, whereas AI used on commercial females produces animals with only 50% of their genes from the elite sector. However, with current MOET techniques each donor female produces relatively few embryos, at least compared to the number of doses of semen which can be produced by each male. Partly for this reason, but also because of the more complex techniques involved, MOET remains far more expensive than AI, and so less attractive as a means of dissemination. Some newer techniques which have the potential to allow the wider use of ET are discussed next.

In vitro production of embryos

Over the last few decades a great deal of effort has gone into developing techniques for the *in vitro* maturation, fertilisation and culture of eggs from farm animals as well as humans. *In vitro* literally means in glass, as opposed to in the body. The rationale for this research in humans is to allow certain types of infertility to be overcome (e.g. that due to blocked Fallopian tubes), or to allow the use of donor semen or eggs in cases of complete infertility of one partner. The main aim of developing these techniques in farm animals is to allow the use of the thousands of eggs present in the ovaries of female animals at birth, most of which never develop to the point of ovulation [3, 19, 22, 59].

One of the earliest intended uses of *in vitro*-produced embryos was to improve the beef merit of calves from dairy or suckler cows by creating a supply of embryos with ¾ or ⅞ beef genotype. Initially, the main source of eggs was the ovaries of slaughtered beef heifers. Companies were established in several countries, including the UK and Eire, to collect eggs from beef heifers with a high proportion of continental beef breeds in their genetic make-up, and to produce embryos from these by maturing them and then fertilising them with semen from high merit proven bulls. These embryos were then marketed for transfer into beef suckler cows or dairy cows. Transfers were made either singly, or to create twins either by transferring an *in vitro*-produced embryo into cows already carrying a natural embryo, or by transferring two *in vitro*-produced embryos. Despite a ready supply of ovaries from slaughtered heifers, early techniques produced few transferable embryos per ovary. Also, some *in vitro* culture techniques are implicated in the birth of very large calves, generally with associated calving difficulties [31]. Partly for these reasons, some of the original companies are no longer in business.

All purebred cows are slaughtered or die of natural causes at some stage, but most are no longer prime candidates for selection by the time this happens, so recovering eggs *post mortem* is not so beneficial. (Though the technique may still be of interest for recovering oocytes from particularly valuable purebred animals. For example, in the UK the breeding company Genus offers such a service, and produces an average of about nine high quality, transferable embryos per donor.) Hence, the method outlined above is particularly suitable for the dissemination of genetic improvement. The majority of donors are likely to be crossbred animals of unknown genetic merit, with high performance in commercial beef traits. Despite the lack of predictions of genetic merit of donors, selection on the basis of breed or crossbred type, or crude selection on phenotype, should be sufficient to ensure that their merit for beef production is better than that of most dairy cows and many suckler cows. Also, the technique requires semen from only a small number of very highly selected bulls to fertilise eggs, so it allows very high male selection intensities and accuracies to be achieved. This will usually compensate for possible deficiencies in the selection procedures for donor females. With improvements in the techniques involved, and selection of appropriate parents, the *in vitro* production of embryos could improve both quality and uniformity of beef production in future.

More recently techniques have been developed to allow the recovery of unfertilised eggs directly from the ovaries of live cows. These involve collection of eggs through an ultrasonically guided needle inserted into the ovary, usually via the vagina [30]. This type of recovery is called ***in vivo* aspiration of oocytes**, or **ovum pick up (OPU)**. It has several potential advantages compared with recovery of eggs from slaughtered cows, or conventional embryo recovery techniques:

- Purebred animals of high genetic merit can be used as donors, so the technique is of potential benefit in genetic improvement and not just in dissemination.

- The technique can be applied to produce embryos in a more planned manner than with *post mortem* recovery.

- OPU allows collection of oocytes from younger donors, compared to conventional embryo recovery techniques.

- It is possible to collect oocytes from donors in the early stages of pregnancy.

- Eggs can be collected from donors on a weekly basis, allowing tens or potentially hundreds of embryos to be produced from the same donor. (In the UK the Genus OPU service produces an average of 7.3 oocytes and 1.7 transferable embryos per collection.)

Because it is an invasive technique it does require skilled veterinary input. However, there appears to be no evidence of injury to the donor, nor any adverse effects on subsequent reproductive performance. This method of embryo production still depends on *in vitro* culture systems which, at the time of writing, can still lead to the production of oversized calves.

OPU is being used in several countries to produce a higher number of transferable embryos than can be produced by conventional MOET. Increasing the yield of transferable embryos in this way could have a major impact on the rate of gain achievable in MOET breeding schemes, possibly allowing rates of gain up to 34% above those possible in conventional progeny testing schemes [35]. Also, high yields of *in vitro*-produced embryos would make some of the methods of controlling inbreeding more practical (e.g. equalising family sizes, use of factorial mating designs). Embryos produced by OPU are likely to be used in genetic improvement programmes in elite herds, rather than for mass dissemination of improvement to the commercial sector, unless the yield of transferable embryos increases dramatically, or it is combined with other new reproductive technologies such as embryo cloning.

In theory, *in vitro*-produced sheep embryos would have many of the advantages already mentioned for cattle embryos. However, the techniques are less well developed for sheep, and the lower value of the end product is likely to limit applications in sheep breeding and especially in commercial sheep production.

Semen and embryo sexing

Sexing semen has been the holy grail of reproductive technologists for decades. Most of the methods tested have used real or apparent physical differences between sperm bearing X and Y chromosomes as the basis for separation. For example, separation methods aimed at exploiting differences in mass, surface charge, antigenic properties and buoyant density of X- and Y-bearing sperm have been investigated, but success rates and repeatability have usually been low [11, 57]. In the last few years a more reliable technique for semen sexing has been developed. The technique uses the fact that sperm bearing the X and Y chromosomes differ slightly in their DNA content. Using a technique called **flow cytometry** it is possible to detect these differences in size, and to sort sperm into two groups accordingly. The technique involves staining sperm with a fluorescent stain, and using differences in fluorescence when the cells pass through a laser beam to identify and sort X- and Y-bearing sperm. Currently this technique is slow, especially when high purity of sorted samples is required. As a result, it can be used to sort only small quantities of semen, such as those required for *in vitro* fertilisation. Small numbers of calves and lambs of the expected sex have been produced from sorted semen [12, 13]. It is likely that developments in the technique will allow faster sorting and higher accuracy of sorting in the near future, and this in turn will allow conventional AI with sexed semen.

Several approaches have been taken to developing techniques for sexing embryos [57, 63]. These are based on removing a small number of cells from embryos at the 16-cell stage or later stages of development. In some cases the sex of the embryo can be determined by direct observation of a preparation of chromosomes using a microscope (**karyotyping**), to confirm the presence of either two X chromosomes or one Y and one X chromosome. Alternative approaches include the use of an immunological test for the H-Y antigen produced only by male embryos, using sex-linked differences in enzyme activity, or using some of the molecular techniques described later in this chapter to detect sequences of DNA specific to the Y chromosome [63]. The last method is already being used on a small scale commercially, and is the most promising method for wider scale application in future.

In genetic improvement programmes, sexing of either semen or embryos could be useful in increasing the selection intensity applied to females by producing more of them. This appears to be of little value when performance records are available on both sexes, and where the total number of animals tested is fixed, as an increase in the number of females implies a reduction in the number of males, and therefore a reduction in male selection intensities. However, when performance recording is sex-limited, as with milk production, there may be benefits in creating more animals of the recorded sex. This may be especially beneficial in nucleus schemes, if they already have access to high genetic merit males which have been accurately tested in a separate population [9, 53].

The development of a cheap, reliable technique for sexing semen in large enough quantities for conventional AI could lead to major improvements in the dissemination of genetic improvement and in the efficiency of animal production. In meat production systems there would be interest in using 'female' semen to breed replacement animals from the highest merit females in the herd or flock. Since males usually show higher growth rates and leaner carcasses, there may be a preference for 'male' semen to produce animals for slaughter. However, so-called once-bred heifer systems, using only female semen, may produce the highest overall efficiency [50]. In these systems each heifer leaves a replacement female calf before being slaughtered for beef production. In conventional dairy cattle systems, female semen could be used to breed replacements from the best cows, releasing a higher proportion of the herd for mating to male semen from a beef breed, or for crossing to a beef bull.

Similarly, the development of techniques for the production of large numbers of cheap, sexed embryos could lead to the use of high merit female embryos to produce replacements, and cloned male embryos to produce slaughter animals, or the development of once-bred heifer systems based only on female embryos. In dairy herds elite female embryos could be used to produce replacements, and male embryos of high beef merit and a proven record for calving ease could be used in the majority of cows. Also, improved techniques for embryo multiplication could increase the benefits and practicality of crossbreeding. In cattle and sheep, crossbreeding is sometimes impractical because the low reproductive rate of these species means that many purebreds are needed to produce a regular supply of F1 females. In some cases there are too few breeds with high genetic merit in the traits of interest to sustain a rotational crossing scheme. However, improved reproductive techniques could allow a plentiful supply of F1 (or other) embryos to produce replacement females from a relatively small population of elite animals of the component pure breeds. This could lead to replacement policies in cattle and sheep similar to those already employed by conventional means in the commercial tier of the pig and poultry industries, where most replacement females are purchased crossbreds.

Embryo cloning

The production of groups of identical embryos from a single original embryo is often termed **embryo cloning**, or **embryo multiplication**. This can be achieved in one of two ways at present. The first involves physically splitting embryos into two halves (or four quarters). If the original embryo was at an early stage of development, the resulting halves need to be inserted into an empty ***zona pellucida*** or 'egg shell', before transfer. If the embryo is at a later stage of development, the halves can be transferred without this procedure [63]. Splitting into more than four parts is not successful, as there are too few cells for normal development. Also, repeated bisection of embryos after further culture is not successful, because it leads to too great a disparity between the physical composition and true chronological stage of development of the embryos. Embryo bisection can be employed to increase the yield of transferable embryos, particularly when donors are rare or very valuable, or in MOET nucleus breeding schemes where a target number of embryos is needed from each donor.

A larger number of identical embryos can be produced by the procedure of **nuclear transfer**. This involves removing the *zona pellucida* from a 16-, 32- or 64-cell embryo, and separating the identical cells within. Each of the resulting cells can be inserted into an unfertilised egg from which the nucleus containing the single copy of chromosomes has been removed (see Plate 60). An electric current causes fusion of the introduced embryonic cell and the unfertilised egg, and the new embryo then develops as if newly fertilised [63]. The advantage of this technique is that it produces larger numbers of identical animals, although the techniques involved are more complex. Until recently, nuclear transfer had only been accomplished using early embryos, or cells from early embryos. However, in 1996 it was reported that viable cloned embryos, and subsequently, live cloned lambs, had been produced by transferring cells which were originally derived from embryos, but had been cultured and multiplied in the laboratory ([7, 8]; see Plate 61). The ability to multiply embryonic cells in the laboratory, and achieve successful development of resulting embryos following nuclear transfer, obviously gives the potential for far greater numbers of identical animals to be produced from a single pair of elite parents. It also creates potential opportunities for the rapid multiplication of embryos which have been genetically modified in some way (e.g. by gene transfer).

In 1997 it was reported that a viable lamb had been produced following nuclear transfer from a cultured cell originating in an adult ewe – the first time that a mammal has been produced from a cell derived from an adult ([60]; see Plate 61). Clearly this presents new opportunities for **adult cloning**, since clones could be derived from a single animal of proven performance, rather than from embryos with high average, but variable merit produced by two proven parents.

A possible complication with nuclear transfer is that it is not just nuclear DNA which is transferred. As outlined in Chapter 2, within the cytoplasm of eggs and embryos there is DNA in the mitochondria, which is inherited exclusively from the female producing the eggs. There have been reports that this mitochondrial DNA, which gets passed from mother to daughter only, has a small but significant effect on milk production [18]. If these findings are confirmed, and the issue is still hotly debated, then there may be variation between clones as a result of variation in mitochondrial DNA in the unfertilised recipient eggs.

In genetic improvement programmes, cloning could be used to produce many animals of the same genotype in order to improve the accuracy of evaluation, or to allow evaluation of traits normally measured post-slaughter on some members of the cloned group. This would involve implanting some embryos from each cloned line to produce animals for testing, and freezing others to allow subsequent use (or further cloning) of the best tested cloned lines in breeding or dissemination programmes. Cloning is expected to be of value in breeding programmes under certain circumstances, for example in dairy cattle nucleus schemes if cloned lines are tested in a separate population. However, if cloning is considered only in the context of closed breeding schemes, with fixed numbers of animals tested, then the expected benefits generally diminish or disappear, as keeping more identical animals means that fewer different families can be kept, and so selection intensities will be reduced [15, 53].

Cloning could allow the more effective use of non-additive genetic variation. Conventionally, selection for quantitative traits within breeds attempts to identify animals with the highest additive genetic merit. It is difficult to make use of non-additive genetic variation in selection since particular combinations of non-additive genes which lead to the high merit of parents get split up and mixed between one generation and the next. However, comparison of cloned lines created from different parents would allow selection of those lines with the best combination of additive and non-additive genetic merit. Since the animals produced from the same cloned line have identical genotypes, the favourable genotype could be recreated exactly by further cloning. Similarly, cloned lines could be produced from crosses between breeds to exploit heterosis if this is important for the traits of interest. Clones with the optimum proportion of each parent breed could be produced, so minimising the problems seen with rotational crossing, which inevitably produces many animals with sub-optimal proportions of the different breeds involved.

While the benefits of cloning in genetic improvement may be limited, the potential of the technique to accelerate dissemination of genetic improvement to commercial herds or flocks is great. For this potential to be realised, reliable and cost-effective methods for cloning will be required. Also, improved and cost-effective techniques for delivery will be needed, including reliable methods for freezing cloned embryos and subsequent non-surgical transfer of these. Possible disadvantages include the risk of producing very large numbers of animals which subsequently turn out to be susceptible to disease, or to be unsuitable to new production methods. However, it should be possible to reduce such risks by maintaining reasonable genetic diversity in the conventionally bred elite populations, and producing several or many separate cloned lines from these.

Cloning may enable the production of animals which are more closely tailored to particular markets. Also, within each of these market types, cloning could help commercial farmers to produce a more uniform product, which is often a requirement of modern food retailers.

The current or potential future value of these reproductive technologies in genetic improvement programmes, and in the dissemination of genetic improvement, is summarised in Table 9.1.

Reproductive technology	Value in genetic improvement programmes	Value in dissemination of improvement to commercial tier
Artificial insemination	••••	••••
Multiple ovulation and embryo transfer	•••	•
In vitro production of embryos	•••	•••
Sexing of semen and embryos	••	•••
Cloning of embryos	••	••••

Table 9.1. The current or potential future value of reproductive technologies in genetic improvement programmes, and in the dissemination of genetic improvement. The technologies are rated on a scale from one to four dots, with one dot representing little or no value, and four dots representing high value.

Molecular genetic technologies

To date, most selection in livestock has been practised with little or no knowledge of what is happening at the DNA level. Selection has been on the effects of the genes, rather than directly on the genes themselves. Some traits are controlled by a single gene, and have a large visible effect, e.g. coat colour, polledness, double muscling. In these cases, there is a fairly close association between the gene and its effects, although non-additive gene action, such as dominance, can still cause difficulties in determining the true genotype of an animal. However, for most traits of economic importance, the performance of animals is affected by their genotype at many different loci. Also, in most cases, the effects of these genes are influenced by environmental effects on the traits of interest, so the link between particular genes and performance is obscured even further.

Despite these problems, very effective methods have been developed to identify animals with favourable predicted breeding values, as discussed in Chapter 5. These work by removing as many environmental influences as possible, and using clues from the performance of candidates for selection, and their relatives, to estimate the additive effect of all loci affecting the trait of interest. They are good at predicting the average genetic merit of offspring, but cannot distinguish between them until the offspring get performance records of their own, or performance records from their progeny. These techniques have been applied with considerable success in most livestock species, as outlined earlier in this book. However, some new molecular technologies offer the opportunity to improve on current methods. The aim in this section is to review briefly some of these technologies which are beginning to have an impact on the genetic improvement of livestock, or which may have such an impact in the future. The three related molecular technologies considered are genome mapping, the use of molecular markers in selection, and gene transfer.

Phenotypic markers of variation at the DNA level, such as variation in coat colour, the presence or absence of horns and variation in animal size and shape, have probably been used in livestock selection since domestication began. More recently, variation in blood groups and milk proteins have been used as markers of variation at the DNA level. However, it is only in the last decade or so that we have had the tools to measure this variation directly at the DNA level. These tools, known as **molecular genetic markers**, are basically alternative segments of DNA at a particular site on a chromosome which are identifiable by means of a laboratory test. Molecular genetic markers have the potential to refine animal selection programmes in some cases, and change them more dramatically in a few cases. They are also of value in checking the identity and parentage of animals, checking the origin of animal products, improving our understanding of evolution, and paving the way to a fuller understanding of the structure and function of genes, and their modification or transfer. It is quite possible to find an association between a particular molecular marker and an animal characteristic of interest, without knowing which chromosome, or which part of a chromosome, the marker is located on. However, the search for useful molecular markers is being guided increasingly by the burgeoning information coming from genome mapping projects. Hence, in this section genome mapping is discussed before outlining the use of molecular markers in selection. These subjects are followed by an outline of gene transfer. Each of these technologies depends on fairly recent advances in molecular and cell biology, so the most important of these advances are summarised briefly first of all. More details can be found in the texts listed at the end of the chapter.

Developments in molecular and cell biology

As outlined in Chapter 2, the existence and gross structure of chromosomes has been known since the early 1900s, and the chemical structure of DNA and its role in inheritance have been known since the 1950s. However, it is only relatively recent advances in molecular and cell biology which have allowed the identification and location of individual functional genes or other sequences of DNA on chromosomes, and provided the molecular tools to assist conventional selection procedures, and to allow gene transfer. Briefly, these advances include:

The discovery of restriction enzymes. Bacteria produce a large number of enzymes called restriction enzymes, which are capable of 'cutting up' sequences of DNA. These enzymes act like microscopic scalpels, to protect the bacteria by attacking viral or other foreign DNA [42, 52]. Different restriction enzymes recognise different sequences of four to six base pairs in DNA, and make their cuts whenever they come across these sequences of bases in a strand of DNA. For example, one commonly used restriction enzyme recognises the sequence GAATTC, and cuts the strand between the G and the adjacent A. (The letters are abbreviations for the four bases in DNA: adenine (A), thymine (T), guanine (G) and cytosine (C), as explained in Chapter 2). This property of restriction enzymes has been widely used in the study of mammalian DNA. Treating (or **digesting**) identical sequences of DNA with the same enzyme(s) produces identical sets of fragments, each of a specific size. Conversely, treating different sequences of DNA with the same restriction enzyme produces different sets of fragments of different sizes. Hence, these enzymes are useful in detecting similarities and differences in the DNA sequences of different animals (see Figure 9.1).

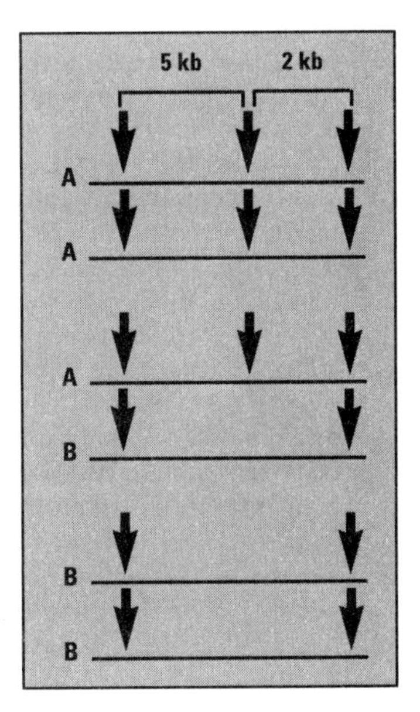

Figure 9.1 Differences in the DNA sequences of different animals can be detected using restriction enzymes. The diagram shows fragments of DNA from each member of a pair of chromosomes from three different animals. Each DNA fragment has either two or three restriction sites – sequences of 4 to 6 bases recognised by, and cut by, a particular restriction enzyme. Treatment of these fragments with the appropriate restriction enzyme will produce only DNA fragments 7 kilobases (kb) long when only two restriction sites are present, but will produce fragments 5 and 2 kb long when there are three restriction sites present. Hence, three genotypes are distinguishable. Animals with only two restriction sites present on the strand of DNA from each chromosome (genotype BB), produce only fragments which are 7 kb long. Those with three restriction sites present on both strands of DNA, produce fragments 5 and 2 kb long (genotype AA). Those with two restriction sites on one strand and three on the other produce fragments 7, 5 and 2 kb long (genotype AB) [61].

The development of techniques to allow DNA fragments of different lengths to be separated and detected. Fragments of DNA of different lengths (e.g. resulting from digestion with restriction enzymes) can be separated by applying a sample of the fragments of mixed sizes to one end of a thin film of gel and applying an electric current to the gel (this technique is called **electrophoresis**). The fragments then move across the gel at different rates, depending on their size. Large fragments move slowly, and small fragments move quickly. The gel can be stained or treated in a variety of ways to show up a series of bands, with each band representing DNA fragments of a particular size. When different samples (e.g. DNA from different animals) are applied side by side to the same gel, a series of bands can be detected in a column arising from each sample. By comparing the patterns of bands across columns, it is possible to tell whether or not different samples contain DNA fragments of a similar length (see Figure 9.2 and Plate 62).

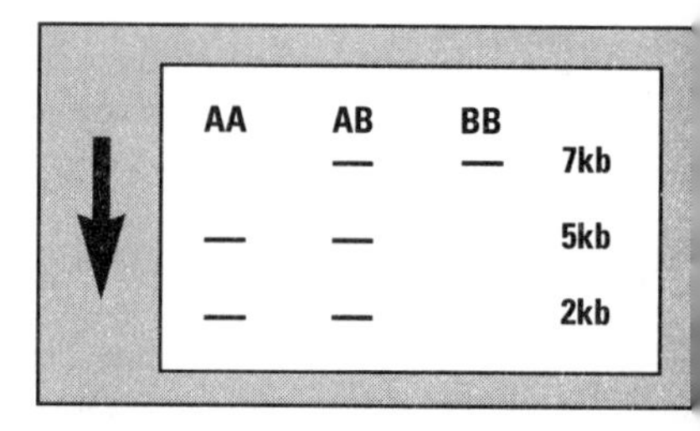

Figure 9.2 Differences in the length of DNA fragments can be detected by applying DNA samples from different animals to a film of gel and applying an electric current to the gel to separate fragments of different length. This is called gel electrophoresis – smaller fragments move further away from the point at which the original sample was applied to the gel. The gel is then stained or treated in some other way (e.g. using a DNA probe), to allow the positions of bands, which represent fragments of different size, to be detected. The diagram shows the pattern of bands expected if samples of DNA from the three genotypes shown in Figure 9.1 were treated with restriction enzymes, and then subjected to gel electrophoresis [61].

The development of techniques to multiply sequences of DNA rapidly. There are two methods for the rapid multiplication of DNA. One, known as **cloning DNA**, is a method of multiplying DNA in living cells. In this case the mammalian DNA fragments which are to be multiplied can be incorporated into the DNA from a **vector**, such as a plasmid – a small self-replicating circle of double-stranded DNA which exists independently in some bacterial cells. Incorporation of the foreign DNA is achieved by treating the vector DNA with the same (or a similar) restriction enzyme used to create the fragments of mammalian DNA. Essentially, this means that the mammalian DNA fits into the gap left in the vector DNA. These **recombinant DNA molecules** can then be introduced into a rapidly reproducing host (a bacterium in the case of plasmid vectors) where they are reproduced along with the host cell DNA [42, 52]. After many cycles of replication, many copies of the original DNA fragment can be extracted from the host cells. Similar techniques exist to allow multiplication of fragments of foreign DNA by incorporating them into yeast chromosomes.

The other method, which has become very widely used since its discovery in the mid-1980s, involves DNA multiplication in a test tube rather than in living cells. This is called the **polymerase chain reaction**, or **PCR** for short. Polymerase is an enzyme involved in the normal replication of DNA in the cell. However, sequences of DNA can be multiplied in the laboratory, by adding a heat-stable version of this enzyme, together with two **primers** (short fragments of DNA which start off the replication of a specific sequence of DNA), and a supply of the four DNA bases, A, T, G and C. Heating up this mixture denatures the DNA in the fragments of interest, causing the strands to separate. Subsequent cooling of the mixture causes the primers to 'home in' on complementary sequences of DNA, and to initiate the process of copying the sequences of interest on each of the two separated strands. Each cycle of heating and cooling produces a doubling of the number of copies of the original DNA sequence, so in a matter of hours it is possible to produce sufficiently large quantities of DNA for characterisation. Other important properties of PCR are that: (i) it multiplies *specific* sequences of DNA which depend on the primers used; (ii) it is most effective for relatively short DNA sequences and (iii) only very small amounts of DNA are needed initially, e.g. from hair follicles.

The development and automation of techniques to determine the sequence of bases on short strands of DNA. A number of techniques exist to enable the sequence of bases on a strand of DNA to be determined. These involve producing fragments of DNA which differ in size from each other by one base only. These are then ordered by size, and the identity of the base at the end of the fragment is determined [42, 52]. The process is still slow and costly, so it is only practical to use it for relatively short fragments of DNA.

The development of DNA probes and labelling techniques. DNA **probes** are single-stranded molecules of DNA of a specific base sequence. Like all strands of DNA, under the right circumstances these probes will bind to complementary sequences or fragments of DNA (as outlined in Chapter 2). Hence, they can be used to detect the presence of a complementary DNA strand in samples of DNA, e.g. obtained by 'blotting' from a gel in a manner equivalent to blotting ink with absorbent paper (see Plate 62), or in intact chromosomes. Probes can be 'labelled', which means that they can be made either radioactive or, more recently, fluorescent. This allows them to be readily detected later, when they have bound to a complementary DNA sequence. The use of sophisticated microscopes, cameras and computer-aided analysis systems allows rapid location of fluorescently labelled probes.

The development of markers of variation at the DNA level. Any identifiable segment of DNA in the genome (the entire genetic code of animals) which shows variation between animals can be used as a marker. Hence, markers may be all or part of a 'functional' gene, or they may be a part of the rest of the genome which does not code directly for the production of a protein. Several different types of marker have been developed, based on some of the methods described above, for detecting variation at the DNA level. These markers are of value in genome mapping, marker assisted selection, parentage verification and product identification. The two types of marker currently of most value in livestock genome mapping, and likely to be of value in marker assisted selection, are outlined briefly below. However, new marker techniques are being developed rapidly. See [42] and the genetics journals listed in the Further Reading section for details of some other techniques of potential value.

As mentioned earlier, there is variation between animals in the number and position of sites at which a given restriction enzyme will cut the DNA. This variation is termed **restriction fragment length polymorphism (RFLP)**. (Polymorphism literally means 'different forms'.) RFLPs are often used now to detect variation in the DNA sequence of different animals by: (i) using PCR to multiply a fragment of DNA containing restriction sites from the animals of interest; (ii) digesting the separate multiplied samples with one or more restriction enzymes and (iii) separating the digested fragments from each animal in adjacent columns of a gel (the fragments of different size move to different positions on the gel) and (iv) distinguishing different genotypes from the size of DNA fragments produced, by staining or use of a labelled probe as described above (see Figures 9.1, 9.2 and 9.4 and Plate 62). RFLPs are often located in functional genes, which increases their value as potential markers. However, the major limitation to the use of this type of marker is that in many cases a given RFLP only involves two different variants or alleles, so animals can be assigned to only a small number of classes or genotypes [2]. (The terms allele and locus are still used for markers even though the marker may not be a functional gene.)

Throughout the genome there are many regions where the same sequence of base pairs is repeated several times, end to end. These repeat sequences are termed **minisatellites** or **microsatellites**, depending on the size of the sequence of base pairs which gets repeated. The repeated sequences in minisatellites are 10 to 60 base pairs long, while those in microsatellites are only 2 to 5 base pairs long [2, 42]. The number of times a given sequence of bases is repeated varies between animals. For example, at a particular point on equivalent chromosomes from different animals the two bases AC may be repeated anywhere between 16 and 22 times (a microsatellite). Because of this variation, these repeat sequences can be used as markers (see Figure 9.3).

The main advantage of microsatellite markers is the large number of variants or alleles present at a particular site or locus, which means that animals can be assigned to many different classes or genotypes (see Plate 63). Hence, at the time of writing, microsatellite markers are the markers of choice in livestock genome mapping and they are likely to be of most value in marker-assisted selection. Detecting variation at a microsatellite locus involves: (i) the use of restriction enzymes to remove intact the repeated sequences of bases from samples of DNA

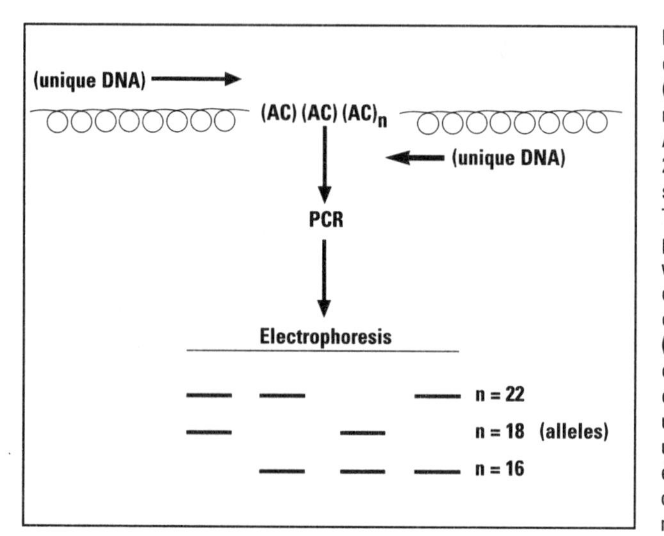

Figure 9.3 Diagram illustrating the use of variation in the number of repeats (n) at a microsatellite locus as a marker. In this example the base pair AC may be repeated either 16, 18 or 22 times (only three AC repeats are shown in the diagram, to save space). That is, there are three alleles at this particular locus; individual animals will have either two copies of any one of these alleles (homozygotes), or a copy of two different alleles (heterozygotes). Fragments of DNA containing the microsatellite locus can be removed from a DNA sample using restriction enzymes, multiplied using PCR, and then separated by electrophoresis to reveal the genotype of individual animals at a particular microsatellite locus [61].

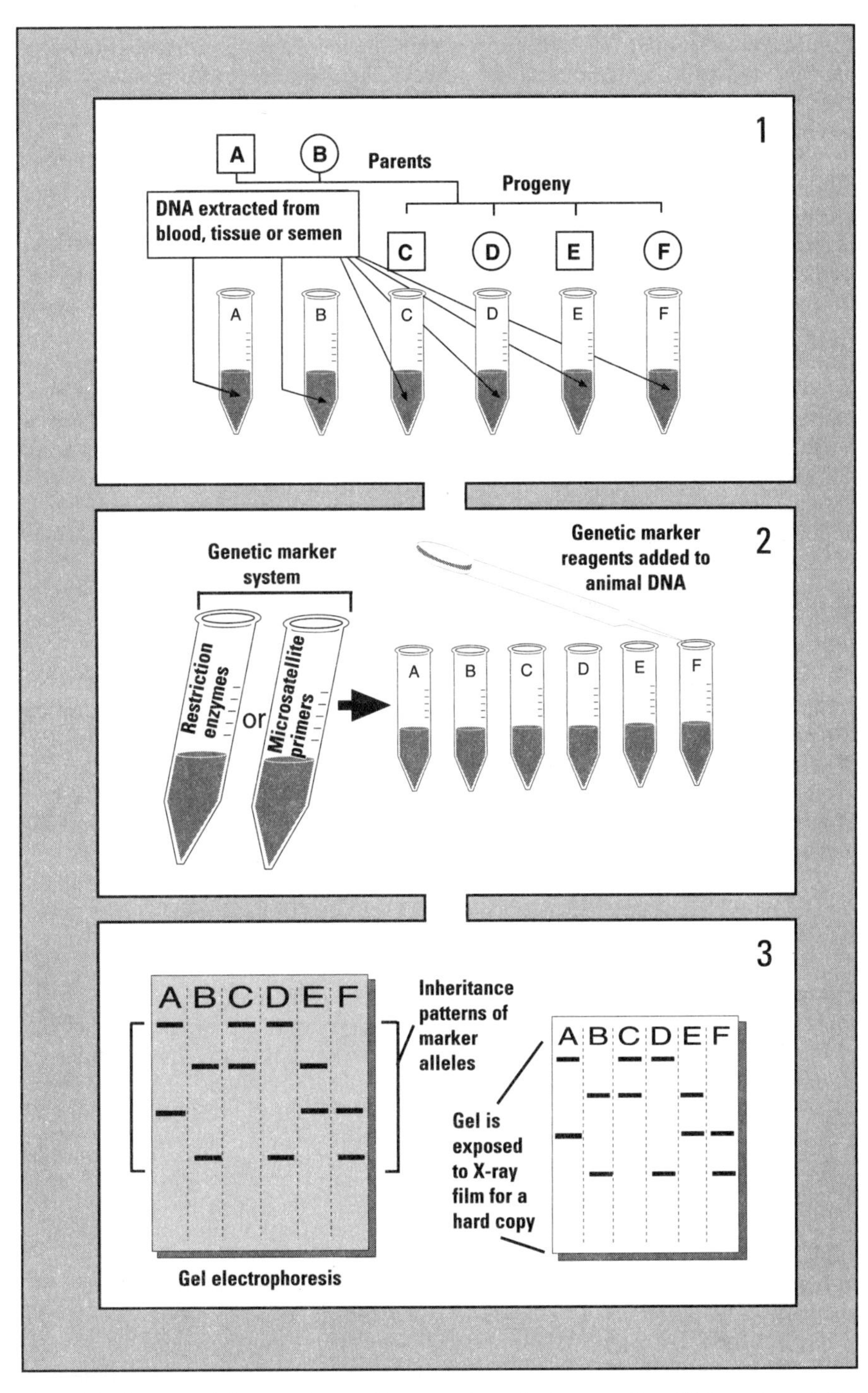

Figure 9.4 Summary of the steps involved in genotyping animals using either RFLP or microsatellite markers [47].

from different animals; (ii) PCR to multiply these sequences from the different samples and (iii) separation and detection of the different sized fragments from each of the samples in a similar way to that described earlier (see Figures 9.3 and 9.4). These procedures are becoming increasingly automated.

Minisatellites are useful in checking the parentage of animals, since each animal produces a virtually unique pattern of bands, known as a **DNA fingerprint**, after application of some of the techniques outlined above. They are less useful as markers in mapping or selection because fragments of DNA are produced from different loci, so it is not usually possible to detect variation among alleles at single loci.

Some important features of RFLP and microsatellite markers which make them particularly useful are that: (i) they are codominant – that is, each of the homozygous and heterozygous genotypes which are found at a marker locus can be distinguished by the laboratory test, unlike the situation with some other markers (including minisatellites) where heterozygotes are indistinguishable from one of the homozygotes; (ii) because they are DNA-based they can be applied to animals of both sexes; (iii) because they are DNA-based they can be applied to animals at any age and (iv) those which are based on PCR need only small initial amounts of DNA from the animals of interest.

Genome mapping

Chromosomes are made up of long double strands of DNA. The informative or coding parts of the DNA are the paired bases adenine (A), thymine (T), guanine (G) and cytosine (C). Functional genes are sequences of these bases which form the codes for the production of proteins. However, functional genes are thought to account for only about ten per cent of the total DNA (the whole genome). The rest of the DNA is either involved in controlling when and in which tissues particular genes are switched on, or its function is unknown at present.

The ultimate purpose of a **genome** (or **gene**) **map** is to describe the location of functional genes, or other sequences of DNA, on each of the chromosomes of the species concerned. The enormity of this task will be easier to appreciate from some statistics on the genomes of mammals. The genomes of mammals are estimated to be composed of 3,000 million base pairs, and to contain at least 100,000 genes, averaging about 3,000 base pairs each. If the complete sequence of base pairs for one of these mammalian species was known and was written in a series of books, it would fill 200 volumes, each of 1,000 pages. Each volume would be bigger than the telephone directory of most large cities and it would be less interesting, as it would consist only of sequences of the four letters A, T, G and C. For comparison, the genomes of fruit flies, yeast and the bacterium *E. coli* would take up ten books, one book and 300 pages respectively [52]. Despite the enormity of the task of mapping the whole genome of any mammalian species, a great deal of research activity is already underway. Large regions of the genomes of humans and the mouse, in its role as a model for humans and other mammals, and parts of the genomes of each of the main domestic livestock species have been mapped already. The high degree of international collaboration between laboratories, to avoid duplication of effort, is greatly speeding progress in mapping the genomes of each of these species.

So what use is a genome map anyway? The main driving force for the production of the human genome map is that it enables the investigation, and potentially the treatment, of a wide range of human diseases with a genetic component. Also, comparison of the genome maps produced for different species is providing a powerful tool for the study of evolution. Similarly, genome maps can be used to make more informed decisions about which breeds, strains or individuals should receive highest priority in conservation of genetic resources. In farm livestock and crop species, genome maps can be useful in making informed choices of markers for economically important traits to accelerate conventional selection programmes, and for locating genes of potential interest for gene transfer. These applications are discussed in more detail in later sections.

Types of genome map

A genome map describes the order of genes, or other sequences of DNA, and the spacing between them, on each chromosome. There are different types of map which have different scales, or levels of **resolution**. Just as with road maps, genome maps with different levels of resolution are useful for different purposes. For instance, if you want to drive from Edinburgh to London, a low resolution route planner map is far more useful for planning the journey at a glance than a high resolution street map. Once you get to London, the street map is far more useful than the route planner. The ultimate level of resolution for a genome map is the full DNA sequence of the genome. Producing a map with this level of resolution is the eventual aim of the Human Genome Project but, at the time of writing, this is still a few years away from being achieved. Maps with this level of resolution over the whole of the genome may never be affordable, or necessary, in farm livestock. However, such high resolution mapping may be useful in particular regions of interest in the genomes of farm animals. Also, the availability of high resolution maps from the human and mouse genome mapping projects is often helpful in livestock genome mapping projects, as explained below. Maps fall into two categories: **linkage** or **genetic maps**, and **physical maps**. These two kinds of map, which are described below, are highly complementary.

Linkage maps

A linkage map shows the **recombinational** or **genetic distance** between pairs of markers on the same chromosome. This is achieved by measuring the frequency with which the markers separate from one generation to the next, as a result of recombination (the crossing over of equivalent segments between the paternally- and maternally-derived members of a pair of chromosomes; see Chapter 2 for details). The recombinational or genetic distance between markers is measured in units called centimorgans (cM). Two markers are said to be one centimorgan apart if they are separated as a result of recombination one per cent of the time. In general, the more closely linked any two markers are on a chromosome, the less frequently they will be separated as a result of recombination. However, there are parts of chromosomes where recombination occurs more frequently (hot spots), and parts where it occurs less frequently (cold spots), so the association between recombination rate, or genetic distance, and physical distance is not perfect. (Also,

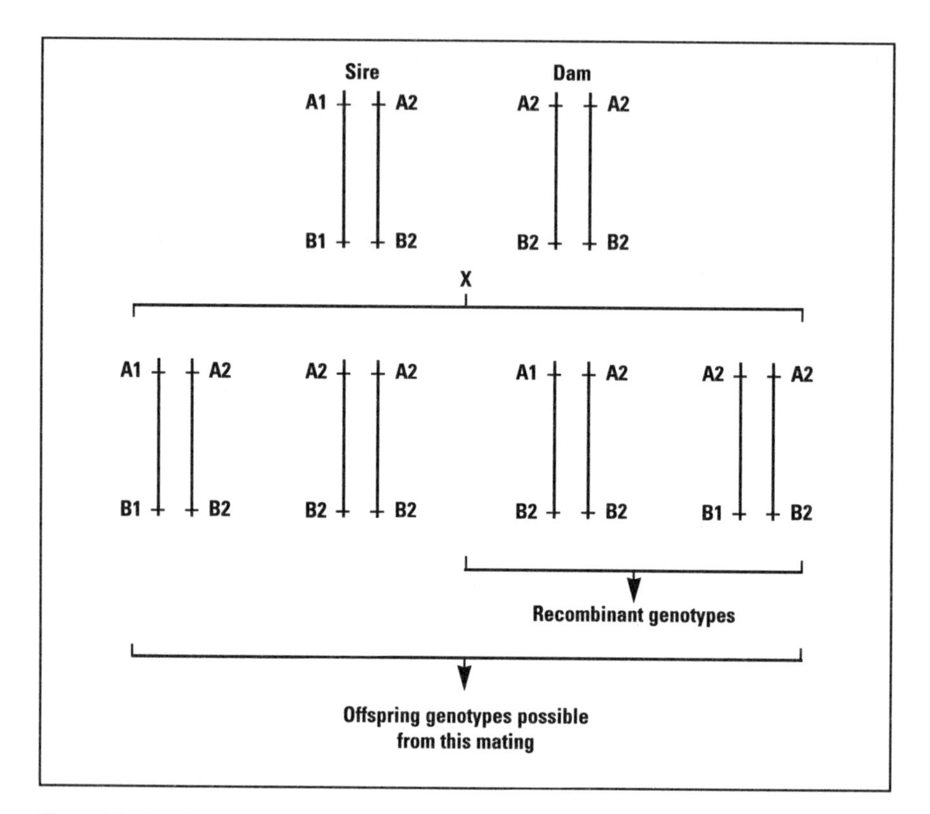

Figure 9.5 A schematic diagram illustrating measurement of the recombination rate between two marker loci, A and B. In this example, the sire is heterozygous at both marker loci, and he is mated to a dam who is homozygous at both marker loci. Four genotypes of offspring are possible from this particular mating. Two of these offspring genotypes (A1A2B2B2 and A2A2B1B2) are only possible as a result of recombination between the two marker loci during sperm production in the sire. The recombination rate is the frequency with which these recombinant offspring genotypes are produced. (Detecting offspring genotypes which are the result of recombination, and hence measuring the recombination rate, is only possible for matings between certain parental genotypes.) The genetic distance between markers is measured in units called centimorgans (cM). Two markers are said to be one centimorgan apart if they are separated as a result of recombination 1% of the time. In general, the more closely linked any two markers are on a chromosome, the less frequently they will be separated as a result of recombination [52].

the relationship between map distance and recombination rate only holds over relatively short distances. Over longer distances, double recombination occurs. That is, a segment of chromosome containing one of the markers crosses over twice, so that the original association between markers is restored, and the fact that recombination has occurred is undetected. There are formulae which adjust for this and more accurately reflect the relationship between longer map distances and recombination rate.) In mammals, 1 centimorgan is roughly equivalent to 1 million base pairs. In other words, markers which separate one per cent of the time are probably about 1 million base pairs apart, those which separate two per cent of the time are probably about 2 million base pairs apart, and so on. This is illustrated in Figure 9.5.

To create a linkage map, it is necessary to have:

- **Variation (polymorphism) at a series of marker loci among animals in the population concerned.**

- **A test to detect the presence of each of the alternative marker genotypes at these loci**. In the past, recombination rates could only be measured between genes which had easily measurable effects in the animals, e.g. blood groups, enzyme types, coat colour, feathering patterns. (The first linkage maps for poultry were produced about 60 years ago, based on the large number of mutations causing variation in colour, feathering and other physical features, which had been identified by poultry enthusiasts, breeders and researchers [1].) However, new molecular techniques mean that any measurable sequence of DNA can act as a marker, and recombination rates between any two of these can be estimated. As mentioned earlier, microsatellite markers show the highest degree of variability or polymorphism, and have become widely used in livestock genome mapping.

- **DNA samples from two or more generations of related animals (or appropriate samples for other types of analysis, e.g. blood group or enzyme analysis)**. These are required in order to track the transmission of markers from one generation to the next. Small quantities of DNA can be obtained from blood, other body tissues, hair roots, milk or a mouth swab [26]. Often, crosses between divergent breeds are used to create populations for genome mapping. For instance the European Pig Gene Mapping Project uses crosses between Chinese breeds such as the Meishan and European breeds like the Large White, and crosses between wild boar and current commercial breeds. In poultry, crosses between Red Jungle Fowl and commercial strains or crosses between broilers and layers have been used. Using divergent crosses increases the probability of detecting genetic variation at any potential marker site. That is, breeds which diverged a long time ago, or have been selected for different traits, are more likely to have different alleles at any particular locus than breeds or lines which are closely related or have a similar selection history. Hence, first cross (F1) animals from matings between divergent breeds or lines are more likely to be heterozygous at any marker locus than crosses between similar lines. When the F1s are mated to each other recombination occurs, which allows genetic distance to be estimated between markers.

Physical maps

Linkage maps provide information on the genetic distance between linked genes or markers, but not necessarily on their exact location. Even the chromosome to which particular linkage groups belong was unknown in many cases until fairly recently. In contrast, the aim of physical maps is to locate genes or markers on

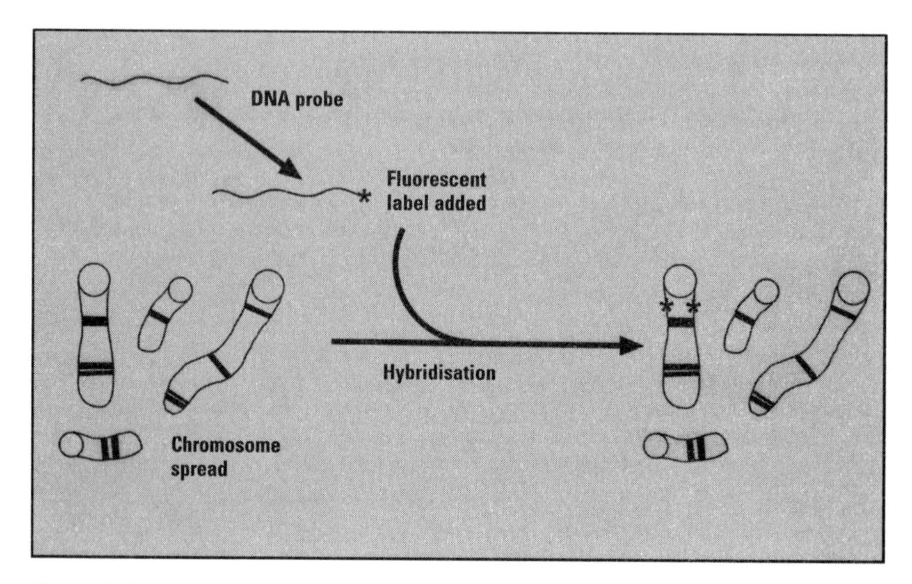

Figure 9.6 Diagram illustrating the use of fluorescent *in situ* hybridisation (FISH) to detect the position of a gene on a chromosome. First a DNA probe for the gene of interest is prepared and fluorescently labeled. This is then allowed to hybridise with a preparation of chromosomes on a slide. The location of the fluorescently labelled probe, and hence the location of the gene of interest, can then be detected using sophisticated microscopes, cameras and computer-aided analysis systems [61].

individual chromosomes, or specific regions of these.

The lowest resolution physical map is called a **chromosomal** or **cytogenetic map**. This is based on the distinctive banding patterns seen when stained chromosomes are observed under a light microscope. The light and dark bands occur as a result of differences in the amounts of the DNA bases adenine/thymine and guanine/cytosine in different regions of the chromosome. On each chromosome, these bands have been given internationally agreed numbers for ease of reference (see Figure 9.8).

Assigning genes or markers to particular chromosomes depends on the ability to separate chromosomes. Two main methods are used. The first involves sorting chromosomes according to their size when they flow singly past a laser beam and collecting them in separate tubes (the method is similar to that described earlier for separating sperm bearing X- and Y-chromosomes [52]). This works well for species in which the chromosomes are quite different in size (e.g. humans, pigs). Another method, which has been used in cattle and sheep genome mapping, is the use of **somatic hybrid cell lines**. This involves fusing cells from the animal of interest with rodent immortalised cell lines. These are cell lines which originated in tumours in rodents. They are used widely in research in cellular and molecular biology, because of their ability to grow and replicate rapidly during culture in the laboratory. The resulting hybrid cells contain chromosomes from both the rodent species concerned and the species being mapped, for example, cattle. During successive rounds of replication, the chromosomes from the species of interest are progressively eliminated from the hybrid cells, and eventually some of the hybrid cells contain only a few, one, or parts of one chromosome from the species of

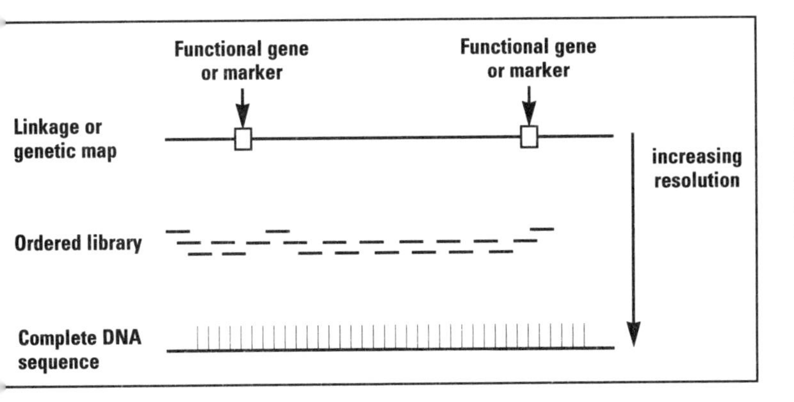

Figure 9.7 Some of the different types of map used in human and livestock genome mapping, and their levels of resolution [4].

interest. Cells containing different chromosomes can then be separated and propagated for further analysis.

Once a culture of cells containing a known chromosome has been established, the presence or absence of particular genes can be established by looking for the gene product (e.g. a species-specific protein) in cultures of cells. When different species-specific proteins are detected, this indicates that the genes responsible for these products all occur on the same chromosome. Alternatively, and more commonly today, the presence of particular DNA sequences in different hybrid cell cultures can be detected using labelled DNA probes.

Labelled probes are also used in **fluorescence *in situ* hybridisation (FISH)**, which is one of the most precise techniques for determining the physical location of a gene [2]. With this method the labelled probe is allowed to **hybridise** with chromosomes from the animal of interest which have been prepared on a slide (i.e. the probe binds to its complementary DNA sequence on the chromosome). The particular chromosome to which the fluorescently labelled probe has bound can then be identified (see Figure 9.6 and Plate 64). Simultaneous use of different probes with different labels also allows the order of genes to be established. Fluorescence *in situ* hybridisation has been used in both cattle and sheep genome mapping, as well as in other livestock, human and mouse genome mapping.

The techniques outlined above are those of most importance in livestock genome mapping. While the aims of livestock genome mapping are far more modest than those of human genome mapping, some of the more sophisticated techniques employed in human genome mapping are being used to a limited extent in livestock. These include the use of mapping methods which map only expressed genes (i.e. coding sequences of DNA), rather than mapping expressed genes and non-coding sequences of DNA indiscriminately. Also, the creation and cloning of overlapping fragments of DNA in yeast or bacteria (**yeast artificial chromosomes** (**YACs**) or **bacterial artificial chromosomes** (**BACs**)) is being used in livestock genome mapping. Sequencing fragments of DNA maintained as YACs or BACs produces a permanent **library** of fragments which can be recognised, but whose order on the chromosome is still unknown. However, the overlapping parts of many fragments can be identified from this library, allowing them to be ordered as they occur on the chromosome. Producing maps with this level of resolution is very laborious, so only small parts of the genomes of farm livestock are being mapped in this way at the moment. However, the approach is useful in find-

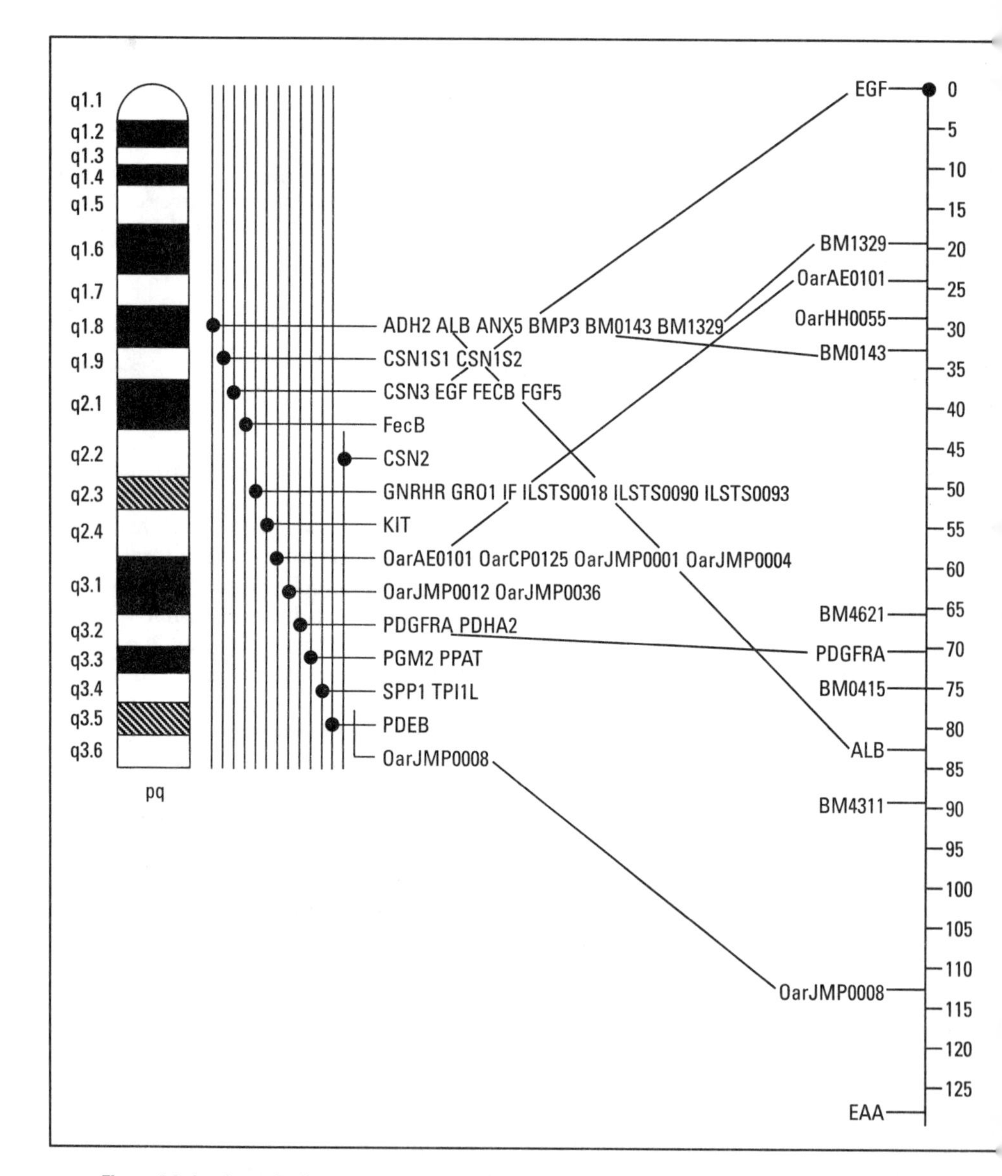

Figure 9.8 A schematic diagram showing some of the genes or markers present on sheep chromosome number 6. The letters and numbers to the left of the chromosome in this diagram are the agreed reference numbers for the particular bands seen on chromosome 6 after staining. The abbreviations in the middle of the diagram refer to physically mapped genes or markers (e.g. FeCB is the abbreviation for the Booroola fecundity gene). The labels for these genes are attached to vertical lines which indicate the likely location of these genes or markers – in most cases in this example the vertical lines span the whole chromosome, indicating that the genes are known to be on chromosome 6, but their exact location was unknown at the time the map was drawn. The 'scale' on the right hand side of the diagram is a linkage map showing the genetic distances (in centimorgans) between a group of genes or markers which are now known to occur on chromosome 6 (there are other linkage maps available for this chromosome which are not shown for reasons of clarity). (Redrawn from a map produced from the SheepBase World Wide Web site using Anubis mapping software.)

ing the sequence of bases in specific regions, for example between two markers which are flanking a gene of interest. Figure 9.7 illustrates some of the different types of map used and their levels of resolution.

Creating a consensus map

As indicated earlier, the various mapping techniques have different levels of resolution. One of the main challenges for groups involved in mapping is to collate information produced by these different mapping techniques to produce an agreed or **consensus map**. This is simplest when collaborators have been mapping the genomes of the same animals from so-called **reference families**. However, information from other groups working on the same species can be incorporated into maps as well. One approach used in producing consensus maps is to locate on the physical map a few key RFLP or microsatellite marker loci from different linkage groups. Once the location of these key markers is established, then the location of other linked markers can be inferred [5]. (By analogy, if you were presented with four different maps of Britain, all showing the same two major cities, but different minor ones, on different scales, some drawn roughly and others with care, it would be possible to produce a more complete single map by placing all minor cities relative to the two major ones common to all maps.)

From the maps already produced there appears to be a surprisingly high degree of similarity or **homology** between parts of the genome of different mammalian species. DNA sequences which appear to have remained the same in different species through evolution are said to be **conserved**. Although the same conserved sequences are often located on equivalent chromosomes in different species, their order within the chromosome is not always the same, so information from one species still needs to be used cautiously in another species. But, as a result of the similarities, progress in mapping the genome in one species can be of direct benefit to mapping in other species. In particular, efforts in mapping livestock species have benefited greatly from genome mapping work in the human and the mouse.

Status of livestock genome maps

There have been international collaborative groups and some independent groups working on genome mapping in all major livestock species for several years now. As a result, linkage maps have been published for each of the livestock species, and large numbers of loci have been placed on physical maps. At the time of writing, linkage maps for cattle and sheep are available with several hundred markers and several hundred physically mapped sites, but these numbers are growing almost daily. Several genes with important effects have been located, including those responsible for the genetic diseases BLAD (bovine leukocyte adhesion deficiency) and DUMPS (deficiency of uridine monophosphate synthase) in cattle [61], and that controlling resistance to scrapie in sheep [28]. Also, markers have been detected which are closely linked to the genes causing Weaver syndrome, polledness and double muscling in cattle, and the Booroola gene in sheep [1]. Figure 9.8 shows the location of some of the genes or markers present on sheep chromosome number 6.

One of the major challenges of genome mapping in any species is the storage, retrieval and use of the vast quantity of data produced. Special databases have been developed for the mapping activities in each of the major livestock species, often in collaboration with those responsible for the human genome database. Some of these databases are for the use of the people working on individual projects, but there are also public domain databases which allow anyone interested to check on the current status of maps. For those with access to the Internet, up-to-date information on sheep and cattle genome maps can be obtained from the following sites:

http://dirk.invermay.cri.nz (SheepBase sheep gene map database)

http://locus.jouy.inra.fr/cgi-bin/bovmap/intro.pl (BovMAP cattle database; other cattle databases can be accessed from this site)

These sites also contain links to databases for other species. Alternatively, databases for all livestock species can be accessed from the Roslin Institute site in the UK:

http://www.ri.bbsrc.ac.uk

or the National Agricultural Library site in the US:

http://probe.nalusda.gov

Use of molecular markers in selection

Two main applications of molecular markers are considered in this section. The first is the use of markers to assist in the introduction of a single gene with a favourable effect from one population to another. For example, there may be interest in introducing a single gene controlling polledness or resistance to a disease to a different breed, cross, or family. The usual route for introduction of a gene of this sort is to cross the current preferred breed to a breed carrying the gene of interest, and then backcross to the original preferred breed for several generations. Molecular markers may be useful in identifying animals carrying the favoured allele, and so help speed up the recovery of the original genotype. This type of application is called **marker-assisted introgression (MAI)**. The second application is the use of markers to accelerate selection for a particular trait within a population. In this case the aim of using markers is simply to boost the improvement made by conventional means, by homing in on genes already present in the breed concerned. This use of markers is called **marker-assisted selection (MAS)**.

In both marker-assisted introgression and marker-assisted selection the gene of interest may be either a single gene with a large effect (major gene), such as those affecting coat colour, polledness and double muscling, or it may be one of many genes affecting a quantitative trait of interest, such as milk yield and quality, growth rate or wool production. These loci which affect quantitative traits are termed **quantitative-trait loci (QTL)**, or **economic-trait loci (ETL)**. In populations which have had effective improvement programmes for many generations it is most likely that marker assisted selection will be for QTLs rather than major genes, since major genes with large favourable effects are likely to have been fixed in this population already.

In addition to these two applications, molecular markers can provide a more definitive test of ancestry. For instance, the probability of excluding an incorrect parentage can increase from one in a few thousand using twelve traditional bloodtype markers, to one in several million by using between five and ten DNA-based markers [26]. This makes the techniques very powerful for pedigree verification in livestock (and of great value in forensic science). If the techniques become cheap enough, they could also prove valuable as an alternative to pedigree recording at birth in difficult or very extensive management systems (e.g. animals kept under range conditions). Similarly, the techniques could be used to confirm the breed of animals or their products. This may prove popular if trends towards branding animal products on their breed of origin continue (e.g. several major retailers in the UK sell beef specifically labelled from Aberdeen Angus crosses, or traditional beef breeds, lamb from traditional hill breeds and milk from specified breeds including the Channel Island and Ayrshire breeds).

Molecular markers could also be useful in ensuring that breeds or individuals carrying particularly rare or useful genes are included in conservation programmes. A related application of the techniques would be in the control of inbreeding in selection or conservation programmes. Inbreeding coefficients estimate the average level of inbreeding of offspring from a particular mating. In fact there will be some variation around this average because, as a result of chance, some offspring receive more and some receive fewer genes than average from the common ancestor. Molecular markers could be used to identify those offspring that receive fewer genes than average from the common ancestor, and so are less inbred than average for that group of sibs [55].

The choice of markers and methods of establishing a link between a marker and a trait of interest are outlined next, before giving more detailed examples of the use of markers in MAI and MAS.

Choice of markers

The ideal marker for both MAI and MAS would be the gene responsible for the trait of interest, or a shorter sequence of DNA within this gene. For example, the ideal marker for selecting for polledness would be the gene responsible for polledness itself. This means that selection on the marker always delivers the expected genotype. However, most markers are not the genes of interest themselves, they just happen to be located in the same region of the chromosome as the gene of interest. In these cases it is important to recognise that the marker and the gene of interest will sometimes become separated from each other from one generation to the next because of recombination. The more closely linked the marker gene and the gene of interest, the less often this will happen, and the more useful the marker is. Hence, it is very important to establish the nature of the association between the marker and the gene of interest. For MAI of a gene between two very different breeds it should be relatively easy to find a close

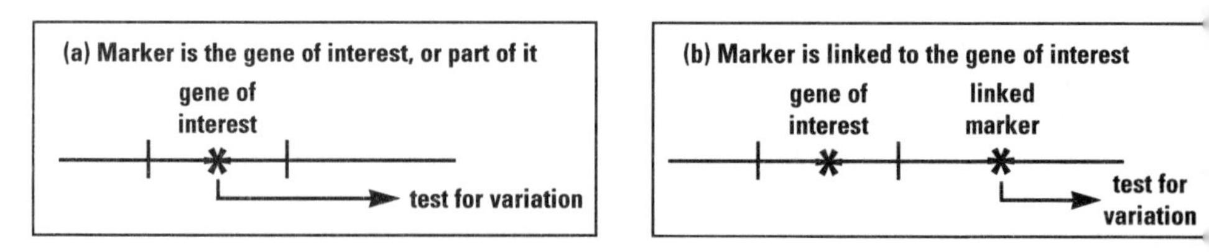

Figure 9.9 Two possible uses of markers. (a) In the first case a test is available for the gene of interest directly (i.e. the marker is the gene responsible for the trait of interest, or it is a shorter sequence of DNA within this gene). (b) In the second case the marker is not the gene of interest, but it is closely linked to this. In the initial generations of crossing between divergent breeds or lines it should be possible to find markers in which one marker genotype is always associated with favourable performance, since the two breeds are heterozygous at many loci. However, following many generations of selection in offspring from these initial crosses, this linkage will break down because of recombination. The more closely linked the marker and the gene of interest, the longer it will take for this association to break down [61].

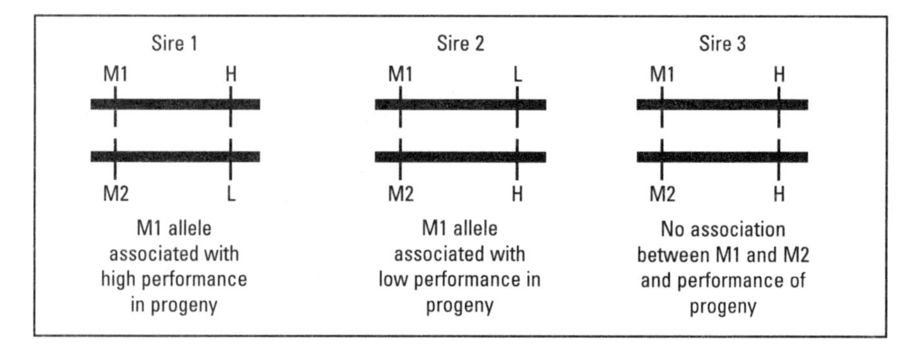

Figure 9.10 Marker and QTL genotypes for three sires within a breed, and the expected association between marker genotype and performance in progeny of these three sires. One marker allele may be associated with favourable performance in some families, and unfavourable performance in others. This arises because the two alleles have been segregating in the population for many generations, so there have been many opportunities for recombination. Hence for MAS within a breed it is important that markers are identified and used within families [24].

marker which will remain associated with the trait of interest for long enough to be useful, because the two breeds are likely to have different genotypes at many loci. This situation is illustrated in Figure 9.9. However, within a population the same marker allele may be associated with favourable performance in one family and unfavourable performance in another. This arises because the two genes have been segregating in the population for many generations, so there have been many opportunities for recombination. Hence, for MAS it is important that markers are identified and used within families. This situation is illustrated in Figure 9.10.

There are three main approaches to choosing markers, in order to investigate whether or not they are linked to traits of interest:

- **The 'shotgun' approach**. This involves choosing a set of markers at random, and checking for associations with animal performance. This approach is becoming less common as information on the location of markers on genome maps improves.

- **The map-based approach**. This involves choice of markers which are evenly spaced over the genome, to maximise the chances of having at least one marker reasonably close to the gene of interest. Much of the mapping activity in livestock has been aimed at getting adequate coverage of the genome to allow this method of identifying markers for QTLs. All livestock genome maps are now of sufficiently high resolution to show markers at 10–20 cM intervals over most of the genome (this represents about 125 to 250 evenly spaced markers in total). This is thought to be sufficient to allow detection of linkage with QTLs which have a reasonably large effect. The chances of detecting associations between markers and QTLs are increased by looking for associations with pairs or groups of markers. Similarly, the risks of wasting selection effort in MAS because the association between a marker and the trait of interest has broken down can be reduced by using two or more markers (**flanking markers**), on either side of the gene for the trait of interest [25, 34].

- **The candidate gene approach**. In this case a known mapped gene may be chosen because it has a similar function to the gene of interest and may be linked to it. For example, a mapped gene with a major effect on disease resistance in one species would be an ideal candidate as a marker for a similar disease in another closely related species. Similarly, within a species, a candidate gene may be chosen because it has a known function linked to the trait of interest (e.g. a gene coding for a hormone affecting milk yield may be investigated as a marker for yield).

Establishing a link between a marker and a trait of interest

To establish an association between a marker and the trait of interest, DNA samples are needed from at least two generations of animals – parents and offspring – and performance records are needed on offspring. The DNA samples are used to track the movement of the markers from parents to offspring and to look for associations between particular marker genotypes and performance in the offspring. This is illustrated in Figure 9.11. In this example two breeds of cattle of different body size are crossed. There is a hypothetical marker linked to a QTL affecting body size. The large breed is homozygous for one marker genotype, and the small breed is homozygous for the other marker genotype. Animals in the F1 cross are of intermediate body size, and have one copy each of the two marker alleles (i.e. they are heterozygous for the marker). Until now there was no way of detecting any linkage between size and the marker. However, in the F2 generation there is segregation of the marker and body size, so the association between them can be detected.

Resource populations to detect useful marker associations have been set up in many countries. These usually involve crosses between divergent breeds or lines, as with the populations used in mapping, to increase the efficiency of detecting useful markers. For example, a cattle resource population has been established at the Roslin Institute in Edinburgh from crosses between the Charolais and the Holstein Friesian breeds to look for useful markers of growth, carcass, milk

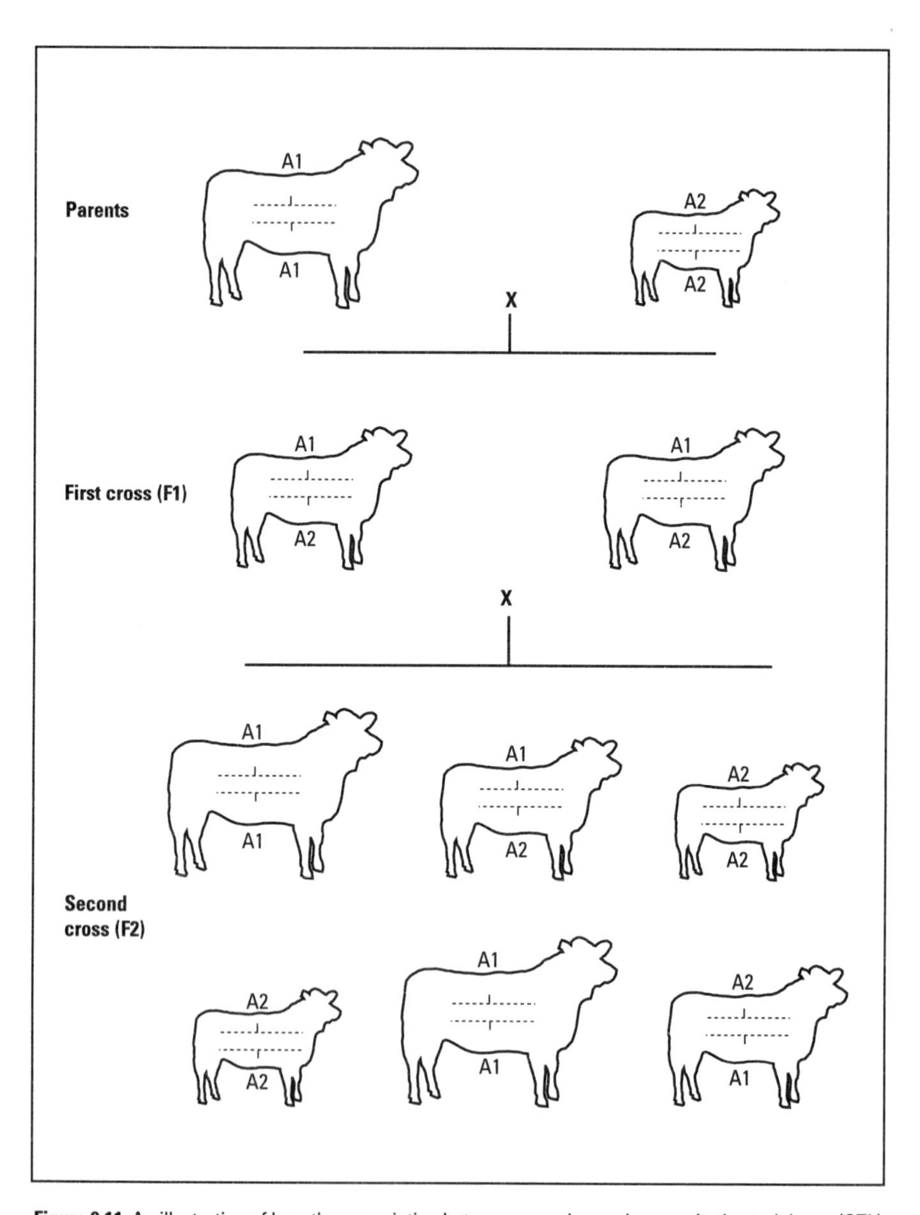

Figure 9.11 An illustration of how the association between a marker and a quantitative trait locus (QTL) can be detected. In this hypothetical example two breeds of cattle of different body size are crossed. There is a marker linked to a QTL affecting body size. The large breed is homozygous for one marker genotype (A1A1), and the small breed is homozygous for the other marker genotype (A2A2). Animals in the F1 cross are of intermediate body size, and have one copy each of the two marker alleles (A1A2). At this stage there is no way of detecting any linkage between body size and the marker – the breeds may differ in marker genotype simply by chance, and it may not be linked to body size. However, in the F2 generation there is segregation of both the marker and the QTL affecting body size, so the association between them becomes apparent. On average, animals of A1A1 marker genotype are larger than those of A1A2, which are larger than those of A2A2 genotype. So the QTL affecting body size is linked to the marker [26].

production and other characteristics. In New Zealand and Australia, crosses have been made between experimental sheep selection lines, originally selected for divergent growth, leanness, wool production or parasite resistance, to help in the search for useful markers. Once potential markers are found, their value will still need to be confirmed in the commercial populations in which they are to be used, but it is more efficient to concentrate the initial search on wider crosses. This may also detect genes segregating in one population which would be of value if introgressed to current commercial populations, e.g. genes affecting disease resistance.

In dairy cattle populations marker relationships are being detected in what are called **granddaughter designs**. These involve the use of bulls (the grandsires) which are heterozygous for the marker gene. Hence, the marker is segregating

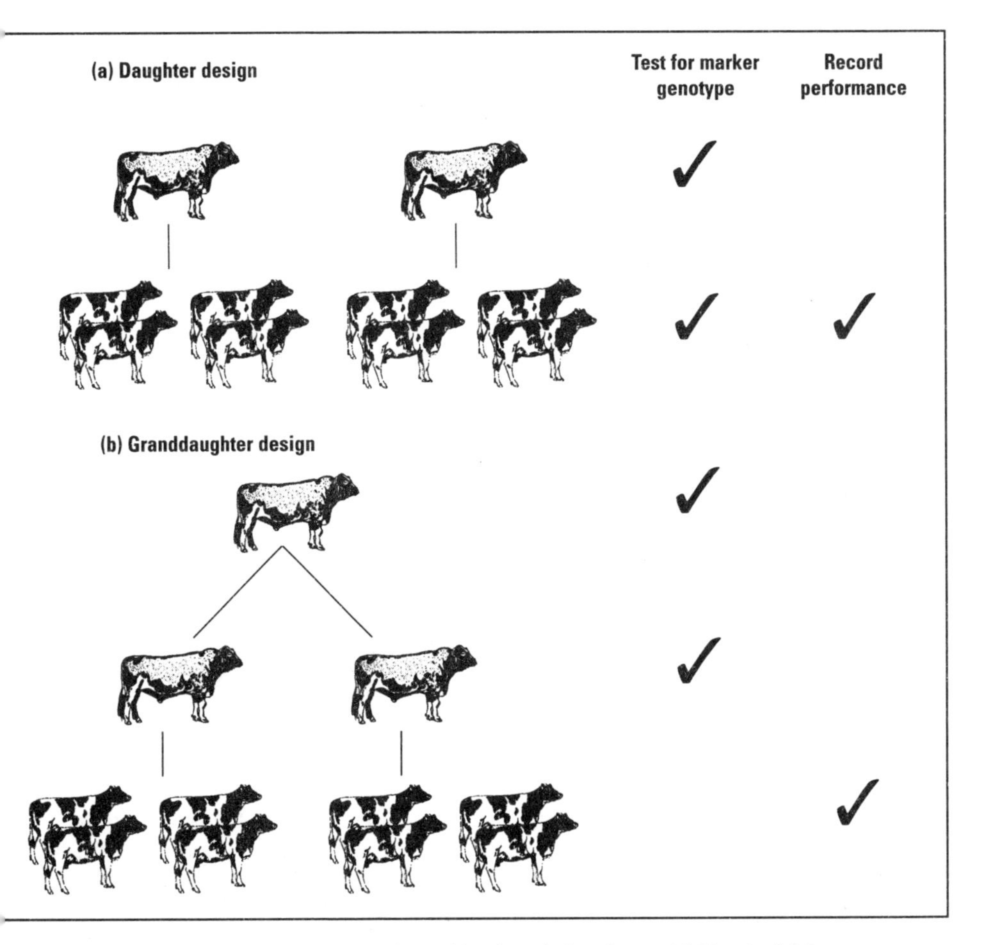

Figure 9.12 A comparison of daughter and granddaughter designs for establishing the link between a marker and a trait of interest in dairy cattle (or other animals where the trait of interest is sex limited). (a) In daughter designs marker genotypes are determined for sires and daughters, and performance records are obtained from daughters. (b) In granddaughter designs marker genotypes are determined for grandsires and sires, and genetic merit of sires is predicted from their daughters' performance records.

among their sons. That is, some sons get a copy of the marker allele linked to the favourable QTL effect, and others get a copy of the other marker allele linked to the unfavourable QTL effect. Since many of the traits of interest (e.g. milk, fat and protein production) are only expressed in females, the genetic merit of sons must be determined from their daughters' performance. These cows are granddaughters of the original sires, hence the name of the design. So, the relationships between marker genotypes and production are established using the sons' marker genotypes and their predicted transmitting abilities (PTAs) for production. This is illustrated in Figure 9.12.

Having demonstrated associations between markers and QTLs, how is this information to be used routinely? By the time the associations have been tested, sons of the original bulls already have progeny test results, and the marker information is redundant as far as these sons are concerned. While it would be relevant for other grandsons of the original sire, it is likely that progeny testing schemes will be sampling the next generation of young bulls well before the marker information is available. However, there are two ways to use the information earlier. The first is to continue to use the marker associations in the next generation of the same family, i.e. for screening great grandsons. The risks of recombination breaking the association between the marker and the QTL are higher in later generations. However, the practical consequences of using the marker in error to select among full sibs are minimal, because full sibs without daughter milk records have identical PTAs, since these are based solely on their parents' PTAs and they have the same parents. At the moment, sons from large full sib groups resulting from embryo transfer on contract mated cows, or in MOET schemes, get chosen for progeny testing largely at random, or on the basis of appearance. So markers could have a potentially beneficial effect with negligible risk when they are used to discriminate between full sibs without performance records of their own.

An alternative is to use the production records from large numbers of daughters of the original bull to look for marker-QTL associations. This is called a **daughter design**. In this case, brothers of the cows could be selected for progeny testing on the basis of their marker genotype, as well as on their predicted PTA (see Figure 9.12). The problem with this approach is that it requires very large daughter groups and a lot of genotyping to establish reliable links between markers and QTLs. It is also of value only in the short term, as the preferred marker genotype becomes fixed after several generations of selection.

An example of marker assisted introgression

The introduction of the polled gene from a polled breed of cattle to a horned breed provides a good example of marker assisted introgression. In most countries, hornless cattle are preferred to horned cattle, because of the reduced risk of injury to humans and other cattle. However, many popular breeds are horned, so calves need to be dehorned at a young age. This is costly and may compromise the welfare of the calf in the short term, even if it improves it in the long term. In most Western breeds of cattle, the polled allele (P) is dominant over the recessive horned allele (p) [26]. In some breeds both the horned and polled types are present. Also, occasional mutations to the polled type occur within breeds which are normally horned, and animals with this mutation could be useful in programmes to introduce polledness. However, since mutations are rare, polled

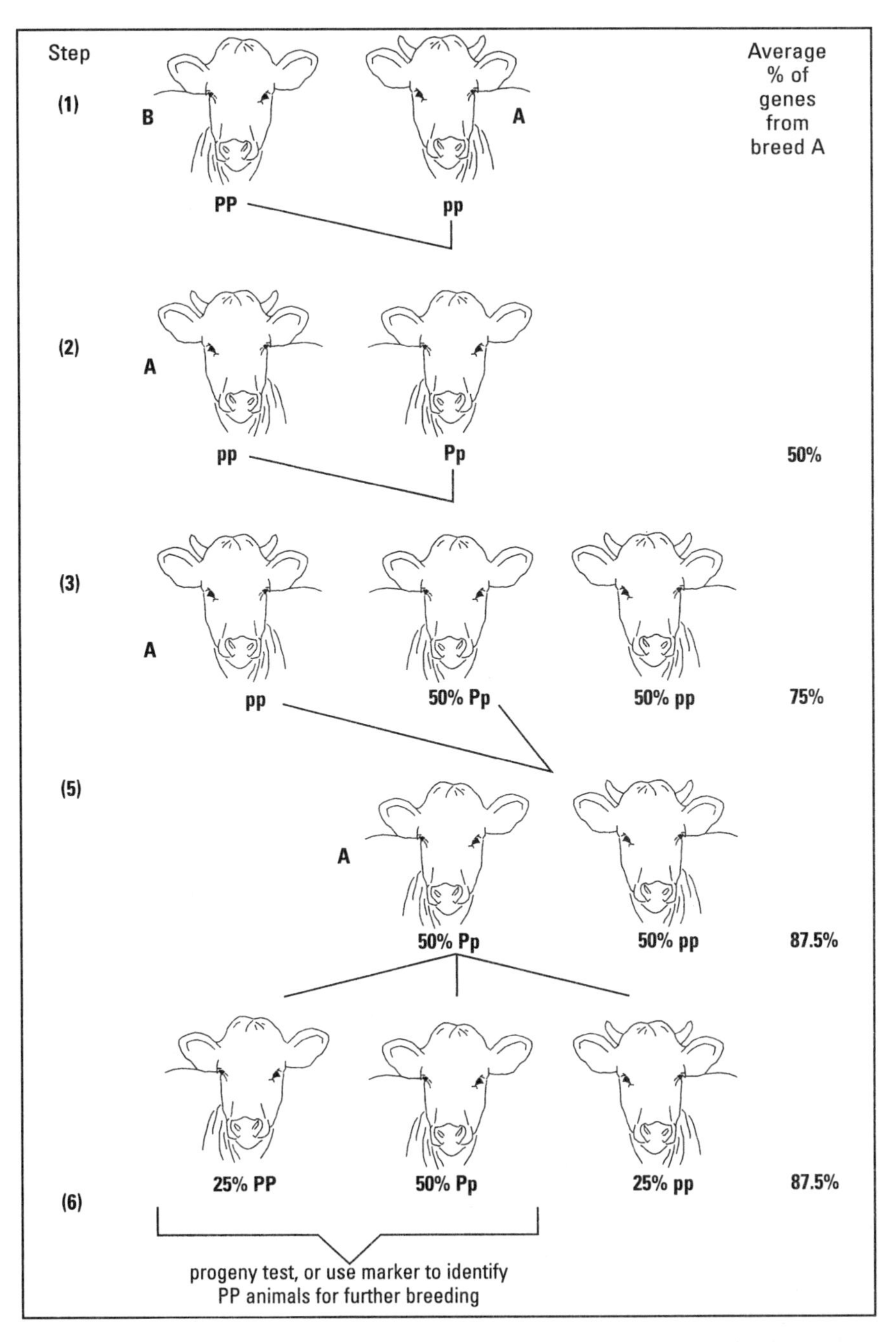

Figure 9.13 The steps involved in a backcrossing programme to introduce the polled gene from breed B to a horned breed, breed A. The diagram shows only two generations of backcrossing to breed A after the initial cross between breeds A and B. This produces offspring with an average of 87.5% of their genes from breed A, which are then interbred. Further generations of backcrossing could be undertaken if a higher proportion of breed A was required.

strains of breeds which are normally horned can be established by crossing to animals from a polled breed followed by back-crossing to the original breed. This could be achieved without the use of markers by the following steps, which are also illustrated in Figure 9.13:

1. Mate horned cows of the preferred breed A to bulls of a polled breed B. All the resulting calves will be heterozygous polled animals (Pp), with 50% of their genes from breed A and 50% from breed B.

2. Mate the F1 crossbred females back to bulls of the preferred breed A. This will result in calves of two genotypes with respect to polledness; on average half of the calves will be heterozygous polled animals (Pp), and half homozygous horned animals (pp). The calves will now average 75% breed A genes and 25% breed B genes.

3. Backcross only the polled (Pp) animals to breed A again. About half the calves will be heterozygous polled and half will be homozygous horned animals, as before. The calves now average 87.5% breed A genes and 12.5% breed B genes.

4. Repeat step three until the proportion of genes from the preferred breed is high enough (e.g. overall performance is close to that of pure breed A animals, or the proportion of breed A genes is high enough to allow pedigree registration).

5. Mate graded-up heterozygous polled animals to each other. This will produce three genotypes of calves: 25% PP (homozygous polled – the desired genotype), 50% Pp (heterozygous polled), and 25% pp (horned) on average.

6. Progeny test polled bulls by mating them to horned cows. Those which are homozygous polled will never produce horned calves. On average, heterozygous polled bulls will produce horned calves from 50% of matings to horned cows. About seven polled calves (and no horned calves) would need to be produced from matings to horned cows to be 99% sure of detecting a heterozygous polled bull.

7. Select and breed from homozygous polled bulls only.

Step six in the above procedure is very time consuming and expensive, but it could be completely abolished by using a molecular marker to distinguish between homozygous and heterozygous polled animals. Similarly, in breeds with both horned and polled animals present already, markers could be used to identify homozygous polled animals (this would be an example of MAS rather than MAI).

In steps two to five above it was possible to eliminate 50% or 25% of the animals (the horned ones) from the grading-up programme by visual inspection. MAI could be of even more value if the aim is to introgress genes for a trait which cannot be readily observed in the animals, or is expensive or difficult to measure. Also, markers in other regions of the genome, apart from the one being introduced, can be used to identify and exclude animals with a high proportion of genes from the least desirable breed in the backcrossing programme.

Disease resistance genes, or genes affecting traits seen in only one sex, or post slaughter, could all be potential candidates for MAI. This approach is being used by some pig breeding companies in crosses between Chinese and European breeds. The aim is to introgress genes conferring high litter size from, for example, the Meishan breed while maintaining the higher genetic merit for growth and carcass traits of the European breeds.

Although MAI is more efficient than alternative programmes without molecular markers, all grading-up programmes of this sort are time consuming and costly. Also, there is often little scope to select for traits of economic importance, other than the one being introgressed. So, before embarking on a MAI programme it is important to check whether the economic benefits at the end of the programme are likely to exceed those which could have been obtained by continuing conventional selection [55].

An example of marker assisted selection

A good example of the use of MAS is in the selection of dairy bulls for progeny testing by some companies in the US. The aim is to improve the accuracy of identifying high merit young bulls for progeny testing. This involves the use of markers at the loci controlling the secretion of the milk proteins κ-casein and β-lactoglobulin. There are two alleles segregating at each locus, abbreviated to A and B in both cases. The B allele at the κ-casein locus has favourable effects on milk processing properties, so there may be benefits in direct selection on genotype. However the genotypes at these loci also serve as linked markers within sire families for QTLs affecting the production of milk, fat and protein yields and fat:protein ratios [10].

This is illustrated in Table 9.2 which shows the mean PTAs for sons of three

κ-casein genotype	No. sons and their mean PTAs	β-lactoglobulin genotype		
		AA	AB	BB
AA	No. sons	7	20	11
	PTA Milk kg	458.0	592.0	427.0
	PTA Protein kg	19.3	18.9	14.0
	PTA Fat kg	22.3	23.3	22.3
AB	No. sons	7	19	20
	PTA Milk kg	754.0	680.0	601.0
	PTA Protein kg	22.3	19.7	18.4
	PTA Fat kg	19.1	19.5	23.1
BB	No. sons	6	4	3
	PTA Milk kg	732.0	690.0	781.0
	PTA Protein kg	26.2	18.2	23.2
	PTA Fat kg	18.2	16.0	19.0

Table 9.2 Mean PTAs for the sons of three Holstein Friesian bulls heterozygous at both the κ-casein and β-lactoglobulin loci [10].

double heterozygous Holstein Friesian bulls (i.e. these grandsires were heterozygous at both the κ-casein and β-lactoglobulin loci). Sons with the genotypes AA for κ-casein and BB for β-lactoglobulin have the poorest PTAs for milk and protein yield, and the highest (i.e. worst in many Western milk markets) fat:protein ratio. Conversely, the genotypes BB for κ-casein and AA for β-lactoglobulin had the highest PTAs for protein, and lower (i.e. better) fat:protein ratios. The most favourable combination of genotypes will depend on the current milk pricing policy in the country concerned. However, based on US prices at the time, the sires with the five 'best' genotypes had PTAs which were 108 kg milk and 1.9 kg protein higher than the overall average, and had PTAs for protein 4.0 kg higher than their PTAs for kg fat [10].

A larger study on markers for QTLs in dairy cattle was completed more recently. This involved fourteen elite Holstein Friesian sire families in the US, with a total of 1,500 progeny tested sons [17]. This study showed evidence of chromosomal regions associated with effects of up to +8kg on protein production and +300 kg milk on milk production. The search for useful markers is also in progress in several other countries, and the use of markers in dairy cattle breeding may become commonplace over the next few years.

It is estimated that using linked markers of QTLs in dairy and other breeding programmes could increase annual response by up to 30% [29, 33, 38]. Whether or not there are benefits from MAS, and the scale of potential benefits, will depend on the size of the QTL effect (this, and hence the value of MAS, may be overestimated in some studies), the strength of the linkage between the marker and the QTL, and the rates of change possible by conventional means. Even when benefits are expected from the use of MAS, much work remains to be done on the optimal integration of MAS into conventional selection programmes.

If markers can be found for traits which are difficult to measure, such as meat quality, other carcass traits and reproductive performance, then the approaches to selection outlined above could have wide application in sheep and beef cattle breeding, as well as in dairy cattle breeding. Markers also have a potentially valuable role in selecting for disease resistance. Already they are widely used in dairy cattle breeding to screen for genetic diseases in candidates for progeny testing. There are similar applications in formal schemes or individual breeder's programmes to select for increased resistance to scrapie in the Shetland, Swaledale, Suffolk, Texel and several other breeds of sheep in Britain. In each of these cases the genetic disease, or genetic control of susceptibility, is largely or totally controlled by a single gene [28]. However, markers could also be useful in selecting for diseases affected by several or many genes, in the same way that they are used to assist selection at QTLs.

Gene transfer

The molecular genetic technologies discussed so far are all used to manipulate the existing genetic variation within a breed or in crosses between breeds. The introgression of a new gene is time consuming because of the need to go through several generations of backcrossing to get rid of the unwanted part of the genome of the 'new' breed. Potentially the same effect could be achieved by direct transfer of the relevant gene from a new breed to the existing favoured breed or strain.

(Although the gene would need to be of major economic importance to justify the cost of gene transfer.) Physically inserting the DNA coding for a gene with a desired effect into the DNA of another animal is termed **gene transfer**, and the animals receiving the foreign DNA are called **transgenic animals**. The high level of conservation of DNA between species means that there are prospects for the transfer of genes not only within species, but also between species, and potentially even between different classes of organism.

There are four main steps involved in creating transgenic animals:

- **Identification and multiplication of a gene with a significant and desirable effect**. This is a major limitation to gene transfer at the moment. There are a vast number of genes present in the mammalian genome, but for only a small fraction of these is the function and exact location in the genome known at present. When a gene of interest is identified, multiple copies can be produced using one of the techniques outlined earlier.

- **Introducing the DNA coding for the desired gene to the preferred breed or line**. This is usually achieved experimentally in mammals by direct injection of several hundred copies of the foreign DNA sequence into the nucleus of an early stage embryo (see Plate 65). This sounds a fairly crude approach, but it appears to work, albeit with low efficiency. For example, usually less than one per cent of manipulated embryos successfully develop into transgenic young, though not all transgenic offspring express the transgene [45, 58]. Alternative methods include the use of vectors, such as retroviruses, which carry copies of the cloned DNA coding for the new gene. This method has been used to create transgenic poultry. Transgenic animals may have different numbers of copies of the foreign gene incorporated into their genome, and they may be incorporated in different places, which can affect the level of expression. The success rate for creating transgenic animals could be improved if these factors could be controlled.

  Another potential method for introducing foreign DNA is via **stem cells**. These are cells, such as those found in early embryos, which are capable of developing into any organ or tissue. When inserted into another normal, developing embryo, these foreign cells multiply and become incorporated in the organs and tissues of the resulting animal. Animals with tissues derived from two sets of parents are termed **chimeras**. At the time of writing it is possible to create cultures of mice stem cells in the laboratory, but not those from farm livestock. If this becomes possible then it may be feasible to introduce foreign DNA to these stem cell cultures, to check that incorporation has been successful, and then to transfer stem cells to developing embryos to create chimeras carrying the modified DNA. While these animals would express the foreign DNA in tissues incorporating the stem cells, they would only transmit the foreign DNA to their offspring if the stem cells were incorporated into reproductive organs. Also, it should be possible to make specific changes to genes in embryo stem cells, rather than simply adding repeat copies of genes as with DNA injection [42].

The recent developments which have enabled the production of cloned animals from cultured cells could also be of benefit in gene transfer. If specific genetic changes are made to cells in culture, then cloned transgenic animals could be produced. All of the cells of such animals would include the genetic modification, thus avoiding the problems associated with chimeras.

- **Regulating expression of the introduced gene**. It is clearly vitally important that the right genes get switched on in the right tissues, and at the right time. Genes can be coupled to a variety of controlling sequences of DNA which cause the gene to be expressed in response to different hormones. But understanding of the regulation of expression is far from complete. In some experiments, genes with a very specific effect, such as secretion of a human milk protein by sheep, have been transferred successfully. That is, the desired effect was achieved, and there were no apparent undesirable effects [66]. However, in other experiments involving transfer of growth hormone genes into pigs and sheep, there were highly undesirable side effects, including serious physiological and anatomical problems such as lameness, arthritis and infertility [44]. It is clear that much more will need to be known about incorporation, and regulation of expression of genes if these problems are to be avoided. This knowledge could also lead to methods of regulating the expression of existing genes, without introduction of foreign DNA.

- **Confirming transmission of the transferred gene to the next generation of animals**. If the aim of gene transfer is to create improved animals for breeding rather than for the production of modified products themselves, then it is important to confirm that the desired effects are observed in offspring, as well as in the original modified animals. It appears that most transgenic animals which actually express the transferred gene do pass on this new gene to their offspring.

The most likely application of gene transfer technology currently is in the production of novel pharmaceutical proteins in milk, as mentioned above [58]. There are several proteins used in the treatment of human diseases which are very difficult and costly to obtain by other means, but which are being, or could be, produced in relatively abundant quantities in the milk of transgenic animals. Applications of this type are likely to become more widespread. This is partly because it appears to be relatively simple to identify such genes, and feasible to transfer them without running into problems of gene expression and regulation. Also, providing there are no serious side effects for the animals, producing transgenics to cure human disease stands a better chance of public acceptance than creating transgenics to improve agricultural productivity. However, agricultural applications which are of direct benefit to animals as well as humans, such as improving the disease resistance of animals, may be publicly acceptable also. These issues are discussed in more detail in the next section.

Ethical implications of new technologies

Many fields of human endeavour lead to ethical dilemmas, and agriculture is certainly no exception. While there are ethical dilemmas in many areas of agriculture, the application of new technologies, particularly in animal production, is perhaps the area where there is most public concern. Views on the use of animals by humans range from those who regard any use of animals as immoral, to those who feel any use of animals in the interest of humans is justified. However, the majority of people accept the use of animals for a range of purposes, including food production, providing that the animals are treated humanely [39].

It is still difficult to decide whether particular treatments of animals are humane, and some form of ethical analysis is often used to help reach rational decisions. There are two main schools of thought in ethical analysis [37]. The first is **consequentialism** – that the ethics of certain actions can be judged solely by the consequences, or that 'the ends justify the means'. Applied to animal production, this approach would allow any treatment of animals to be justified, providing that the benefits for humans were high enough. The second is **deontology**, which states that rights and duties are of overriding importance, irrespective of consequences. Applied to animal production, this approach would rule out certain procedures, regardless of the benefits. In practice, it is difficult to solve ethical dilemmas using only one of these approaches [37].

The recent report of the Banner Committee provides one of the most relevant, and balanced, assessments of ethical issues in animal breeding. The Banner Committee was formed by the UK government "to consider the ethical implications of emerging technologies in the breeding of farm animals; to advise on the adequacy of the existing legal and other safeguards in those areas; and to make recommendations" [39]. The committee used both the approaches outlined above to produce a framework for ethical discrimination among new breeding technologies in farm animals. The members of the committee had expertise in animal genetics, veterinary medicine, farming, law, ethics, philosophy, animal welfare and consumer affairs. The committee considered the technologies of artificial insemination, superovulation and embryo transfer, *in vitro* production of embryos, sexing, cloning and genetic modification by gene transfer. To assist in its deliberations, the committee sent a consultation letter to a wide range of organisations in the food and agriculture industries, animal welfare, consumer and religious organisations. In it they asked whether the organisation had intrinsic objections to the use of these technologies, or views on their effects on welfare, genetic diversity, social or economic life, the potential environmental risks from the use of the new technologies, and the use of patent law to protect commercial interests in genetically modified animals. Additionally, the committee visited practitioners of the techniques, and sought the views of a number of expert witnesses. They presented their recommendations early in 1995.

The crux of their report is that the humane use of animals respects three principles:

"(a) Harms of a certain degree and kind ought under no circumstances to be inflicted on an animal.

(b) Any harm to an animal, even if not absolutely impermissible,

> nonetheless requires justification and must be outweighed by the good which is realistically sought in so treating it.
>
> (c) Any harm which is justified by the second principle ought, however, to be minimized as far as is reasonably possible." [39].

In other words, there are some procedures which should simply never be used on animals, no matter what the potential benefits. There are others which could be used, even if there is potential harm to the animal. If there is any harm, this needs to be justified by the benefits expected, and steps need to be taken to make sure that the harm is minimised. These principles recognise that a cost:benefit analysis alone is not sufficient to justify the use of harmful procedures. The report also emphasises that harm does not just mean physical harm, but includes treatment of animals which might be considered degrading or disrespectful of the nature of the animal. The committee applied the three principles to each of the breeding technologies mentioned above to help discriminate among them on ethical grounds. They recognised that some individuals and groups disagree with any use of animals for food production, but applied the principles from the majority view that the humane use of animals for this purpose is acceptable.

The first principle would exclude the use of technologies which are regarded as intrinsically objectionable. Many of the organisations consulted had objections to the use of new breeding technologies. These included the views that use of these technologies treated animals as "raw materials", disrespected the "integrity of nature", or allowed practitioners to "play God". However, the committee felt that the uses of the new technologies were not *automatically* objectionable on these grounds, since many of them simply accelerate the selection process possible by natural means. Even those which directly modify the genetic make-up by gene transfer can still respect the essential nature and wellbeing of the animal. As an example they cited three hypothetical uses of gene transfer. The first aimed at increasing the protein content of cows' milk, the second aimed at producing only female chicks from an egg-laying line of poultry, and the third aimed at improving feed efficiency in pigs by reducing their sentience and responsiveness. The committee reasoned that neither of the first two of these aims was intrinsically objectionable, since they did not affect the animals' "defining characteristics, nor threaten the achievement of its natural ends or good". However, they argued that the third use would be intrinsically objectionable since it was designed specifically to modify the normal behaviour of pigs.

The committee felt that the use of DNA from different species to create transgenic animals could be considered in the same framework, since in most cases this was aimed at the addition of a trait without significantly modifying the animal's natural characteristics, rather than creating a new type of animal. For example, the transgenic sheep mentioned earlier have an additional human protein secreted in their milk, but in all other respects they are normal sheep. (The use of DNA across species does create problems for people with dietary restrictions for religious or other reasons. Other committees have considered this issue and suggested appropriate labelling of foods from modified animals to ensure freedom of choice for consumers.) While there are already government-appointed committees in the UK (as there are elsewhere) which make recommendations on farm animal welfare, or on the use of animals in experiments, the Banner Committee

Breeding technology	Cattle	Sheep and goats	Pigs	Poultry
Semen collection	•	•	•	•
Artificial insemination	•	X	•	(•)
Superovulation	(•)	(•)		
Embryo transfer	(•)	X	X	
Ovum pick up (OPU)	(?)			
In vitro fertilisation	(•)			
Sexing	(•)			
Cloning	(•)			
Genetic modification	(•)			

Table 9.3 Summary of the findings of the Banner Committee on the welfare implications of new breeding technologies. • indicates that the committee considered there are no serious welfare concerns with the use of the technique; (•) indicates that the techniques do not inherently compromise welfare, but there are related concerns, e.g. the production of abnormally large calves following some *in vitro* embryo culture procedures, concerns on the choice of appropriate animals as embryo recipients or possible intrinsic objections to the use of gene transfer in some cases; X indicates concerns over welfare implications of this technique. (?) indicates insufficient information available to gauge the welfare consequences [39].

recommended the creation of a standing committee with a broader remit to consider ethical questions related to current and possible future uses of animals.

The report emphasised that while there may be no intrinsic objections about the 'ends' or aims of a particular technology, there would be objections if the 'means' to that end seriously compromise animal welfare. For that reason the welfare implications of each of the technologies was examined. The findings are summarised in Table 9.3.

Broadly speaking, the committee felt there were no strong welfare concerns about semen collection and AI in any livestock species, with two exceptions. The first was the use of laparoscopic AI in sheep. They recommended that this should remain a veterinary procedure, and be performed under a code of practice covering analgesia, and the health, maturity and suitability of the animals, and that it should be restricted to use in disease control or recognised breed improvement programmes.

Techniques requiring the use of non-therapeutic surgery were considered unacceptable for routine use by the Banner Committee, as they have been by other groups in the veterinary profession in Britain and several other countries. While this is a reasonable prior choice of boundary, it appears that laparoscopy itself may compromise welfare no more than some routine handling operations on sheep, in so far as this can be measured by blood hormones which act as indicators of stress [27]. However, the procedure causes pain in humans for some days afterwards, and it is probable that it is painful for sheep too, hence the recommendation on analgesia. Although a great deal of research effort is going into finding alternative methods of AI in sheep, as yet there are no alternatives which produce consistently high conception rates with frozen semen. Also, there are concerns about the welfare implications of some of the alternative non-surgical methods of AI.

The second welfare concern was over the use of AI in poultry. The concern here was not with the technique itself, but due to the fact that in some strains of turkey its use is necessary because heavily muscled birds are incapable of natural mating. This was considered intrinsically objectionable. These strains have been created in the past by conventional selection, which highlights the fact that ethical concerns are not restricted to the use of new breeding technologies. (For example many people would regard as intrinsically objectionable the high incidence of Caesarean sections in homozygous double-muscled cattle, or the congenital defects common in some dog breeds due to selection for particular breed characteristics. In both cases the dramatic changes in the form of the animals have been brought about by conventional selection.)

The committee reached similar conclusions on the use of superovulation, embryo transfer and other embryo technologies to those reached for AI. Where non-surgical procedures can be used for embryo recovery or transfer, they were not regarded as inherently bad for welfare. Where surgical procedures are required, such as in sheep and pigs, the committee recommended similar restrictions to those for laparoscopic AI in sheep. There were concerns that although several of the techniques did not inherently compromise welfare, some associated procedures could do so. For instance, some embryo culture procedures are thought to be implicated in the birth of abnormally large calves from *in vitro*-produced embryos. Similarly, the use of recipients of an inappropriate breed, cross or degree of maturity may increase the incidence of difficult calvings following embryo transfer, and so compromise welfare. Hence the recommendations include consideration of these wider risks to welfare, as well as risks from the technologies themselves.

In a similar vein, a welfare checklist has been proposed to help deliberations on the use of new technologies in cattle breeding [56]. This includes assessing: (i) pain and fear associated with the technique itself, or its immediate consequences; (ii) periparturient problems associated with the technique (e.g. oversize calves associated with some *in vitro*-produced embryos, or calving difficulties associated with twinning); (iii) physical or psychological problems demonstrable in all or most of the genetically modified offspring (such as those seen in the modified pigs mentioned earlier) and (iv) increased incidence of disease or disability only demonstrable by observation of relatively large populations over several generations (e.g. increased incidence of mastitis or lameness). A two-stage review process was proposed so that problems of types (i) to (iii) would need to be solved before a technique was approved for wider use, but a second stage of controlled use on a wider scale would be needed to detect problems of the sort outlined in (iv).

The Banner Committee did not regard genetic modification in its own right as a threat to animal welfare, although there may be consequences which affect welfare, such as the high incidence of lameness and arthritis in the modified pigs and sheep mentioned earlier. Both the creation of genetically modified organisms, and breeding from them, are controlled procedures already in the UK, requiring a licence from the Secretary of State. Before a licence is granted, the likely adverse effects on the animal have to be weighed against the likely benefits of the modification. Hence possible threats to welfare should be considered already and the committee supported the existing controls. However, they recommended that if consideration of adverse effects does not already encompass intrinsic objections, such as damage to the natural integrity of the animal, it should be broadened to do so. Additionally, the committee recommended greater uniformity across species in the legislation governing some of the technologies, and made suggestions on training of operators to safeguard animal welfare.

There has been much debate on the issue of patenting genetically modified animals, and the techniques involved in producing them. Those in favour often argue that the protection offered by patents is essential in encouraging biotechnology companies to invest in research and development. Those against often argue that it is immoral to patent modified animals or associated techniques, since this encourages the view that animals are simply industrial commodities. The issue was considered by the Banner Committee, which concluded that patents were not objectionable in principle, but that consideration of patent applications should include moral criteria as well as technical ones. They also suggested that the threat to small producers from very widely drawn claims to patent protection be monitored.

On the issue of genetic diversity, the group felt that the new techniques were likely to have only a slightly greater impact than existing techniques, and since it is in breeders' interests to maintain diversity, this was not seen as a major threat. However, they made several recommendations including: (i) the formation of a UK register of breeds to record population sizes; (ii) a survey to measure diversity within and between breeds using molecular markers and production traits; (iii) the construction of a biodiversity database and (iv) the establishment of a genome bank for cryogenic storage of gametes and embryos to enable reintroduction of genes lost from populations.

The committee also considered possible socio-economic impacts of the new breeding technologies. These included the impact of new techniques on food surpluses, the effect on the viability of small farms, their effect on third-world agriculture, and the risk of alienating consumers. While recognising the potential risks in these areas, the committee felt that existing legislation was sufficient to deal with these. They cited the ban in the EU on the use of the hormone bovine somatotropin to increase milk yields in dairy herds as an example of the use of this legislation to deal with perceived socio-economic threats. They suggested that those involved in developing and using new breeding technologies need to be sensitive to public concerns, open to debate on the issues involved, and supportive of reasonable systems for regulation, provision of information and labelling. The UK government has accepted most of the committee's recommendations, but reserved judgement on the need for a standing committee on ethical issues related to the use of animals, and on the need for a gene bank, and defended existing legislation on patents.

The report of the Banner Committee is particularly relevant to the discussion in this chapter. However, other approaches have been proposed elsewhere. For instance, an ethical matrix has been proposed to help evaluate the impact of new biotechnologies from the perspective of the animals, producers, consumers and the 'biota' or wider environment [37]. (See also [14].) In each case, three ethical principles are considered: respect for wellbeing, autonomy and justice. Applying these principles to animals, wellbeing equates to welfare, animal autonomy includes freedom to express normal behaviour, and justice equates to respect for the integrity and nature of the animal. In the case of consumers, wellbeing would encompass any health risks posed by food produced by new biotechnological processes, autonomy would encompass consumer choice, and justice would encompass product price, which affects the ability of different groups to benefit from the product. While this approach is interesting, there is a risk that it leads to such wide thinking that the practical problems remain unsolved.

The vast majority of this book has concentrated on the technical aspects of cattle and sheep breeding. However, it is entirely appropriate that analyses of the technical aspects of a breeding scheme be accompanied by an analysis of the ethical implications. Many of the breeding techniques and selection goals described throughout the book are likely to be neutral with respect to animal welfare. But, based on experience from pig and poultry breeding, and to a lesser extent dairy cattle breeding, the sustained pursuit of some breeding goals, such as higher output alone, may have unfavourable consequences on welfare. For example, selection for milk yield in dairy cattle is associated with a higher incidence of mastitis. However, with a better understanding of the genetics of the production traits and diseases involved it is possible to turn this situation around, and develop breeding goals which limit the deterioration in resistance to disease, and preferably improve it. It is important not to forget that many of the new breeding technologies provide opportunities for great good in this respect, as well as the possibility of harm. For instance, the use of acceptable reproductive technologies could allow more rapid progress in selection for disease resistance or calving ease. Similarly, gene transfer may allow the rapid introduction of genes conferring disease resistance or polledness from one breed to another.

By their very nature, ethical problems are difficult to solve. On their own admission, the framework proposed by the Banner Committee is not easy to apply in all circumstances. However, their framework and recommendations provide an important foundation for a more open dialogue in what is clearly an area of great public concern. In the longer term, this must help to restore public confidence in farming and science, currently at a low ebb in many industrialised countries, while allowing maximum benefit to be derived from those technologies which are considered acceptable.

Summary

- New reproductive and molecular genetic technologies could lead to more effective genetic improvement programmes in livestock, or to more rapid dissemination of improved genes from elite to commercial sectors of the livestock industries. While most of these technologies are relevant in all species, many are likely to be of particular value in ruminants.

- Artificial insemination is already widely used in cattle breeding but, largely because of technical difficulties, it is not yet as widely used in sheep breeding. AI can allow higher male selection intensities, shorter male generation intervals and higher accuracy of selection, as well as providing genetic links across herds or flocks. Hence, it is a highly effective method for increasing rates of genetic improvement. It is also an extremely useful method for dissemination.

- Multiple ovulation and embryo transfer (MOET) potentially offers similar benefits in selection of females to those offered by AI in males. However, in practice the benefits are often smaller. As a result of this and its relatively high cost, the technique is largely confined to specialised breed improvement schemes (often involving elite nucleus populations), to facilitating international trade in genetic material and to accelerating multiplication of newly introduced breeds.

- *In vitro* production of embryos can dramatically increase embryo yield compared with conventional MOET. This could accelerate rates of improvement in breeding schemes, especially when oocytes are collected from live elite donors (ovum pick up). However, it could also allow more effective dissemination to the commercial sector, for example when oocytes are collected from slaughtered beef heifers of high performance, for transfer to dairy or suckler cows of lower beef merit.

- While the use of reproductive technologies such as MOET can substantially increase rates of genetic improvement, rates of inbreeding often increase proportionately more. However, much of this gain can often be achieved with minimal increase in inbreeding by modifications to the design of the breeding scheme.

- In most circumstances sexing of either semen or embryos appears to be of little value in accelerating genetic improvement. However, the development of a cheap, reliable technique for sexing semen in large enough quantities for conventional AI could lead to major improvements in the dissemination of genetic improvement and in the efficiency of animal production. Semen or embryo sexing on a smaller scale could still allow more effective dissemination if it is coupled with *in vitro* production of embryos.

- Cloning appears to be of fairly limited value in accelerating genetic improvement, but the potential of the technique to accelerate dissemination of genetic improvement to commercial herds or flocks is great. For this potential to be realised, reliable and cost-effective methods for cloning will be required. Also, improved and cost-effective techniques for delivery will be needed.

- To date, most selection in livestock has been practised with little or no knowledge of what is happening at the DNA level. Selection has been on the effects of the genes, rather than directly on the genes themselves. However, some new molecular technologies offer the opportunity to improve on current methods. Genome mapping, the use of molecular markers in selection, and gene transfer are already having, or are likely to have, an impact on livestock improvement.

- These molecular technologies are based on relatively recent advances in molecular and cell biology including: (i) the discovery of restriction enzymes which are capable of 'cutting up' sequences of DNA; (ii) the development of techniques to allow DNA fragments of different lengths to be separated and detected; (iii) the development of techniques to multiply sequences of DNA rapidly; (iv) the development of techniques to determine the sequence of bases on short strands of DNA; (v) the development of DNA probes and labelling techniques; and (vi) the development of markers of variation at the DNA level.

- Any identifiable segment of DNA in the genome, which varies between animals, can act as a marker. Hence, markers may be all or part of a 'functional' gene, or they may be part of the rest of the genome which does not code directly for the production of a protein. Several different types of marker have been developed, based on some of the methods described above, for detecting variation at the DNA level. These markers are of value in genome mapping, marker assisted selection (MAS) parentage verification and product identification.

- The two types of marker currently of most value in livestock genome mapping and of most potential value in marker assisted selection are restriction fragment length polymorphisms (RFLPs) and microsatellites. Microsatellite markers are usually highly polymorphic. As a result they are now the markers of choice in livestock genome mapping and they are likely to be of most value in marker assisted selection in the immediate future. However, other types of marker are being developed rapidly, and these may be of value also.

- Some important features of RFLP and microsatellite markers which make them particularly useful are that: (i) they are codominant; (ii) because they are DNA-based they can be applied to animals of both sexes and at any age and (iii) those which are based on PCR need only small initial amounts of DNA from the animals of interest.

- The purpose of a genome map is to describe the location of functional genes or other sequences of DNA (e.g. markers) on the chromosomes. There are different types of map which have different levels of resolution. The ultimate level of resolution for a genome map is the full DNA sequence of the genome. However, maps with this level of resolution over the whole of the genome may never be affordable, or necessary, in farm livestock.

- Livestock genome maps can be useful in making informed choices of markers for economically important traits in order to accelerate conventional selection programmes, and for locating genes of potential interest for gene transfer. Genome maps are also useful in deciding which breeds, strains or individuals should receive highest priority in conservation programmes.

- Maps fall into two categories: linkage maps and physical maps. Linkage maps show the genetic distance between pairs of markers on the same chromosome. Physical maps show the location of genes or markers on individual chromosomes.

- Research groups have been working on genome mapping in all major livestock species for several years now, often in collaboration with one another. As a result, linkage maps have been published for each of the livestock species, and large numbers of loci have been placed on physical maps.

- Molecular markers may speed up introduction of a gene of interest from one breed to another (marker-assisted introgression – MAI). Compared to conventional backcrossing programmes, markers can help by: (i) identifying backcrosses carrying the particular favoured allele from the new breed and (ii) among animals with one or two copies of this allele, identifying those with least of their genome from the new breed.

- Markers may also accelerate selection for traits of economic importance within a population (marker-assisted selection – MAS). In this case the aim of using markers is simply to boost the improvement made by conventional means, by homing in on genes already present in the breed concerned. The gene of interest may be either a single gene with a large effect, or it may be one of many genes at a locus affecting a quantitative trait (a QTL).

- The ideal marker for both MAI and MAS would be the gene responsible for the trait of interest, or a shorter sequence of DNA within this gene. However, most markers are not the genes of interest themselves, they just happen to be located in the same region of the chromosome as the gene of interest.

- There are three main approaches to choosing markers in order to investigate whether or not they are linked to traits of interest: (i) the 'shotgun' approach, where markers are chosen at random; (ii) the map-based approach, using markers which are evenly spaced over the genome and (iii) the candidate gene approach, where a known mapped gene may be chosen as a marker because it has a similar function to the gene of interest.

- To establish an association between a marker and the trait of interest, DNA samples are needed from at least two generations of animals – parents and offspring – and performance records are needed on offspring. The DNA samples are used to track the movement of the markers from parents to offspring, and to look for associations between particular marker genotypes and performance in the offspring.

- In dairy cattle populations, marker relationships are being detected in what are called granddaughter designs. These involve grandsires which are heterozygous for the marker gene, so the marker is segregating among their sons. The genetic merit of sons is predicted from their daughters' production and the association between genetic merit for production and markers is then investigated. Alternatively, daughter designs use the production records from large numbers of daughters of a bull to look for marker-QTL associations.

- Physically inserting the DNA coding for a gene with a desired effect into the DNA of another animal is termed gene transfer, and the animals receiving the foreign DNA are called transgenic animals. The high level of conservation of DNA between species means that there are prospects for the transfer of genes not only within species, but also between species, and potentially even between different classes of organism.

- Gene transfer usually involves: (i) identification and multiplication of a gene with a significant and desirable effect; (ii) introducing the DNA coding for the desired gene to the preferred breed; (iii) regulating expression of the introduced gene and (iv) confirming transmission of the transferred gene to the next generation of animals.

- Currently gene transfer is used most often in the production of novel pharmaceutical proteins in milk. It appears to be relatively simple to identify genes coding for these proteins, and feasible to transfer them without running into problems of gene expression and regulation. Also, providing there are no serious side effects for the animals, producing transgenics to cure human disease stands a better chance of public acceptance than creating transgenics to improve agricultural productivity.

- In many countries there is public concern over the application of new technologies in animal production. Most people accept the use of animals for a range of purposes, including food production, provided that the animals are treated humanely. However, it is often difficult to decide whether or not a particular treatment is humane.

- The UK government has accepted the recommendations of a committee it established to review this area (the Banner Committee). These recommendations were that the humane use of animals respects three principles: (i) that some treatments of animals are so harmful that they should never be permitted under any circumstances; (ii) that if a harmful treatment is permitted, the harm it causes must be justified by the good being sought and (iii) that steps would be taken to minimise any harm which is justified by the second principle.

- The first principle would exclude the use of technologies which are regarded as intrinsically objectionable. However, the committee felt that the use of new breeding technologies was not *automatically* objectionable on these grounds, since many of them simply accelerate the selection process possible by natural means. Even those which directly modify the genetic make-up by gene transfer can still respect the essential nature and wellbeing of the animal. While there may be no intrinsic objections about the 'ends' or aims of a particular technology, there would be objections if the 'means' to that end seriously compromise animal welfare.

- It seems appropriate that analyses of the technical aspects of a breeding scheme be accompanied by an analysis of the ethical implications. In many cases breeding techniques and selection goals will be neutral with respect to welfare. In others, a better understanding of the genetics of production traits and disease, or other 'welfare traits', should allow the development of breeding goals which limit the deterioration in disease or welfare traits, and preferably improve them.

- The framework and recommendations of the Banner Committee provide an important foundation for a more open dialogue in what is clearly an area of public concern. In the longer term, this must help to restore public confidence in farming and science, currently at a low ebb in many industrialised countries, while allowing maximum benefit to be derived from those technologies which are considered acceptable.

References

1. Archibald, A.L., Burt, D.W. and Williams, J.L. 1994. 'Gene mapping in farm animals and birds: an overview.' *Proceedings of the 5th World Congress on Genetics Applied to Livestock Production*, Vol. 21, pp. 5–12.
2. Archibald, A. and Haley, C. 1993. 'Mapping the complex genomes of animals and man.' *Outlook on Agriculture*, 22:79–84.
3. Betteridge, K.J., Smith, C., Stubbings, et al. 1989. 'Potential genetic improvement of cattle by fertilization of fetal oocytes *in vitro*.' *Journal of Reproduction and Fertility, Supplement*, 38:87–98.
4. Billings, P.R., Smith, C.L. and Cantor, C.R. 1991. 'New techniques for physical mapping of the human genome'. *FASEB Journal*, 5:28–34.
5. Broad, T.E. and Hill, D.F. 1994. 'Mapping the sheep genome: practice, progress and promise.' *British Veterinary Journal*, 150:237–52.
6. Caballero, A., Santiago, E. and Toro, M.A. 1996. 'Systems of mating to reduce inbreeding in selected populations.' *Animal Science*, 62:431–42.
7. Campbell, K.H.S., McWhir, J., Ritchie, W.A. and Wilmut, I. 1996. 'Sheep cloned by nuclear transfer from a cultured cell line.' *Nature*, 380:64–6.
8. Campbell, K.H.S. and Wilmut, I. 1997. 'Totipotency or multipotentiality of cultured cells: applications and progress.' *Theriogenology*, 47:63–72.
9. Colleau, J.J. 1991. 'Using embryo sexing within closed mixed multiple ovulation and embryo transfer schemes for selection on dairy cattle.' *Journal of Dairy Science*, 74:3973–84.
10. Cowan, C.M. 1994. 'Use of genetic marker technology to select AI bulls.' *British Cattle Breeders Club Digest* No. 49, pp. 45–8.
11. Cran, D.G. and Johnson, L.A. 1996. 'The predetermination of embryonic sex using flow cytometrically separated X and Y spermatozoa.' *Human Reproduction Update*, 2:355–63.

12. Cran, D.G., Johnson, L.A., Miller, N.G.A., et al. 1993. 'Production of bovine calves following separation of X- and Y-chromosome bearing sperm and *in vitro* fertilization.' *Veterinary Record*, 132:40–1.

13. Cran, D.G., McKelvey, W.A.C., King, M.E., et al. 1997. 'Production of lambs by low dose intrauterine insemination with flow cytometrically sorted and unsorted semen.' *Theriogenology*, 47:267.

14. de Boer, I.J.M., Brom, F.W.A. and Vorstenbosch, J.M.G. 1995. 'An ethical evaluation of animal biotechnology: the case of using clones in dairy cattle breeding.' *Animal Science*, 61:453–63.

15. de Boer, I.J.M., Meuwissen, T.H.E. and van Arendonk, J.A.M. 1994. 'Combining the genetic and clonal responses in a closed dairy cattle nucleus scheme.' *Animal Production*, 59:345–58.

16. Dingwall, B. and McKelvey, B. 1993. 'Artificial insemination and embryo transfer technology for sheep and goats.' *World Agriculture*, 1993, pp. 78–81.

17. Georges, M., Nielsen, D., Mackinnon, M., et al. 1995. 'Mapping genes controlling milk production: towards marker assisted selection in livestock.' *Genetics*, 139:907–20.

18. Gibson, J.P., Freeman, A.E. and Boettcher, P.J. 1997. 'Cytoplasmic and mitochondrial inheritance of economic traits in cattle.' *Livestock Production Science*, 47:115–24.

19. Gordon, I. 1994. *Laboratory Production of Cattle Embryos*. CAB International, Wallingford.

20. Gordon, I. 1996. *Controlled Reproduction in Cattle and Buffaloes*. CAB International, Wallingford.

21. Gordon, I. 1997. *Controlled Reproduction in Sheep and Goats*. CAB International, Wallingford.

22. Gordon, I. and Lu, K.H. 1990. 'Production of embryos *in vitro* and its impact on livestock production.' *Theriogenology*, 33:77–87.

23. Grundy, B., Caballero, A., Santiago, E. and Hill, W.G. 1994. 'A note on using biased parameter values and non-random mating to reduce rates of inbreeding in selection programmes.' *Animal Production*, 59:465–8.

24. Haley, C.S. 1995. 'Livestock QTLs – bringing home the bacon?' *Trends in Genetics*, 11:488–92.

25. Haley, C.S. and Knott, SA. 1992. 'A simple regression method for mapping quantitative trait loci in line crosses using flanking markers.' *Heredity*, 69:315–24.

26. Haley, C., Williams, J., Woolliams, J. and Visscher, P. 1996. 'Gene mapping and its place in livestock improvement.' *British Cattle Breeders Club Digest* No. 51, pp. 5–15.

27. Haresign, W., Williams, R.J., Khalid, M. and Rodway, R. 1995. 'Heart rate responses and plasma cortisol and β-endorphin concentrations in ewes subjected to laparoscopy and its associated handling procedures.' *Animal Science*, 61:77–83.

28. Hunter, N., Moore, L., Hosie, B.D., et al. 1997. 'Association between natural scrapie and PrP genotype in a flock of Suffolk sheep in Scotland.' *Veterinary Record*, 140:59–63.

29. Kashi, Y., Hallerman, E. and Soller, M. 1990. 'Marker assisted selection of candidate bulls for progeny testing programmes.' *Animal Production*, 51:63–74.

30. Kruip, T.A.M. 1994. 'Oocyte retrieval and embryo production *in vitro* for cattle breeding.' *Proceedings of the 5th World Congress on Genetics Applied to Livestock Production*, Vol. 20, pp. 172–9.

31. Kruip, T.A.M. and den Daas, J.H.G. 1997. '*In vitro* produced and cloned embryos: effects on pregnancy, parturition and offspring.' *Theriogenology*, 47:43–52.

32. Land, R.B. and Hill, W.G. 1975. 'The possible use of superovulation and embryo transfer in cattle to increase response to selection.' *Animal Production*, 21:1–12.

33. Lande, R. and Thompson, R. 1990. 'Efficiency of marker assisted selection in the improvement of quantitative traits.' *Genetics*, 124:743-56.

34. Lander, E.S. and Botstein, D. 1989. 'Mapping Mendelian factors underlying quantitative traits using RFLP linkage maps.' *Genetics*, 121:185–99.

35. Lohuis, M.M. 1993. *Strategies to improve efficiency and genetic response of progeny test programs in dairy cattle*. Doctoral thesis, University of Guelph, Canada.

36. Luo, Z.W., Woolliams, J.A. and Simm, G. 1994. 'An assessment of present and future effectiveness of embryological techniques in ruminants.' *AgBiotech News and Information* 6:13N–18N.

37. Mepham, T.B. 1996. 'Ethical impacts of biotechnology in dairying.' In C.J.C. Phillips (ed.), *Progress in Dairy Science*, CAB International, Wallingford. pp. 375–95.

38. Meuwissen, T.H.E. and van Arendonk, J.A.M. 1992. 'Potential improvements in rate of genetic gain from marker-assisted selection in dairy cattle breeding schemes.' *Journal of Dairy Science*, 75:1651–9.

39. Ministry of Agriculture, Fisheries and Food. 1995. *Report of the Committee to Consider the Ethical Implications of Emerging Technologies in the Breeding of Farm Animals*. HMSO, London.

40. Nicholas, F.W. 1979. 'The genetic implications of multiple ovulation and embryo transfer in small dairy herds.' Paper presented at the *Annual Conference of the European Association for Animal Production*, Harrogate.

41. Nicholas, F.W. 1996. 'Genetic improvement through reproductive technology.' *Animal Reproduction Science*, 42:205–14.

42. Nicholas, F.W. 1996. *Introduction to Veterinary Genetics*. Oxford University Press, Oxford.

43. Nicholas, F.W. and Smith, C. 1983. 'Increased rates of genetic change in dairy cattle by embryo transfer and splitting.' *Animal Production*, 36:341–53.

44. Pinkert, C.H., Dyer, T.J., Kooyman, D.L., et al. 1990. 'Characterization of transgenic livestock production.' *Domestic Animal Endocrinology*, 7:1–18.

45. Pursel, V.G., Pinkert, C.A., Miller, K.F., et al 1989. 'Genetic engineering of livestock.' *Science*, 244:1281–8.

46. Robinson, J.J. and McEvoy, T.G. 1993. 'Biotechnology – the possibilities.' *Animal Production*, 57:335–52.

47. Schook, L.B. 1995. *Mapping the Pig Genome. A Practical Primer.* Minnesota report 234-1995, Minnesota Agricultural Experiment Station, University of Minnesota.

48. Smith, C. 1984. 'Rates of genetic change in farm livestock.' *Research and Development in Agriculture*, 1:79–85.

49. Smith, C. 1986. 'Use of embryo transfer in genetic improvement of sheep.' *Animal Production*, 42:81–8.

50. Taylor, St.C.S., Moore, A.J., Thiessen, R.B., et al. 1985. 'Efficiency of food utilization in traditional and sex-controlled systems of beef production.' *Animal Production*, 40:401–40.

51. Toro, M.A., Silio, L., Rodrigañez, J., et al. 1988. 'Inbreeding and family index selection for prolificacy in pigs.' *Animal Production*, 46:79–85.

52. US Department of Energy. 1992. *DOE Human Genome Program. Primer on Molecular Genetics.* US Department of Energy, Office of Energy Research, Office of Health and Environmental Research, Washington DC.

53. Villanueva, B. and Simm, G. 1994. 'The use and value of embryo manipulation techniques in animal breeding.' *Proceedings of the 5th World Congress on Genetics Applied to Livestock Production*, Vol. 20, pp. 200–7.

54. Villanueva, B., Simm, G. and Woolliams, J.A. 1995. 'Genetic progress and inbreeding for alternative nucleus breeding schemes for beef cattle.' *Animal Science*, 61:231–9.

55. Visscher, P.M. and Haley, C.S. 1995. 'Utilizing genetic markers in pig breeding programmes.' *Animal Breeding Abstracts*, 63:1–8.

56. Webster, J. 1994. 'New breeding technologies: ethical and animal welfare issues.' *British Cattle Breeders Club Digest* No. 49, pp. 41–4.

57. White, K.L. 1989. 'Embryo and gamete sex selection.' In L.A. Babiuk and J.P. Phillips (eds), *Animal Biotechnology: Comprehensive Biotechnology. First supplement*, Pergamon Press, Oxford, pp. 179–202.

58. Wilmut, I., Archibald, A.L., Harris, S., et al. 1990. 'Methods of gene transfer and their potential use to modify milk composition. *Theriogenology*, 33:113–23.

59. Wilmut, I., Haley, C.S. and Woolliams, J.A. 1992. 'Impact of biotechnology on animal breeding.' *Animal Reproduction Science*, 28:149–62.

60. Wilmut, I., Schnieke, A.E., McWhir, J., et al. 1997. 'Viable offspring derived from fetal and adult mammalian cells.' *Nature*, 385:810–13.

61. Womack, J.E. 1996. 'The bovine gene map.' In C.J.C. Phillips (ed.), *Progress in Dairy Science*, CAB International, Wallingford. pp. 89–103.

62. Woolliams, J.A. 1989. 'Modifications to MOET nucleus breeding schemes to improve rates of genetic progress and decrease rates of inbreeding in dairy cattle.' *Animal Production*, 49:1–14.

63. Woolliams, J.A. and Wilmut, I. 1989. 'Embryo manipulation in cattle breeding and production.' *Animal Production*, 48:3–30.

64. Wray, N.R. and Goddard, M.E. 1994. 'Increasing long-term response to selection.' *Genetics Selection Evolution*, 26:431–51.

65. Wray, N.R. and Goddard, M.E. 1994. 'MOET breeding schemes for wool sheep. 1. Design alternatives.' *Animal Production*, 59:71–86.

66. Wright, G., Carver, A., Cottam, D., et al. 1991. 'High level expression of active human α-antitrypsin in the milk of transgenic sheep.' *Biotechnology*, 9:830–4.

Further Reading

Animal Genetics (Journal of the International Society for Animal Genetics)

Gibson, J.P. and Smith, C. 1989. 'The incorporation of biotechnologies into animal breeding strategies.' In L.A. Babiuk and J.P. Phillips (eds), *Animal Biotechnology: Comprehensive Biotechnology. First supplement*, Pergamon Press, Oxford. pp. 203–31.

Hafs, H.D. (ed.) 1993. 'Genetically modified livestock: progress, prospects and issues.' *Journal of Animal Science*, 71, Supplement 3.

Mammalian Genome (Journal of the International Mammalian Genome Society)

Nicholas, F.W. 1996. *Introduction to Veterinary Genetics*. Oxford University Press, Oxford.

Proceedings of the World Congresses on Genetics Applied to Livestock Production

Schook, L.B. 1995. *Mapping the Pig Genome. A Practical Primer*. Minnesota report 234-1995, Minnesota Agricultural Experiment Station, University of Minnesota, St. Paul, Minnesota.

Smith, C. and Smith, D.B. 1993. 'The need for close linkages in marker-assisted selection for economic merit in livestock.' *Animal Breeding Abstracts*, 61:197–204.

Theriogenology

Trends in Genetics

van Arendonk, J.A.M. and Bovenhuis, H. 1996. 'The application of genetic markers in dairy cow selection programmes.' In C.J.C. Phillips (ed.), *Progress in Dairy Science*, CAB International, Wallingford. pp. 105–23.

Table showing values of selection intensity for different proportions of animals selected[1] (Source [2]; see Chapter 4).

	No. in population						
No. selected	**5**	**10**	**15**	**20**	**25**	**30**	**35**
1	1.163	1.539	1.736	1.867	1.965	2.043	2.107
2	0.829	1.270	1.492	1.638	1.745	1.829	1.898
3	0.553	1.065	1.311	1.469	1.584	1.674	1.748
4	0.291	0.893	1.162	1.332	1.455	1.550	1.628
5		0.739	1.032	1.214	1.345	1.446	1.527
6		0.595	0.916	1.110	1.248	1.354	1.439
7		0.457	0.809	1.016	1.161	1.271	1.360
8		0.318	0.708	0.928	1.081	1.196	1.288
9		0.171	0.611	0.846	1.006	1.126	1.222
10			0.516	0.767	0.936	1.061	1.160
11			0.422	0.692	0.869	0.999	1.102
12			0.328	0.619	0.805	0.941	1.047
13			0.230	0.547	0.743	0.884	0.994
14			0.124	0.476	0.683	0.830	0.944
15				0.405	0.624	0.777	0.895
16				0.333	0.566	0.726	0.848
17				0.259	0.508	0.676	0.803
18				0.182	0.451	0.627	0.758
19				0.098	0.394	0.579	0.714
20					0.336	0.530	0.672
21					0.277	0.483	0.629
22					0.216	0.435	0.588
23					0.152	0.387	0.546
24					0.082	0.338	0.505
25						0.289	0.464
26						0.239	0.423
27						0.186	0.382
28						0.131	0.340
29						0.070	0.298
30							0.255
31							0.210
32							0.164
33							0.115
34							0.062
35							
36							
37							
38							
39							
40							
41							
42							
43							
44							
45							
46							
47							
48							
49							
50							

[1] Estimates of the selection intensity for numbers between the rows or columns of this table may be obtained by a linear approximation.

Table cont.

	No. in population						
No. selected	**40**	**45**	**50**	**75**	**100**	**150**	**200**
1	2.161	2.208	2.249	2.403	2.508	2.649	2.746
2	1.957	2.008	2.052	2.217	2.328	2.478	2.580
3	1.810	1.864	1.911	2.084	2.201	2.357	2.463
4	1.694	1.750	1.799	1.980	2.101	2.263	2.372
5	1.596	1.654	1.705	1.893	2.018	2.185	2.297
6	1.510	1.571	1.624	1.818	1.947	2.118	2.233
7	1.434	1.497	1.552	1.752	1.884	2.060	2.177
8	1.365	1.430	1.487	1.693	1.828	2.007	2.127
9	1.301	1.369	1.427	1.638	1.777	1.960	2.082
10	1.242	1.312	1.372	1.588	1.730	1.916	2.040
11	1.187	1.258	1.320	1.542	1.686	1.876	2.002
12	1.134	1.208	1.271	1.498	1.645	1.838	1.966
13	1.084	1.160	1.225	1.457	1.607	1.803	1.932
14	1.037	1.114	1.181	1.418	1.571	1.770	1.901
15	0.991	1.071	1.139	1.381	1.536	1.738	1.871
16	0.947	1.029	1.099	1.346	1.504	1.708	1.843
17	0.904	0.988	1.060	1.312	1.472	1.680	1.816
18	0.862	0.949	1.022	1.279	1.442	1.652	1.790
19	0.822	0.911	0.986	1.248	1.413	1.626	1.766
20	0.782	0.873	0.951	1.217	1.386	1.601	1.742
21	0.744	0.837	0.916	1.188	1.359	1.577	1.720
22	0.706	0.802	0.882	1.160	1.333	1.554	1.698
23	0.668	0.767	0.849	1.132	1.308	1.531	1.677
24	0.631	0.733	0.817	1.105	1.283	1.509	1.656
25	0.595	0.699	0.786	1.079	1.259	1.488	1.636
26	0.558	0.665	0.754	1.053	1.236	1.467	1.617
27	0.522	0.633	0.724	1.028	1.214	1.447	1.598
28	0.486	0.600	0.693	1.003	1.192	1.428	1.580
29	0.450	0.568	0.663	0.979	1.170	1.409	1.563
30	0.414	0.535	0.634	0.956	1.149	1.390	1.545
31	0.378	0.503	0.604	0.932	1.128	1.372	1.529
32	0.341	0.471	0.575	0.910	1.108	1.354	1.512
33	0.304	0.439	0.546	0.887	1.088	1.337	1.496
34	0.267	0.407	0.517	0.865	1.069	1.320	1.480
35	0.228	0.375	0.488	0.843	1.050	1.304	1.465
36	0.188	0.342	0.459	0.821	1.031	1.287	1.450
37	0.147	0.309	0.430	0.800	1.012	1.271	1.435
38	0.103	0.276	0.401	0.779	0.994	1.255	1.421
39	0.055	0.242	0.372	0.758	0.976	1.240	1.407
40		0.207	0.343	0.738	0.958	1.225	1.393
41		0.171	0.313	0.717	0.941	1.210	1.379
42		0.133	0.283	0.697	0.923	1.195	1.366
43		0.093	0.253	0.677	0.906	1.181	1.352
44		0.050	0.221	0.657	0.889	1.166	1.339
45			0.189	0.637	0.873	1.152	1.327
46			0.156	0.617	0.856	1.138	1.314
47			0.122	0.598	0.840	1.125	1.301
48			0.085	0.578	0.824	1.111	1.289
49			0.046	0.559	0.808	1.098	1.277
50				0.539	0.792	1.085	1.265

Table cont.

	No. in population			
No. selected	**75**	**100**	**150**	**200**
51	0.520	0.776	1.072	1.253
52	0.501	0.760	1.059	1.242
53	0.481	0.745	1.046	1.230
54	0.462	0.729	1.034	1.219
55	0.443	0.714	1.021	1.208
56	0.423	0.699	1.009	1.197
57	0.404	0.684	0.997	1.186
58	0.384	0.669	0.985	1.175
59	0.365	0.654	0.973	1.165
60	0.345	0.639	0.961	1.154
61	0.325	0.624	0.949	1.144
62	0.305	0.609	0.937	1.133
63	0.285	0.594	0.926	1.123
64	0.265	0.580	0.914	1.113
65	0.244	0.565	0.903	1.103
66	0.223	0.551	0.892	1.093
67	0.202	0.536	0.881	1.083
68	0.180	0.521	0.869	1.073
69	0.158	0.507	0.858	1.064
70	0.135	0.492	0.847	1.054
71	0.112	0.478	0.837	1.044
72	0.087	0.463	0.826	1.035
73	0.061	0.449	0.815	1.026
74	0.032	0.434	0.804	1.016
75		0.420	0.794	1.007
76		0.405	0.783	0.998
77		0.391	0.773	0.989
78		0.376	0.762	0.980
79		0.361	0.752	0.971
80		0.346	0.742	0.962
81		0.332	0.731	0.953
82		0.317	0.721	0.944
83		0.302	0.711	0.936
84		0.286	0.701	0.927
85		0.271	0.691	0.918
86		0.256	0.680	0.910
87		0.240	0.670	0.901
88		0.224	0.660	0.893
89		0.208	0.650	0.884
90		0.192	0.640	0.876
91		0.176	0.631	0.868
92		0.159	0.621	0.860
93		0.142	0.611	0.851
94		0.124	0.601	0.843
95		0.106	0.591	0.835
96		0.088	0.581	0.827
97		0.068	0.572	0.819
98		0.048	0.562	0.811
99		0.025	0.552	0.803
100			0.542	0.795

Table cont.

	No. in population			No. in population
No. selected	150	200	No. selected	200
101	0.533	0.787	151	0.414
102	0.523	0.779	152	0.407
103	0.513	0.771	153	0.400
104	0.504	0.763	154	0.392
105	0.494	0.755	155	0.385
106	0.484	0.748	156	0.378
107	0.474	0.740	157	0.370
108	0.465	0.732	158	0.363
109	0.455	0.724	159	0.356
110	0.445	0.717	160	0.348
111	0.436	0.709	161	0.341
112	0.426	0.702	162	0.333
113	0.416	0.694	163	0.326
114	0.406	0.686	164	0.318
115	0.397	0.679	165	0.311
116	0.387	0.671	166	0.303
117	0.377	0.664	167	0.296
118	0.367	0.656	168	0.288
119	0.357	0.649	169	0.280
120	0.348	0.641	170	0.273
121	0.338	0.634	171	0.265
122	0.328	0.626	172	0.257
123	0.318	0.619	173	0.249
124	0.308	0.612	174	0.242
125	0.298	0.604	175	0.234
126	0.287	0.597	176	0.226
127	0.277	0.590	177	0.218
128	0.267	0.582	178	0.210
129	0.257	0.575	179	0.202
130	0.246	0.567	180	0.194
131	0.236	0.560	181	0.185
132	0.225	0.553	182	0.177
133	0.215	0.546	183	0.169
134	0.204	0.538	184	0.160
135	0.193	0.531	185	0.152
136	0.182	0.524	186	0.143
137	0.171	0.516	187	0.134
138	0.160	0.509	188	0.125
139	0.148	0.502	189	0.116
140	0.137	0.495	190	0.107
141	0.125	0.487	191	0.098
142	0.113	0.480	192	0.089
143	0.101	0.473	193	0.079
144	0.088	0.465	194	0.069
145	0.075	0.458	195	0.059
146	0.062	0.451	196	0.048
147	0.048	0.444	197	0.038
148	0.033	0.436	198	0.026
149	0.018	0.429	199	0.014
150		0.422		

Glossary of technical terms

accuracy (of predicted breeding values or transmitting abilities)
The correlation between predicted and true breedin values or predicted and true transmitting abilities.

accuracy (of selection) (r)
The correlation between the selection criterion (e.g. a index) and the breeding goal.

additive correction factors
Amounts which are added to, or subtracted from, th performance records of animals which belong t particular classes (e.g. singles or twins) to adjust fo non-genetic effects when predicting genetic merit.

additive genetic standard deviation (sd_A)
Square root of the additive genetic variance.

additive genetic variance (or variation) (V_A)
Variance (or variation) in a trait due to the combine effects of genes with additive action. Variance (c variation) in breeding values.

allele
A particular sequence of bases at a given locus or site on chromosome. There may be many alternative sequence possible at a particular locus in the breed as a whole, bu individual animals either carry copies of two differen alleles (one on each member of the pair of chromosome concerned – these animals are heterozygotes), or tw copies of the same allele (these animals are homozygotes) The alternative sequences of bases at a particular site ar termed alleles or genes interchangeably.

artificial insemination (AI)
Deposition of semen into the reproductive tract of a femal animal, either via the vagina, or directly into the uteru using a laparoscope. The semen is often extended to allow wider use, and frozen to allow long-term storage.

artificial selection
Selection of animals by humans, rather than, or in additio to, natural selection. Like natural selection, artificia selection acts on the many differences between individua animals, but the choice is based on characteristics of perceived benefit to the breeder, rather than on fitness fo a particular natural habitat, environment etc.

autosomes
All chromosomes apart from the sex chromosomes. Genes on these chromosomes are said to be autosomal.

backcross
Animal resulting from backcrossing.

backcrossing
Mating of crossbred animal back to one of the parent breeds.

bacterial artificial chromosomes (BACs)
Bacterial chromosomes containing fragments of mammalian (or other) DNA. These are used to replicate DNA fragments for further study.

se (chemical)
The chemical substances which, when paired, make up the 'rungs' of a DNA 'ladder'. There are four different bases in DNA (adenine, thymine, guanine and cytosine).

se (population)
Group of animals with unknown parents in genetic evaluations, whose PBVs are set equal to zero, or other group of animals with PBVs set to zero (e.g. those born in a particular year).

st linear unbiased prediction (BLUP)
A statistical procedure for predicting animal breeding values. Can be applied under several sets of assumptions or models which account for different relationships between animals. BLUP estimates environmental effects and predicts breeding values simultaneously, and so disentangles genetics from management and feeding more effectively, and produces better predictions of breeding value than other methods.

eed
A recognised group of interbreeding animals within a domestic species. Often animals of a breed are of fairly uniform appearance, but it some cases animals are considered to belong to a breed purely by virtue of their geographical location.

eed society
An organisation formed to promote a particular breed, and to record details of ancestry (and sometimes performance also) of animals of the breed, registered by members.

reeding goal (or objective)
The characteristic(s) which selection is intended to improve, e.g. carcass lean content, margin between milk production and feed costs.

reeding value (BV)
Additive genetic merit of an animal. Double the expected deviation in progeny performance from the population mean. An animal's true breeding value cannot be measured directly, but it can be predicted from various sources of information including the animal's own performance, and that of its relatives.

andidate gene
A gene which potentially contributes to variation in a trait of interest. For example, a gene with a known or suspected link to the trait of interest may be chosen as a marker (e.g. a gene coding for a hormone known to affect milk yield, or growth rate).

entromere
Part of a chromosome. Located near the middle of the chromosome, or at one end. During the initial phases of duplication of chromosomes, the two copies remain joined at the centromere.

hromosome
Structures found in the nuclei of all cells, which are made up of DNA, and on which genes are sited. Chromosomes occur in pairs in the nuclei of body cells, and singly in gametes (sperm and eggs).

cloning (DNA)
Multiplication of DNA by incorporation into the DNA o vector (such as a plasmid – a small self-replicati circle of double stranded DNA which exists independent in some bacterial cells). The resulting 'recombinant' DN molecules are then introduced into a rapidly reproduci host (a bacterium in the case of plasmid vectors) whe they are reproduced along with the host cell DNA. Aft many cycles of replication, many copies of the origin DNA fragment can be extracted from the host cells.

cloning (of embryos)
The production of groups of identical embryos either b physically splitting embryos or by transferring culture cells into eggs from which the nucleus has been remove (nuclear transfer).

co-dominance
Type of gene action in which heterozygotes sho characteristics of each of the homozygous types (e.g. with roan coat colour in Shorthorns).

codon
A sequence of three bases in mRNA which codes for th production of a particular amino acid (the building block which proteins are made from).

collateral relatives
Relatives from the same generation as the animal of interes e.g. sibs, cousins.

combining ability
Ability of particular breeds to produces high-perfomin offspring, as a result of high heterosis, when they a crossed.

complementarity
An outcome of some crossbreeding schemes, in which th overall efficiency of a production system is improved b crossing breeds which have high genetic merit in differen traits, e.g. use of large, lean terminal sire breeds on dam breeds of intermediate size and good maternal performance

complete dominance
Type of gene action in which the presence of one allel completely masks the effect of another allele at the sam locus (e.g. the black coat colour gene in cattle).

composite
A new (synthetic) breed formed by crossing two or mor breeds, then selecting within the new population.

consensus map
A genome map, using data from several studies, showin locations of linkage groups and physically mapped loci.

conserved DNA
DNA sequences which appear to have remained the same in different species through evolution.

contemporary animals/contemporary groups
Animals which have been treated in a similar way, e.g born over a relatively short period of time, on the same farm, and fed and managed similarly.

›rrelated response (to selection)	The response measured in one trait following selection on another.
›rrelation coefficient (r)	A value between –1 and +1 which measures the direction and strength of the association between two characteristics, or between the same characteristic at different times. Positive values indicate that as one trait increases, the other generally increases too (e.g. live weight and fatness). Negative values indicate that as one trait increases, the other generally decreases (e.g. milk yield and milk protein %). There are several types of correlations but two of the most widely used in animal breeding are the phenotypic correlation, or correlation between measurements on animals, and the additive genetic correlation, or correlation between breeding values.
ɔvariance	A measure of the association between traits, used in its own right, and in calculating correlation coefficients. Covariances between two traits are calculated in a similar way to calculating the variance of a single trait.
rossbred (animal)	Animal with parents of different breeds or strains.
rossbreeding	Mating parents of two or more different breeds, strains or species together.
ulling	To remove an animal from a herd or flock, e.g. because it has a poor predicted breeding value.
aughter design	A method of detecting associations between markers and traits of interest (usually sex-limited traits), where marker genotypes are determined for sires and daughters and performance records are obtained from daughters.
leoxyribonucleic acid (DNA)	The chemical substance which chromosomes and genes are made of. DNA has a ladder-like structure with pairs of four different bases (adenine, thymine, guanine and cytosine) making up the 'rungs'. This structure, together with the fact that these bases always pair in the same way, provides a very reliable mechanism for copying DNA during cell division. Genes or alleles are simply sequences of bases on one side of a DNA molecule. Different genes or alleles at a particular locus are sequences which differ slightly – often only in one base pair.
liploid	The state in which the chromosomes occur in pairs. This is the case for all body cells, except the gametes, e.g. humans have 23 pairs, cattle have 30 pairs and sheep have 27 pairs.
direct breeding value	The genetic merit of an animal for its direct (as opposed to maternal) genetic influence on a trait (e.g. its influence on weaning weight via genes for growth as opposed to genes for uterine capacity and milk production).

DNA fingerprint	The virtually unique pattern of bands produced on a gel a a result of the different alleles present at minisatellite lo in a DNA sample from an individual animal. Used to chec parentage.
DNA probe	A single-stranded molecule of DNA used to detect th presence of a complementary DNA strand in samples c DNA or in intact chromosomes. Probes can be 'labelle by making them either radioactive or fluorescent.
economic-trait locus (ETL)	See quantitative-trait locus.
economic value (of a trait)	The marginal profit resulting from a genetic change of on unit in that trait, e.g. the increase or decrease in profi resulting from a change of 1 kg in live weight compared t the current average value, with no change in other traits i the breeding goal. Used to apportion emphasis to differen traits in selection indexes.
electrophoresis	Application of an electric current to a film of gel on whic DNA samples have been placed, to separate fragments o DNA of different size and charge.
elite or nucleus breeders/herds/flocks	Breeders/herds/flocks at the top of the improvemen pyramid supplying breeding stock to other breeders.
environmental correlation	A measure of the extent to which environmenta conditions that are favourable for one character ar favourable or unfavourable for a second character (rang from –1 to +1).
environmental variance (or variation) ($\mathbf{V_E}$)	Variance (or variation) in a trait due to non-genetic effects including differences in feeding and management etc remaining after adjusting records of performance.
epistasis	Epistasis occurs when the presence of an allele at one locus masks the effect of an allele at another locus (i.e there is an interaction between alleles at different loci). Visible examples include interactions between alleles a different coat colour loci in several species.
estimated breeding value (EBV)	See predicted breeding value.
estimated transmitting ability (ETA)	See predicted transmitting ability.
expected progeny difference (EPD)	See predicted transmitting ability.
fixed base	A fixed group of animals (e.g. those born in a particular year) from which deviations in breeding value may be expressed. The group usually remains fixed for several evaluations.
fluorescence *in situ* hybridisation (FISH)	A technique for determining the physical location of a gene using a fluorescently labelled DNA probe.

mete — A sex cell – a sperm in the male, and an unfertilised egg (ovum) in the female.

ne — A sequence of bases on one side of a DNA molecule, which codes for the production of proteins. These proteins either have a structural role (e.g. in muscle) or serve as signals controlling body functions (e.g. enzymes). The alternative forms of the sequence of bases at a particular site are termed genes or alleles interchangeably. The word gene is also used to describe the site on a chromosome where a particular gene occurs. More properly, this is termed a locus.

ne frequency — The proportion or percentage of genes (alleles) of each type at a particular locus, in a herd, flock or other group of animals. Must add up to 1 or 100% for all alleles at this locus.

ne map — See genome map.

ne transfer — Insertion of DNA from one animal into the genome of another. This transfer may be between individuals of the same or different breeds, species or even classes of organism.

eneration interval (**L**) — The weighted average age of parents when their offspring are born.

enetic correlation — A measure of the direction and strength of the association between breeding values for two characters e.g. live weight and fat depth (range from –1 to +1).

enetic drift — Chance change in gene frequencies.

enetic evaluation — The prediction of breeding values.

enetic groups — Groups of animals of different average merit e.g. from different countries. In genetic evaluations base animals are assigned to different genetic groups to avoid bias in predicting breeding values.

enetic links — Links between contemporary groups provided by related animals. These links are necessary in order for BLUP to estimate environmental effects and predict breeding values simultaneously, and to enable comparison of PBVs across groups, herds, flocks or years.

enetic trend — The change in mean predicted breeding value in a population of animals over time.

enome — The entire genetic code of an animal contained in the haploid or diploid set of chromosomes.

enome map — A description of the location of functional genes, or other sequences of DNA, on the chromosomes of the speci concerned. Genome maps are made up of linkage or gene maps, and physical maps.

genomic imprinting

Type of gene action in which the expression of a gene offspring depends on which sex of parent contributed (e.g. the callipyge gene affecting muscularity in sheep).

genotype

The particular combination of genes or alleles which animal inherits.

genotype x environment interaction

Occurs when genotypes do not rank the same in differe environments, or when the advantage to a particul genotype in one environment is smaller or greater than another.

genotype frequency

The proportion or percentage of animals of each genotype a herd, flock or other group. Must add up to 1, or 100% over all possible genotypes.

grading up

Repeated mating of females (usually), and subsequent their female offspring, to males of a new breed.

granddaughter design

A method of detecting associations between markers an traits of interest (usually sex-limited), where marke genotypes are determined for grandsires and sires an genetic merit of sires is predicted from their daughters performance records.

group breeding scheme

Cooperative breeding scheme with nucleus of elit animals screened from members' flocks or herds Recording and selection are concentrated in the nucleus which then produces breeding stock for members.

haploid

The state in which chromosomes occur singly, rather tha in pairs. This is the case in gametes e.g. the sperm or egg of humans, cattle and sheep carry 23, 30 and 27 singl chromosomes.

heritability (**h^2**)

The proportion of superiority of parents in a trait (i.e. the proportion of the selection differential) which, on average is passed on to offspring. Or, the additive genetic variatio in a trait (the variation in breeding values), expressed as proportion of the total phenotypic variation in that trait.

heterosis (or hybrid vigour)

The advantage in performance of crossbred animals above the mid-parent mean.

heterozygote/heterozygous

Animals which carry copies of two different alleles at a particular locus (e.g. RW).

homology (of DNA sequence)

The high degree of similarity between parts of the genome of different mammalian species.

mozygote/homozygous

Animals which carry two copies of the same allele at a particular locus (e.g. RR).

nprovement lag

The difference in genetic merit between tiers of a livestock industry.

breeding

The practice of mating related animals. This may be done deliberately, as in linebreeding. It is also an inevitable consequence of long-term selection in a closed population.

breeding coefficient (**F**)

A measure of the amount of inbreeding, defined as the probability that two alleles at any locus are 'identical by descent'.

breeding depression

The decline in performance (especially in traits associated with functional fitness, such as reproductive rate and disease resistance) which is a consequence of inbreeding.

complete penetrance

Occurs when a character is entirely controlled by a single gene, but not all animals of a given genotype show the expected phenotype (e.g. malignant hyperthermia in pigs).

ndependent culling levels

A method of selection for more than one trait. Involves setting minimum qualifying standards, or thresholds, in each trait of interest. To qualify for selection, animals must then surpass the qualifying standard in each trait.

ndex coefficients (or weights) (**b**)

The weighting factors applied to the phenotypic measurements of one or more traits on an animal, or its relatives, to derive a single index score on which to base selection.

ndex measurements

The set of measurements of performance on candidates for selection, to which index coefficients are applied to calculate an index score. May be the same as traits in the breeding goal if these can all be measured in both sexes, and on the live animal, or they may be different from the goal traits, e.g. if these are only available on one sex, or are measured after slaughter.

ndex selection

Selection on an overall score (or index) of genetic merit which combines information on several traits of importance, or from different classes of relatives.

individual heterosis

Heterosis as a result of the individual animal (rather than its sire or dam) being crossbred.

introgression

Introduction of a single gene for a favourable characteristic to an existing breed by crossing to a new breed carrying the gene, and then backcrossing to the original breed for several generations. This is followed by mating of males and females from the final backcross generation to produce animals which are homozygous for the introgressed gene.

in vitro production of embryos

Laboratory production of embryos, following collection unfertilised eggs from the ovary of a slaughtered female from a live donor via ovum pick up. These eggs are fir matured, then fertilised and cultured further before transfe

linebreeding

The practice of deliberately mating closely relate animals. This was widely used by early improvers of far livestock. It is far less common in large animal breedin today because of the risks of producing non-viable lines a a result of inbreeding depression.

linkage

Occurs when two loci are close together on the sam chromosome. Because segments of the chromosome ar involved in crossing over, genes which are closer together o the same chromosome will tend to cross over together mor often than genes which are far apart on the same chromosom This can lead to associations between characters.

linkage map

A map showing the recombinational or genetic distanc between pairs of markers on the same chromosome. Thi is achieved by measuring the frequency with which th markers separate from one generation to the next, as result of recombination.

locus (pl. loci)

The site ('address') of a gene on a chromosome.

major gene

A single gene with a large effect, such as those affectin coat colour, polledness and double muscling.

marker

Any variable and measurable characteristic of an anima which can potentially be used as a predictor of performance or can be used to create a linkage map (e.g. genotype at a molecular marker locus, blood group, milk protein type).

marker-assisted introgression (MAI)

The use of markers to speed up introduction of a gene o interest from one breed to another by: (i) identifying backcrosses carrying the particular favoured allele from the new breed, and possibly; (ii) amongst animals with one or two copies of this allele, identifying those with least of their genome from the new breed.

marker-assisted selection (MAS)

The use of markers to accelerate selection by identifying animals likely to have a favourable genotype for the trait of interest. The gene of interest, to which the marker is linked, may be a single gene with a large effect, or it may be one of many genes affecting a quantitative trait (a QTL).

maternal breeding value

The genetic merit of an animal for its maternal (as opposed to direct) genetic influence on a trait (e.g. its influence on weaning weight via genes for uterine capacity and milk production rather than via genes for growth).

aternal heterosis — Heterosis in reproduction and other maternal traits as a result of the breeding female being crossbred.

eiosis (reduction division) — The type of cell division which occurs during production of the gametes (sperm and eggs). During meiosis the cells which become sperm or eggs receive only a single (haploid) set of chromosomes. Also, recombination leads to a mixing of segments of the maternally-derived and paternally-derived chromosomes.

essenger RNA (mRNA) — RNA is a substance very similar to DNA. Messenger RNA is a type of RNA which transfers the code for the production of a protein from the DNA in the nucleus of a cell, to the ribosomes ('protein factories') in the cytoplasm of the cell.

icrosatellite — A region in the genome where the same sequence of base pairs is repeated several times, end to end. The repeated sequences are two to five base pairs long. The number of times a given sequence of bases is repeated varies greatly between alleles. Because of this variation, microsatellites are very useful as markers in genome mapping and for marker assisted selection.

inisatellite — A region in the genome where the same sequence of base pairs is repeated several times, end to end. The repeated sequences are 10 to 60 base pairs long. The number of times a given sequence of bases is repeated varies between alleles. Minisatellites are useful in checking parentage.

itochondria — Small structures in cells responsible for generating energy to 'fuel' the activities of the cell – the cell's 'power stations'.

itochondrial (cytoplasmic) inheritance — Transmission of small amounts of genetic information from mothers to daughters via mitochondrial DNA.

itosis (multiplication division) — The type of cell division which occurs during normal growth and development. Each of the 'daughter' cells produced receives a copy of the complete original (diploid) set of chromosomes.

nodel — Set of assumptions used in predicting breeding values (or estimating genetic parameters) which differ in sophistication. The name of the model indicates the relationships used and the animals for which BVs are predicted e.g. sire model BLUP predicts BVs for sires from their progeny records, animal model BLUP predicts BVs for all animals included in the evaluation, using all relationships between them.

nolecular marker — Any identifiable segment of DNA in the genome. Markers may be all or part of a 'functional' gene, or they may be a part of the rest of the genome which does not code directly for the production of a protein. RFLPs and microsatellit are commonly used molecular markers in livesto genome mapping, and are potentially useful in mark assisted selection.

multiple ovulation and embryo transfer (MOET)
A series of reproductive techniques includi superovulation of a donor female, mating, recovery of t resulting embryos, and transfer of fresh or frozen embry to recipient females.

multiplicative correction factors
Multiplication factors which are used to adju performance records of animals which belong particular classes (e.g. singles or twins) to adjust f non-genetic effects when predicting genetic merit.

multiplier breeders/herds/flocks
Breeders/herds/flocks in the middle of the improveme pyramid buying breeding stock form elite breeders abov and supplying pure or crossbred breeding stock commercial producers below.

multi-trait across country evaluation (MACE)
A method for predicting breeding values of animals usi information from different countries. Based on multi-tra BLUP evaluation, but rather than evaluating man different traits at once, MACE treats records on the sam trait from different countries as if they were different trait This allows for differences between countries in th heritability of the trait of interest, and allows for differen genetic correlations between this trait recorded in differen countries.

multi-trait (BLUP) evaluation
Prediction of breeding values for several trait simultaneously. This improves the accuracy of th evaluation, if the traits are correlated, provided tha accurate estimates of these correlations are used in th evaluation.

multi-trait selection
Selection to alter several traits, e.g. by using a selectio index.

mutation
Chromosomal mutations involve changes in the number o structure of chromosomes. Gene or point mutations occu when mistakes are made in copying DNA which are no corrected by repair enzymes. If gene mutations occur i the reproductive organs, they lead to a change in the geneti code of offspring. Gene mutations are an important sourc of genetic variation.

natural selection
The process by which animals, and other forms of life, which are best suited to their particular environment o niche, stand a higher chance of surviving and reproducing than those that do not ('survival of the fittest'). Natura selection acts on the many differences between individua animals.

on-additive genetic variance (or ariation) (V_{NA})	Variance (or variation) in a trait due to the combined effects of genes with non-additive action.
clear transfer	Transfer of cells from an embryo, or from cell culture, into an unfertilised egg from which the nucleus has been removed. A method of producing clones.
icleus (cell)	Part of a cell containing the chromosomes.
icleus (herd, flock or population)	A group of highly selected animals; animals of high genetic merit.
icleus breeding scheme	A breeding scheme in which recording and selection are concentrated on an elite group of animals. These may be in a central herd or flock, or dispersed over several herds or flocks.
verdominance	Type of gene action in which the performance of heterozygous animals is more extreme than that of both homozygous types (e.g. the Inverdale fertility gene in sheep).
vum pick up (OPU)	Collection of eggs from donors through an ultrasonically guided needle inserted into the ovary.
artial dominance	Type of gene action in which the performance of heterozygotes is intermediate to that of the two homozygotes, but is closest to that of animals homozygous for one of the alleles concerned (e.g. the Booroola gene with respect to litter size).
aternal heterosis	Heterosis in reproductive performance as a result of sires being crossbred.
edigree	A record of the ancestry of an animal.
edigree (registered) animal	An animal whose ancestry is known, or more usually, whose ancestry is recorded by a breed society or similar organisation.
edigree (stud) breeder	A breeder of purebred animals who usually registers details of these, especially ancestry, with a breed society or similar organisation.
pedigree index	An index combining information on ancestors' performance/breeding values.
pedigree selection	Selection on (breeding values derived from) ancestors' information.
performance test	A comparison of (and subsequently selection amongst) animals based on their own performance.
permanent environmental effects	Non-genetic effects which influence an animal's

	performance throughout its life.
phenotype	Outward appearance (e.g. coat colour), or performance of animals measured in some way (e.g. milk yield). In some cases the phenotype is determined solely by the genotype (e.g. coat colour). In many other cases it is the result of both genetic and non-genetic (environmental) factors.
phenotypic correlation	A measure of the direction and strength of the association between observed performance, or phenotype, in two characters e.g. live weight and fat depth measured on the same animals (range from –1 to +1).
phenotypic standard deviation (**sd_P**)	Square root of the phenotypic variance.
phenotypic variance (or variation) (**V_P**)	Variance (or variation) in a trait due to the combined effects of genes and the environment.
physical map	A map showing the physical location of genes or markers on individual chromosomes.
pleiotropy	Pleiotropy occurs when an allele at a single locus influences more than one characteristic of the animal (e.g. the double muscling gene in cattle affects both muscularity and fertility).
polymerase chain reaction (PCR)	A method of rapidly multiplying sequences of DNA ‘in a test tube’ rather than in living cells.
predicted breeding value (PBV)	Prediction of the additive genetic merit of an animal. May be based on various sources of information including the animal’s own performance, and that of its relatives. Double the expected deviation in progeny performance from the population mean.
predicted transmitting ability (PTA)	Prediction of the additive genetic merit of an animal. May be based on various sources of information including the animal’s own performance, and that of its relatives. The expected deviation in progeny performance from the population mean (i.e. half the predicted breeding value).
progeny test	A comparison of animals (and subsequently selection amongst them) based on the performance of their progeny.
purebred (straightbred)	Animal with both parents of the same breed.
purebreeding (straightbreeding)	The practice of mating animals belonging to the same breed.
quantitative-trait locus (QTL)	A locus affecting a quantitative trait such as milk yield and quality, growth rate or wool production.
recombinant DNA	DNA which is combined from different sources, e.g. from a mammal and a plasmid.

combination — The crossing-over of segments of the maternally-derived and paternally-derived members of a pair of chromosomes during meiosis.

cords in progress — Early lactation milk records (especially from heifers) used to predict total lactation yields for genetic evaluations. This allows earlier selection of breeding animals, and hence shorter generation intervals than selection on completed lactations.

ference family — A family of animals used to produce DNA samples for genome mapping.

gression (coefficient) (b) — A statistic which measures the direction and strength of association between two characters, or between the same character at different times, or between the same character on related animals. Regression coefficients measure the rate of change in one character per unit change in another character, in the units of measurement.

lationship matrix — A table showing the expected proportion of genes in common in a group of animals.

liability (or repeatability) of predicted reeding values (or transmitting abilities) — The squared accuracy of PBVs, i.e. the squared correlation between predicted and true breeding values (or transmitting abilities).

epeatability — The correlation between repeated records from the same animal – a measure of the value of repeated records in selection.

esource population — Population of animals used in genome mapping or to detect useful associations between markers and traits of economic importance.

esponse (to selection) (**R**) — The change in the mean performance in a population of animals as a result of selection. Usually measured per annum or per generation.

estriction enzymes — Enzymes produced by bacteria, which are capable of 'cutting up' sequences of DNA. Different restriction enzymes 'recognise' different sequences, usually of four to eight base pairs in DNA.

estriction fragment length olymorphism (RFLP) — Variation between animals in the number and position of sites at which a given restriction enzyme will cut the DNA. RFLPs can be used to detect variation in the DNA sequence of different animals, and hence they are used as markers in genome mapping.

ibosomes — Small structures in the cytoplasm of cells responsible for assembling amino acids to produce proteins, using instructions from genes carried by messenger RNA. Ribosomes are the cell's 'protein factories'.

rolling base	A group of animals (e.g. the most recent age gro included in an evaluation) from which deviations breeding value may be expressed. This group usual changes at each evaluation.
rotational crossing	This involves the use of the same two or three (or mor breeds in rotation in a crossbreeding programme.
seedstock/stud breeder/herd/flock	See elite breeder/herd/flock.
segregation	The separation and chance sampling of paternally- an maternally-derived genes during meiosis in the production a sperm or egg.
segregation ratios	The ratios of different offspring genotypes expected fro matings between particular genotypes of parent.
selection between breeds or strains	Substituting one breed or strain for another.
selection criterion (criteria)	The measurement(s) on which selection is based (e.g. liv weight at 400 days of age, fleece weight).
selection differential (**S**)	The difference between the mean performance of selecte animals (i.e. those identified to become parents of the nex generation) and the overall mean of the group of anima from which they were selected.
selection index (**I**)	An overall score of genetic merit which combine information on several measured traits, or from differen classes of relatives. The emphasis on each trait in the inde usually depends on the strength of its association with trait in the breeding goal, and their relative economic value.
selection intensity (**i**)	The superiority of animals selected (the selectio differential) expressed in standard deviation units.
selection limit or plateau	The point at which no further response to selection can be achieved, as a result of exhaustion of genetic variation in the trait under selection.
selection within breeds or strains	Choosing better parents within a particular breed or strain.
semen (or embryo) sexing	Sorting of semen (or embryos) by a variety of methods according to whether they are carrying X or Y chromosomes.
sibs (siblings)	Animals with one or both parents in common, i.e. brothers or sisters. Full sibs have both parents in common, half sibs have only one parent in common.
sib test (sib selection)	A comparison of animals (and subsequently selection between them) based on the performance of their siblings.

andem selection — A method of selection for more than one trait. Involves selection for one trait for one or more generations, followed by selection for a second trait for one or more generations, possibly followed by selection on more traits, eventually returning to selection on the first, and so on.

emporary environmental effects — Non-genetic effects which have a relatively short impact on animal performance, e.g. affecting only one lactation, or one weight record.

erminal sire — Sire, strain or breed selected for specialised meat production characteristics.

est day records — Actual milk records obtained on sampling days (usually at monthly intervals) which are usually then used to predict full lactation yields of milk, fat and protein in milk recording schemes.

transfer RNA — RNA is a substance very similar to DNA. Transfer RNA is a type of RNA involved in transporting free amino acids in the cell to ribosomes for assembling proteins.

transgenic (animal) — A genetically modified animal whose genome contains 'foreign' DNA, e.g. from another animal, possibly of a different breed or species.

triplet — A sequence of three bases in DNA which codes for the production of a particular amino acid (the building blocks which proteins are made from), or signals where a gene stops or starts.

two-stage selection — Selection between animals initially on one trait (or set of traits) and then, between the remaining animals, on a second trait (or set). Two stage selection is usually practised when a second trait (or set of traits) becomes available later in life, or when it is too expensive or difficult to measure on all animals.

variable expressivity — Type of gene action in which animals show the expected phenotype, but to varying degrees (e.g. mulefoot in cattle and pigs).

yeast artificial chromosomes (YACs) — Yeast chromosomes containing fragments of mammalian (or other) DNA. These are used to replicate DNA fragments for further study.

Index of terms